Comprehensive Guide on Organic and Inorganic Solar Cells

Solar Cell Engineering Series

Comprehensive Guide on Organic and Inorganic Solar Cells

Fundamental Concepts to Fabrication Methods

Edited by

Md. Akhtaruzzaman

Solar Energy Research Institute (SERI), The National University of Malaysia (@Universiti Kebangsaan Malaysia), Bangi, Malaysia

Vidhya Selvanathan

Universiti Kebangsaan Malaysia, Bangi, Malaysia

Academic Press is an imprint of Elsevier
125 London Wall, London EC2Y 5AS, United Kingdom
525 B Street, Suite 1650, San Diego, CA 92101, United States
50 Hampshire Street, 5th Floor, Cambridge, MA 02139, United States
The Boulevard, Langford Lane, Kidlington, Oxford OX5 1GB, United Kingdom

British Library Cataloguing-in-Publication Data
A catalogue record for this book is available from the British Library

Library of Congress Cataloging-in-Publication Data
A catalog record for this book is available from the Library of Congress

ISBN: 978-0-323-85529-7

For Information on all Academic Press publications
visit our website at https://www.elsevier.com/books-and-journals

Publisher: Candice Janco
Acquisitions Editor: Lisa Reading
Editorial Project Manager: Aleksandra Packowska
Production Project Manager: Poulouse Joseph
Cover Designer: Miles Hitchen

Typeset by MPS Limited, Chennai, India

Contents

List of contributors

Md. Akhtaruzzaman
National University of Malaysia, Bangi, Malaysia; Solar Energy Research Institute, Universiti Kebangsaan Malaysia, Bangi, Malaysia

Nowshad Amin
Institute of Sustainable Energy, Universiti Tenaga Nasional (@UNITEN; The Energy University), Kajang, Malaysia

Nadrah Azmi
College of Engineering, Universiti Tenaga Nasional, Jalan Ikram-UNITEN, Kajang, Malaysia; Institute of Sustainable Energy (ISE), Universiti Tenaga Nasional, Jalan Ikram-UNITEN, Kajang, Malaysia

Puvaneswaran Chelvanathan
Solar Energy Research Institute (SERI), The National University of Malaysia, Bangi, Malaysia; Graphene & Advanced 2D Materials Research Group (GAMRG), School of Engineering and Technology, Sunway University, Jalan Universiti, Bandar Sunway, Malaysia

A.K. Mahmud Hassan
Universiti Kebangsaan Malaysia, Bangi, Malaysia

Mohammad Ismail Hossain
Department of Materials Sciences and Engineering, City University of Hong Kong, Kowloon, Hong Kong, P.R.China; Department of Applied Physics, Hong Kong Polytechnic University, Kowloon, Hong Kong, P.R. China; Department of Electrical and Computer Engineering, University of California, Davis, CA, United States

Mohammad Aminul Islam
Department of Electrical Engineering, Faculty of Engineering, Universiti Malaya, Jalan Universiti, Kuala Lumpur, Malaysia

Makoto Karakawa
Nanomaterials Research Institute, Kanazawa University, Kanazawa, Japan

Dietmar Knipp
Geballe Laboratory for Advanced Materials, Department of Materials Science and Engineering, Stanford University, Stanford, CA, United States

Muhammad Ammar Bin Mingsukang
Center for Ionics, Faculty of Science, Department of Physics, University of Malaya, Kuala Lumpur, Malaysia

Masahiro Nakano
Nanomaterials Research Institute, Kanazawa University, Kanazawa, Japan

Jean-Michel Nunzi
Nanomaterials Research Institute, Kanazawa University, Kanazawa, Japan; Department of Physics, Engineering Physics and Astronomy, Department of Chemistry, Queens University, Kingston, ON, Canada

Wayesh Qarony
Department of Applied Physics, Hong Kong Polytechnic University, Kowloon, Hong Kong, P.R. China; Materials Sciences Division, Lawrence Berkeley National Laboratory, Berkeley, CA, United States; Department of Electrical Engineering and Computer Sciences, University of California Berkeley, CA, United States

Md. Khan Sobayel Bin Rafiq
Solar Energy Research Institute, Universiti Kebangsaan Malaysia, Bangi, Malaysia

Kazi Sajedur Rahman
Solar Energy Research Institute, Universiti Kebangsaan Malaysia (The National University of Malaysia), Bangi, Malaysia

Muhammad Rizwan
Department of Chemistry, Faculty of Science, The University of Lahore, Lahore, Pakistan

Vidhya Selvanathan
Universiti Kebangsaan Malaysia, Bangi, Malaysia; National University of Malaysia, Bangi, Malaysia

Md. Shahiduzzaman
Nanomaterials Research Institute, Kanazawa University, Kanazawa, Japan

Kamaruzzaman Sopian
Solar Energy Research Institute, Universiti Kebangsaan Malaysia, Bangi, Malaysia

Tetsuya Taima
Nanomaterials Research Institute, Kanazawa University, Kanazawa, Japan

Yuen Hong Tsang
Department of Applied Physics, Hong Kong Polytechnic University, Kowloon, Hong Kong, P.R. China

Ashraf Uddin
School of Photovoltaic and Renewable Energy Engineering, University of New South Wales, Sydney, NSW, Australia

Yulisa Binti Mohd. Yusoff
Institute of Sustainable Energy, Universiti Tenaga Nasional (@The National Energy University), Kajang, Malaysia

Preface

Electricity has become such a quintessential part of humanity in today's world. To understand how vastly our energy demands have expanded, just think of the number of times you flip a switch per day. With the continuous growth of industrial sectors coupled with the latest innovations in the fields of transportation, agriculture, medicine, and science, it is expected that our electricity demand will increase by around 80% by 2050. So, on the one hand, we have extremely high energy demands required to fuel different facets of our daily life, and on the other, there is scarcity of continuously depleting fossil fuels that serve as our primary source of energy. Taking this ideas together, we desperately need a reliable, sustainable, and inexhaustible source of energy to keep our world running, which we believe can the sun. Serving as a natural nuclear reactor, the amount of energy that can be derived from the sunlight available on earth at a given point of time can equal up to 173,000 TW. Simply put, 1 hour of sunlight exposure on earth can generate a sufficient number of photons to theoretically fuel the global energy requirement for an entire year. Such an unlimited abundance of energy has inspired decades of research to efficiently understand the science behind the art of deriving electrical energy from sunlight, also known as photovoltaics. The field of photovoltaics has continuously evolved over the years where new technologies are ventured and creative solutions are proposed every day. The first generation of solar cells was dominated by silicon-based materials, limiting the improvisations to techniques and processes. However, the introduction of thin-film solar cells paved the way for the inclusion of several inorganic materials as primary components in the device. Subsequently, with the inclusion of organic materials in third-generation solar cells, almost every element in the periodic table finds unique utility as a solar cell material. Hence, this book attempts to address the evolution of solar cell technologies beginning from the second-generation inorganic material-based thin-film photovoltaics, followed by organic solar cells and, finally, the latest hybrid organic–inorganic approaches. The content of this book is intended to be a comprehensive guide discussing the theoretical background as well as practical knowledge required for its fabrication (including material selection and experimental techniques). In each subsection, the working principle and architecture of a category of solar technology are presented, followed by a dissection of every component within the architecture. Prerequisite characteristics for material selection of each component are then discussed with relevant examples from current literature. Subsequently, crucial experimental procedures for the fabrication of these devices are introduced, which will help the audience to visualize practical applications of the technology.

CHAPTER

Principle of photovoltaics

1

Nowshad Amin

Institute of Sustainable Energy, Universiti Tenaga Nasional (@UNITEN; The Energy University), Kajang, Malaysia

1.1 Introduction

In this chapter, the characteristics and amount of the sun's energy as the main input source of solar photovoltaic (PV) energy will be discussed to show how enormous an energy bank is safely placed millions of kilometers away from us. Then, solar PV fundamentals together and solar cell classification will be introduced for better comprehension of sunlight to electricity conversion. Solar PV cells are electricity generators that differ from more well-known hydroelectric-, diesel- or nuclear reactor-based generators. Energy conversion occurs in a unique way based on the semiconductors' quantum effect, abolishing the need of any heat or mechanical parts as seen in conventional electricity generators. The tremendous growth of solar PV technology over the past few decades has helped the levelized-costs-of-generating-electricity making it very cost competitive among all available electricity-generating sources as shown in Fig. 1.1, which is mainly derived from the rigorous efforts of researchers supported by renewable energy policies around the world as per International Energy Agency (IEA) report. Countries are now including solar PV technology as one of the key contributors in their long term future energy roadmap owing to the confidence gained over the past decade.

1.2 Solar energy

The sun's energy reaches the earth in the form of an electromagnetic wave passing through outer space. The sun is known to be a nuclear fusion reactor and its radiation spectra can be replicated by a perfect black-body heated at 6000K. Considering an average distance of 150 million kilometers between the sun and earth, the incident light on the earth's surface is an assortment of plane electromagnetic waves of various frequencies. The radiation spectra of the sunlight that arrives on the earth's surface is shown in Fig. 1.2, along with the incident light before entering the atmospheric belt of the earth. The solar radiation is attenuated by at least 30% during its passage through the earth's atmospheric belt, where

Comprehensive Guide on Organic and Inorganic Solar Cells. DOI: https://doi.org/10.1016/B978-0-323-85529-7.00015-3

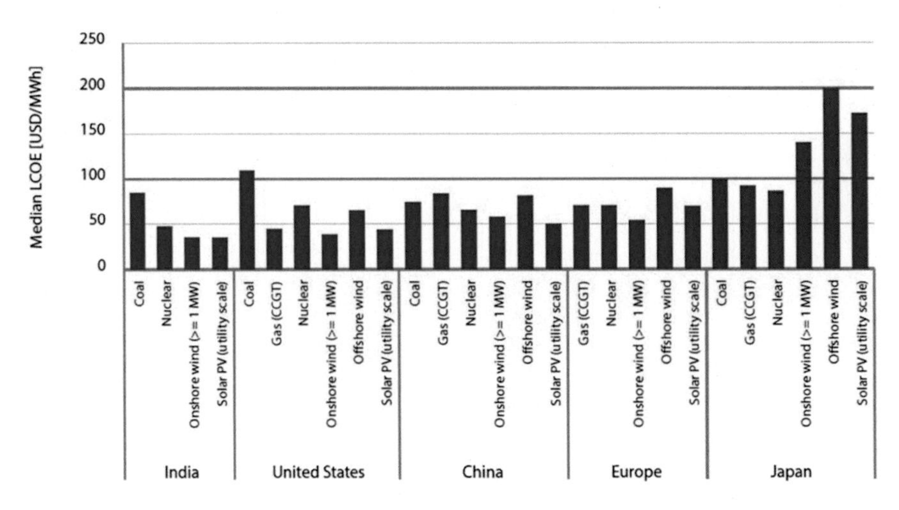

FIGURE 1.1

Median LCOE technology costs by region (data taken from IEA 2020 report).

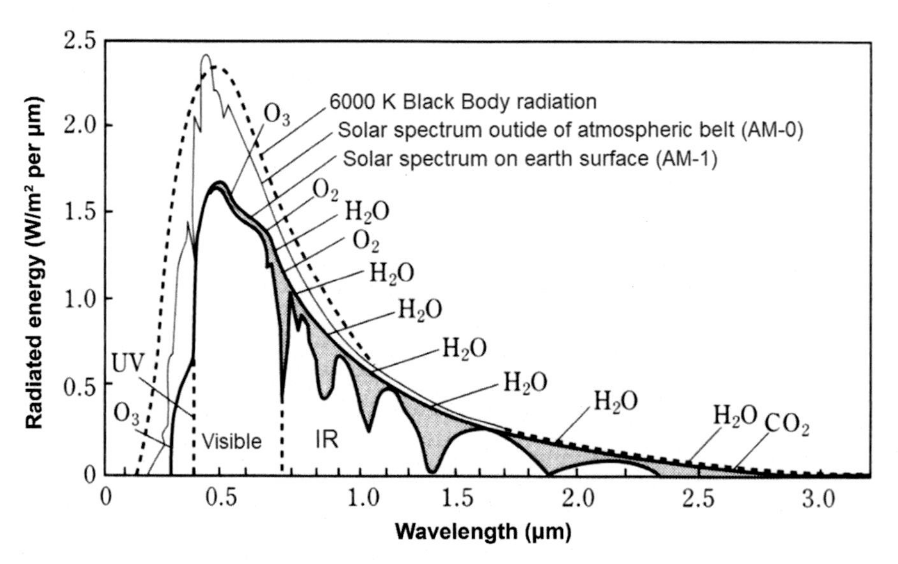

FIGURE 1.2

Solar energy spectra for three different air mass conditions.

most of the energetic ultraviolet or blue waves are reflected by the similarly-sized aerosols in outer space (this is the reason the sky is blue). In addition, the water moisture in the surrounding atmosphere absorbs a substantial amount of incoming solar energy, thereby reducing the amount of sunlight reaching the surface.

To simulate the radiation spectrum of the sun at various places, we use 6000K black-body radiation at the origin (sun), then 5700K black-body radiation for the spectrum on the earth's surface. If we quantify the amount of energy radiated from the sun's surface to electrical power, it equals to 3.8×10^{23} kW, which reaches an energy density of about 1.4 kW/m^2 outside of the earth's atmospheric belt. This energy density is termed as a solar constant. However, the sunlight (energy density) that falls at a location varies depending on the latitude, time, and weather. Again, seasonal variation also changes the sunlight's air-volume in a particular place, which is called Air Mass (AM). It is the shortest and considered unity when the sun is directly overhead and depicted as AM1. AM is 0 in outer space, but for other places on earth, it is derived by the following equation, where θ is the angle of the sun from straight overhead.

$$\text{Air mass} = \frac{1}{\cos\theta} \tag{1.1}$$

Therefore, when the sun is 45 degrees from center, the radiation is AM1.5. One convenient way to estimate AM is to measure the length of the shadow that is cast by any vertical object of height h as shown in the equation below.

$$\text{Air mass} = \sqrt{\left(1 + \left(\frac{s}{h}\right)^2\right)} \tag{1.2}$$

Therefore, the total radiated energy of the sun reaching the earth can be derived from the product of the solar constant with the projected surface area of the earth. One quick example shows that when taking the earth's diameter of 6356 km, the total incident solar energy becomes 177×10^{12} kW. This enormous amount of energy is the total energy from the sun that can be used for various purposes, of which 30% is reflected back to the space. Among the remaining 70% that reaches the earth, around 47% is utilized to keep the earth warm enough for living creatures. Then, the remaining 23% is stored in either seawater or ice with a part of it being used for cloud or rain creation. Among all of these, the energy of energy that is used for wind, wave, or convection energy is only 0.2%, equivalent to 0.37×10^{12} kW. Furthermore, the energy needed for the growth of plants and animals on earth, the photosynthesis of biological systems, is only 0.02% (equivalent to 400×10^{8} kW). Nonetheless, these small amounts of solar energy are what is really important for the conservation of the global ecosystem.

To summarize the source of solar energy, it can be concluded that terrestrial sunlight varies dramatically and unpredictably in availability, intensity, and spectral composition. On clear days, the length of the sunlight's path through the atmosphere and the optical AM are important parameters. The indirect or diffuse component of sunlight can be particularly important for less than ideal conditions. Reasonable estimates of global radiation (direct plus diffuse) received annually on horizontal surfaces are available for most regions of the world. There are uncertainties that can be caused by local geographical conditions and approximations involved in the conversion to radiation on inclined surfaces.

1.3 Photovoltaic effect

PV systems comprise the technology to convert sunlight directly into electricity without additional fuel. The term "photovoltaic" is derived from the Greek language. "Photo" means light and "voltaic" means electricity. Charged carriers are produced based on the photo-conduction phenomenon upon incident light on any semiconductor. The light generated carriers will polarize in regions of positively- and negatively-charged particles due to possible diffusion or drift occurring as a result of asymmetric spatial distribution or built-in potential of a p-n junction device and create electromotive forces termed as the PV effect. The PV effect due to spatial non-uniformity of charged carriers has the Dember and photo-electromagnetic effect, but they do not possess a direct relationship with solar PV cells. Therefore, only the PV effect related to the interfacial electric field of the semiconductors will be discussed (Fig. 1.3).

The PV effect is a key to solar energy conversion, where electricity is generated from light energy. Owing to quantum theory, light is regarded as packets of energetic particles called photons, whose energy depends only on light frequency. The energy of visible photons is sufficient to excite electrons and bound into solids up to higher energy levels where they are freer to move. Meanwhile, there are internal electric fields due to the electron affinity and Fermi level between the semiconductor and attached subsequent materials in cases of semiconductor p-n junctions, crystal borders, and semiconductor interfaces or surfaces. Hence, when

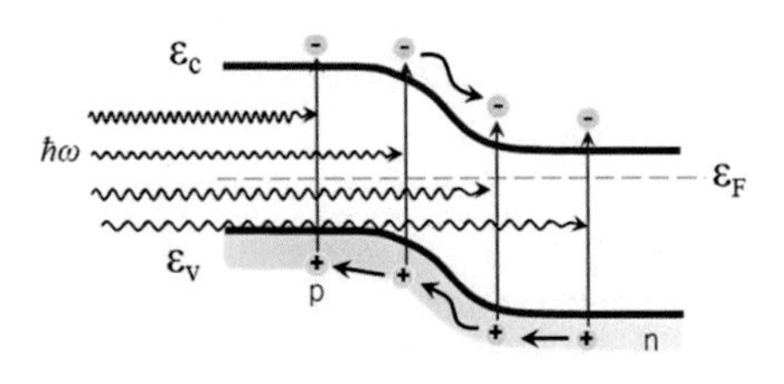

(A) Photovoltaic effect in p-n junction

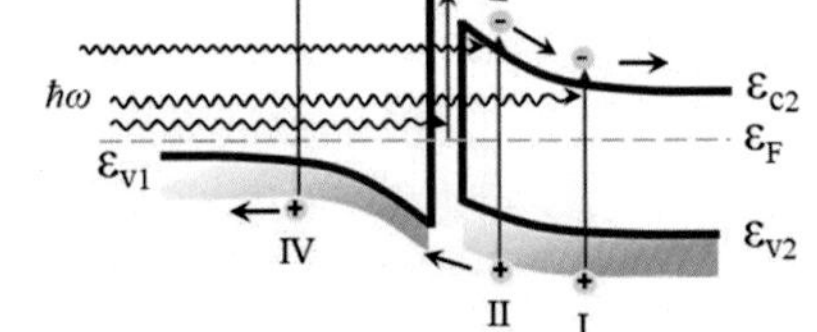

(B) Photovoltaic effect in hetero-junction

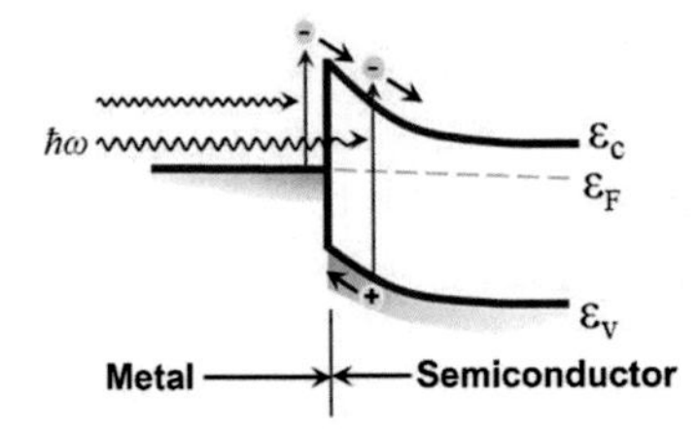

(C) Photovoltaic effect in Schottky barrier contact

FIGURE 1.3

Photovoltaic effect in various semiconductor junctions and interface.

incoming photons strike such regions, electron-hole pairs are generated that drift in opposite directions to create charge polarization and eventually electromotive force due to light irradiation known as the PV effect. It is only for this PV effect that any semiconductor p-n junctions, heterojunctions, or even Schottky barrier junctions experience interface potential that can create solar cells. On the other hand, many materials possess PV properties but fail to provide ample current due to higher internal resistance for their crystalline structure and are then used as optical sensors. Binary semiconductors, such as ZnO, CdS, ZnSe in sintered form, are used as optical detectors.

1.4 Fundamentals of solar cells

To generate a PV effect, an inbuilt electric potential must exist in semiconductors. The potential that exists between interfaces of materials helps in generating this phenomenon. In general, there is a contact potential between two electronically asymmetric materials. Examples include storage batteries that can generate contact potentials between lead or graphite and electrolytes or thermocouples that use the contact potential differences of two contacted metals. Similarly, contact potential is induced depending on a combination of either semiconductor and metal or semiconductor and semiconductor. The p-n junctions are engineered with good reproducibility with stable combinations of materials that use inherent or inbuilt electric fields in the interfaces (Fig. 1.4).

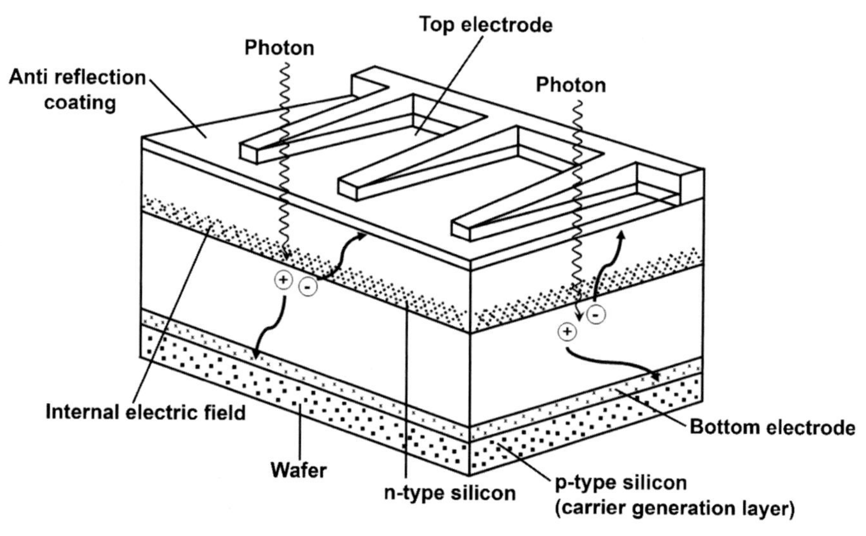

FIGURE 1.4

Structure of a typical crystalline silicon solar cell.

A solar cell is generally a p-n junction with built-in asymmetry that separates the light excited electrons away from the junction region to be extracted out to external circuits. The effectiveness of the solar cell depends on the choice of light absorbing materials as well as supporting materials for the carrier to be efficiently collected through external terminals. To explain basic solar cell fundamentals, only silicon-based solar cells will be inferred in this section. Pure silicon as a group IV semiconductor is undoped, but it must doped with materials that can create either positive (p) type or negative (n) type polarity in silicon. Usually it is done by incorporating group V materials such as P to make n-type silicon and group III materials such as B to make p-type silicon. Silicon doping enhances its electrical conductivity that can easily be controlled or manipulated up to certain limit, which is also called valence electron control typically used in transistor or integrated circuits. If both n-type and p-type silicon are crystallographically joined, then the interface is called the p-n junction. Initial carrier diffusion on both sides of the interface creates a depletion region or space charge region as mobile charges are exhausted due to recombination in the region. This is the reason the internal electric field exists thereby restricting the enhancement of the electric field and pushing any charge carrying particles, for example, electron and hole, on both sides of the p-n junction. A present, all electronic devices such as the diode, transistor, LED, or LASER etc. utilize an internal electric field as the main working principle that originates from the interface potential.

Practically used solar cells are essentially large area p-n junctions that use the interface electric field for the PV effect. A simple silicon solar cell schematic is shown in Fig. 1.3. As seen, the n-type layer is disproportionately thin to allow the incoming light to immediately reach the junction area. As light is illuminated on a solar cell, photons of different wavelengths hit the semiconductor surface, which, in this case, is the n-type silicon region. Only a fraction of the photons are converted into electrical energy, since only photons with energy equal to or greater than the energy bandgap of the semiconductor (Si) are absorbed. Photon absorption leads to the generation of an electron-hole pair, also known as EHP generation. The majority-carrier concentrations (the total number of electrons in an n-type semiconductor or the total number of holes in a p-type semiconductor) are unaffected by photon-assisted carrier generation, as newly generated EHP concentrations are insignificant compared to the majority-carrier concentrations. However, minority-carrier concentrations (the total number of electrons in a p-type semiconductor or the total number of holes in an n-type semiconductor) increase significantly. This change upsets the equilibrium condition between the diffusion force and electrostatic force, resulting in the PV effect. Electrons originating from the p region eventually diffuse into the depletion region, where the potential energy barrier at the junction is lowered, allowing current to flow and establish a voltage at the external terminals. Holes created in the n-doped region travel in the opposite direction to the p-doped side. The generation of charges depending on the incoming photon flux as well as movement of charges in the above manner results in the amount of current density existing along the terminals.

1.5 Energy conversion of solar cells

Energy conversion efficiency (or simply conversion efficiency) of a solar cell, η, is the percentile ratio of output electrical power from the terminals of solar cell to the incident sunlight energy over the same area of the cell. Therefore, the conversion efficiency η can be expressed as the following equation (Fig. 1.5).

$$\eta = \frac{\text{Electrical power output from a solar cell, } P_{out}}{\text{Incident sunlight energy, } P_{in}} \times 100\% \tag{1.3}$$

However, a more difficult procedure is needed to define or derive this conversion efficiency as the solar cell's ultimate performance index. In other words, the conversion efficiency of the same solar cell changes with changes of the incoming light spectra. Moreover, the output energy changes depending on the load connected to the cells even under the same incident light. Therefore, in accordance with IEC TC-82, it is decided to consider the incoming light source under AM 1.5 with incident power of 100 mW/cm^2 for any terrestrial solar cells at maximum power output conditions, which is also termed as nominal efficiency (η_n).

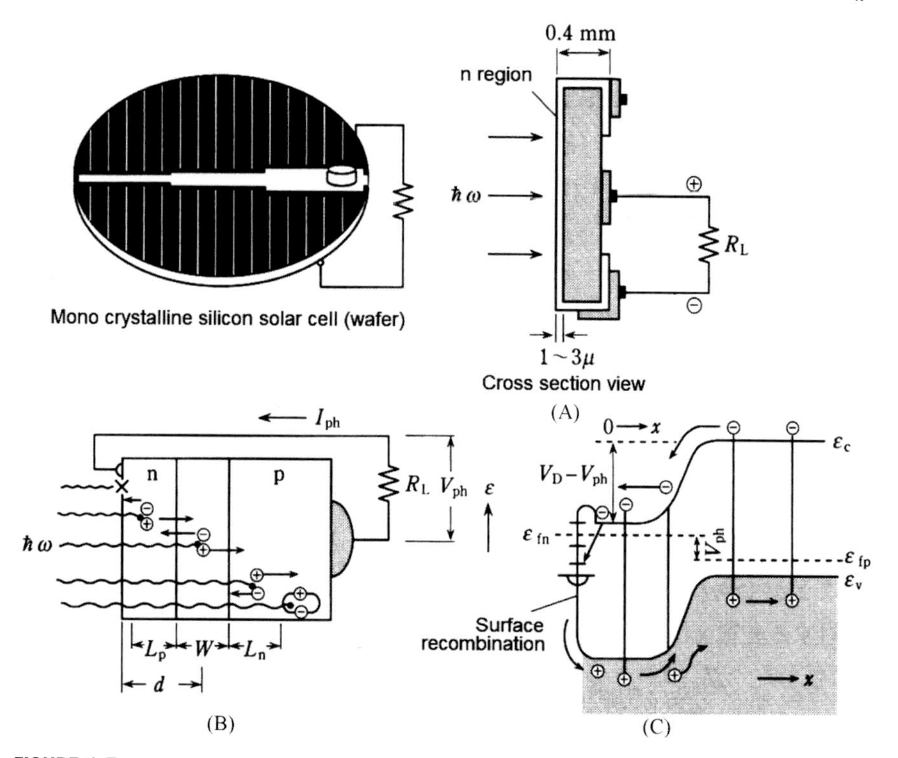

FIGURE 1.5

Working principle of a silicon solar cell (A) cross section of the solar cell, (B) enlarged view of p-n junction and (C) energy band gap diagram showing carrier flow.

The conversion efficiency measured under this condition is usually described in the specifications of solar panels. For publication purposes, we show this value of conversion efficiency. In the derivation of nominal efficiency, it is important to derive the relationships among maximum output power voltage $V_{\max}$ (or V_{mp}), maximum output power current $I_{\max}$ (or I_{mp}), open circuit voltage V_{oc}, and short circuit current density J_{sc} from the measured output characteristics of the solar cell.

When light enters from the surface of a p-n junction type solar cell, the junction exists at a distance, d, from the surface. Considering minority carrier diffusion lengths of both p and n regions as L_n and L_p, the optical absorption coefficient α in respect to the wavelength λ and the opto-electronic quantum efficiency γ, the carrier generation $g(x)$ at a distance from surface x, is proportional to the optical absorption $d\Phi/dx$ as shown in the following equation.

$$g(x) = \gamma\varphi_0\alpha e^{-\alpha x} \tag{1.4}$$

Here, Φ_0 is the optical flux density of the wavelength λ on the surface ($x = 0$). However, in reality, most of the incident light flux generated by the carriers is extinct due to recombination around the surface, $x = 0$, which is known as surface recombination loss. The carriers (both electron and hole) that contribute to the PV effect come from the carriers that are collected due to diffusion from the neutral region border until the length of the minority carrier diffusion (L_n and L_p) on both sides of the depletion region. To determine total photo current, the current in the n-region must be found by integrating the product of $g(x)$ and $\exp[-(d-x)/L_p]$ from $x = 0$ to d and the current for the p-region and add both components. Consequently, the total current for the monochromatic incident light at the time terminals of both sides are shorted can be found by the following equation.

$$\frac{dI_{sc}(\lambda)}{d\lambda} = \gamma A\alpha\lambda\left\{\frac{L_p}{1-\alpha L_p}\left[e^{-\alpha d} - e^{-d/Lp}\right] + \frac{L_n e^{-\alpha d}}{1-\alpha L_n}\right\} \tag{1.5}$$

In reality, the junction depth is considered much less than the light penetration depth and αL_n, $\alpha L_p \ll 1$, and create the carrier collection effect, so Eq. (1.5) can be simplified as below.

$$\frac{dI_{sc}(\lambda)}{d\lambda} = \gamma A\alpha \cdot \lambda\left(L_n + L_p\right)\mathrm{e}^{-\alpha d} \tag{1.6}$$

Fig. 1.6 shows an example that compares both the calculated data and experimental data for the spectral response of the silicon solar cell, considering $d = 2\ \mu\text{m}$, $L_n = 0.5\ \mu\text{m}$ and $L_p = 10\ \mu\text{m}$.

In the case of p-n junction type solar cells, the conversion efficiency would be higher when the overlapping increases between the calculated spectral response distribution as shown above and the incident sunlight distribution shown in Fig. 1.2. However, the cutoff wavelength on the longer wavelength side of the spectral response distribution of any material depends on the energy bandgap, whereas the shape of the spectrum depends on the geometrical dimensions of the device, the constants such as L_p, L_n, μ_p, μ_n, and the light absorption coefficient spectra $\alpha(\lambda)$ as shown in Fig. 1.7. Hence, the theoretical limit of the maximum

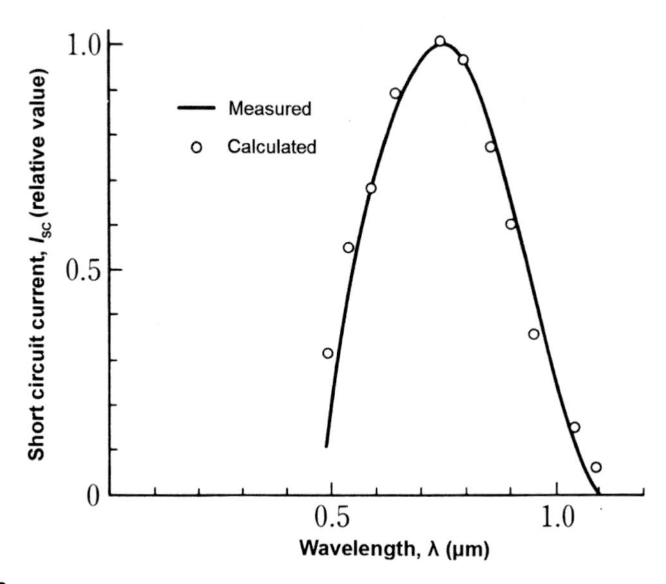

FIGURE 1.6

Spectral response of a silicon solar cell.

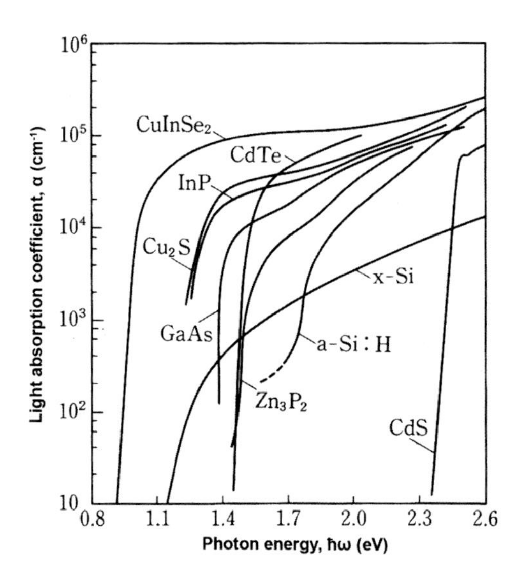

FIGURE 1.7

Light absorption coefficient spectra of semiconductors used in solar cells.

achievable conversion efficiency, $\eta_{\max}$, depends on these parameters. If we take the incident photon density of the solar radiation as a function of λ, $\Phi(\lambda)$ and the electronic charge as q, then the actually measured short circuit current I_{sc} will be as below.

$$I_{sc} = \int_0^{\infty} I_L(\lambda)d\lambda = qA\gamma(L_n + L_p)\int_0^{\infty} \varphi(\lambda)\alpha e^{-\alpha d}d\lambda \tag{1.7}$$

The short circuit current I_{sc} flows from the n-type region to the p-type region as seen in Fig. 1.5. The relationship between the terminal voltage V and flowing current I is the solar cell current-voltage characteristics as shown in the following Eq. (1.8), where the p region is regarded as positive.

$$I = I_0\left[\exp\left(\frac{qV}{nkT}\right) - 1\right] - I_{SC} \tag{1.8}$$

Here, I_0 is the reverse saturation current of the diode. Fig. 1.8 shows the output characteristics of the solar cell. Following the equation above, a voltage is produced per the light intensity when the terminals are in an open state. This voltage is termed as open circuit voltage, V_{oc}, as it becomes the following Eq. (1.9), where $I = 0$ (open circuit) in Eq. (1.8).

$$V_{oc} = \frac{nkT}{q}\ln\left(\frac{I_{sc}}{I_0} + 1\right) \tag{1.9}$$

As shown in Fig. 1.8, the maximum power point P_m is defined when the solar cell is connected to the optimum load R_L and the corresponding voltage and

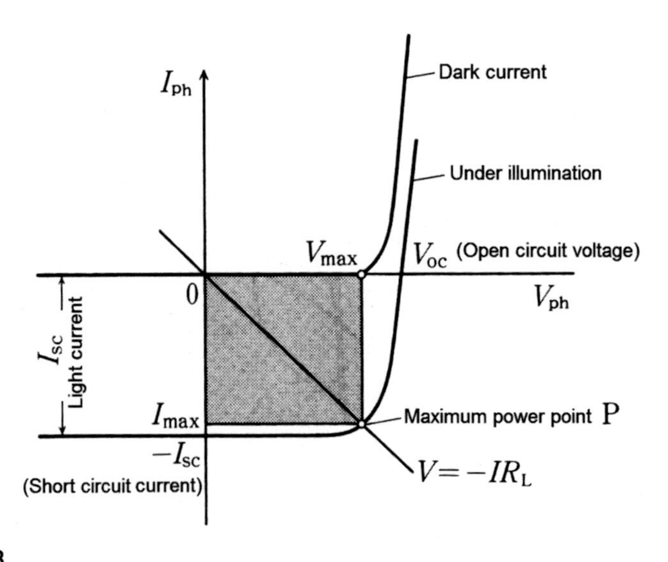

FIGURE 1.8

Current-voltage characteristics of the solar cell.

current are denoted by V_{max} (or V_{mp}) and I_{max} (or I_{mp}). The areas calculated by V_{max} and I_{max} are equivalent to the output power. Hence, following Eq. (1.8), P_{out} as the generalized output power of any solar cell is shown in Eq. (1.10).

$$P_{out} = V \cdot I$$

$$= V \cdot \left\{ Isc - I_0 \left[\exp\left(\frac{qV}{nkT}\right) - 1 \right] \right\} \tag{1.10}$$

From Fig. 1.7, it is also clear that at the maximum power point P_{max} becomes,

$$\frac{dP_{out}}{dV} = 0 \tag{1.11}$$

Hence, from the above two equations, the maximum power-point voltage V_{max} satisfies the following relationship.

$$\exp\left(\frac{qV_{max}}{nkT}\right)\left(1 + \frac{qV_{max}}{nkT}\right) = \left(\frac{I_{sc}}{I_0}\right) + 1 \tag{1.12}$$

Meanwhile, the maximum power point current I_{max} can be expressed as the following.

$$I_{max} = \frac{(I_{sc} + I_0) \cdot qV_{max}/nkT}{1 + (qV_{max}/nkT)} \tag{1.13}$$

In reality, the nominal energy conversion efficiency of any solar cell is measured under a solar spectra replicated light source that usually shows AM 1.5 and 100 mW/cm^2 for terrestrial application solar cells and AM 0 and 100 mW/cm^2 for space application solar cells. For instance, if all the parameters such as P (V_{max}, I_{max}) as well as V_{oc}, J_{sc} are found from a terrestrial application solar cells output characteristics measurement, the nominal conversion efficiency η_n for light exposed area S (cm^2) can be derived from the following equation.

$$\eta_n = \frac{V_{max} \cdot I_{max}}{P_{in}S} \times 100(\%)$$

$$= \frac{V_{oc} \cdot J_{sc} \cdot FF}{100(\mathrm{mW/cm^2})} \times 100(\%)$$

$$= V_{oc}(V) \cdot J_{sc}(\mathrm{mA/cm^2}) \cdot FF(\%) \tag{1.14}$$

where

$$\mathrm{FF} = \frac{V_{max} \cdot J_{max}}{V_{oc} \cdot J_{sc}} \tag{1.15}$$

Here, *FF* is the curve fill factor (FF) or, simply, FF, which is the ratio between equivalent area of the maximum output power (P_{max}) and the product of $V_{oc} \times J_{sc}$ as shown in Fig. 1.8, which also expresses the junction quality of the solar cell device.

Hence, the conversion efficiency of a solar cell is directly proportional to the I_{sc}, V_{oc}, and the FF during the performance evaluation where the input power (P_{in}) exposed to the solar cell is used as 1 kW/m^2 or 100 mW/cm^2.

1.6 Equivalent circuit of solar cells

The equivalent circuit of the solar cell has output characteristics as shown in Eq. (1.8), which is generally described with two components, the p-n junction diode's rectifying component and the constant current source component I_{sc} that depends on the intensity of incoming light. Apart from that, there are series resistances R_s that limit the terminal current and parallel resistances R_{sh} that facilitate the leakage current of the p-n junction part. All of these are shown in the equivalent circuit of the solar cell in Fig. 1.9. As shown in the figure, the solar cell terminal current and voltage are related by Eq. (1.16).

$$I = I_{sc} - I_0\left[\exp\left\{\frac{q(V + R_S I)}{nkT}\right\} - 1\right] - \frac{V + R_s I}{R_{sh}} \tag{1.16}$$

As shown from Fig. 1.9, for a particular solar cell at the time of lower light intensity and within the lower range of I_{ph}, the diode current I_d and the leakage current V_d/R_{sh} become almost equal. The solar cell equivalent circuit equation becomes Eq. (1.17), where it is more affected by R_{sh} than R_s.

$$I = I_{sc} - I_0\left[\exp\left(\frac{qV}{nkT}\right) - 1\right] - \frac{V}{R_{sh}} \tag{1.17}$$

On the contrary, in the case of higher light intensity, $I_d \gg V_d/R_{sh}$, the effect of R_s becomes more prominent than R_{sh} and the equation simplifies to (1.18).

$$I = I_{sc} - I_0\left[exp\left\{\frac{q(V + R_s I)}{nkT}\right\} - 1\right] \tag{1.18}$$

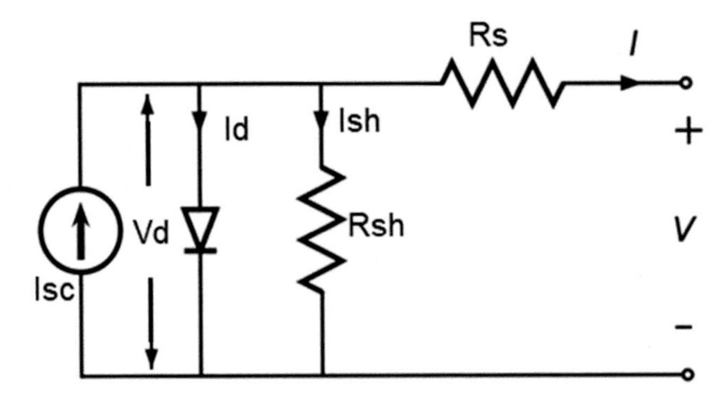

FIGURE 1.9

Equivalent circuit of the solar cell.

R_s does not have a significant effect on the open circuit voltage V_{oc}, whereas the short circuit current I_{sc} drastically decreases. Meanwhile, R_{sh} does not show any impact on I_{sc} but leads to a decrease in V_{oc}.

It is important to understand how the series resistance R_s affects the output current and can be shown with a simple example. In the case of silicon p-n junction solar cell, let us consider a short circuit current density J_{sc} of 30 mA/cm^2, I_o of 50 pA/cm^2, $n = 1$ at the incident power of 100 mW/cm^2. The calculation result of the R_s as the output parameter is based on Eq. (1.16), is shown in Fig. 1.10. In this case, the corresponding conversion efficiency is shown with respect to R_s values, whereby the shunt or parallel resistance is taken to be infinity to cause the leakage current to be 0. As seen, the solar cell output characteristics are largely affected by R_s. All of these can be explored in many of the available solar cell performance related simulation software. If we execute the same calculation taking R_{sh} as the variable parameter, the results are found as shown in Fig. 1.11. As seen from the figure, R_{sh} seems to affect the light-induced current comparatively less but V_{oc} has been largely affected. In basic silicon p-n junction solar cells, the V_{oc} is around 0.51 V. However, today's silicon solar cells are more complex in structure with back surface field as well as anti-reflection coatings with textured structures able to improve both V_{oc} and J_{sc} to reach over 25% conversion efficiency. Figs. 1.10 and 1.11 can be considered as the most important basic characteristics for R_s and R_{sh} as derived from the basic solar cell output characteristics calculation, which practically helps to understand the impacts of R_s and R_{sh} that may arise from various fabrication processes.

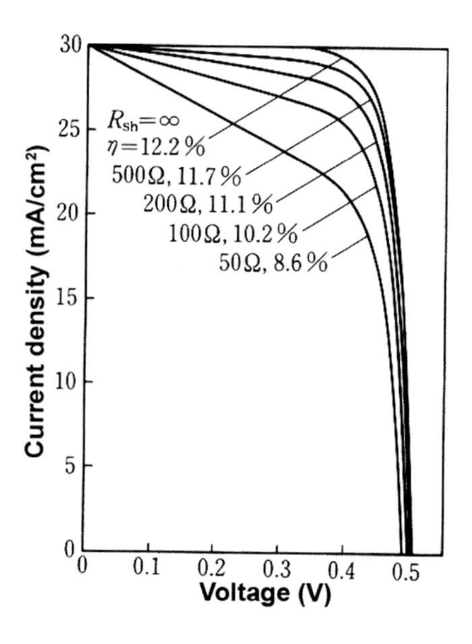

FIGURE 1.10

Output characteristics of a silicon solar cell at variable series resistances.

1.7 Collection efficiency

The carrier collection efficiency of a solar cell is an essential term that measures the spectra of generated carriers upon incident ideal sunlight and is then converted into the spectral response via the quantum effect (photon to electron energy transformation). It can be derived from the total number of carriers within the p-n junction upon solving the diffusion equation of the generated minority carrier. To provide effective carrier collection, wider bandgap semiconductors are used as window layers from where incident light enters either into heterojunction or heterointerface solar cells as shown in Fig. 1.12. Most

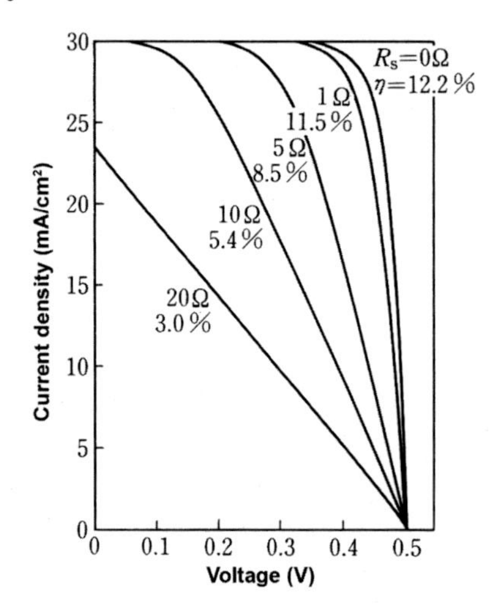

FIGURE 1.11

Output characteristics of a silicon solar cell at variable shunt resistances.

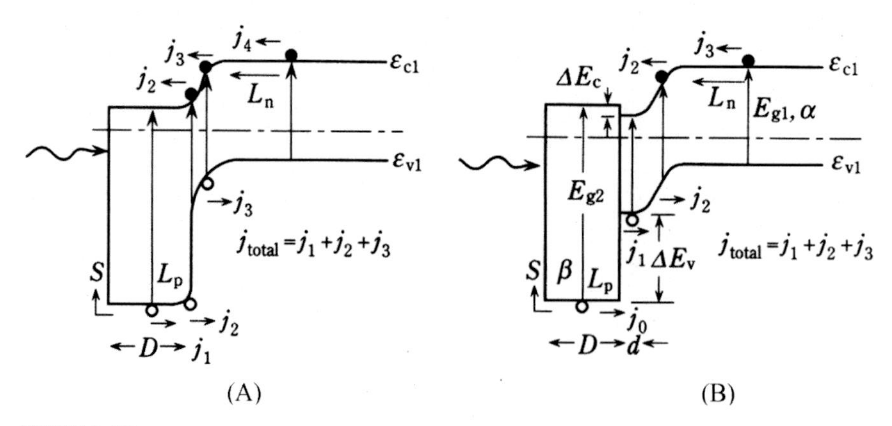

FIGURE 1.12

Band profile of solar cells with a wider window layer; (A) hetero junction solar cell and (B) hetero interface solar cell.

ideal heterojunction solar cells expect both the p and n regions to have the same doping concentration and that there is no inter-diffusion of carriers between them. However, in reality, the doping concentration is deliberately kept higher to reduce the resistance in the surface layer, which causes most of the hetero-junction solar cells to become hetero-interface solar cells. The basic difference in the carrier collection for both homo-junction and hetero-junction solar cells lies in the following fact. At a distance from the surface in both cases, the light intensity and collected carriers are found to be more improved for hetero-junction solar cells. However, structural design optimization is important for the materials to achieve higher conversion efficiency in hetero-junction solar cells, considering the complex interrelationship among the materials' bandgap and optical and electrical properties as well as matching the solar radiation energy spectra.

1.8 Theoretical limit of efficiency

In recent years, although the conversion efficiencies of solar PV devices have reached quite comparable values, a common question always arises as to why the efficiencies are not higher in comparison to other electrical power generating machines, for example, diesel generators. As we have learned some basics about conversion efficiency terminologies and equations in the above sections, we can now understand the sets of parameters involved in the efficiency measurement of the solar PV cells unlike conventional generators. Let's now look into the theoretical limit of solar cells, which involves two important facts: the materials used to absorb the sunlight and the incoming solar spectra. Ultimately, the overlapping between these two spectra, that is, absorption spectra of the main materials and the sunlight spectra, will decide the optimal limit of the solar cell device's performance. Fig. 1.13 shows some of the solar cell materials'

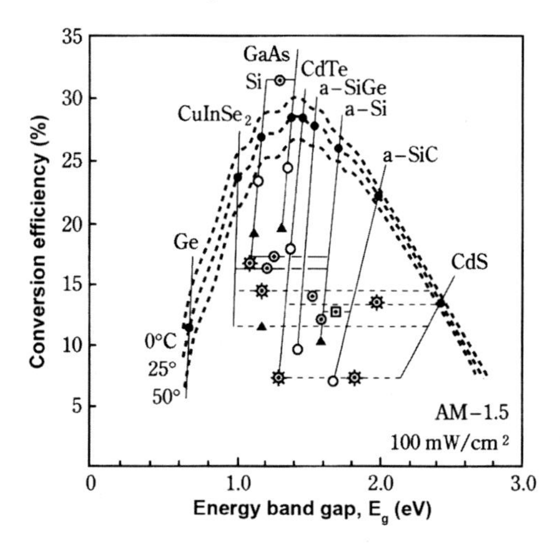

FIGURE 1.13

Theoretical efficiency limit of various kinds of solar cells at room temperature.

theoretical limits derived from the absorption spectra under AM-1.5, 100 mW/cm^2 incident light and practical data on the conversion efficiency observed in some materials, such as GaAs-based solar cells (theoretical limit is 28.5%). However, the calculation for the theoretical limit uses a simple structure for any light absorbing material, which may not be appropriate for all research-scale or commercially available solar cell. In the case of crystalline silicon solar cells, the theoretical limit for a single junction structure is 27%, whereas now many different configurations show greater than 26% efficiency. On the other hand, low cost amorphous silicon solar cells have theoretical limits around 25.5%; however, we have seen data in the 13%–15% range with some associated light-soaking degradation. Other thin film solar cells (TFSCs), such as CIGS (Cu-In-Ga-Se) or CdTe, also show 28% theoretical limits as found from different studies.

Nonetheless, the theoretical conversion efficiency limit is not greater than 30% for a single light absorbing material as discussed above. With the rigorous efforts of researchers around the world over the past few decades, the differences between the theoretical limits and practically achieved solar cells have moved closer with innovative structural engineering and light trapping technologies. The following discussion may help to understand the loss any solar cell incurs when exposed to sunlight spectra. At first, the conformity (overlapping) between the incident solar spectra and the spectral response of any solar cell that comes from the solar cell material's light absorption quality are very crucial. As mentioned earlier, the spectral response of any solar cell depends on the absorbing material's properties, such as bandgap and junction depth. In principle, the theoretical limit depends on the material property factor/constant that is bandgap, which exposes the inability of the light (photon) collection by the material as the resulting mismatch between the solar spectrum and spectral response of the material. More clearly, these are losses that correspond to light moving through the solar cell without being absorbed and the light that is reflected or scattered from the surface of the solar cell. Apart from these collection losses, the other overlapping part can be fully utilized for photon to electron excitation or solar cell effects in spite of several loss mechanisms. Among the losses that limit the collection efficiency of solar cells, they can be classified in the following categories:

1. Light reflection loss from the surface even though collectible by spectral response
2. Surface recombination loss of the carriers generated by the collected light (photon)
3. Bulk recombination loss of the carriers in the semiconductor bulk
4. Series resistance loss due to the Joule heat produced by the internal carrier flow (current)
5. Voltage factor loss due to the polarized electric field by the light generated carriers limited by the p-n junction's diffusion voltage V_d, where photons equivalent to bandgap energy encounter loss of $h\nu - qV_{oc}$ ($qV_{oc} < qV_d < E_g$)

The key factor to improving the efficiency of solar cells depends on the above mitigation of loss mechanisms using various innovative techniques.

1.9 Classification of solar cells

Solar cells can be classified in many ways, such as the generation or based on the main light absorption materials in the physical structure. Although solar cells were first designed for very specialized uses such as in spacecrafts and satellites, they can now be found in everyday use, such as wristwatches to central power stations. The development of the solar cell originates from the work of French physicist Alexander Edmond Becquerel in 1839. He discovered the PV effect while experimenting with a solid electrode in electrolyte solution. He observed that when light fell upon the electrode, voltage developed. The discovery of photoconductivity in selenium led to the fabrication of the first selenium solar cell by W.G. Adams in 1877. In 1883, the first true solar cell that was only around 1% efficient was built by Charles Fritts who coated the semiconductor selenium with a very thin and transparent layer of gold to form the junction. Another metal semiconductor junction solar cell, which was made of copper and semiconductor copper oxide was demonstrated in 1927. By 1930, both the selenium cell and copper oxide cell were being employed in light-sensitive devices, such as photometers for photography. These early solar cells had energy-conversion efficiencies of less than 1%. This standoff was finally overcome with the development of the Si solar cell patented by Russell Ohl in 1941.

The modern age of solar power technology arrived in 1954 when Bell Laboratories, experimenting with semiconductors, accidentally found that silicon doped with certain impurities was very sensitive to light. In 1954, three American researchers, namely G.L. Pearson, Daryl Chapin, and Calvin Fuller, demonstrated a 6% efficient silicon solar cell when used in direct sunlight, which increased to 14% by 1958 and 28% by 1988. In the same year, the first thin film Cu_2S/CdS heterojunction solar cell with 6% efficiency was reported by Reynolds. In 1956, Jenny reported a GaAs solar cell with 4% conversion efficiency. In 1963, D.A. Cusano fabricated the first thin film CdTe solar cell based on CdTe/Cu_2Te heterojunction with 6% efficiency. Bonnet and Rabenhorst reported a thin film CdTe/CdS solar cell in 1972 with 6% efficiency. In 1974, S. Wagner et al. reported a thin film $CuInSe_2$/CdS heterojunction solar cell with 12% conversion efficiency. However, the 1980s and 1990s have been a period in which public and governmental support for PVs has been underemphasized, but significant activity in the research community has continued. Generally speaking, solar PV technology can be classified into three generations, as explained in subsequent sections and shown in Fig. 1.14.

1.9.1 First generation solar cells

First generation solar cells are made of semiconducting p-n junctions consisting of silicon. Silicon cells have a high efficiency averaging 20%, but very pure

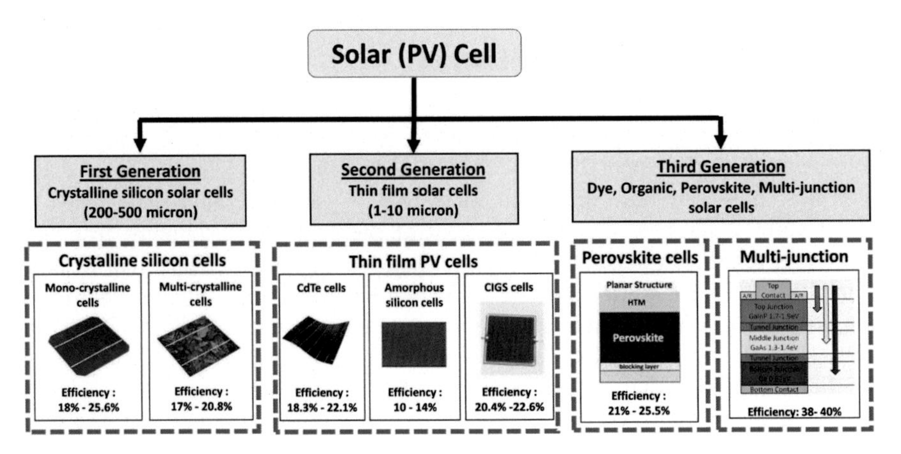

FIGURE 1.14

Solar cell classification.

silicon is needed and the price is high when compared to the power output. These solar cells are manufactured from pure silicon and their theoretical efficiency can reach a maximum of 33%. First generation solar cells account for over 90% of commercial production and include both mono and multicrystalline silicon. However, in recent years, a "directionally solidified (DS) wafer" (also called a "cast mono" wafer) has been developed that is different than the usual Czochralski growth process and has slowly replaced mainstream mono-crystalline silicon. The manufacturing processes associated with the production of first generation solar cells are still inherently expensive; hence, these cells may take 5–6 years to pay back their initial investment.

1.9.2 Second generation solar cells

Second generation solar cells are based on reducing the cost of first generation cells by employing thin film technologies. Thus, TFSCs have been regarded as a potential low cost, high efficiency solar cell which can replace Si solar cells in the PV market. The obvious advantages of TFSC over Si solar cells are low temperature processing techniques, low material usage, a variety of deposition processes and its compatibility with low cost substrates. Combining all of these advantages, low cost PV cells are achievable and suit the criteria for large scale applications, such as terrestrial deployment. Types of solar cells that fall under the TFSC category are cadmium telluride (CdTe), copper-indium-gallium-diselenide (CIGSe), copper-zinc-tin-sulfide (CZTS) and amorphous silicon (a-Si) solar cells. CdTe, CIGS, and amorphous silicon TFSCs entered the commercialization stage over 20 years ago.

1.9.3 Third generation solar cells

Third generation solar cells have been developed to enhance the average opto-electronic conversion performance of second generation technology while maintaining low production costs. The goals are to promote thin-film solar cells that use novel approaches to obtain efficiencies in the range of 30%–60%. Dye-sensitized solar cells once flourished to some extent for low cost approaches with great expectations. In recent years, Gratzel and his co-workers achieved 10.9% conversion efficiency using organometal lead halide perovskites as an alternative to dye-sensitized solar cells. The structure of the device is a simple heterojunction thin film. Organo metallic lead halide perovskite ($CH_3NH_3PbI_3$) is currently a leading thin film material for its numerous beneficial properties to realize low cost and high efficiency solar cells. $CH_3NH_3PbI_3$ (bandgap: 1.5 eV) has a high absorption coefficient of over $10^5\ cm^{-1}$, which means that all the potential photons of sunlight with energy greater than the bandgap can be absorbed within the 400-nanometer thick $CH_3NH_3PbI_3$ absorber layer. On the other hand, there are a few more approaches that can improve the efficiency such as spectrum splitting (multi junction solar cells), incident spectrum modification (by using concentrators), multiple electron-hole pair generation by a single photon and some others. Technologies associated with third generation solar cells include multi-junction PV cells, tandem cells, and nano-structured cells to pick up better incident light and convert excess thermal energy to improve voltages or carrier collection.

Nonetheless, there are various PV cells that show different light absorbing materials, structures, and associated fabrication methods. Table 1.1 shows the highest confirmed research scale cell results. These highest confirmed efficiencies have been published under the title of "Progress in Photovoltaics" since 1993 that provide an authoritative summary of the current state-of-the-art. The latest version, 56, was published in 2020.

1.10 Efficiency measurement

Solar cell efficiency is always misinterpreted or speculated for yield. Efficiency is measured from the solar cell's current voltage characteristics at defined standard illumination conditions, but the yield is the outdoor performance of any solar cell or panel for its promised output at standard conditions. Again, solar cell current-voltage characteristics are influenced not only by incident light intensity but also the spectra used for the measurement to replicate sunlight. Therefore, careful consideration is needed to minimize the error margin in the performance measurement. Measuring the characteristics under perfect standard conditions or at a given condition needs special skill. Solar cell efficiency measurement depends on the precise spectral content of the sunlight that fluctuates with the AM, moisture content, turbidity, etc. It is really not appropriate to use natural sunlight as the

Table 1.1 Confirmed non-concentrating terrestrial PV cell efficiencies measured under the global AM1.5 spectrum (1000 W/m^2) at a cell temperature of 25°C (IEC 60904–3: 2008 or ASTM G-173–03 global).

Classification	Eff. (%)	V_{oc} (V)	J_{sc} (mA/cm^2)	FF (%)	Description/reference
Silicon					
Si (Crystalline)	26.7 ± 0.5	0.738	42.65	84.9	Kaneka, n-type rear IBC
Si (DS: Directionally Solidified)	24.4 ± 0.3	0.7132	41.47	82.5	Jinko Solar, n-type
III-V Cells					
GaAs (thin-film)	29.1 ± 0.6	1.1272	29.78	86.7	Alta Devices
GaAs (multicrystalline)	18.4 ± 0.5	0.994	23.2	79.7	RTI, Ge substrate
InP (crystalline cell)	24.2 ± 0.5	0.939	31.15	82.6	NREL
III-V Cells					
GaAs (thin-film)	29.1 ± 0.6	1.1272	29.78	86.7	Alta Devices
GaAs (multicrystalline)	18.4 ± 0.5	0.994	23.2	79.7	RTI, Ge substrate
InP (crystalline cell)	24.2 ± 0.5	0.939	31.15	82.6	NREL
Thin-Film Chalcogenide					
CIGS (Cd Free)	23.35 ± 0.5	0.734	39.58	80.4	Solar Frontier
CdTe (cell)	21.0 ± 0.4	0.8759	30.25	79.4	First Solar, on glass
CZTSSe (cell)	11.3 ± 0.3	0.5333	33.57	63.0	DGIST, Korea
CZTS (cell)	10.0 ± 0.2	0.7083	21.77	65.1	UNSW
Amorphous /microcrystalline Si					
Si (amorphous)	10.2 ± 0.3	0.896	16.36	69.8	AIST
Si (microcrystalline)	11.9 ± 0.3	0.550	29.72	75.0	AIST
Perovskite					
Perovskite cell	21.6 ± 0.6	1.193	21.64	83.6	ANU
Dye-sensitized					
Dye (cell)	11.9 ± 0.4n	0.744	22.47	71.2	Sharp
Organic					
Organic (cell)	15.2 ± 0.2	0.8467	24.24	74.3	Fraunhofer ISE
Multijunction Devices					
Five-junction cell (bonded) (2.17/1.68/1.40/1.06/0.73 eV)	38.8 ± 1.2	4.767	9.564	85.2	Spectrolab, two terminal

Edited from Green, M. A., Dunlop, E., Hohl-Ebinger, J., Yoshita, M., Kopidakis, N., & Hao, X. (2020). Solar cell efficiency tables (version 57). Progress in Photovoltaics: Research and Applications, *2020; 1–13 (Green et al., 2020).*

input light as natural light largely varies depending on the location, atmospheric conditions, seasonal variation, incident angle, luminance, and spectral irradiance. Here, the most recommended efficiency measurement method will be introduced, which uses replicated sunlight. In this case, the standard sunlight distribution to which measurements are referenced is the AM1.5 distribution of Fig. 1.1. Solar simulators are employed as recommended light sources as close as AM1.5 spectral distribution using a filtered xenon lamp, ELH lamp, and even LEDs nowadays. Whatever the light source used, the illumination source must give a collimated beam of uniform intensity at the test plane with stability at the time of measurement within the specified limit.

A typical experimental setup for measuring current-voltage characteristics is shown in Fig. 1.15. It is more desired to have a four-point probe measurement scheme to keep current and voltage probes separate to the device under testing to eliminate the effects from series resistance as well as contact resistances. The PV cell should be kept at 25°C throughout the measurement time with complete shutdown of other light interferences except the solar simulator light. The lamp intensity can be calibrated using a reference cell measured under standard conditions elsewhere. The current-voltage characteristics can then be measured by using variable load resistance to the device under test, that is solar PV cell.

There is another kind of important measurement of solar cells, which is the spectral response of the cell. It is the direct comparison of the output of a cell with the calibrated spectral response. A steady-state source of monochromatic light from a monochromator can be used as the simplest method. It gives the carrier generation or simply current response of the solar cell for each light wavelength with the closest intensity of sunlight replication. The ultimate spectral response of the solar cell would match with the short circuit current density, if properly measured in both methods.

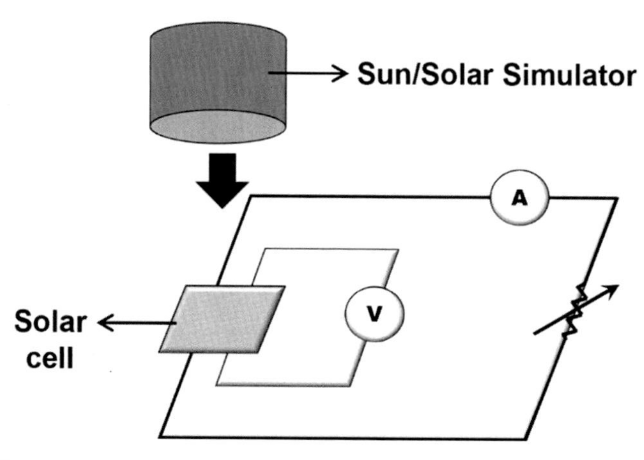

FIGURE 1.15

A typical experimental setup for measuring current-voltage characteristics.

1.11 Summary

Sunlight is continuously pouring down, free of cost, which can more than sustain our existence on earth. Apart from making livable conditions on earth, solar energy can be harnessed as the most efficient form, that is, electricity to cater all our needs using a solar cell. Sunshine that reaches the earth varies drastically and unpredictably in terms of availability, intensity, and spectral composition; however, solar cells of various available materials can convert substantial amounts of sunlight through a phenomenon called the PV effect. Around 98.98% of energy that can be collected on earth comes from solar energy and the remaining 0.02% comes from geothermal energy. As discussed in this chapter, the solar energy sent to earth is 10,000 times more than mankind's average usage. Solar PV devices were realized based on the discovery of the PV effect in the 19th century, but momentum has slowed over the past 70 years. Compared with other energy sources, solar PV energy systems do not require moving parts and silently produces clean energy free of GHG emissions with minimal maintenance. Expansion or scale-merit is better than any others due to the addition of solar panels for any expansion. Based on known physics and the adoption of novel technology for the materials and structures, solar cells have evolved dramatically in recent times having light to electricity conversion efficiencies increase from 6% (single cell) to nearly 50% (tandem). Novel PV materials are being researched extensively, and larger solar farms are being established around the world. A majority of countries have optimistically incorporated around 50% of their annual energy sources to be solar PV energy for their energy roadmap by 2050. It is hoped that the fundamental discussion in this chapter assists the readers to a level that they can easily follow the subsequent chapters on other topics related to PV energy technologies.

Acknowledgments

The author wishes to thank the Ministry of Higher Education of Malaysia (MOHE) for providing the research grant with the code of LRGS/1/2019/UKM-UNITEN/6/2 to support our research activities on solar photovoltaic energy. Appreciation is also give to the Universiti Tenaga Nasional (UNITEN) for providing substantial support on solar energy related R&D over the years.

References

Green, M. A., Dunlop, E., Hohl-Ebinger, J., Yoshita, M., Kopidakis, N., & Hao, X. (2020). Solar cell efficiency tables (version 57). *Progress in Photovoltaics: Research and Applications*, *2020*, 1–13. doi:10.1002/pip.3371.

Further reading

Green, M. A. (1998). *Solar cells: Operating principles, technology and systems applications*.

Hamakawa, Y., & Kuwano, Y. (1994). *Solar enegy engineering: Solar cells*. Baifukan Publications.

Manabe, S., & Stouffer, R. J. (1980). *Journal of Geophysical Research*, *85*(1980), 5529.

Nelson, J. (2008). *The physics of solar cells*. Imprerial College Press.

Thekackara, M. P. (1970). The solar constant and the solar spectrum measured from a research aircraft. NASA Technical Report No. R-351.

CHAPTER

Organic solar cells 2

Ashraf Uddin

School of Photovoltaic and Renewable Energy Engineering, University of New South Wales, Sydney, NSW, Australia

2.1 Introduction and working principles

The development of a sustainable future energy option is one of the most important challenges for the human race. The solar energy components of the energy needs will continue to grow significantly as pressure mounts to generate power in a clean and renewable way. Photovoltaic (PV) energy is the most rapidly growing sector of the electricity generation industry worldwide. Among the many alternative energy sources, PV technology shows high potential and cost-effectiveness that can efficiently convert sunlight into electricity and fulfill the global energy demand upon large area deployment. The PV effect, which is the physical phenomenon responsible for converting light to electricity, was first observed in 1839 by French physicist, Edmund Becquerel (1839). Becquerel observed a voltage generated when one of two identical electrodes in a weak conducting solution was illuminated.

PV energy has immense potential for many terrestrial applications, including bulk utility-based power generation in solar plants and building-integrated PV (BIPV) and building-applied PV in smart buildings. The interest in PV systems continues to grow with more countries investing on PV systems to meet the growing energy demand.

The world PV market is currently dominated by inorganic crystalline silicon solar cells, which constituted approximately 90% of the market in 2019. One of the main challenges in the silicon PV industry is the high fabrication and panel installation costs. Therefore, many studies have been conducted by the PV research community to overcome the cost issues with next generation PV technologies. Heliatek, based in Germany, is actively seeking the application of OSCs because the production of organic solar cells (OSC) and the use of materials are much less expensive than crystalline silicon solar cells. The company is experimenting with organic PV modules on concrete walls outside the building, as well as on glass and metal.

Organic photovoltaic (OPV) devices have received a great deal of attention, mainly due to their several advantages, especially their low cost, lightweight, and

Comprehensive Guide on Organic and Inorganic Solar Cells. DOI: https://doi.org/10.1016/B978-0-323-85529-7.00006-2

simple fabrication on flexible substrates. OSCs have received widespread attention and will be an excellent alternative for many different market segments. The simplicity of materials, flexibility and cost are issues that cannot be solved by silicon solar cells in the future. OSCs are lightweight, adaptable to color, and can be produced on transparent and flexible surfaces at low cost. This makes them attractive for markets where other technologies cannot compete, such as in building and consumer product integration. However, the efficiency and long-term operational stability of OSCs have not yet reached the same levels as their inorganic PV counterparts.

OPV devices must meet two basic requirements to be competitive: a PCE of more than 18% and a an average lifetime greater than 12 years. Recently, through collaborative development of high-performance photoactive materials, an understanding of the membrane nanostructure morphology, and device structural optimization, PCE has steadily increased due to iterative upgrades of materials and techniques, reaching a record 18% PCE (Liu et al., 2020), the threshold required for commercial viability (Liu et al., 2020). However, high-performance implementation has little impact on the technology if the resulting device life does not meet process requirements. Many environmental factors affect the performance of OSCs, including heat, light, humidity, and oxygen exposure, causing the performance of OSCs to be significantly affected during operation. Therefore, increasing the longevity and commercialization of OSCs remains a significant challenge. Thermal degradation is a critical factor that inevitably leads to performance degradation when solar panels are operating outdoors, as solar panels typically reach 65°C–85°C in sunlight. Recent studies of several high-efficiency OPV devices have shown that even after concise running times, the efficiency can be severely compromised.

A most important factor in producing high efficiency OSCs is the donor/acceptor morphology in the bulk heterojunction (BHJ) active layer, as shown in Fig. 2.1. The donor/acceptor morphology is formed in the active layer after spin-coating of the blended solution. The scale of the donor/acceptor phase separation depends on spin-coating process parameters such as speed, temperature, solvent, and material solubility. The enhancement of charge separation and transport in OSCs device structure could be expected to be substantially higher if the donor/acceptor morphology can be controlled on a molecular scale. Compared to inorganic semiconductors, the physical

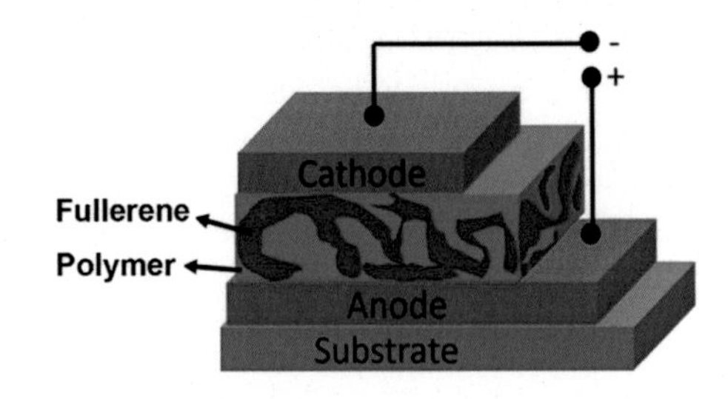

FIGURE 2.1

A schematic diagram of a bulk-heterojunction organic solar cell device structure.

understanding of OSCs materials and devices are still lacking because of the fundamental differences in the optoelectronic properties of organic materials. The scope of organic semiconductor material and device design are limited by this lack of understanding. Among those limitations, the differences of OPV devices compared to inorganic semiconductor-based devices are: (1) the organic semiconductor materials are electronically disordered, dispersing the rates of charge transfer and transport processes; (2) the organic semiconductor is usually not doped, thus precluding the conditions that allow charge dynamics to be linearized in the description of device physics; (3) the active layers are often heterogeneous, either as multicomponent films or because of nonuniform molecular ordering; (4) the charges and excited states are localized on individual molecules or molecular segments, with the result that charge and energy transport processes are relatively slow; and (5) the dielectric permittivity is low, leading to stronger space charge effects.

2.1.1 Device performance characteristics

Solar cells are semiconductor devices that convert light into electricity. Under illumination, light flux is shined on the solar cells. When the photon energy is equal to or greater than the bandgap of the material, the photon is absorbed by the material, one electron is excited into the conduction band (C_B), and one hole is left in the valence band (V_B). The excited electron-hole pairs must be separated and collected to extract the energy. The most common structure of the solar cell is a p-n junction. By creating the doping concentration bias, the electron-hole pairs can be separated and then be collected. The p-n junction can be formed by single bulk material with differential doping concentrations as shown in Fig. 2.2.

Another commonly used structure is the heterojunction solar cell. These solar cells usually have a p-i-n structure. Three materials with different doping behaviors are integrated into the devices. The electron transports through the electron

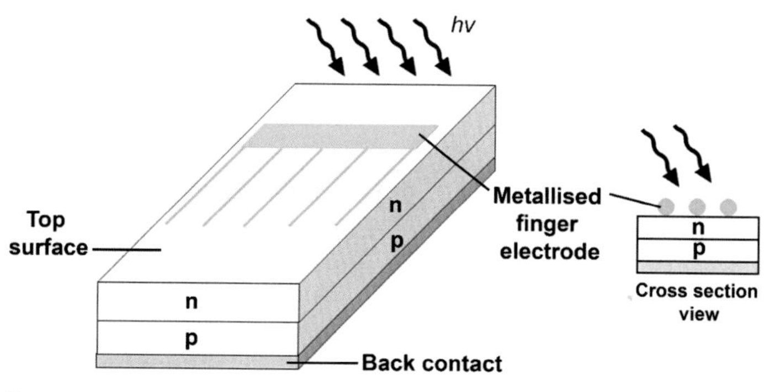

FIGURE 2.2

Schematic diagram of a p-n junction type silicon solar cells.

transportation layer and is then collected by the cathode. Holes transport through the hole transportation layer and are collected by the anode.

Under illumination, the solar cell can build potential due to the excitation of electrons and holes. The current density-voltage (J-V) curve of a solar cell is the J-V characteristics of the diode shifted by the light-generated current as shown in Fig. 2.3. In the dark, carriers are injected into the circuit by applying a forward bias voltage. Under illumination, the J-V curve shifts to the fourth quadrant as the carriers are generated and is described by an established model using the Shockley equation:

$$J = J_0\left[\exp\left(\frac{\mathrm{qV}}{\mathrm{nkT}}\right) - 1\right] - J_L \tag{2.1}$$

where J is current density, J_0 is dark saturation current density, J_L is light generated current density, q is elementary charge, V is voltage, n is ideality factor, k is Boltzmann's constant, and T is the absolute temperature. The short circuit current J_{sc} is defined as the current density through the solar cell at zero applied bias ($V = 0$) under illumination. Additionally, it is associated with the external quantum efficiency (EQE). The relationship is expressed as: (Feng, Liu, & Yu, 2014)

$$J_{sc} = \frac{q}{hc}\int_{\lambda_{min}}^{\lambda_{max}} EQE \times P_{in}(\lambda) \times d\lambda \tag{2.2}$$

where λ, h, and c denote wavelength, Planck's constant, and the speed of light, respectively. In general, the EQE as a function of wavelength is acquired by

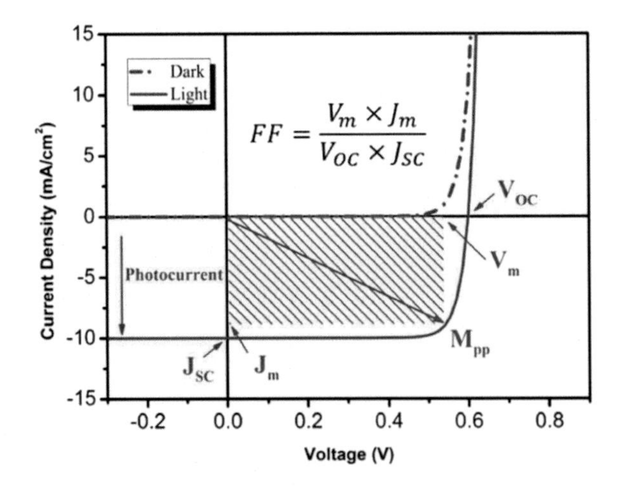

FIGURE 2.3

Current density—voltage (J-V) characteristics of a photovoltaic device under illumination and in the dark. Short-circuit current density J_{sc}, open-circuit voltage V_{oc}, current density J_m, and voltage V_m at the maximum power point and fill factor are shown.

Reproduced with permission from Qi, B., J. Wang, Fill factor in organic solar cells, Physical Chemistry Chemical Physics, 15 (2013) 8972–8982, © 2013, The Royal Society of Chemistry (Qi and Wang, 2013).

comparing the number of photogenerated collected electrons to the number of incident photons at a specific wavelength. The EQE of a fully functional OPV device is the product of four types of efficiencies, each corresponding to a step in the charge generation process. These include the efficiency of light absorption (η_A), exciton diffusion (η_{ED}), charge separation (η_{CS}), and charge collection (η_{CC}), giving:

$$EQE = \eta_A \times \eta_{ED} \times \eta_{CS} \times \eta_{CC} \tag{2.3}$$

At open circuit conditions, the value of V_{oc} can be expressed as the following equation when there is no current passing through the cell:

$$V_{oc} = \frac{nkT}{q} \ln\left(\frac{J_L}{J_0} + 1\right) \tag{2.4}$$

In BHJ OSCs, the V_{oc} can be described by a linear fit between the highest occupied molecular orbital (HOMO) level of the donor and the lowest unoccupied molecular orbital (LUMO) level of the acceptor. Compared with inorganic solar cells, OSCs suffer from a large V_{oc} loss. The origin of the V_{oc} loss has been intensively studied (Elumalai and Uddin, 2016). An empirical loss factor of 0.3 V is introduced to determine the maximum voltage of the OPV device (Cowan, Roy, & Heeger, 2010; Scharber et al., 2006):

$$V_{oc} = \left(\frac{1}{e}\right)\left(\left|E_{HOMO}^{Donor}\right| - \left|E_{LUMO}^{Acceptor}\right|\right) - V_{loss} \tag{2.5}$$

Despite the empirical evidence developed by Scharber et al. (2006), an experimental investigation discovered a linear relationship between the V_{oc} and effective bandgap (E_g) via analysis of charge transfer absorption by means of Fourier transform photocurrent spectroscopy. An equation was developed to determine the V_{oc} upon applying a linear fit to various OSCs:

$$V_{oc} \approx \frac{E_g}{e} - 0.43V \tag{2.6}$$

Due to the intricate mechanism governing the operation of OSCs, multiple factors can influence the V_{oc} (Elumalai and Uddin, 2016). Recently, Brebels et al. reported that increasing the dielectric constants could potentially mitigate the inherent V_{oc} restriction imposed by loss originating from the D-An energetic offset (Brebels, Manca, Lutsen, Vanderzande, & Maes, 2017). The authors thus concluded that the V_{oc} is correlated to the binding energy of the charge transfer excitation, i.e. the dielectric constant:

$$V_{oc} = \frac{E_{LUMO}^{Acceptor} - E_{HOMO}^{Donor} - E_b}{q} - C \tag{2.7}$$

where the E_b is the exciton binding energy, C is the constant related to temperature and illumination, and q is the elementary charge.

The fill factor (FF) is the product of J_{sc} and V_{oc} in conjunction with the maximum power point (MPP) of both current-density and voltage, representing the ratio of the squareness of the solar cell as depicted in Fig. 2.3. The shape of the square characterizes how efficient the photogenerated carriers can be extracted to corresponding electrodes. The ideal condition for FF is 100%, when the J-V curve in which the MPP coincided with J_{sc} and V_{oc} is a rectangle. The FF can be expressed as:

$$FF = \frac{J_{MP} \times V_{MP}}{J_{sc} \times V_{oc}} \tag{2.8}$$

where J_{MP} and V_{MP} denote current-density and voltage at MPP, respectively. FF is defined as the ratio of the maximum power to the product of V_{OC} and J_{SC}. FF describes the quality of the devices. A schematic J-V diagram is shown in Fig. 2.3, where the maximum power output is $J_{mp} \times V_{mp}$. It is usually linked to two major resistive losses: shunt resistance and series resistance. Shunt resistance describes the difficultly of current leaking inside the device, therefore the larger the shunt resistance, the better FF value. Series resistance is defined by the resistive loss when the current flows from the devices. A smaller series resistance is preferable in solar cells to decrease resistive loss. A circuit diagram of a solar cell including the shunt resistance and series resistance is shown in Fig. 2.4.

Based on the above-mentioned three PV parameters, a mathematical formula was utilized to describe the efficiency of a solar cell:

$$PCE = \frac{J_{SC} V_{OC} FF}{P_{in}} \tag{2.9}$$

where PCE is power conversion efficiency and P_{in} is the incident light power density. P_{in} is standardized at the AM 1.5 spectrum with an irradiation intensity of 1000 W/m^2, while the solar cell is operating at a temperature of 25°C. The PCE of the solar cell is heavily dependent on variations on those PV parameters (J_{sc}, V_{oc} and FF), in which light absorption of organic materials, combinations of

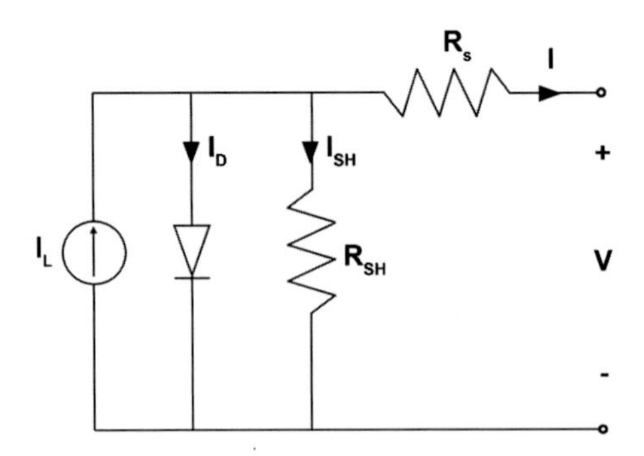

FIGURE 2.4

The equivalent circuit diagram of a solar cell includes the shunt resistance R_{SH} and series resistance R_S. The larger the shunt resistance R_{SH} is, the less current leaked within the device. The smaller the series resistance R_s is, the smaller the resistive loss.

polymer/fullerene, band alignment of donor and acceptor with neighboring layers, morphology of active layers, charge transport, and recombination at the inference of active layer/hole transporting layer (HTL) or active layer/electron transporting layer (ETL) influences those parameters.

2.1.2 Single layer organic solar cells

The simplest structure of the single active layer OSC is shown in Fig. 2.5A. It is made by sandwiching an organic electronic material between two metallic electrodes. The electrodes are normally a layer of wide bandgap material indium tin oxide (ITO) with high work function and a layer of low work function metal such as silver, calcium, or aluminum. An electric field is developed in the device structure by the difference of work function between the two electrodes. When the light photon is absorbed in the organic active layer, an electron-hole pair is generated. An electron is transferred to the LUMO and leaves holes in the HOMO to form excitons. The work function differences between the two electrodes help to split the exciton (electron-hole pairs), thereby attracting electrons to the positive electrode and holes to the negative electrode.

The structure of a single layer OSC is to place a thermo-evaporative organic molecular layer between two electrodes with different functions (transparent electrode/organic photosensitive semiconductor/electrode), so that current-carrying electrons can be easily moved between the two electrodes because the two electrodes have different work functions. In this case, the Schottky barrier will form between the lower function electrode and polymer. The quantum efficiency of single layer OSC is low due to the low carrier mobility of 10^{-3} cm^2/Vs in the organic electronic material. This shows that the charge generated by the light

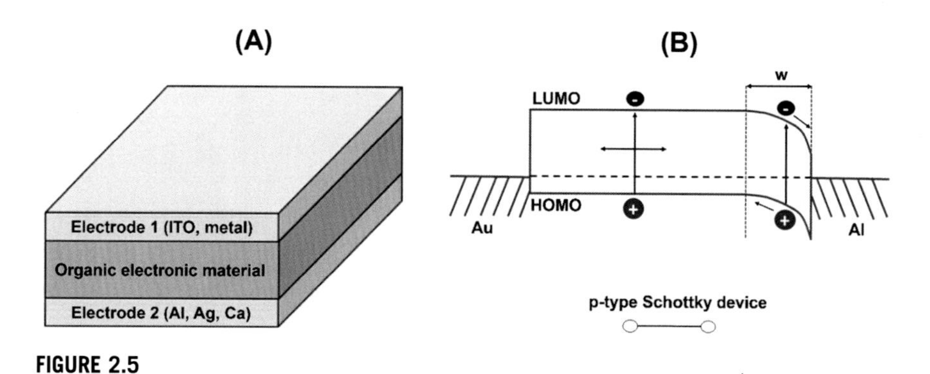

FIGURE 2.5

(A) Schematic diagram of one single layer device structure of organic solar cell. (B) Energy bend diagram of the device structure with a Schottky contact at the Al side, therefore only light induced carriers with the W width region can be separated, which is described as exciton diffusion limited.

energy in the organic semiconductor requires more time to collect from the electrode. The low mobility of the charge will increase the charge recombination, which then reduces the efficiency of the OPV cell. The formation of excitons also results in a lower efficiency of OSCs.

2.1.3 Bilayer organic solar cells

The structure of bilayer OSCs contains two organic electronic layers sandwiched between two conductive electrodes, as shown in Fig. 2.6A. Electrostatic forces are generated at the interface between the two organic layers due to differences of electron affinity (EA) and ionization energies. The organic materials are selected to make the electrostatic forces large enough that excitons can be split more efficiently than the single layer PV cell. One organic layer is called a donor layer, which is normally a smaller bandgap material that can easily absorb light to produce electron-hole pairs. Another organic layer is called the acceptor layer which can easily accept electrons from the donor layer. The electron acceptor layer has higher EA and ionization potential (IP). This bilayer OSC structure is also called a planar donor-acceptor heterojunction.

A bilayer heterojunction OSC is achieved by the stacking of two different organic materials contacting two metal electrodes. The difference in EA will contribute to the separation of carriers when the working functions of the metal contacts match the HOMO of the donor material and the LUMO of acceptor material, respectively.

The electron of the acceptor will transfer from the HOMO to LUMO when:

$$I_D - A_A - U_C < 0$$

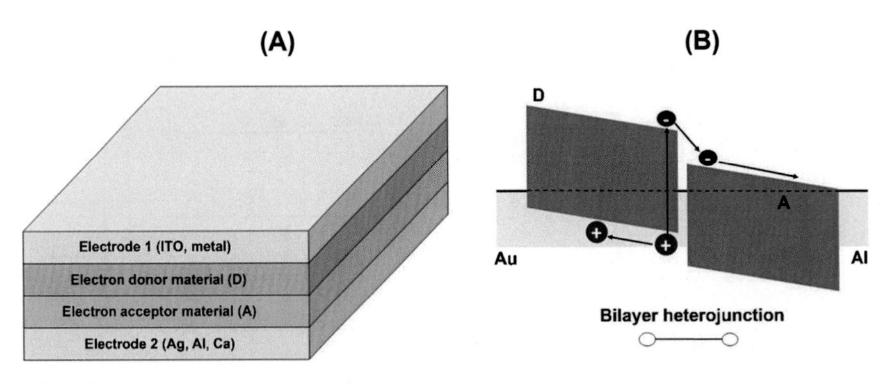

FIGURE 2.6

(A) Schematic diagram of a bilayer heterojunction organic solar cell, where D and A refer to donor and acceptor, respectively. (B) Energy band diagram of a bilayer heterojunction solar cell. The HOMO of the donor will contact the metal of the higher working function, and the LUMO of the acceptor will contact the metal of the lower working function.

where I_D stands for the ionization potential of the donor, A_A is the EA of the acceptor, and U_C is the Coulombic interaction. The excitation energy for the donor electron to transfer to the acceptor LUMO is from illumination. One advantage of the bilayer heterojunction is the monomolecular charge transport. After the separation of excitons, n-type material (acceptor) will support the transport of electrons, while p-type material (donor) will support the transport of holes. By the monomolecular charge transport, electrons and holes are effectively separated and the possibility of recombination is significantly reduced. Bilayer hetero unction solar cells show linear dependency of light-induced current on illumination. It can by synthesized from casting or evaporation of material on the substrate or the other material.

Excitons are effectively dissociated at the donor-acceptor interface, and a double-layered OPV cell is formed by inserting a receptor layer between the donor layer and cathode. However, until now, double-layer OSCs have obtained PCE values much smaller than inorganic PV cells, mainly because the organic semiconductor exciton diffusion length is short, generally around 10–20 nm. Compared to the single-layer concept, the two-layer concept allows for greater efficiency because it also allows for the reduction of the distance that excitons diffuse before dissociation, thus significantly reducing the charge carrier recombination. The main advantage of a two-layer OSC is that the carrier moves in the appropriate organic material phase after exciton dissociation, which reduces charge recombination.

2.1.4 Bulk-heterojunction organic solar cells

The introduction of a BHJ as an active layer in OSCs has significantly improved the performance of polymer PV cells. It forms a large heterojunction in the active layer by both the donor and acceptor electronic materials and then allows the two phases to separate so that the two components self-assemble into a porous network connection. This structure is advantageous over single and bilayer structures because the morphology of the film is controllable, allowing for a dissociation site for all generated excitons. In fact, using exciton diffusion length order receptors and donor nanodomains, excitons always find a donor-acceptor interface before recombination. BHJs of conjugated polymers and fullerene derivatives represent an exciting approach to solid-state OSCs. The BHJ is the most efficient structure utilized today. In such a configuration, the organic active layer is a complex diffusive interface in the form of a bi-continuous interpenetrating network between the two materials, and the surface area of the interface can be utilized to its greatest extent.

BHJ solar cells often need to mix donor and acceptor molecules at the nanoscale, because many prototype conjugated polymers are used in the nanometer range. Excellent charge transfer is premised on ordered donor and acceptor domains, where such an arrangement yields the best results. The results show that a domain width of the exciton diffusion length of 2 times thicker can simultaneously promote the active generation of electric charge. The preparation of such a nanopattern is complicated, and the device fabrication often relies on the phase separation of the donor material and the acceptor material during the formation of the absorber film. Still, despite all the recent improvements, the efficiency of these devices has not been high enough for the commercially applications.

2.1.4.1 Operating principles of bulk heterojunction

In OSCs, light absorption creates a bound exciton and then, current can be produced by the dissociation of the exciton. BHJ OSCs are generally comprised of donor and acceptor materials, leading to many spatially distributed heterojunctions rather than planar heterojunction. Compared to the planar structure, the BHJ structure enables more excitons to be generated at the donor-acceptor (D-A) interface due to increased contact area and reduces the average distance for an exciton to travel to the interface. This structure requires energetic offsets between the donor and acceptor to attain high J_{sc} and FF.

These unique properties have given rise to all the main differences in terms of operating mechanism between organic and inorganic solar cells. For OPV devices, photogeneration of bound electron-hole pairs due to photon absorption largely hinges on the polymeric and acceptor materials. The detailed elucidation of the operating mechanism of BHJ OSCs is illustrated in Fig. 2.7. The photoactive layer sandwiched between two dissimilar electrodes absorbs a photon resulting in the creation of an exciton, which is a coulombically bound electron and hole pair. In the organic active layer, this photoinduced exciton generated in the donor material is dissociated at donor-acceptor interfaces to produce the photocurrent. The driving force required to overcome the exciton's high binding energy ($0.3 \sim 1$ eV) is provided by the energy offset in the LUMO between the donor and

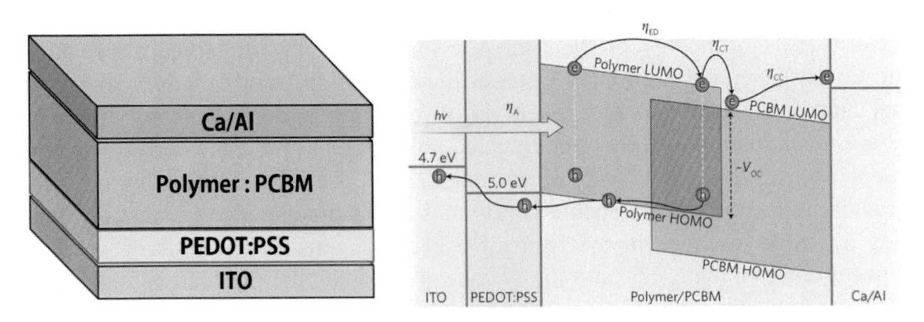

FIGURE 2.7

Schematic diagram of the bulk heterojunction organic solar cells (OSC) structure and energy band diagram and the operating principles of an OSCs.

Reproduced with permission from Ref. Li, G., Zhu, R., & Yang, Y. (2012). Polymer solar cells. Nature Photonics, 6, *153., © 2012, Springer Nature (Li, Zhu, and Yang, 2012).*

acceptor material. Thus, the exciton diffuses to the donor-acceptor interface where the bound electron-hole pair is separated into free electron and hole before deactivating to the ground state. After that, the positively charged hole and negatively charged electron are collected by the corresponding cathode electrode and anode electrode, respectively, thus providing the photocurrent.

Only a fraction of the incident light is absorbed with photon energies greater than or equal to the bandgap energy (E_g). The bandgap is subject to the semiconductor material, which is defined by the energy level between the HOMO and LUMO, typically in the range of 1 to 3 eV. It is worth mentioning that the molecular orbital stands for one-electron wave functions associated with the energies of the HOMO and LUMO. However, it is important to note that the molecular orbital cannot be measured, whereas the IP and EA can be determined experimentally. Therefore, the HOMO and LUMO levels are defined via theoretical calculations, which can be considered as the vertical IP and vertical EA, respectively. Unlike metals (no bandgap) and insulators (wide bandgap), the E_g of semiconductors is in an intermediate range of 0.5–4.0 eV. An E_g of 1.1 eV is able to absorb 77% of the solar irradiation, whereas most organic semiconductors materials have bandgaps greater than 1.5 eV, which reduces the number of solar photons to be captured to about 50%. It is obvious that the utilization of low bandgap polymers in OSCs increases the possible harvesting of the photon and subsequently improves the J_{sc}.

2.2 Normal and inverted device structure configurations

Typically, OSCs consist of a thin organic layer embedded between two electrodes where at least one side of the device is transparent, allowing for light to pass through, as shown in Fig. 2.8. The thin photo-active layer has been developed on substrates like glass or flexible polyethylene terephthalate (PET) with a highly transparent conducting layer of ITO. This ITO conductor not only allows light to access the active layer but also collects charge carriers from the active layer. Two types of device architectures are commonly used, one has four layers on the substrate and the other has five layers, enabling selective transport of carriers to the corresponding electrodes. For the normal device structure, a superior efficiency of over 6% was attained with P3HT:phenyl C61 butyric acid methyl ester ($PC_{61}BM$) blend, in which a conducting layer of poly(3,4-ethylenedioxythiophene)-poly(styrenesulfonate) (PEDOT:PSS) was inserted between the ITO and the active layer using solution deposition techniques. This layer has been treated as a HTL that blocks off electrons to access the anode electrode, as well as serving as a buffer layer which smooths out the ITO and seals the active layer from oxygen.

Traditionally, the final layer of this normal structure ends up with an electron accepting electrode that has low work function, such as aluminum (Al) or silver (Ag).

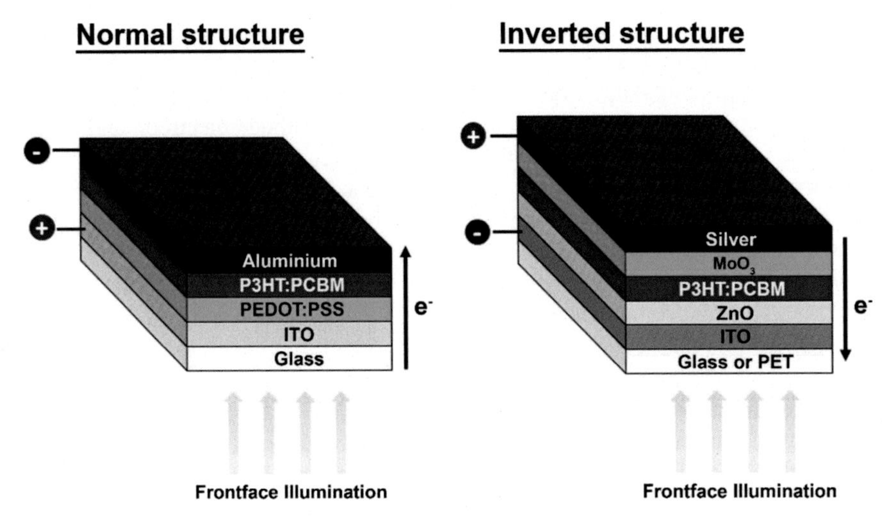

FIGURE 2.8

The normal device structure comprises spin-coated PEDOT:PSS as hole transporting layer (HTL) and P3HT:PCBM as active layer and evaporated Al as electrode. The inverted structure contains five layers instead of four layers. The device consists of spin-coated ZnO as ETL, P3HT:PCBM as the active layer, and evaporated MoO_3 and Ag as the HTL and electrode, respectively.

However, sometimes, an electron transport layer [i.e. ZnO or lithium fluoride (LiF) or tris(8-hydroxyquinoline) aluminum (Alq3)] has been utilized in conjunction with the metal electrode to improve the PCE. With this normal structure, the light enters the photoactive layer through the substrate, also known as a back-contact device. However, air stability has been a major issue for this normal structure. With exposure to air, PEDOT:PSS can absorb moisture present in the atmosphere and release the acidic protons etching away the ITO layer. Apart from that, the degradation mechanisms have been unequivocally revealed in previous studies employing low work function metallic electrode, where the oxidation of aluminum allows the penetration of oxygen and moisture to the active layer, resulting in significant degradation for OSCs. To overcome these shortcomings, one alternative approach is to use an inverted device structure, whose direction of charge transport is opposite the normal structure, because of the inverse order of the layers. A comparison of two types of device structures is shown in Fig. 2.8. The inverted structure leads to better air stability due to the elimination of PEDOT:PSS, where the acidic and hygroscopic nature of PEDOT:PSS is detrimental for not only the ITO, but also the metal electrode. Among alternative HTL materials, molybdenum trioxide (MoO_3) has been demonstrated to have outstanding optical and electrical properties, when substituted for PEDOT:PSS. Additionally, a more stable and high work function metal, such as silver (Ag) or gold (Au), can be employed as the anode electrode, thus preventing fast degradation of the device.

2.3 Key factors behind organic photovoltaic cell efficiency

Several key factors influence OPV cell efficiency. Donor/acceptor materials can be divided into three categories based on their optical bandgap: (1) narrow bandgap material; (2) medium bandgap material; and (3) wide bandgap material. Usually, narrow optical bandgap materials have a wider absorption range of solar spectra. Other key factors include the electrical, optical, and structural properties of the organic photoactive materials. They also include interface engineering to enhance the carrier collection probability at the electrodes. To reduce the ohmic contact resistance at the electrode (cathode and anode) interfaces, additional carrier transport layers, such as electron transport and hole transport layers with suitable work functions to match the energy levels of donor and acceptor materials (Fig. 2.9), are generally inserted between the BHJ active layer and electrodes. These electron and hole transport layers also block the carriers to reach the opposite electrodes. In addition, selecting the appropriate device structure is crucial.

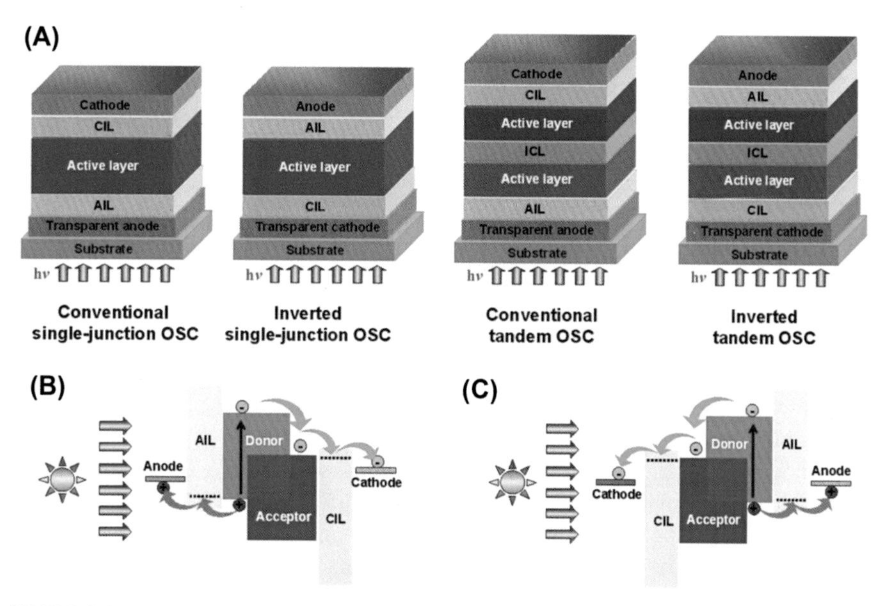

FIGURE 2.9

(A) Schematic of the device structures of conventional and inverted single-junction OSCs with an anode interface layer (AIL) and cathode interface layer (CIL). (B) and (C) are the schematic bang energy level diagram and charge-transporting processes in conventional and inverted OSCs, respectively.

Reproduced with permission from Ref. Zhigang, Y., Jiajun, W., & Qingdong, Z. (2016). Interfacial materials for organic solar cells: Recent advances and perspectives. Advanced Science, 3, *1500362, © 2016, John Wiley and Sons (Zhigang, Jiajun, and Qingdong, 2016).*

For high efficiency OSCs fabrication, different methods have been adopted by molecular engineers to strategically design small molecules. The modulation of end groups is one of the most noteworthy methods (constitutional or functionalities units at the extremity of a macromolecule or oligomer) for easy alignment of molecular energy level. As an example, Ding and his coworkers (Xiao, Jia, & Ding, 2017) developed a low-bandgap acceptor small molecule CO*i*8DFIC, by assembling an electron-donating CO-bridged ladder-type unit (CO*i*8) with two electron-accepting end groups (DFIC). The highly efficient acceptor ITIC was modulated with thienyl-fused indanone end-groups. A new small molecular acceptor (SMA) (ITCC) was designed with a higher lying LUMO level and more compact π−π stacking distance that led to an impressive V_{oc} of ~1 V and improved FF. Halogenation, specifically fluorination, is the another repeatedly used end group modulation technique. This is implemented by using electron-deficient elements such as F- or Br- atoms into the end-capping groups (Wang et al., 2018). Fluorination can simultaneously downshift the HOMO and LUMO levels of the modulated molecules without causing strong steric hindrance. The fluorinated molecules also exhibit improved intramolecular interactions due to the noncovalent interactions, which, in turn, enhance the crystallinity of molecules charge transport. The higher polarization and a reduced Coulombic potential between holes and electrons are also observed in fluorinated molecules. However, these highly efficient fluorinated SMAs usually require multiple-step synthesis processes that can critically restrict material accessibility and require device upscaling for mass production.

Small molecule donors (SMDs) are used in high performance solution processed OSCs at a smaller scale, especially in tandem cells as a high bandgap front cell donor material, to avoid the overlap of absorption spectra and subsequent low J_{sc}, which are typically observed in tandem solar cells. SMDs have easier energy level tunability and reduced batch-to-batch variation (Li et al., 2016). The molecular structures of the fullerene and small molecule nonfullerene acceptors (NFA) used in high efficiency OSCs are shown in Fig. 2.10.

In novel polymer design, alkyl side chain-engineering plays a key role. The alkyl side chains of conjugated polymers can influence the PV properties for aggregation behavior, including stacking/packing and intermolecular interactions in the solid state. Liu et al. (2017) have designed and synthesized a series of wide bandgap copolymers with different alkyl side chains on benzodithiophene (BDT) donor- and benzodithio-phene-4,8-dione (BDTDO) acceptor- units. They have found that the length of the alkyl side chains impacts the optical, electrical, and PV properties and the molecular packing and crystallinity of polymers. Large steric hindrance and the decreased planarity of the main backbone can be caused by the longest alkyl substituents on the BDTDO unit and lead to decreased π−π stacking interactions.

The best PV performance with a moderate bulky side chain on BDTDO unit are observed with high PCEs over 10% in both fullerene- and NFA-based OSCs.

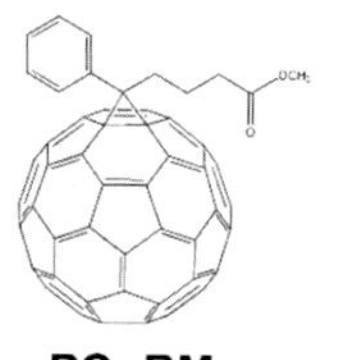

$PC_{61}BM$

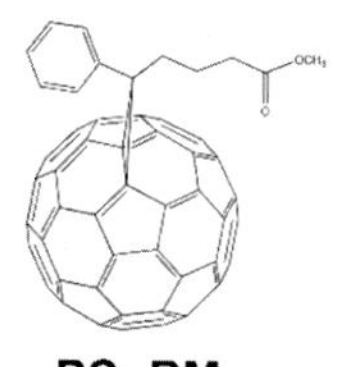

$PC_{71}BM$

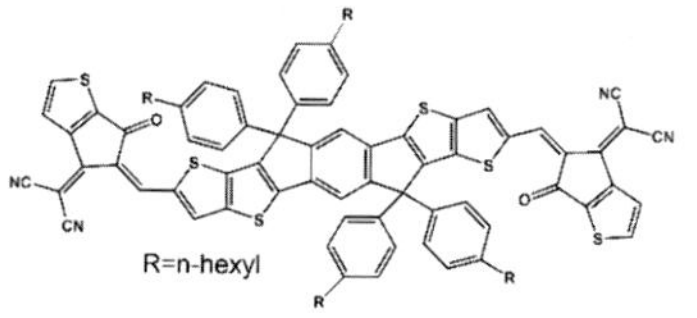

R=n-hexyl

ITCC

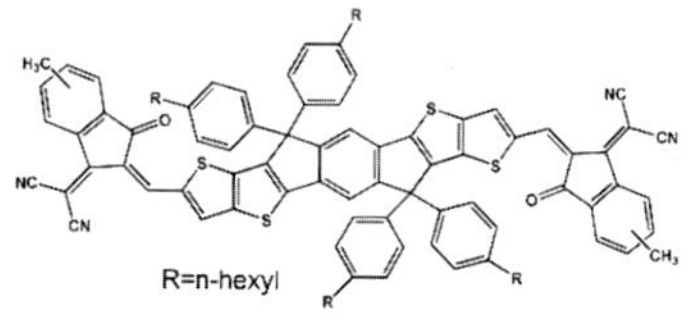

R=n-hexyl

IT-M

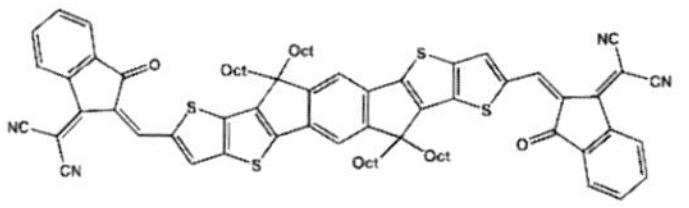

C8-ITIC

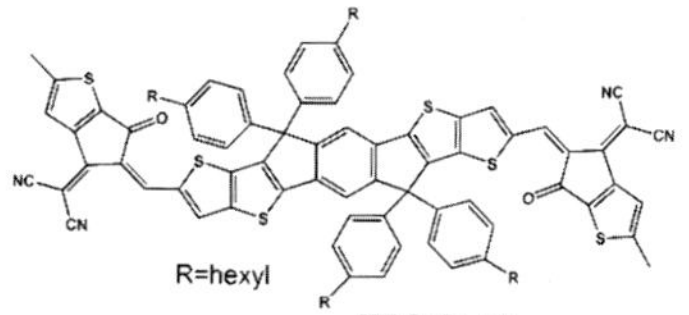

R=hexyl

ITCC-M

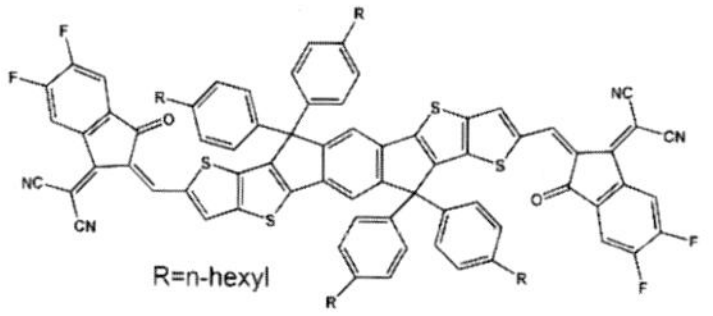

R=n-hexyl

IT-4F

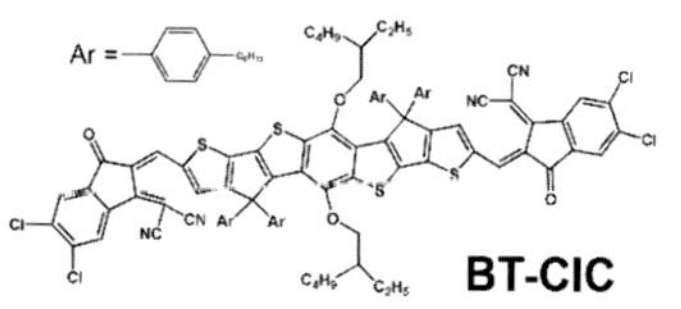

BT-CIC

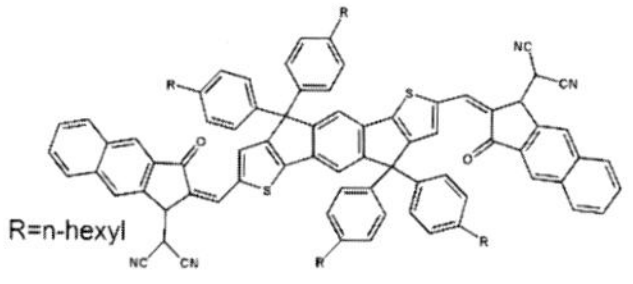

R=n-hexyl

IDTN

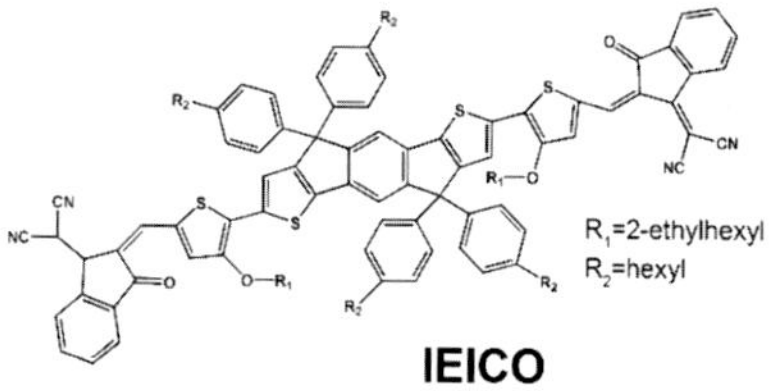

R_1=2-ethylhexyl
R_2=hexyl

IEICO

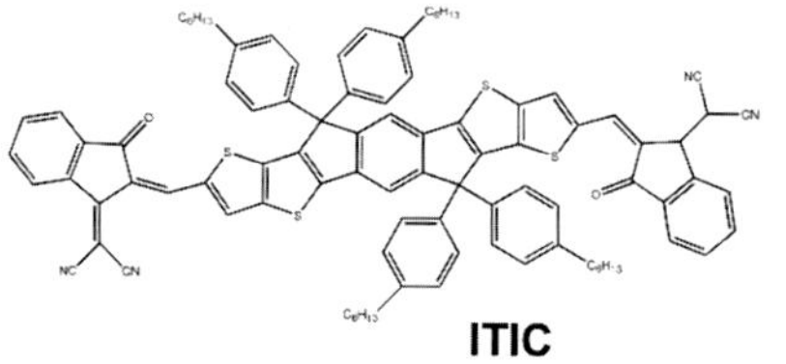

ITIC

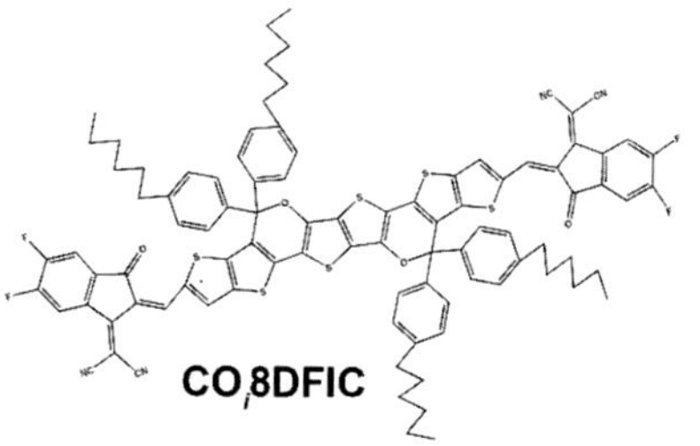

CO$_i$8DFIC

(*Continued*)

In a novel polymer, BDTS-TDZ alkylthio side chains demonstrated stronger intermolecular interactions. For efficient charge transport, the corresponding energy levels of the polymer should be lowered, and better crystallinity and faces on the packing properties are required by the insertion of sulfur atoms in the alkylthio side chains. The molecular structures of polymer and SMD materials used in high efficiency OSCs are shown in Fig. 2.11.

An efficient way to improve OSCs is to design the narrow bandgap organic materials with a wide range of absorptions of the solar spectra. Material selection for the high performance OSCs should be based on complementary absorption between the donor and acceptor pair for the wider absorption range. As an example, if an organic acceptor material with a light absorbing peak at 800 nm must achieve a maximum absorption range, this acceptor material can be paired with a donor material with a light absorbing peak at 500 nm. NFA materials have an advantage of an easily tuneable absorption spectrum and a more intense absorption compared with fullerene materials. NFA materials have more flexible choices for donor and acceptor pairs for BHJ OSCs.

Moreover, energy level alignment of the donor-acceptor pair also plays an important role in improving the performance of OSC devices. By selecting the donor with a low HOMO level and an acceptor with a high LUMO level, the open-circuit voltage (V_{oc}) of device can be enhanced. In a BHJ device, the offset of HOMO and LUMO levels between acceptor and donor is the driving force for the dissociation of excitons. For efficient exciton dissociation, a large HOMO and LUMO offset is needed for high efficiency OSC devices. However, low V_{oc} and high energy loss are associated with a large HOMO/LUMO level offset. Thus, for designing a high-performance OSC device, it is essential to find a balance between efficient exciton dissociation and low energy loss. A driving force higher than 0.3 eV is needed for the most efficient fullerene OSCs. However, in a recent development of NFA-based devices, efficient exciton dissociation was exhibited even with a donor-acceptor band edge offset lower than 0.3 eV. The nonfullerene-based device indicates a larger potential to achieve high efficiency with high V_{oc}.

The donor-acceptor morphology is another key aspect for improved device performance. The molecular geometries of the donor and acceptor materials are mainly related to the morphology of the blended layer. It can be explained in four aspects, including domain size, phase separation, molecular orientation, and π-π stacking. In fullerene-based devices, the spherical geometry of the fullerene

FIGURE 2.10

Acceptor materials—molecular structures of the fullerene and small molecule NF acceptor materials for high efficiency organic solar cells. The represented molecules are phenyl-C61-butyric acid methyl ester (PCBM), carbonoxygen-bridged unit (COi8), difluoro-substituted IC (DFIC), indacenodithieno[3,2-b]thiophene (IT), 2-(3-oxo-2,3-dihydroinden-1-ylidene)malononitrile (INCN), inda-cenedithiophene (IDT), and benzodithiophene (BDT).

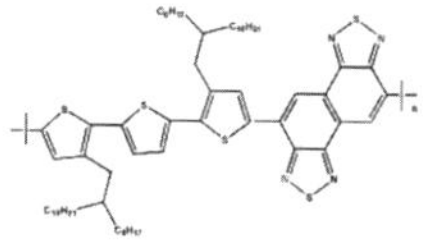

PNTT 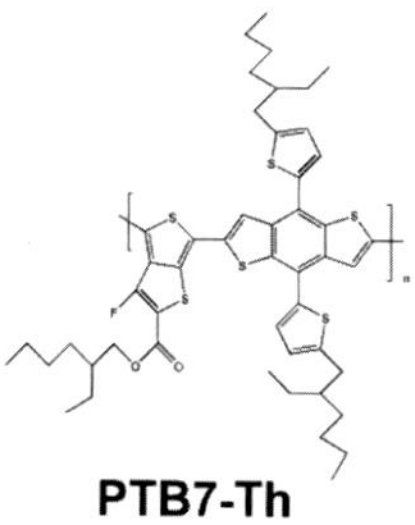 **PTB7-Th** 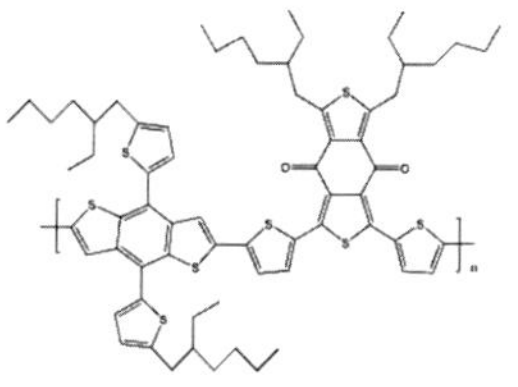 **PBDB-T**

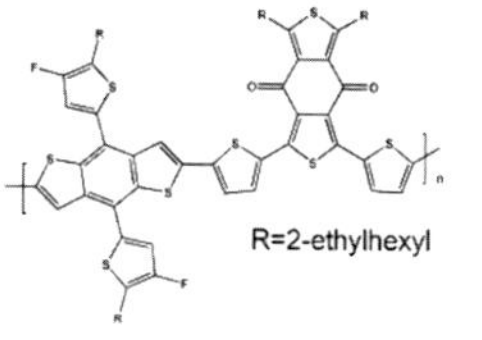

R=2-ethylhexyl

PBDB-TF

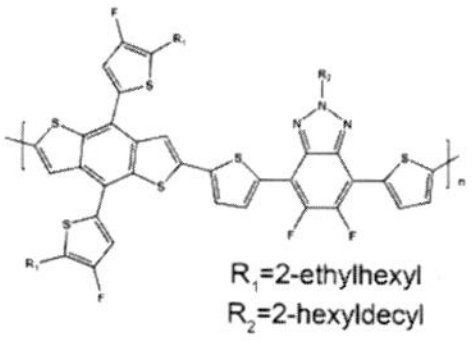

R_1=2-ethylhexyl
R_2=2-hexyldecyl

PBTA-TF

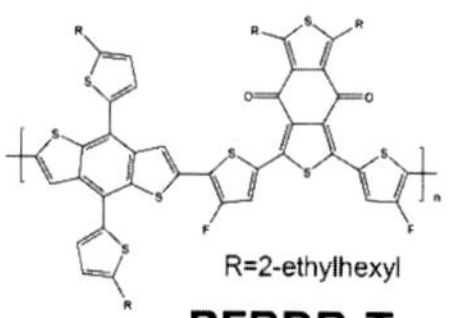

R=2-ethylhexyl

PFBDB-T

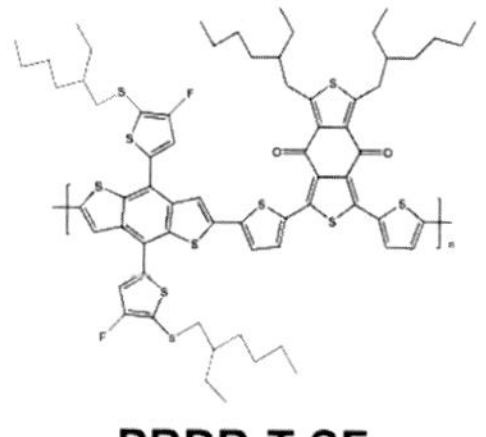

PBDB-T-SF

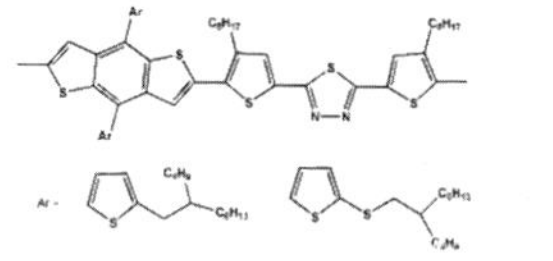

PBDT-TDZ **PBDTS-TDZ**

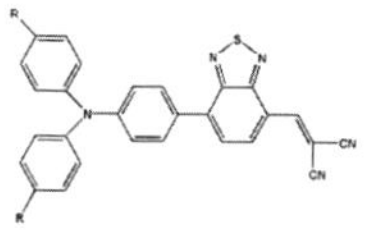

DPDCPB: R=H
DTDCPB: R=CH_3

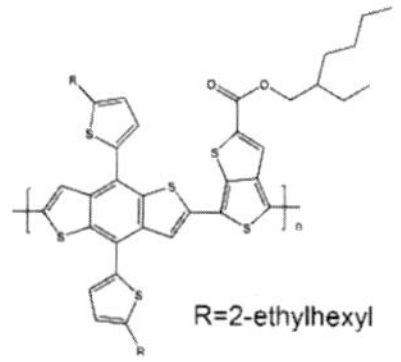

R=2-ethylhexyl

PBDTTT-E-T

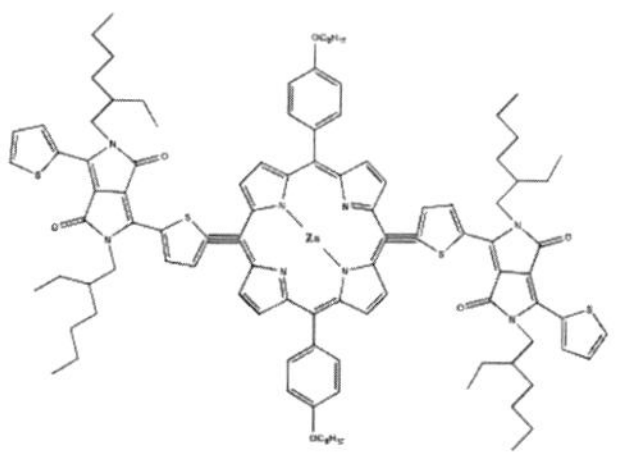

DPPEZnP-O

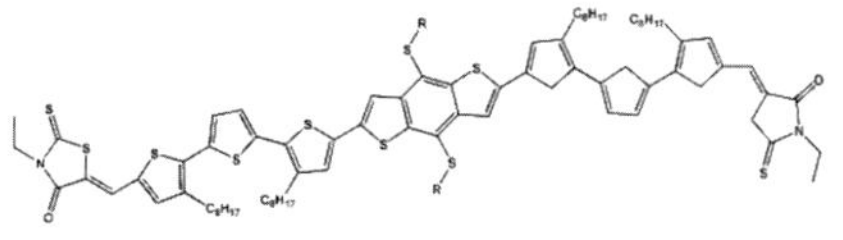

DR3TSBDT

(*Continued*)

acceptor is perfect for forming effective phase separation, while most of the NFAs in the device show a relatively planar structure with poor phase separation ability which can cause good miscibility between donor and acceptor. The design of nonplanar nonfullerene materials to improve the crystallinity of donor or acceptor materials can be adopted to obtain a good phase separation in the OSCs. In the donor-acceptor blended layer, the formation of a large domain size and large crystals should be avoided to reduce the interfacial area and inefficient exciton dissociation. Molecular orientation is another key factor to consider for high efficiency devices. The face-on molecular orientation is favorable for high-efficiency devices in the blended layer for the effective vertical charge transport channel to increase the carrier transport ability. In the molecular structure, effective intermolecular π-π stacking is needed to attain higher carrier transport in the active layer.

Recently, newly designed NFAs and PTB7-Th donor OSCs were reported. Most of the designed perylenediimide (PDI)-based NFAs work well with donor PTB7-Th for its better film morphology with a strong face-on stacking and complementary wide absorption spectrum. It is important to mention that acceptors, IDT-2BR1, Ta-PDI, oo-PDI, and CO$_i$8DFIC based-devices all exhibit an efficiency higher than 8%. Donor PTB7-Th has also shown high compatibility with some of the low bandgap acceptors like ITIC, and ITIC-Th gives an efficiency up to 8.5%. The highest efficiency of 12.16% is achieved with a PTB7-Th: COi8DFIC based-device with an inverted structure as shown in Fig. 2.12. It is expected to achieve more high-efficiency OSCs to reach the commercialization requirement based on PTB7-Th donor material and widely compatible donor with novel efficient acceptors.

2.4 Ternary strategy of organic solar cells

The ternary strategy is a unique method for OSCs that can significantly improve device performance. For the single junction OSC, the intrinsic narrow absorption range of both two donor and acceptor materials can limit the utilization of the solar spectrum by cells, this strategy that is using the third material that has a complementary absorption profile was designed to enlarge the total absorption.

FIGURE 2.11

Donor materials—molecular structures of the polymer and small molecule donor materials for high efficiency OSCs. The representative building blocks of the polymers are benzothiadiazole (BT), benzo[1,2-b:4,5-b′]dithiophene (BDT), thienothiophene (TT), 3-fluorothieno[3,4-b]thiophene-2-carboxylate (FTT), benzo[1,2-c:4,5-c′]dithiophene-4,8-dione (BDD), 1,3,4-thiadiazole (TDZ), alkoxycar-bonyl-substituted thieno[3,4-b]thiophene (TT-E), benzotriazole (BTA), and diketo-pyrrolopyrrole (DPP).

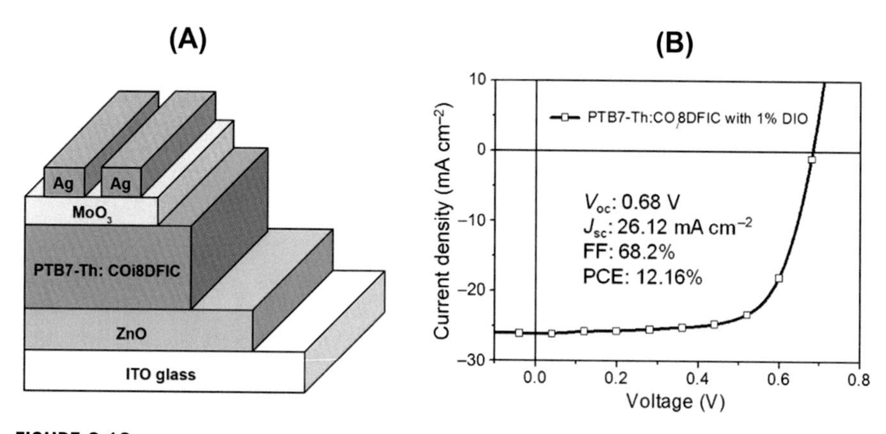

FIGURE 2.12

(A) The bulk heterojunction inverted device structure with the highest power conversion efficiency. (B) J-V curve for the PTB7-Th:COi8DFIC-based organic solar cell.

Reproduced with permission from Ref. Xiao, Z., Jia, X., & Ding, L. (2017). Ternary organic solar cells offer 14% power conversion efficiency. Science Bulletin, 62, *1562–1564, © 2017, Elsevier B.V. (Xiao et al., 2017).*

Unlike other strategies, the ternary strategy still retains the simplicity of the fabrication process of OSCs, and it has attracted broad attention (Lin, Wright, Veettil, & Uddin, 2014; Xu et al., 2018). Normally, based on the additional material used, the ternary OSC can be divided into two types, including the donor-donor-acceptor and donor-acceptor-acceptor type. Initially, the third component is rationally selected to increase the light-absorbing ability as well as the J_{sc} of the device. However, more research has found that the additional component may also enhance other parameters of a device like V_{oc} and FF. This may occur because the third component has additional positive impacts on charge transportation, the morphology of the BHJ, and exciton dissociation.

Since different components have different energy levels and different morphology in the blend, the working mechanisms of different ternary solar cells are also different. In general, the working mechanism of the ternary solar cell can be divided into three categories, including the charge transfer mechanism, the energy transfer mechanism, and the parallel linkage transfer mechanism as shown in Fig. 2.13. In the charge transfer mechanism, the third component provides an additional pathway for the exciton dissociation and carrier transportation. To utilize this mechanism to create a ternary device, selecting materials with a cascade energy level alignment is commonly applied. In the energy transfer mechanism, the third component acts more like an energy transfer agent that can absorb light and transfer the photogenerated excited states to the corresponding component (donor for the third component is the acceptor and the acceptor for the third component is a donor). Benefitting from that, the corresponding part in this ternary device can generate more free charge carriers, which indicate a higher J_{sc}. In the parallel linkage transfer mechanism, the third component acts as an alloy sensitizer that will mix with one of the original materials and form a

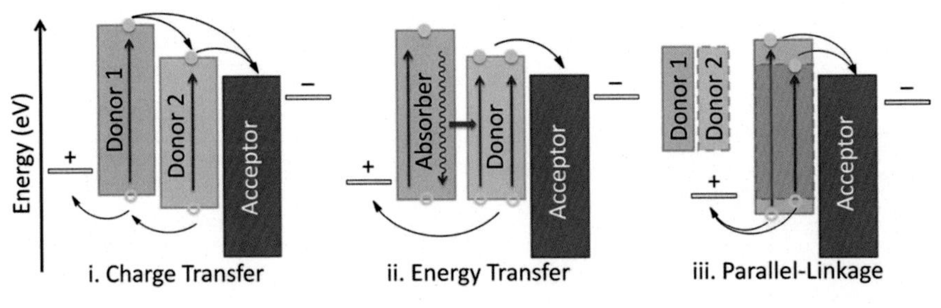

FIGURE 2.13

Three working mechanisms of ternary blend organic solar cells devices.

Reproduced with permission from Journal of Physical Chemistry Letters. *2013, 4 (11), 1802–1810, © 2013, American Chemical Society.*

special alloy phase. No charge transfer will occur between the two donors or acceptors. Instead, the ternary device is working more like two individual OSCs that the charge transfer tunnels can only form at the two interfaces of the donor and acceptor. Although these three working mechanisms are quite different, they usually are not independent. In many cases, these three mechanisms can exist simultaneously.

2.5 Electron transparent layer

Recently the most widely used ETL in OSCs is zinc oxide (ZnO). The LUMO of electron acceptors aligns well with the ZnO Fermi level (Lin, Miwa, Wright, & Uddin, 2014; Mahmud et al., 2017). Moreover, ZnO also increases the shunt resistance of the device by blocking the hole because of its high IP. ZnO is a wide bandgap material and transparent to visible light. It can absorb the UV light and works as a UV light filter for the photoactive layer. Based on the conduction and V_B positions, metal oxide semiconductors (MOSs) can be p-type or n-type for the OSCs. The MOS V_B is required to match the polymer HOMO level for a p-type contact. The C_B of the n-type MOS is required to match the acceptor LUMO level to work as an electron transfer layer.

Wide bandgap interface materials also serve as a carrier's barrier by improving the carrier selectivity contacts. The main roles of ETL materials are as follows:

1. Align/adjust the energetic barrier height between the adjacent electrodes and photoactive layer.
2. Serve as a selective contact materialization for holes or electrons.
3. Control the polarity of the device to make it a normal or inverted device structure as shown in Fig. 2.8.
4. Prohibit a physical or chemical reaction between the active layer and electrode.
5. Serve as an optical spacer in the device structure.

The cathode interfacial layer with low-work function metal (top electrode) and the bottom adjoining the transparent conductive oxide (TCO) electrode are based on the normal or inverted device structure. In the BHJ, active layer alkali metals or related compounds are used for the ohmic contact to the electron acceptor. Normally, lithium fluoride (LiF) and cesium carbonate (Cs_2CO_3) are used to reduce the low work function of the electrodes. They exhibited enhanced the V_{OC} and good electron injection properties for enhanced device efficiency. However, oxidation of alkali metal compounds over a long period of time leads to the degradation of device stability. The effective alternative is to use transition MOSs, such as ZnO and TiO_2, with work functions corresponding to the fullerenes LUMO levels. These two MOSs are well known for optical transparency, solution processibility, and their chemical resistance to oxygen and moisture. Effective alternatives of n-type semiconducting oxides with low work-function metals are used as cathode contacts. These oxides demonstrate the suitability in both conventional and inverted device structures.

For the fabrication of inverted structure OSCs, ZnO colloids or nanoparticles are spin coated as an electron transport layer on TCO substrates. ZnO nanostructures can be obtained from precursor solutions containing zinc salts through solution-processed methods such as sol–gel, a hydrothermal process or solvothermal. ZnO NPs are used as ETLs in OSCs devices. A PCE value of 4% has been demonstrated in inverted poly (3-hexylthiophene) (P3HT) cells, as shown in Fig. 2.14 using ZnO ETL (Liang, Zhang, Jiang, & Cao, 2015). There are many defect states in the synthesized ZnO nanostructures. A large fraction of dangling

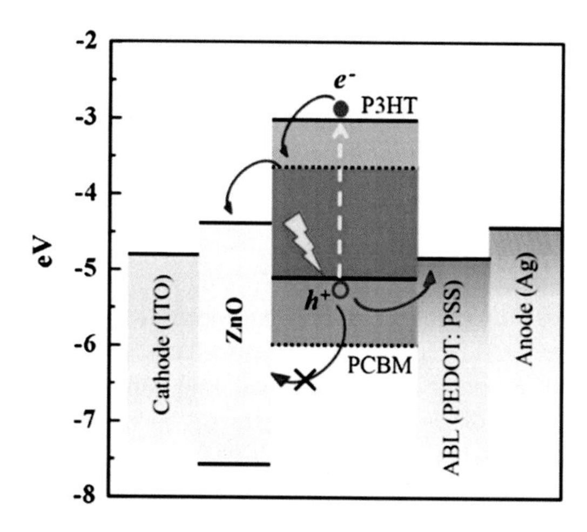

FIGURE 2.14

Schematic of energy levels for each layer of an inverted P3HT:PCBM device.

Reproduced with permission from Ref. Liang, Z., Zhang, Q., Jiang, L., & Cao, G. (2015). ZnO cathode buffer layers for inverted polymer solar cells. Energy & Environmental Science, 8, *3442–3476, © 2015, The Royal Society of Chemistry.*

bonds exists in the small diameter of ZnO NPs, generating a high density of defect states in the bandgap. The removal of these localized defect states in the band gap of charge transport layers is very important for the operational stability of the device, as well as the enhanced device performance of inverted OSCs. Therefore, to improve the conductivity of ZnO NPs, UV light exposure is required to improve the conductivity of ZnO NPs by means of photodoping and defect filling. It is also found that UV-ozone (UVO) treatment passivates the surface defect states. PDTG-TPD:PC71BM-based OSC devices with ZnO NPs as ETL exhibited enhanced device performance with a PCE of 8% using UVO treatment. However, it is not enough to optimize the device performance with this UV exposure method. It is possible to achieve optimum device performance using P3HT:PCBM-based devices by increasing the ZnO nanostructure crystallinity with annealing. Moreover, ZnO nanoparticles across ZnO nanowires were planted to increase the electron lifetime, improve charge collection, and decrease recombination at the corresponding electrodes. It has shown that the carrier lifetime increased twofold by incorporating electrospun ZnO nanowires.

In solution processed ZnO film, multiple studies have ascribed the morphological change to the variation in postthermal conditions, subsequently affecting PV performance of OPV devices. The ZnO layer uses a sol-gel method with static and dynamic annealing conditions. It was claimed that the P3HT:$PC_{61}BM$ film-based device using nano-ridged ZnO exhibited a PCE of 4%, which is approx. 25% higher than the device processed with a planar ZnO. This conclusion seems to be contradictory to other findings, where the performance of a PTB7:$PC_{71}BM$-based device is insensitive to the morphology (nano-ridged and planar) of the ZnO, as long as the optimum thickness of the active layer remains constant. The multiple ZnO ETLs for various polymeric systems suggest that the properties of the interface between active layer and ETL can be significantly varied.

Appropriate interface engineering requires specific materials for OSCs for this purpose. Metallic electrodes with transparent conducting oxides (TCO) poor ohmic contacts arise due to the (1) the formation of interfacial dipoles (2) mismatched work function or misalignment of energy levels and (3) interfacial trap states. To develop good and efficient ohmic contacts, various charge-extracting interlayers can be employed between the electrodes and the active layer. Transition MOS are considered as potential candidates among the various interfacial materials owing to their superior optical transparency, high environmental stability, and facile synthesis routes. The energy-level alignment between the donor and the acceptor in OSCs devices with ohmic contacts is used to determine the open circuit voltage V_{OC}; if not, it is determined by the contact electrode work function differences. The J_{sc} is determined by the charge collection efficiency across the photoactive layer after the illumination for photogenerated carriers. Device series resistance, shunt resistance, and charge recombination/extraction rate determine the FF of OSCs. Finally, the product of normalized Voc, J_{sc}, and FF of OSCs is determined the power conversion efficiency (PCE) under 1 sun at AM 1.5 G illumination.

The ionization energies and electron affinities are selected for high or low work-function MOSs employed in OSCs along with fullerene derivatives. For highly efficient and stable ohmic contacts, these MOSs provide the basis of energy-level bending, Fermi-level pinning, and vacuum-level shifting at the polymer–electrode interfaces. Furthermore, the use of oxide interface layers avoids the direct contact between the photoactive layer and electrodes, which is also adversely affects the efficient charge collection or high density of carrier traps. Maximizing the V_{OC} of the MOS interfacial layers plays an important role for developing ohmic contacts because of an increase in dark current as well as carrier recombination by lowering the built-in potential.

2.6 Hole transport layer

Normally, the HTL of a conducting polymer (poly 3,4-ethylenedioxythiophene: poly styrenesulfonate, PEDOT:PSS) is used due to its advantages of high work function, high conductivity, and solution processability (Wang et al., 2017). However, there are various issues due to the acidic nature of PEDOT:PSS, such as the degradation of the active layer and the ITO anode etching. To solve these issues, the use of various metal oxides such as MoO_x, WO_x, and NiO_x as the HTL have been explored both in inverted and conventional OSCs (Kanwat and Jang, 2014). These oxides have high conductivity, work function, and transparency. They could be suitable for blocking electrons and collecting holes. Although, organic materials like sulfonated poly(diphenylamine), polyaniline (PANI), or (PANI:PSS), and a variety of oxides, like WO_3, MoO_3 or V_2O_5, were used as HTL either in OSCs or organic light emitting diodes. Most of these investigations were conducted to improve the device efficiency through ETLs or HTLs. These materials were not used to study the device stability. Studies on the influence of these HTL materials and the formation of the interface to the active layer of OSCs on the stability and lifetime of carriers are still rare.

In recent studies, it has been shown that the device stability of inverted structure OSCs is significantly higher compared to conventional device structures, probably due to the use of high work function metal anodes. Normally, high work function metal is less sensitive to degradation than low work function metals used as cathodes. The PEDOT:PSS layer used as a HTL is identified as one of the main degradation sites in OSCs due to the absorption of oxygen and water (Son, Park, Yeon Moon, Ju, & Kim, 2020).

2.7 Electrode materials

To improve the device performance means to increase PCE, the selection of materials for the device structure by band energy diagram matching is very important.

In most OSC device structures, researchers use PEDOT:PSS as the hole transport layer and ITO as anode electrode. As a cathode, Al is one of the most widely used metals. Some authors used a LiF thin layer (< 1 nm) interface between the Al cathode and the photo-active layer. The mechanism of the improvement of device performance by the LiF interface layer is unclear. Some researchers believe that LiF passivates the interface defects in the Al/photo-active layers. There is also speculation that LiF molecules may breakdown, and Li atoms form a new compound in the interface region which enhances the electron extraction to improve device performance. This enhanced electron extraction mechanism can be explained as tunneling, band bending at the Al/photo-active layers, or the presence of interfacial dipoles, which help to lower the Al work function. To explain the interface states, density distributions models require energetic disorder in the bulk and dipole-induced disorder at the interface.

In OSCs, PFN and ZrAcac are typically combined with Al as a cathode to improve the device performance. An Al cathode is not good for the long-term device stability because of the chemical reaction with air. Thick films of PFN and PDIN are not suitable for the roll-to-roll fabrication technology because of their low conductivity. In inverted structure OSCs, various cathode materials are used to modify the contact of metal oxides and the active layer.

2.8 Fabrication techniques

2.8.1 Solution processing

In enhance the cost-effectiveness of organic PV technologies, solution processing is being developed for the deposition of organic semiconductors to replace more complicated vacuum-processed methods. In the context of solution phase processing and the interpenetrating donor-acceptor mixture in organic solvents, various film forming techniques ranging from laboratory scale spin coating to large-scale industrial fabrication techniques such as doctor blading, slot-die coating, and ink-jet printing have been applied for OPV device fabrication. Among them, spin coating is the most studied film forming method with its simple procedure, fast operation, and good reproducibility for laboratory scale devices [our papers]. Applications of spin coating have enabled a simple and fast experimental work-flow from deposition of each layer to the completed device. Fig. 2.15 illustrates a simple operation of the spin coating technique, where the prepared solution is applied to an ITO substrate. Thereafter, acceleration of the substrate to a pre-set spin rate allows formation of uniform thickness and film surface.

The surface morphology and thickness of the film in each active solution and concentration can be easily reproduced with the spin coating method. Precise control of the film formation (such as thickness and morphology) plays a crucial role in determining the performance of OSCs. A detailed study on the correlation between active layer thickness and performance were reported by Huang et al.

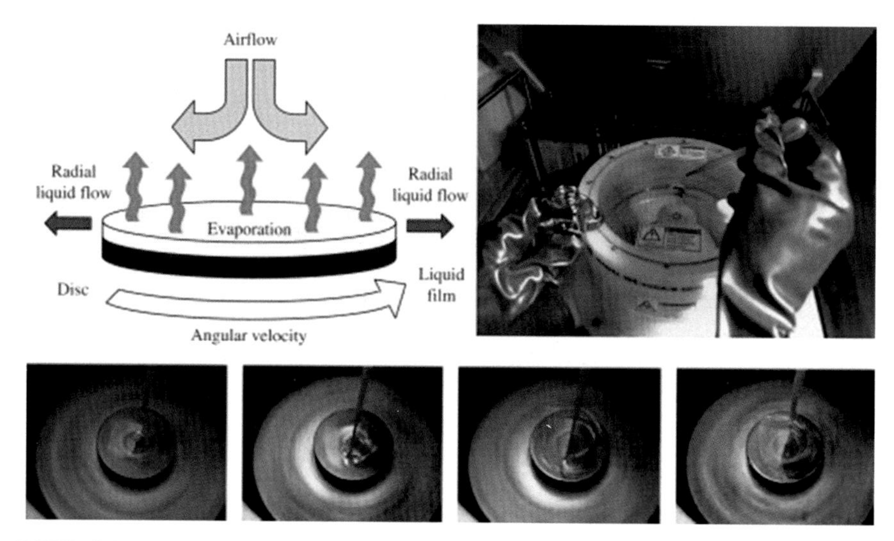

FIGURE 2.15

Schematic diagram of spin coating operation along with a photo of actual spin coating equipment.

Reproduced with permission from Ref. Krebs, F.C. (2009). Fabrication and processing of polymer solar cells: A review of printing and coating techniques. Solar Energy Materials and Solar Cells, 93, *394–412, © 2008, Elsevier B.V. (Krebs, 2009).*

(2016), who demonstrated that poly[4,8-bis(5-(2-ethylhexyl)thiophen-2-yl)benzo [1,2-*b*;4,5-*b*′]dithiophene-2,6-diyl-*alt*-(4-(2-ethylhexyl)-3-fluorothieno[3,4-*b*]thiophene)-2-carboxylate-2,6-diyl] (PTB7-Th)-based devices with an active layer thickness of 150 nm (PCE ~ 9.41%) outperforms a thickness of 230 nm (PCE ~ 8.01%). The hole mobility for the device with 150 nm thickness was higher than that of 230 nm device, which led to enhanced FF and overall efficiency. In a similar report, Gupta et al. investigated the change of morphology of active layers (P3HT:PCBM) on variations in the thickness by varying the rotational speed (Gupta et al., 2013). Carefully varying the thickness led to an increase in PCE from 2.40% to 2.89%. The improved PCE was mainly attributable to a finer distribution with smaller domains. As with previous studies, it is clear that optimum thickness, and morphology of the OSCs can be easily achieved by adjusting the rotational speed. Thus, the spin coating technique has been widely utilized on the laboratory scale for device optimization.

2.8.2 Device fabrication

The prepatterned ITO glasses with an area of 12 mm × 12 mm can be purchased commercially. The chemical materials PTB7-Th, $PC_{71}BM$, ITIC, and IEICO-4F, can be purchased from 1-Materials and Sigma-Aldrich. The chemical materials

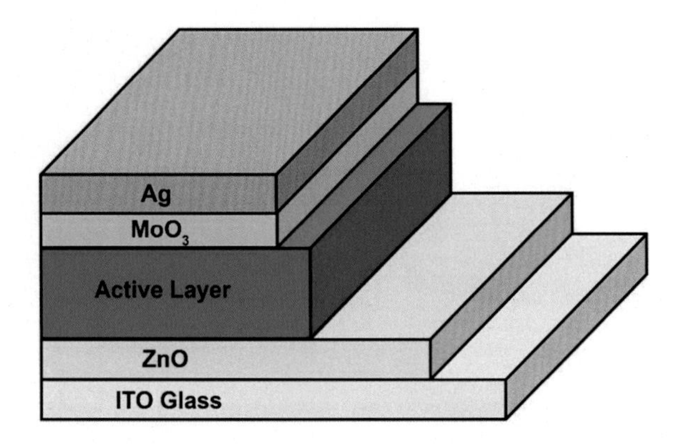

FIGURE 2.16

The schematic of the inverted device structure of organic solar cells.

IDFBR can be purchased from Solarmer. The zinc oxide nanoparticles, reagent alcohol (anhydrous, <0.003% water), chlorobenzene (99.8%), 1,8-diiodooctane, and MoO_3 can be purchased from Sigma-Aldrich. All OSCs can be fabricated in conventional and inverted device structure of ITO Glass/ZnO/active layer/MoO3/Ag, as shown in Fig. 2.16. The ITO glass substrate should be sequentially cleaned through soapy deionized (DI) water, pure DI water, acetone, and isopropanol by 10 min ultrasonication. The ZnO nanoparticle solution (3%) can be made by adding 50 μL zinc dispersion into 1622 μL reagent alcohol. The ZnO nanoparticles solution must be stirred overnight in a N_2 filled glove box and then spin-coated on the top of the cleaned ITO glass with a spin rate of 4000 rpm for the 60 s. After the coated ZnO solution, all samples must be annealed at 120°C on the hot plate inside the N_2 filled glove box for 10 min. Then the active layer solution will be prepared. After the fabrication of the active layer, the coated samples must be placed in a vacuum chamber at a pressure of 10^{-5} Pa. The 10 nm thick film of MoO_3 and 100 nm thick film of silver must be deposited on the sample surface through a shadow mask by thermal evaporation.

2.8.3 Device characterization

The current density-voltage (J–V) characteristics can be measured using a solar cell I–V testing system under an illumination power of 100 mW/cm^2 by an AM 1.5 G solar simulator. During the measurement, the device temperature can be estimated by GM1350 50:1 LCD infrared thermometer digital gun and maintained at around 25°C. An Autolab PGSTAT-30 can be used for the capacitance-voltage (C-V) characterization and electrochemical impedance spectroscopy measurements. These characterizations should be conducted inside the glove box with a frequency range of 106–100k Hz using a frequency analyzer module. A QEX10 spectral response system from PV measurements can be used as a characterization machine for EQE measurements. The optical absorption characterization can be measured by a UV–VIS–NIR spectrometer (Perkin Elmer–Lambda 950). A Bruker Dimension ICON SPM with a scan rate of 0.512 Hz can be used as the atomic force microscope to characterize the film surface morphology. Scanning electron microscopy (SEM) images can be taken using a Nano SEM 230 system. Transmission electron microscopy (TEM) images can be taken using a JEO IF200 system. The micro-photoluminescence (μPL) spectroscopic can be measured by a μPL system. The sample can be excited by a pulsed OPO laser at 532 nm for the PL measurements. For steady-state spectral PL, the signal can be detected by Glacier X TE Cooled CCD Spectrometer with a detection range of 200–1050 nm. For time-resolved PL measurement, the signal can be detected by an id110 VIS 100 MHz Photon Detector operated in free-running mode. In the space charge limited current method, the electron-only and hole-only devices can be fabricated using the device structure of ITO/ZnO nanoparticles/active layer/ZnO nanoparticles/Ag and ITO/PEDOT: PSS/active layer/MoO_3/Ag, respectively. X-ray diffraction (XRD) analysis of the sodium titanate nanofibers can be carried out using a Rigaku Ultima IV multipurpose X-ray diffractometer equipped with Cu Kα radiation and fixed monochromator. For these measurements, an acceleration voltage of 40 kV and 20 mA current can be applied. The XRD patterns can be measured using a continuous scan mode and 2theta (2θ) angle from 5 to 60 degrees with a sampling width of 0.02 degrees and a scan speed of 1 degree/minute.

2.9 Key challenges

To date, there are various approaches that have been involved in enhancing the efficiency of OSCs, such as the synthesis of new materials, interfacial engineering, and morphological optimization (Upama, Mahmud, Conibeer, & Uddin, 2019; Upamaa et al., 2018). Particularly, designing novel acceptor materials has driven a recent surge in research focus. Among them, low bandgap acceptor materials play an important role to improving the PEC of the OSCs and achieving the highest PEC over 18%. The nanoscale phase separation availability of fullerene materials such as [6,6]-pheny-C_{61}/C_{71}-butyric acid methyl ester ($PC_{71}BM$) with superior charge-transporting properties lead the highest fullerene-based OSCs PCE of over 12%. However, these fullerene-based OSCs suffer from several limitations, such as morphology instabilities, fast air degradation, the high production and purification cost, etc. The increase of fullerene-based OSCs PCE is becoming a challenge. Recently, the development of "nonfullerene" acceptor molecules with strong electron-accepting and isotropic electron-transport have become more attractive for the researcher to improve the PCE of OSCs. The PCE of nonfullerene based OSCs has improved dramatically to over 18%. For further improvement of OSCs PCE, the use of NFA materials is a promising investigation area.

Recently, the ternary blend photoactive layer strategy is a unique method for OSCs to improve the device performance for research attention. The ternary blend strategy is very simple for the fabrication of OSCs compared to the tandem solar cell strategy, which is required for sophisticated fabrication process and current matching optimization. Furthermore, the ternary blend strategy has a potential towards large-scale fabrication for commercial application. Initially the ternary strategy was designed to broaden the absorption spectrum by using a third material with a complementary absorption profile since the single-junction OSC plagued by the intrinsic narrow absorption range. The energy level alignment, morphology tuning, energy transfer, miscibility, and crystallization are the key points in the ternary blend strategy to improve the device performance (Duan et al., 2019). Hence, investigating the application of ternary blend strategy in OSCs and its mechanism is another promising method towards higher device performance, especially incorporating a NFA with the ternary strategy shows profound potentials.

For commercial applications, long-term stability of OSCs is also a critical issue together with the improvement of PCE. OSCs have suffered from the performance degradation due to the evolution of optoelectronic and morphological structure because of the unstable nature of organic materials, such as light-induced instability, chemical instability, interfacial degradation, and thermally induced changes (Duan et al., 2020). Most of these instability factors are related to each other and simultaneously lead to degraded device performance. It is difficult to control the device stability from the light, heat, and air induced degradations as OSCs operating under the sun in ambient conditions. Most OSC performance degrades sharply during the early stages of the device operation, which is also termed "burn-in degradation," and can contribute to the major performance loss over the lifetime of OSCs. For instance, the state-of-art poly[(2,6-(4,8-bis(5-(2-ethylhexyl)thiophen-2-yl)-benzo[1,2-b:4,5-b'] dithiophene))-alt-(5,5-(1',3'-di-2-thienyl-5',7'-bis(2-ethylhexyl)benzo[1',2'-c:4',5'-c'] dithiophene-4,8-dione)] (PBDB-T): 3,9-bis(2-methylene-(3-(1,1-dicyanomethylene)-indanone))-5,5,11,11-tetrakis(4-hexylphenyl)-dithieno[2,3-d:2',3'-d']-s-indaceno[1,2-b:5,6-b']dithiophene (ITIC) based nonfullerene and poly[[2,6'-4,8-di(5-ethylhexylthienyl)benzo[1,2-b:4,5-b']dithiophene] [3-fluoro-2[(2-ethylhexyl)carbonyl]thieno[3,4-b] thiophenediyl]] (PTB7-Th):$PC_{71}BM$ based fullerene devices can both lose over 30% of their initial PCE in a few hours under the real operation condition. Therefore, burn-in degradation has become an ineluctable barrier for OSCs to achieve long-time stability for future commercialization. An in-depth understanding of the mechanism behind burn-in degradation has become the precondition to overcome this barrier.

2.10 Recommendations for future research works

Further research on OSCs could be oriented towards the following directions:

1. Incorporating newly developed NFAs to fabricate ternary blend OSCs, seeking for performance and stability improvement. This could be a pathway to boost the device performance of binary OSCs further.

2. Incorporating newly designed materials into the optimized ternary blend OSCs to fabricate quaternary OSCs, seeking performance and stability improvement.
3. Investigating the burn-in degradation process in the ternary OSCs based on two NFAs. The observation may be different from our case study, which focuses on ternary OSCs consisting of both fullerene and NFAs.
4. Developing strategies such as proper encapsulation to further improve the stability of OSCs, especially to suppress the burn-in degradation in OSCs.
5. Developing high-efficiency ternary blend OSCs with large-area using scalable fabrication methods such as blade coating to investigate its scalability towards commercialization.
6. The semitransparent OSC and PSC could be used for visible light sensing applications. For example, these cells could be used as smart displays in electronic products and supply energy to the system.
7. Fabrication of large-area device based on the high efficiency devices achieved during this research to investigate the device scalability.
8. Investigating the stability and robustness of the flexible devices when subject to controlled mechanical stress.
9. Opting for ink-based interfacial layers for fully printable OSCs.

2.11 Conclusions

In this chapter, we have discussed the achievement of high performance OSCs. Organic PVs are extremely attractive candidates for the next-generation low-cost solar cell technologies with mechanical flexibility. However, for OSCs, the efficiency and photo-stability are needed to be improved for commercial applications. These emerging PV technologies is also a great fit for BIPV applications. The objective of this chapter is to provide a basic understanding of OSCs materials and device engineering to improve the device performance in state-of-the-art solar cells.

Acknowledgments

I acknowledge the endless support from my research students and the staff of photovoltaic and renewable energy engineering school (SPREE), UNSW, Sydney.

References

Becquerel, E. (1839). Mémoire sur les effets électriques produits sous l'influence des rayons solaires. *Comptes Rendus*, *9*, 561–567.

Brebels, J., Manca, J. V., Lutsen, L., Vanderzande, D., & Maes, W. (2017). High dielectric constant conjugated materials for organic photovoltaics. *Journal of Materials Chemistry A*, *5*, 24037–24050.

Cowan, S. R., Roy, A., & Heeger, A. J. (2010). Recombination in polymer-fullerene bulk heterojunction solar cells. *Physical Review B*, *82*, 245207.

Duan, L., Guli, M., Zhang, Y., Yi, H., Haque, F., & Uddin, A. (2020). The air effect in the burn-in thermal degradation of nonfullerene organic solar cells. *Energy Technology*, *8*, p1901401.

Duan, L., Zhang, Y., Yi, H., Haque, F., Deng, R., Guan, H., ... Uddin, A. (2019). Trade-off between exciton dissociation and carrier recombination and dielectric properties in Y6-sensitized nonfullerene ternary organic solar cells. *Energy Technology*, p1900924. Available from https://doi.org/10.1002/ente.201900924.

Elumalai, N. K., & Uddin, A. (2016). Open circuit voltage of organic solar cells: An in-depth review. *Energy Environment Science*, *9*, p391–p410.

Feng, W., Liu, J., & Yu, X. (2014). Efficiency enhancement of mono-Si solar cell with CdO nanotip antireflection and down-conversion layer. *RSC Advances*, *4*, 51683–51687.

Gupta, S. K., Sharma, A., Banerjee, S., Gahlot, R., Aggarwal, N., Deepak., & Garg, A. (2013). Understanding the role of thickness and morphology of the constituent layers on the performance of inverted organic solar cells. *Solar Energy Materials and Solar Cells*, *116*, 135–143.

Huang, J., Carpenter, J. H., Li, C. Z., Yu, J. S., Ade, H., & Jen, A. K. Y. (2016). Highly efficient organic solar cells with improved vertical donor–acceptor compositional gradient via an inverted off-center spinning method. *Advanced Materials*, *28*, 967–974.

Kanwat, A., & Jang, J. (2014). Extremely stable organic photovoltaic incorporated with WOx doped PEDOT:PSS anode buffer layer. *Journal of Materials Chemistry C*, p901–p907.

Krebs, F. C. (2009). Fabrication and processing of polymer solar cells: A review of printing and coating techniques. *Solar Energy Materials and Solar Cells*, *93*, 394–412.

Li, G., Zhu, R., & Yang, Y. (2012). Polymer solar cells. *Nature Photonics*, *6*, 153.

Li, M., Gao, K., Wan, X., Zhang, Q., Kan, B., Xia, R., ... Chen, Y. (2016). Solution-processed organic tandem solar cells with power conversion efficiencies >12%. *Nature Photonics*, *11*, 85–90.

Liang, Z., Zhang, Q., Jiang, L., & Cao, G. (2015). ZnO cathode buffer layers for inverted polymer solar cells. *Energy & Environmental Science*, *8*, 3442–3476.

Lin, R., Miwa, M., Wright, M., & Uddin, A. (2014). Optimisation of the sol-gel derived ZnO buffer layer for inverted structure PCPDTBT:PC71BM bulk heterojunction organic solar cells. *Thin Solid Films*, *566*, 99–107.

Lin, R., Wright, M., Veettil, B. P., & Uddin, A. (2014). Enhancement of ternary blend organic solar cell efficiency using PTB7 as a sensitiser. *Synthetic Metals*, *192*, 113–118.

Liu, Q., Jiang, Y., Jin, K., Qin, J., Xu, J., Li, W., ... Ding, L. (2020). 18% efficiency organic solar cells. *Science Bulletin*. Available from https://doi.org/10.1016/j.scib.2020.01.001.

Liu, T., Pan, X., Meng, X., Liu, Y., Wei, D., Ma, W., ... Sun, Y. (2017). Alkyl side-chain engineering in wide-bandgap copolymers leading to power conversion efficiencies over 10%. *Advanced Materials*, *29*, 1604251.

Mahmud, M. A., Elumalai, N. K., Upama, M. B., Wang, D., Soufiani, A. M., Wright, M., ... Uddin, A. (2017). Solution-processed lithium-doped ZnO electron transport layer for efficient triple cation (Rb, MA, FA) perovskite solar cells. *ACS Applied Materials & Interfaces*, *9*, p33841–p33854.

Qi, B., & Wang, J. (2013). Fill factor in organic solar cells. *Physical Chemistry Chemical Physics*, *15*, 8972–8982.

Scharber, M. C., Mühlbacher, D., Koppe, M., Denk, P., Waldauf, C., Heeger, A. J., & Brabec, C. J. (2006). Design rules for donors in bulk-heterojunction solar cells—towards 10% energy-conversion efficiency. *Advanced Materials*, *18*, 789–794.

Son, H. J., Park, H.-K., Yeon Moon, J., Ju, B.-K., & Kim, S. H. (2020). Thermal degradation related to the PEDOT:PSS hole transport layer and back electrode of the flexible inverted organic photovoltaic module. *Sustainable Energy Fuels*, *4*, 1974–1983.

Upamaa, M. B., Elumalaib, N. K., Mahmuda, M. A., Xua, C., Wang, D., Wright, M., & Uddin, A. (2018). Enhanced electron transport enables over 12% efficiency by interface engineering of non-fullerene organic solar cells. *Solar Energy Materials and Solar Cells*, *187*, 273–282.

Upama, M. B., Mahmud, M. A., Conibeer, G., & Uddin, A. (2019). Trendsetters in high-efficiency organic solar cells: Toward 20% power conversion efficiency. *SOLAR RRL*. Available from https://doi.org/10.1002/solr.201900342.

Wang, D., Elumalai, N. K., Mahmud, M. A., Wright, M., Upama, M. B., Chan, K. H., ... Uddin, A. (2017). V2O5-PEDOT: PSS bilayer as hole transport layer for highly efficient and stable perovskite solar cells. *Organic Electronics*, *53*, p66–p73.

Wang, Y., Zhang, Y., Qiu, N., Feng, H., Gao, H., Kan, B., ... Chen, Y. (2018). A halogenation strategy for over 12% efficiency nonfullerene organic solar cells. *Advanced Energy Materials*, 1702870.

Xiao, Z., Jia, X., & Ding, L. (2017). Ternary organic solar cells offer 14% power conversion efficiency. *Science Bulletin*, *62*, 1562–1564.

Xiao, Z., Jia, X., Li, D., Wang, S., Geng, X., Liu, F., ... Ding, L. (2017). *Science Bulletin*. Available from https://doi.org/10.1016/j.scib.2017.10.017.

Xu, C., Wright, M., Ping, D., Yi, H., Zhang, X., Mahmud, M. A., ... Uddin, A. (2018). Ternary blend organic solar cells with a non-fullerene acceptor as a third component to synergistically improve the efficiency. *Organic Electronics*, *62*, 261–268.

Zhigang, Y., Jiajun, W., & Qingdong, Z. (2016). Interfacial materials for organic solar cells: Recent advances and perspectives. *Advanced Science*, *3*, 1500362.

CHAPTER

Introduction of inorganic solar cells

3

Nowshad Amin

Institute of Sustainable Energy, Universiti Tenaga Nasional (@UNITEN; The Energy University), Kajang, Malaysia

Solar photovoltaic cells with reasonable conversion efficiency were first realized using semiconductors, that is, silicon, without classification into any other subcategory. However, beginning in the 1970s and 1980s, researchers around the world began searching for alternatives to silicon, which led to a combination of varieties as seen today. Classifying the main absorption material of solar cells, we may find inorganic and organic materials. The inorganic semiconductor materials used to make photovoltaic cells are comprised of crystalline, multicrystalline, amorphous, and microcrystalline silicon (Si), III–V compounds and alloys such as gallium arsenide (GaAs), chalcogenides such as cadmium telluride (CdTe), and chalcopyrite compounds such as copper indium gallium diselenide (CIGS), copper zinc tin sulfide (CZTS) etc. Since their invention, the key application of Si solar cells has been limited to space-craft vehicle power supplies. However, solar cell technology has benefited significantly from the high standard of silicon technology initially developed for electronics such as transistors and later for integrated circuits. Crystalline silicon has been the leading photovoltaic technology since the development of the first solar cell with reasonable efficiency (6%) at Bell Laboratories in 1954. Single and multicrystalline silicon have dominated the PV market based on their well-established performance stability and high efficiency over the years. Therefore crystalline silicon solar cell technology is the most mature technology in the PV industry today. Silicon solar cells demonstrate high efficiency averaging around 20%. However, very pure silicon is needed and the price is comparatively high compared to the power output. Crystalline silicon solar cells are manufactured from pure silicon and show efficiencies as high as 25% of a theoretical maximum efficiency of 28%. Silicon-based solar cells accounted for over 90% of commercial production in recent years, even though the market share of these solar cells is slowly declining. The manufacturing processes used for the production of silicon solar cells are inherently expensive. However, Si is one of the least absorbing semiconductors used for solar cells as it has an indirect bandgap, while most of the other inorganic semiconductors have a direct bandgap. Therefore at least ten times more crystalline Si is needed to

Comprehensive Guide on Organic and Inorganic Solar Cells. DOI: https://doi.org/10.1016/B978-0-323-85529-7.00005-0

absorb a given fraction of sunlight compared to other semiconductors like GaAs, CdTe, etc. Thicker semiconductor material represents higher material volume and a higher quality material because of the long paths that the high energy electrons excited by the photons must travel before they are delivered to the external circuit to produce suitable current. Therefore some highly light absorbent inorganic semiconductors were investigated to reduce the cost of first generation silicon solar cells by employing thin film technologies. As shown in Fig. 3.1, a number of both inorganic and organic materials are classified based on main solar cell absorber materials.

We know the huge amount of solar irradiation received by the earth makes photovoltaic energy a great and free source of energy. However, the higher cost per watt-peak is a factor that slows the PV industry. This cause has provoked the PV scientific community to search for an alternative technology to yield high efficiency and low-cost solar cells. From this, the concept of Thin Film Inorganic Solar Cell (TFISC or simply TSFC) has been introduced primarily using inorganic light absorbing solid state semiconducting materials. Hence, TFSC is largely synonymous with inorganic solar cells. Thin film technology not only achieves higher efficiency of the solar cell but also contributes to cost reduction in solar cell fabrication. The key advantage of TFSC is the promise of lower costs, since less energy for processing and relatively lower costs for the materials are required, while large-scale production is achievable. The flexibility in the deposition techniques also permits the development and utilization of novel semiconductors, mostly inorganic, which otherwise might be difficult to produce. The deposition of inorganic semiconductors on foreign substrates typically results in polycrystalline or amorphous films with optical and electrical properties that can be considerably different

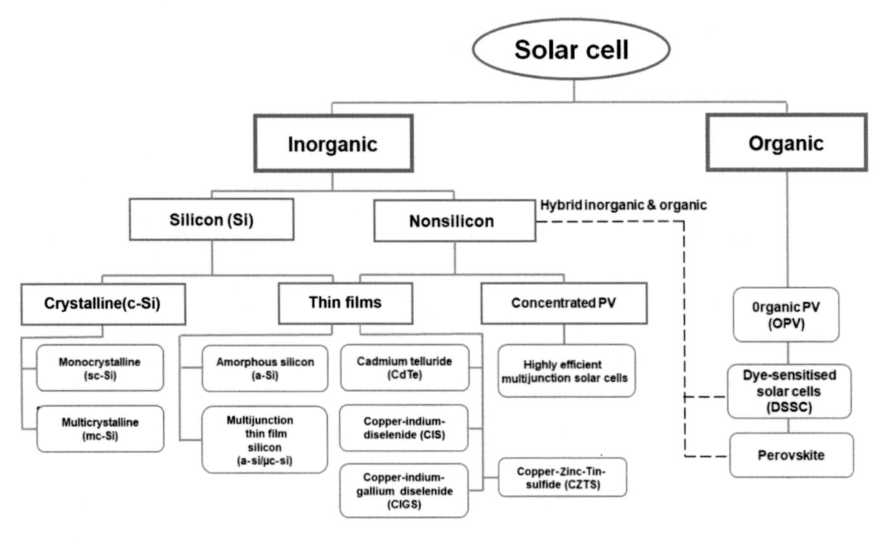

FIGURE 3.1

Possible solar cell classification based on organic and inorganic materials.

from the single-crystal behavior. This is mostly due to the large number of grain boundaries and other lattice defects. However, one of the major problems with TFSC is that in many cases the higher defect density also diminishes the efficiency and stability of the semiconductor compared to the single-crystal cells. Therefore great efforts have been made to realize the influence of lattice defects on the photovoltaic properties of the semiconductor. Moreover, semiconductor material science including fabrication technology on many compound inorganic semiconductors has been evolving over this period. Eventually, TFSCs based on inorganic materials have been regarded as the potential low-cost technology, where high efficiency solar cells could compete with silicon solar cells in the photovoltaic market. The obvious advantages of TFSC over Si solar cells are low temperature processing techniques, low material usage, a variety of deposition processes, and a compatibility with low-cost substrates. Accomplishing all the mentioned advantages, low cost photovoltaic cells are achievable that also fulfills the criteria for large scale applications, such as terrestrial deployment. Several potential photovoltaic materials have been explored during recent decades. Recently, research activities have shifted gradually towards TFSC using polycrystalline compound semiconductors, mostly inorganic, with direct bandgap and high absorption coefficient, which have greater potential to attain high conversion efficiency and higher stability. Types of solar cells that fall under the TFSC category include cadmium telluride (CdTe), copper indium selenide (CIS), gallium arsenide (GaAs), CIGS, CZTS, and amorphous silicon solar cells. CdTe, CIGS, and amorphous silicon thin film solar cells (TFSC) entered the commercialization stage nearly 20 years ago, whereby conversion efficiencies for research scale cells continue to show values greater than 20% with CIGS, CdTe. All TFSCs are contrasting the first generation bulk silicon solar cells in terms of absorber thickness, which is in many cases over 100 times less for thin films than bulk crystalline and poly-crystalline silicon solar cells as shown in Fig. 3.2.

3.1 Cadmium-telluride thin film solar cells

CdTe is a recognized solar cell material due to material advantages and easier methods of thin film deposition to prepare polycrystalline CdTe layers. The maximum theoretical efficiency of the CdTe solar cell corresponding to a band gap of 1.5 eV is about 28%–30%. CdTe has long been known as a leading thin film photovoltaic material due to its near optimal direct bandgap of 1.44 eV and high absorption coefficient. CdTe has a high absorption coefficient and approximately 99% of the incident light is absorbed by a layer thickness of about 1 μm. CdTe has vital advantages as a material for solar cells because of its ability to maintain good electronic properties under thin film form. Over the last 50 years, much effort has been put into developing high efficiency, low-cost thin film polycrystalline CdS/CdTe solar cell devices. CdTe cell production is associated with

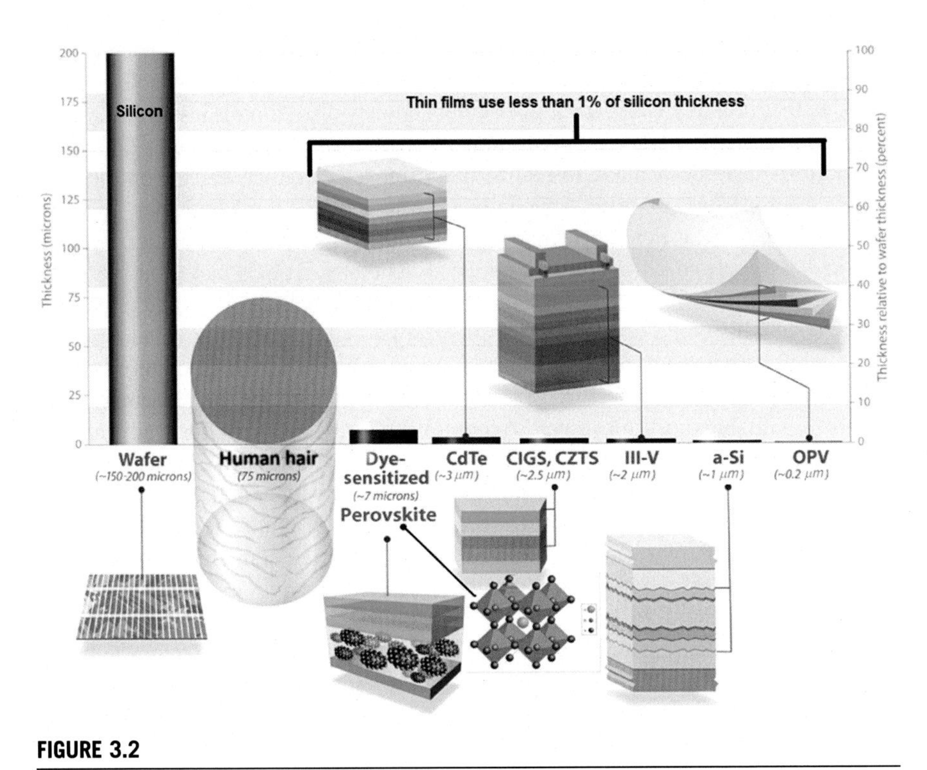

FIGURE 3.2

Technology diversification by inorganic thin film solar cells.

polycrystalline materials and glass, which are potentially much cheaper than bulk silicon. The polycrystalline layers of a CdTe solar cell can be deposited via different deposition techniques such as close-spaced sublimation, physical vapor deposition, RF magnetron sputtering, etc. Electronic properties are one of the most important properties to fabricate high efficiency CdTe solar cell. Basic understandings of CdTe-based solar cell properties are tricky particularly because of the polycrystalline nature. Therefore researchers should concentrate on studies of the fundamental electronic properties of the polycrystalline thin film CdTe and other constituents of the cell relevant to the improvement of the cell/module performance. To expand the stability of CdTe solar cells, more emphasis should be given to the front and back contact of the cells. For a commercially viable product, the module lifetime should be proven to be at least 25 years.

3.2 Copper indium gallium diselenide thin film solar cells

Recently, there has been a curiosity in solar cell technology using advanced polycrystalline thin film absorbers (few micron thickness) instead of Si single crystal

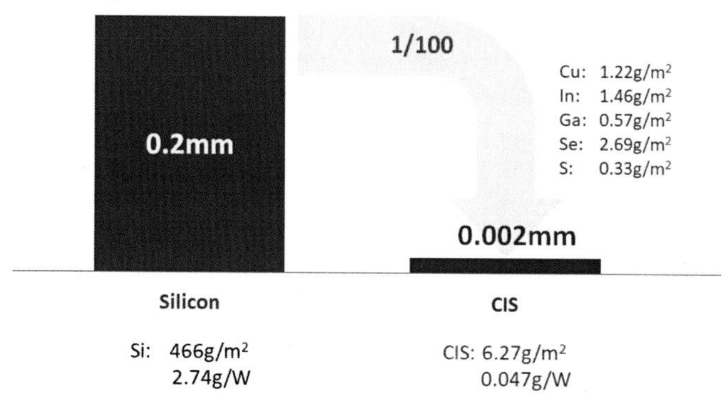

FIGURE 3.3

Higher cost reduction potential of CuInGaSSe solar cells.

wafers (200 μm thick), characterized by less material usage and lower cost with respect to the traditional technology. These materials include chalcopyrite CIGS and kesterite CZTS semiconductors. The success of CIGS TFSC is represented by the appearance of numerous PV companies around the globe. This has added another dimension in the silicon-dominated PV market in terms of lower production costs stemming from low material usage as well as low energy intensive manufacturing process as shown in Fig. 3.3 as a reference. CIGS solar cells have been in mass production since the early 2000s. Despite its scientific potential, its growth and market share have remained very modest over the years. Numerous companies have ventured into this field, each expecting their products will have advantages over other competitors in terms of efficiency and cost. However, the majority of them found that their competitive advantage did not last for long. Thus it is apparent that the mass production of CIGS solar cells is supported by a strong foundation in the science and technology behind it, as proven by the handful of CIGS makers that are still profitable to this day. Research and development in the field of chalcogenide photovoltaics has been pursued over the last five decades. Buoyed by the hopes that the use of polycrystalline TFSC can realize the dream of low cost photovoltaics, early scientists in this field diligently carried out numerous investigations that finally succeeded in bringing the cost of photovoltaics to its all-time low. At the time of writing, Manz AG of Germany declared that the Levelized Cost of Energy of its CIGS modules have undercut those of crystalline PV modules to less than 3 EURct/kWh. Several real-world tests in different parts of the globe using CIGS modules have also proven that CIGS modules are still capable of generating high energy in various climatic conditions, pollution, and irradiation angle. On the scientific front, there have been periods of enthusiastic breakthroughs that resulted in record-breaking efficiencies of both laboratory scale and larger modules of CIGS devices. Interspersed with these breakthroughs are pragmatic periods that indicate that more research must be performed. As of today, CIGS photovoltaic devices achieve the highest energy conversion efficiency among the available thin film technologies. This achievement is considered a miracle by some, given that so many basic material properties are still not well understood. Material scientists have long established a strong correlation between the growth condition of thin films and its properties. Thus empirical science holds the key to further the theoretical understanding on the subject of CIGS solar cells.

3.3 Copper zinc tin sulfide thin film solar cells

To attain terawatt photovoltaics, indium-containing CIGS quaternary material is poised to fall short of the required production due to the insufficiency of In. Therefore quaternary compound Cu_2ZnSnS_4 (CZTS) has been intensively scrutinized recently as an alternative absorber material for TFSC due to its resemblance in material properties with CIGS. In CZTS solar cells, costly indium (In) is substituted by low cost zinc (Zn) and gallium (Ga) with tin (Sn). CZTS is a compound semiconductor of group I–II–IV–VI of the periodic table. As a semiconductor, this material has excellent photovoltaic properties such as a direct-bandgap, high absorption coefficient ($> 10^4$/cm), and optimal bandgap energy of 1.4–1.5 eV which are highly anticipated in a photovoltaic material. Moreover, the CZTS compound comprises abundant (rare-earth free metals), nontoxic, and low-cost materials. From a semiconductor material property point of view, pure sulfide CZTS (x=1) is considered as a better suited photo-absorber for photovoltaic solar energy conversion due to its optimum bandgap of 1.5 eV compared to the selenide-containing CZTSe (x=0) counterpart which has a lower bandgap around 1 eV. Theoretical calculations depict that conversion efficiency as high as 32% is attainable from TFSC containing a CZTS absorber layer of few micrometers. However, the efficiencies attained so far are still much lower than those of CIGS PV devices. This technology is still under R&D stage and yet to reach the commercialization stage. Further extensive optimization of device structure and interfacial and material properties are necessary to further boost the efficiency of laboratory scale CZTS-based devices to more than 15%, and subsequently inspire the translation of the optimized lab scale process to a large area process to confirm a sustainable and successful CZTS thin film module production in the future.

3.4 Novel chalcogenides and emerging photovoltaic technologies

Chalcogenide materials typically display semiconductive properties and are appropriate candidates for solar cells, optoelectronics, sensors, and thermoelectric devices. Metal chalcogenides comprise a large family of 2D materials, which characteristically circumvent the gapless aspect of graphene and enable the light absorption and application as a photodetector. According to the elements involved, metal chalcogenides can be categorized into transition metal and main group metal chalcogenides (MMCs). Transition metal chalcogenides (TMCs) include two subsets such as the well-known TMDs with the form of MX_2 (M=Mo, W as semiconductor; V, Nb, Ta as metal) and the less explored transition metal trichalcogenides (TMTs) with the form of MX_3 (M=Ti, Zr, Hf), where X represents chalcogens (S, Se, Te). A featured property of TMDs is the thickness-dependent bandgap and the indirect to direct bandgap transition when thinning to a monolayer, covering the bandgap range from 1.0 to 2.1 eV. On the other hand, TMT monolayers cover the bandgap from 0.21 to 1.90 eV, with MTe_3 (M=Ti, Zr, Hf) showing metallic properties. Besides TMCs, MMCs in the form of MX, MX_2, and M_2X_3 (M=Ga, In, Ge, Sn; X=S, Se, Te), are also promising candidates for optoelectronic applications. In line with the development of semiconductors, research has also focused on the development of appropriate supports including carbon nanotubes, carbon nanofibers, activated carbon, and graphene that can enhance the activity of semiconductors. Graphene has gained much attention because of its good adsorption and electronic properties. Multijunction solar cells, so far made primarily using the III–V compounds, have obviously confirmed that by minimizing thermalization and transmission losses, very large improvements in efficiency can be achieved over those of single junction cells. These devices find application in generating power for space applications and are used in concentrator systems. The future expansion of multijunction devices using low-cost thin film technologies is particularly promising for producing more efficient and inexpensive devices. Cost reductions will also be substantial when the thin film technologies are directly produced on building materials other than glass, as many materials, for example, tiles and bricks, can be substantially cheaper than glass and have much lower energy contents. There are a few approaches to improve efficiency: spectrum splitting (multijunction solar cells), tandem cells, modifying the incident spectrum (by using concentrators), nanostructured cells to pick up better incident light, multiple electron-hole pair generation by a single photon and some others. Contrary to conventional photovoltaic technologies, concentrated photovoltaic (CPV) devices use lenses and curved mirrors to focus sunlight onto small, but highly efficient, multijunction solar cells mainly based on GaAs or its derivatives. CPV systems frequently use solar trackers and sometimes a cooling system to further increase their efficiency. They are currently more expensive and far less common than conventional PV technologies. Moreover, there are hot carrier solar cells and intermediate bandgap solar cells being investigated at laboratories as the emerging PV technology using inorganic materials.

SUBCHAPTER

Cadmium telluride (CdTe) thin film solar cells 3.1

Kazi Sajedur Rahman

Solar Energy Research Institute, Universiti Kebangsaan Malaysia (The National University of Malaysia), Bangi, Malaysia

3.1.1 Introduction

Semiconductors are a huge part of many modern-day devices like sensors, integrated circuits, energy harvesting devices, optoelectronics, and so on. They can be considered the heart of these applications. Nevertheless, aside from two recognized rudimentary semiconductors, silicon and germanium, large numbers of integrated materials have been used since the microelectronic revolution discovery of the transistor. Nonetheless, many compound semiconductors have been manufactured and grown by several methods by mainstream researchers. Semiconductors are the basic photovoltaic material used in inorganic solar cells. This section deals with a literature review and the structure and characteristics of cadmium telluride (CdTe). Key developments associated with low-cost, high efficiency CdTe thin film solar cells are also provided to give a better and clear representation of how CdTe solar cells have progressed over the years.

3.1.1.1 Cadmium telluride as a solar cell material candidate

The selection of materials for photovoltaic conversion is guided by several considerations as described below:

1. The value of the energy band gap and nature of the band-to-band transitions.
2. The value of the photocarrier lifetime as a function of doping.
3. The capability of the material to be prepared economically in large areas with good electronic properties.
4. The ability to form efficient collecting structures.
5. The ability of the cell to work under concentrated radiation.

Based on this quality criteria, CdTe is a good choice as a solar cell material. Lately, research activities have shifted progressively toward thin film solar cells exploiting compound semiconductors with direct band gaps and high absorption coefficients, which have an enormous potential to achieve high efficiency and stability in contrast to a-Si solar cells. Among them, CdTe is considered a verified

Comprehensive Guide on Organic and Inorganic Solar Cells. DOI: https://doi.org/10.1016/B978-0-323-85529-7.00009-8

thin film solar cell material owing to its interesting properties. CdTe is a binary compound, henceforth it is simpler to reserve the stoichiometry compared to ternary or quaternary compounds like CIS and CIGS. CdTe is advantageous as a solar cell material for its appropriate direct band gap and ability to endure good electronic properties as a thin film. The highest theoretical efficiency of the CdTe solar cell conforming to a band gap of 1.5 eV is approximately 28% to 30% (Rahman et al., 2019). Though the exploration of the CdS/CdTe hetrojunction solar cell began in the early 1970s with about 6% efficiency, the present efficiency of CdTe solar cell has touched 22.1% as testified by First Solar Inc., the leading CdTe thin film-based PV manufacturing company (Rahman et al., 2019). Fig. 3.1.1 exemplifies the energy gap of CdTe around 1.45 eV, which is hypothetically well-matched to solar energy conversion (Nowshad, 2001). Fig. 3.1.2 illustrates that the energy gap of CdTe is direct, leading to absorption coefficient values ($>10^5$ cm^{-1}) much higher than those of silicon (Nowshad, 2001).

3.1.1.2 Basic properties of cadmium telluride

3.1.1.2.1 Physical properties

This part gives a total overview of the entire range of physical properties of CdTe. An orderly attempt has been executed to choose the most consistent values from the current literature. Properties, such as electron and hole mobility and electron affinity, are also mentioned in Table 3.1.1 to provide a better understanding of the material.

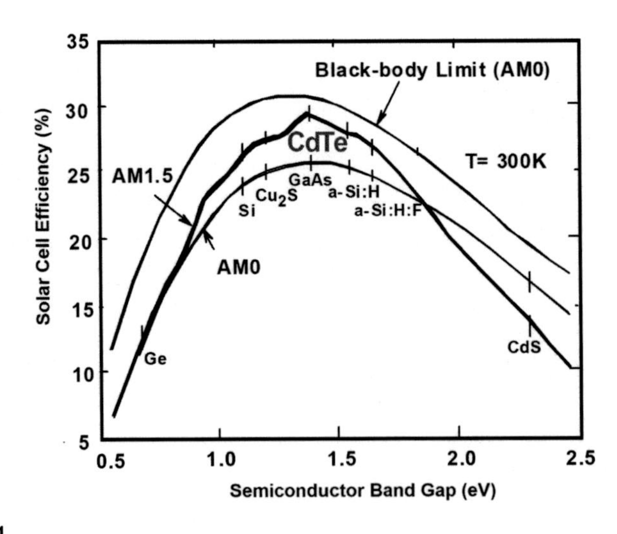

FIGURE 3.1.1

Theoretical efficiency versus energy band gap of photovoltaic devices (Nowshad, 2001).

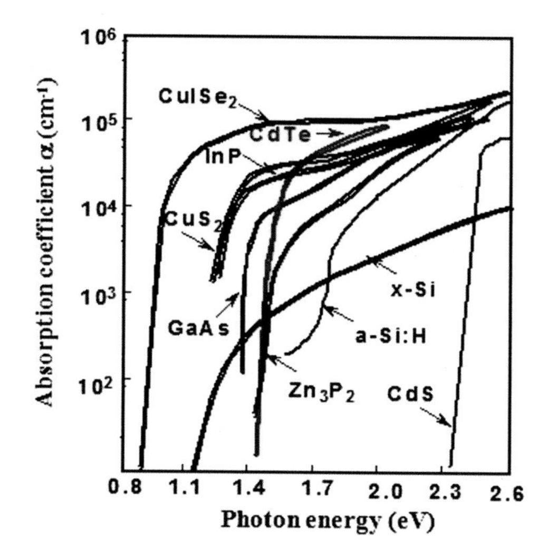

FIGURE 3.1.2

Absorption spectra of photovoltaic materials (Nowshad, 2001).

Table 3.1.1 Physical properties of cadmium telluride (Nowshad, 2001).

Physical properties	Values
Density	5.86 g/cm^3
Melting point	1092°C
Linear expansion coefficient	4.9×10^{-6}/K
Thermal conductivity	0.075 W/cm/K
Dielectric constant	10.3
Lattice constant	6.481 Å Lattice constant have unit
Refractive index	2.91
Band gap at room temperature	1.50 eV
Optical transition type	Direct
Conductivity type	n, p
Electron mobility	1200 cm^2/V.s
Hole mobility	100 cm^2/V.s
Electron affinity	4.28 eV

3.1.1.3 Structural properties

Both Cd and Te are in the fourth row of the periodic table, with atomic numbers of 48 and 52, respectively. Therefore CdTe has a similar equivalent isoelectronic arrangement as Sn, InSb, and AgI. As both HgSe and HgTe are semimetals, CdTe's average atomic number of 50 is the maximum for any IIB–VIA compound semiconductor. Typically, semiconductor materials have tetrahedral bonding in which each atom has four nearest neighbor atoms set in a tetrahedral structure and are held together by collaborating valence electrons, which forms the covalent bonds between atoms. They naturally form cubic or hexagonal structures. The bulk CdTe has a cubic zincblende structure, where each Te atom is tetrahedrally bonded with four nearest neighbor atoms of Cd. The structural properties of CdTe at 300K are summarized in Table 3.1.2.

The stoichiometric ratio of CdTe is typically 1:1 and melts near 1092°C. However, CdTe melts at a lower temperature if the material composition deviates. Single phase CdTe could only be attained if the composition is very close to the 50/50% stoichiometry. In addition to that, CdTe undergoes congruent sublimation, which implies that when the material is heated, the evaporation rates of Cd and Te are roughly the same. As a result, they condense on the substrate at the same rate, preserving the even stoichiometry as the original CdTe source used for deposition. Therefore this unique material quality permits the lump of stoichiometric CdTe to grow into a thin film with almost the same stoichiometric CdTe.

3.1.1.3.1 Thermal and mechanical properties

The thermal and mechanical properties of CdTe at 300K are summarized in Table 3.1.3. The thermal conductivity of 0.075 W/cm/K at 300K for CdTe is the lowest of any tetrahedrally coordinated II–VI compounds except HgSe and HgTe.

3.1.1.3.2 Optical properties

The optical properties of a semiconductor material can be determined by the band gap and quantum efficiency of the corresponding materials. The band gap of CdTe always affects the quantum efficiency of the CdTe solar cell as demonstrated in Fig. 3.1.3.

The quantum efficiency for the CdS/CdTe thin film solar cell remains in the range of 500 to 850 nm. The swift deterioration around 825 nm matches the CdTe band gap, which is close to 1.5 eV (Amin et al., 2017). Consequently, the window material for the CdS/CdTe thin film solar cell must be a wide band gap material

Table 3.1.2 Structural properties of cadmium telluride at 300K (Nowshad, 2001).

Structural properties	Values
Molecules per unit cell	4
Molecular weight	240.00 g
Lattice constant	6.481 Å
Shortest Cd-Te distance	2.806 Å
Concentration of Cd sites	$1.469 \times 10^{22}/cm^{-3}$
Molar volume	$40.99\ cm^3$

Table 3.1.3 Thermal and mechanical properties of cadmium telluride at 300K (Nowshad, 2001).

Thermal and mechanical properties	Values
Crystal structure	Zincblende
Linear expansion co-eff.	$(4.9 \pm 0.1) \times 10^{-6}$ /K^{-1}
Thermal conductivity	0.075 W/cm/K
Heat capacity, Cp	5.9 cal g/atom/K
Isothermal compressibility	3.69×10^{-3} /kbar^{-1}
Microhardness	55–140 kp/mm^2

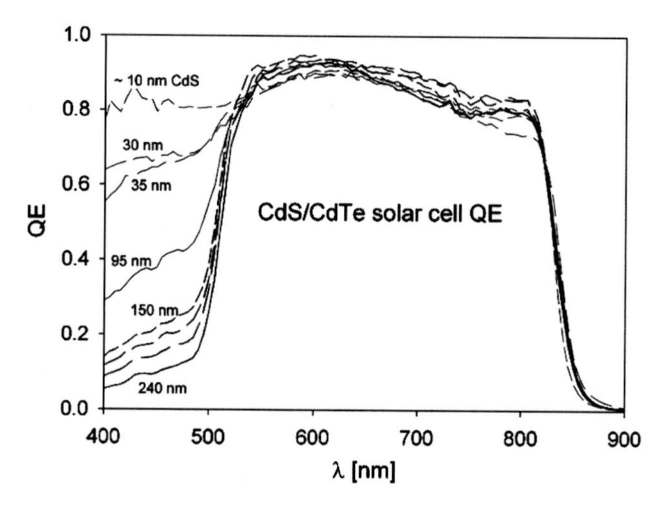

FIGURE 3.1.3

Quantum efficiency of the CdS/CdTe thin film solar cell (Amin et al., 2017).

to capitalize on the irradiance of the solar spectrum enhancing the quantum efficiency and ultimately the device efficiency (Amin et al., 2017).

3.1.1.3.3 Electrical properties

Electrical properties are eventually restricted by the electronic band arrangement deciding the inherent carrier concentration and the lattice vibration modes, which set upper limits on the electron and hole mobility. Nevertheless, electrical properties are driven by impurities, native defects, and the interactions between them. CdTe is a group II–VI compound semiconductor, and the electrical conductivity can be controlled between n-type and p-type. The conductivity change from p-type to n-type can be designated by Cd vacancies. This phenomenon is well documented with respect to "Cd" vapor pressure, where the conductivity type changes from p-type to n-type (Nowshad, 2001). Therefore CdTe conductivity type can be maintained by the partial vapor pressure of the fundamental element, where a higher concentration of "Te" leans toward p-type conductivity and "Cd" richness generates n-type conductivity. The as-grown CdTe thin films are resistive with low carrier concentration and contingent on the growth process and thickness of the film. Nevertheless, doping is normally incorporated in CdTe thin films with dopants such as copper (Cu) or zinc (Zn) to decrease the resistivity and expand the carrier concentration anticipated for the solar cell.

3.1.1.4 Cadmium telluride thin film solar cells

To be a viable contributor to large scale electrical supply, PV material technology should be manufacturable and durable (20-year plus outdoor life) and possess a module efficiency that is 15% or higher to attain PV system cost goals. Polycrystalline materials offer several advantages to achieve these PV material requirements. Over the last 50 years, much effort has been put into developing high efficiency, low-cost thin film polycrystalline CdS/CdTe solar cells. Among group II–VI semiconductor compounds, CdTe is the one of the most promising materials for thin film solar cells because of its low-cost and high efficiency. First, cell production is associated with polycrystalline materials and glass, which are potentially much cheaper than bulk silicon. Second, polycrystalline CdTe layers can be deposited via different techniques such as close-spaced sublimation (CSS), physical vapor deposition (PVD), sputtering, electrodeposition (ED), etc. Third, CdTe has a high absorption coefficient and nearly 99% of the incident light is absorbed by a layer thickness of about 1 μm. CIS and CIGS materials also show almost similar properties as stated above but CdTe has advantages over them based on its electronic properties. The electronic properties are important in fabricating high efficiency solar cells, and CdTe is unique in this respect.

3.1.1.5 Structure of cadmium telluride thin film solar cells

CdTe solar cells typically have a heterojunction configuration because of the short optical absorption length in CdTe and the difficulty of forming a shallow junction with a high conductivity layer. CdTe solar cells are usually fabricated in a superstrate configuration where the light is incident through the glass substrate. The CdS/CdTe solar cell structure can be separated into five major parts: substrate, front contact, window layer, absorber layer, and back contact. Fig. 3.1.4 displays the schematic of the different layers of CdS/CdTe thin film solar cells.

3.1.1.5.1 Substrate

For any type of solar cell, choice of substrate is vital. Basically, CdS/CdTe thin film solar cells are fabricated in superstrate configuration where light comes from the top. Hence, a transparent substrate is required so both reflection and absorption occur on and in the substrate. Additionally, the substrate should survive the

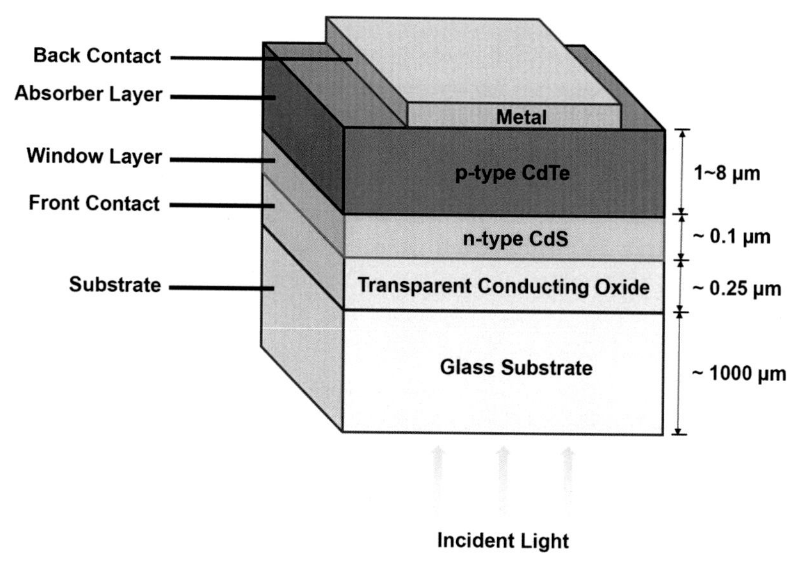

FIGURE 3.1.4

Schematic structure of cadmium sulfide/cadmium telluride thin film solar cells.

deposition process temperature and not contaminate the subsequent layers during the growth process. Therefore considering all the above issues, glass is commonly the best choice as it is highly transparent and inexpensive and can withstand high temperatures. Some common glass types used are soda lime glass, Corning glass, borosilicate glass, flexible ultra-thin glass, etc. The outer side of the glass exposed to sunlight is usually coated with an antireflective coating to enhance the optical performance of the solar cell.

3.1.1.5.2 Front contact

Generally, n^+ transparent conducting oxides (TCOs) have been used as the front contact (Bonnet & Rabenhorst, 1972). Since TCO forms the first layer after the glass substrate, it must be highly transparent over 80% in the visible range and should also have a large optical band gap over 3 eV. Electrically it should have very low resistivity on the order of 10^{-4} Ω-cm and high mobility (>30 cm^2/V. sec) with low free carrier absorption (Chopra, Paulson, & Dutta, 2004). The commonly used TCOs for the front contact of CdS/CdTe thin film solar cells are indium-doped tin oxide (ITO), fluorine-doped tin oxide (FTO), cadmium-doped tin oxide and aluminum-doped zinc oxide. The most widely used material as a front electrical contact for CdS/CdTe solar cells is tin oxide (SnO_2) as it is highly stable under high temperatures. As SnO_2 has too low conductivity for producing a good contact, it is frequently doped either with indium to generate ITO or fluorine to make FTO. Doped SnO_2 is a wide band gap and n-type degenerate semiconductor. The optical and electrical properties of doped SnO_2 depend on the doping material. It has been found that fluorine doping improves the carrier concentration which consequently intensifies the electron mobility (μ) of the SnO_2 film. For undoped SnO_2 films, electron concentration is in the lower 10^{18} cm^{-3} ranges and the mobility, μ, is ~1 cm^2/V-s. However, fluorine-doped films have higher absorption and a lower band gap compared to the undoped films. For fluorine-doped SnO_2 films, the optical absorption coefficient (for an average wavelength between 500 and 900 nm) is ~1200 cm^{-1} compared to ~600 cm^{-1} for undoped film (Balasubramanian, 2004). Therefore ITO or FTO is the material of choice as the front contact depending on the CdS and CdTe deposition processes.

3.1.1.5.3 Cadmium sulfide window layer

Polycrystalline cadmium sulfide (CdS) is reported to be the best heterojunction or n-type partner for the n-CdS/p-CdTe thin film solar cell. CdS is mainly chosen as a window layer because of its high band gap (Eg = 2.4 eV at room temperature) compared to that of CdTe (Eg = 1.45 eV at room temperature). CdS is largely transparent to wavelengths over 520 nm and allows the CdTe absorber layer to collect the photons of lower energy to generate electron-hole pairs under light illumination. Therefore the CdS layer is often referred to as the "window layer" and helps to reduce the interface recombination. Moreover, the lattice mismatch between CdS and CdTe is less than 10% which is considerably low. The thickness of the CdS layer affects the cell performance. Generally, the thickness of the deposited CdS layer is around 100 nm. By reducing the thickness of the CdS layer, some of the light below 510 nm can also cross it to the CdTe providing extra current to the device. Therefore thinning the CdS layer enhances the gain of photons in CdTe. In contrast, for the uniform CdS thin films and diffusion of CdS into CdTe during annealing treatment, it is suggested to maintain the thickness above this limit, because if not, the CdS layer will be completely diffused and the cell is either shunted or causes a low open circuit voltage (V_{oc}). It is found that as the CdS layer thins, the maximum values of both J_{sc} and efficiency are improved but V_{oc} is drastically reduced. To enhance the blue response of the CdS/CdTe solar cells, it is obligatory that the CdS films are sufficiently thin (40–60 nm) and pinhole free to avoid the formation of shunting paths. In addition to the improved blue response, the use of a thin CdS layer also results in a reduction in the V_{oc} and fill factor by 20–50 mV and 5%–10%, respectively. CdS can be deposited by different deposition techniques, such as sputtering, metal organic chemical vapor deposition, chemical bath deposition, thermal evaporation, CSS, molecular beam epitaxy, spray pyrolysis, and so on (Chopra et al., 2004; Nowshad, 2001). CdS layers deposited by low temperature techniques usually require an annealing treatment for recrystallization preceding deposition of the CdTe absorber layer.

3.1.1.5.4 Cadmium telluride absorber layer

CdTe is known as a thin film material owing to its adjacent optimal band gap of 1.45 eV and high absorption coefficient. CdTe is called the absorber layer as it absorbs over 90% of the accessible photons ($h\upsilon > 1.45$ eV) in less than 2 μm thickness (Nowshad, 2001; Rahman et al., 2019). CdTe is p-type doped and the doping level of CdTe is less than CdS and the depletion region is typically in the CdTe layer. Consequently, this is the active region of the solar cell, where most of the carrier generation and collection happens. The fundamental necessities for efficient absorption and photocurrent generation are a homogeneous CdTe film with a thickness ~2 μm, low defect density, bigger grain size (>0.5 μm), and a pinhole-free film. CdTe can be deposited with a wide range of deposition techniques, either chemically or physically (Rahman et al., 2019). PVD has been the conventional approach to grow CdTe, which comprises CSS, sputtering, thermal evaporation, ED, etc. (Arce-Plaza et al., 2018). However, CdTe films grown by CSS at higher temperature >550°C showed improved grain sizes of >2 μm (Amin & Rahman, 2017). Therefore CdTe thin films grown at higher temperatures initiate Cd deficiencies resulting in the material conductivity to be p-type and the grown films are Te-rich and ensure p-type conductivity. The CdS/CdTe stack is typically given a postdeposition $CdCl_2$ heat treatment which enables grain enhancement, reduces the defect density in the films, promotes the interdiffusion between CdS and CdTe layers, and thereby improves solar cell efficiency. However, many associated factors degrade the ideal efficiency and achieved efficiencies are lower. The record efficiency of CdTe thin film solar cell has reached 22.1%, as stated by First Solar, Inc., and the CdTe module efficiency has exceeded 18% (Rahman et al., 2019).

3.1.1.5.5 $CdCl_2$ treatment

It is difficult to achieve high doping concentration for CdTe due to self-compensation from intrinsic defects, for example, vacancies (V_{Cd}, V_{Te}), interstitial defects (T_{ei}), and grain boundaries. Therefore postdeposition annealing treatments are required to activate and/or passivate native defects (Barri et al., 2005). Chloride treatment is a vital technology to enhance the grain size. This treatment is achieved by placing $CdCl_2$ on the back surface of the CdTe layer preceding annealing. $CdCl_2$ application can be done ex situ by either evaporating a $CdCl_2$ thin film on top of CdTe or covering the CdTe sample with a saturated $CdCl_2$/methanol solution (Alnajjar, Alias, Almatuk, & Al-Douri, 2009). Currently, extrinsic doping of the CdTe is not frequently used. However, some researchers believe that oxygen in the treatment acts as a dopant and improves film nucleation or both. Annealing the TCO/CdS/CdTe stack in the presence of chloride at approximately 400°C typically improves the electronic properties of the devices and results in a highly efficient thin film solar cell that is used by most researchers (Alnajjar et al., 2009; Barri et al., 2005). It is experimentally proven that the gains in both V_{oc} and photocurrent generally increase due to the chloride treatment. The exact mechanism is not well understood yet, but grain growth and CdS/CdTe interdiffusion can occur as a result of the chloride treatment. All chloride annealing methods can be conducted at or near atmospheric pressure in oxygen ambient. Thus $CdCl_2$ treatment is indispensable when making efficient CdTe solar cells. $CdCl_2$ treatment enhances the solar cell performance by affecting the following parameters:

1. Recrystallization: It has been observed that the lattice constant rises due to $CdCl_2$ treatment. The lattice constant rises from 6.437 to 6.479 Å for $CdCl_2$ concentrations from 1 to 5 wt.%, as reported by earlier (Barri et al., 2005).
2. Grain growth: $CdCl_2$ treatment eliminates the small grains in submicron levels by increasing the grain size of CdTe layer and consequently increasing the cell performance. It has been reported that the grain size has been improved from 0.1 to >1 μm for a 5 wt.% $CdCl_2$ films (Mullins & Taguchi, 1992).
3. CdTe defect levels: CdTe defect levels reduce with $CdCl_2$ treatment.
4. Type conversion: $CdCl_2$ etching treatment increases the Te concentration at the surface region of CdTe, which converts p-CdTe to p^+-CdTe. This conversion helps to create low ohmic back contact for CdTe solar cell.
5. Current transport: $CdCl_2$ treatment decreases the series resistance of CdTe thin film solar cells by reducing the back-contact barrier height, thereby enhancing the current transport of the solar cell.

3.1.1.5.6 Interdiffusion between cadmium sulfide and cadmium telluride

Interdiffusion between the CdS and CdTe interface and forming a new alloy CdS_xTe_{1-x} is well recognized. It can be formed by either S diffusion to CdTe or Te diffusion to CdS (Enríquez & Mathew, 2004). The primary cause for good quality CdTe/CdS junction formation depends on CdTe and CdS miscibility and a reaction between these CdTe and CdS during the complete cell fabrication process guides the formation of two ternary compounds as the interfacial layer (Ferekides et al., 2004). Due to the formation of an S-rich layer of $CdTe_yS_{1-y}$ and Te-rich layer of CdS_xTe_{1-x} at the interface, it is assumed that $CdTe_yS_{1-y}$ and CdS_xTe_{1-x} reduce the interfacial defect density contributing toward highly efficient CdS/CdTe thin film solar cell (Ferekides et al., 2004). One of the feasible ways to achieve the formation of $CdTe_yS_{1-y}$ and CdS_xTe_{1-x} is the postdeposition treatment of the CdS/CdTe thin film solar cell. The interdiffusion process relies on many factors such as the growth temperature, grain structure, and postgrowth annealing of the films. Typically, CdTe and CdS interdiffusion happens due to postdeposition thermal treatment, where the process is improved by incorporating Cl prior to annealing (Terheggen et al., 2003). Nevertheless, mixing during the growth can be controlled by varying the CdTe growth conditions. The consequence of this diffused layer on the performance of CdS/CdTe solar cell is substantial. As an example, V_{oc} can fluctuate as a function of the level of the interdiffusion (Singh & McClure, 2003). Furthermore, it has also been reported that the Te diffusion in the CdS region reduces the band gap of CdS, and S diffusion in the CdTe region reduce the band gap of CdTe (Wood, Lane, Rogers, & Coath, 1999). Therefore the external quantum efficiency and complete performance of the devices may change.

3.1.1.5.7 Back contact

Fabricating a stable and low resistance back contact for p-type CdTe solar cells is still a challenge. Following post-deposition thermal annealing, the back contact is deposited to complete the cell fabrication. An ohmic contact to CdTe entails a metal with a work function greater than 5.7 eV. No metal has yet been found to suitably match with CdTe and, consequently, there exists a wide Schottky barrier between CdTe and the metal back contact. The approach to overcome the naturally existing Schottky barrier is to create a heavily p-doped CdTe surface by chemical etching and applying a buffer layer or Back Surface Field (BSF) layer of high carrier concentration between CdTe and the metal (Ferekides et al., 2004). An efficient BSF layer is an important structural component to attain high efficiency in a solar cell. It reduces the loss of carriers at the back contact by minimizing the carrier recombination, creates a tunnel contact, and reflects the carriers toward the p-n junctions and consequently increases the cell performance. Generally, a Cu-containing back contact is utilized to improve the device performance of n-CdS/p-CdTe solar cells which creates a quasiohmic, nonrectifying contact and additionally dopes the CdTe layer (Aris et al., 2018). These types of cells exhibit good efficiencies in the beginning. However, the efficiency degrades with time due to Cu diffusion to the front contact which causes a shunting effect (Aris et al., 2018). The back contact commonly incorporates two layers: the initial one is a deeply doped layer with a low-loss electrical contact to CdTe and the secondary contact is metal that conveys the current laterally (Aris et al., 2018). The second approach to reducing the barrier height is to form a highly doped p+ region at the CdTe back contact by evaporating elemental Te or by chemical etching with etchants, such as bromine methanol (BrMeOH) or a mixture of nitric acid and phosphoric acid, HNO_3:H_3PO_4 (McCandless & Dobson, 2004; Potlog, Ghimpu, & Antoniuc, 2007). Some commonly used Cu-containing contacts are: Cu/graphite, Cu/Mo, Cu/Au, Cu_xTe/Au, etc (Nawarange & Compaan, 2011). Sometimes, Cu-free contacts are also exploited for CdS/CdTe solar cells such as: ZnTe, Sb_2Te_3/Mo, Au, etc. (Bätzner, Romeo, Zogg, Wendt, & Tiwari, 2001).

3.1.1.6 Deposition techniques of cadmium telluride thin films

CdTe deposition methods include PVD, CSS, vapor transport deposition (VTD), ED, sputtering etc. (Rahman et al., 2019). All CdTe deposition techniques have pros and cons. The physical techniques involve high cost while chemical techniques confirm lower cost with the compromise of efficiency. The physical method assurances bigger grain size than that attained via chemical techniques, which influences the electrical properties. The deposition temperature is one of the vital factors for CdTe growth as most of the available deposition techniques have substrate heating. It is quite noteworthy that even though important distinctions exist, the performances achieved are self-governing to processing and demonstrate the flexibility of CdTe and its prevalent possibilities in photovoltaic innovations.

3.1.1.6.1 Physical vapor deposition

PVD of CdTe is based on the equilibrium between Cd and Te_2 vapors and CdTe solid (McCandless & Sites, 2003), Cd + $1/2Te_2 \Leftrightarrow$ CdTe. Consequently, CdTe can be grown by co-evaporation from basic sources, direct sublimation from a CdTe source, or vapor transport with a carrier gas to transport Cd and Te_2 vapors from either elemental or CdTe sources (McCandless & Sites, 2003). Thermal evaporation is also a PVD process in which a solid material positioned in a boat is heated to the point where it starts to evaporate and then condenses onto a cooler substrate to form a film of that respected evaporated material. CdTe deposition by thermal evaporation is carried out by gradually increasing the current to the desired deposition current. The temperature of the boat increases with increasing current and eventually heating the CdTe source, and when the temperature is high enough, CdTe evaporation occurs. The evaporation time is adjusted in such a way that the thickness of the films is nearly uniform for different deposition parameters such as for higher deposition current, where the deposition time is less and vice versa.

3.1.1.6.2 Sputtering

Sputtering is a PVD system that is attractive for thin film deposition because sputter-deposited films usually have provide better adhesion on the substrate compared to other techniques. Sputtering is based on the theory that particle to particle collisions will involve an elastic transfer of momentum that can be utilized to apply a thin film to the substrate. Generally, CdTe films deposited by sputtering have a composition close to that of the source material. In CdTe deposition via sputtering, Cd and Te mass transfer occur through removal of the CdTe target by Ar^+ followed by diffusion to the substrate and condensation (McCandless & Sites, 2003). Generally, CdTe deposition by sputtering is carried out at a substrate temperature below 300°C and at deposition pressures of $\sim 10-20$ mTorr (McCandless & Sites, 2003).

3.1.1.6.3 Close-spaced sublimation

CSS is perhaps the most straightforward method in PVD where deposition happens by sublimation and the source plate is near the substrate. Semiconductors that evaporate under 800°C can be covered on substrates in both vacuum and atmospheric pressure. The target materials should be solid in either chunk or powder form (Amin & Rahman, 2017). For instance, CdTe can be deposited at around 600°C with a thickness of $1-10\ \mu m$ within 10 minutes of deposition time, which is one of the quickest deposition times among other PVD approaches. As anticipated, CSS and CdTe are closely interconnected due to widespread use of CSS in

CdTe film growth (Amin & Rahman, 2017). The deposition method for CdTe can substantially influence the material properties and device performance. The CSS technique proposes high deposition rates and can be simply scaled for manufacturing purposes. CSS equipment can be created with a wide assortment of alternatives, such as vacuum, temperature, source-substrate spacing, etc. For instance, the source and substrate in CdTe growth are reserved at a close distance of 2 mm, which can also be altered (Amin & Rahman, 2017). Top and bottom heaters can be selected conferring to the temperature required to evaporate either in atmospheric pressure or evacuating to the mTorr range, loading up with gas like Ar or N_2. A good temperature controller is needed and must be programmable in steps, such as increment or hold time. Remarkably, the entire CSS deposition happens in just 10 minutes, without the evacuation or purging time. Sublimation demonstrates the direct phase transition between a solid and gas state. CSS is reliant on the factors such as sublimation process at the surface of the source and substrate, and transport of the gas from the source to the substrate.

The sublimation reaction is stated in the following equation (Amin & Rahman, 2017):

$$2CdTe(s) \leftrightarrows 2Cd(g) + Te_2(g) \tag{3.1.1}$$

Sublimation creates monatomic Cd and diatomic group VI (Te_2 or S_2) vapor which recombine by the opposite reaction on the comparatively cool substrate. Basically, the coating method is a chemical vapor deposition with locally produced vapor (sublimation) and an inverse reaction (deposition) developing no by-products. A schematic illustration of the CSS is presented in Fig. 3.1.5.

The substrate and source, detached by a small distance (1 mm for example) and reinforced by proper holders, are surrounded in a controlled atmosphere in a fused silica tube provided with gas inlet and outlet tubes. The most important process parameters are the source-substrate temperatures, nature of the atmosphere, reaction tube pressure, and composition of the source material (Amin & Rahman,

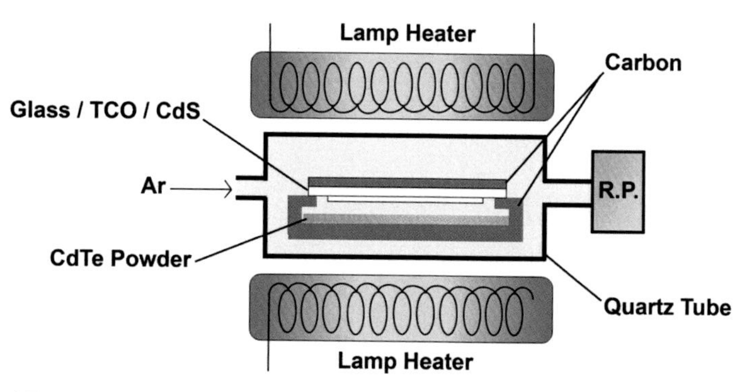

FIGURE 3.1.5

Schematic of close-spaced sublimation (CSS) system.

2017). These parameters are interconnected. For instance, the fractional pressures of Cd and Te_2 in the reaction tube are vital in defining the deposition rate, and these pressures change exponentially with temperature. At a specified source temperature, the sublimation rate rises swiftly as the pressure in the reaction tube reduces from the atmospheric pressure. For fixed source-substrate spacing, high thermal conductivity of the ambient gas increases the substrate temperature, consequently decreasing the growth rate. High deposition rate of CdTe films (up to 10 μm/min) is a superior aspect of the CSS process. Deposition pressures (1–30 Torr), substrate temperature (500°C–600°C), and source temperatures (600°C–800°C) are commonly used for CSS (Amin & Rahman, 2017).

3.1.1.6.4 Vapor transport deposition

Basically, VTD permits very high deposition rate at high substrate temperature and pressure forthcoming 0.1 atm onto moving substrates (McCandless & Sites, 2003). Whereas CSS is diffusion-restricted, VTD works by convective exchange of a vapor stream soaked with Cd and Te to the substrate, where supersaturation of the Cd and Te vapors creates condensation and reaction to form CdTe (McCandless & Sites, 2003). The CdTe source entails a heated chamber comprising solid CdTe in which the carrier gas combines with Cd and Te vapors and is exhausted through a slit over or under the moving substrate at a distance of ~1 cm (McCandless & Sites, 2003). The geometrical arrangement of the source impacts the consistency and use of the vapors in the carrier gas. The carrier gas composition can be changed, as with CSS, to comprise N_2, Ar, He, and O_2. As-deposited VTD films are like CSS films, with closely arbitrary orientation and normal grain size distribution (McCandless & Sites, 2003).

3.1.1.6.5 Electrodeposition

Electrodeposition is a chemical process in which an electric current is utilized in an electrolytic solution with the purpose of ion movement toward the cathode (Arce-Plaza et al., 2018). The material is deposited when the ions arrive on the cathode. This method has numerous benefits; for instance, with this method, it is possible to utilize flexible substrates. Another benefit is that as the substrate is not heated, large area deposition is likely and, consequently, can be industrialized.

3.1.1.6.6 Spray pyrolysis

Spray pyrolysis is a simple chemical technique where the material in solution is crushed by the gas pressure (argon, air, nitrogen, etc.) (Arce-Plaza et al., 2018). For this procedure, it is significant to regulate the solution flow and gas pressure. The crushed solution is sprayed on the hot substrate to generate the film. For this technique, CdTe solution is basically a colloidal system. For this colloidal system, the stabilizing agent is vital as it alleviates the CdTe molecule (Arce-Plaza et al., 2018).

3.1.1.7 Performance prospective of cadmium telluride thin film solar cells

To attain the anticipated breakthrough of photovoltaic technology as a competitive energy source against fossil fuels, both the conversion efficiency and reduction of production cost are essential. Worldwide, researchers are still trying to develop thin film technology for high efficiency, low production cost, and environmental friendliness. Polycrystalline CdTe is the most promising photovoltaic material for the thin film solar cell because of its excellent PV properties. This section covers almost all aspects to improve CdTe thin film solar cell technology, such as starting from the absorber layer, buffer layer, contact layers, etc. Further advances in performance require better understanding of the materials that comprise the solar cell, key interfaces, and device operation models such as those issues listed below.

- Primary research in open-circuit-voltage (V_{oc}) improvement:
  - Interface optimization (front/back contact)
  - Improved carrier lifetime with better CdTe film quality (grain size, grain boundary, defects, fabrication process, etc.) and therefore collection
  - Understanding of doping capability of CdTe and improved carrier concentration.
- Optical property optimization
  - Glass transmission – low iron (Fe) glass
  - TCO absorption
  - Thin window layer, CdS
  - Thicker interdiffusion layer (CdS_x:Te_{1-x}), reflux agent
  - Anti-reflective coating usage
  - Windows less devices or graded junction
  - Back mirror usage.
- Better contacts (front and back):
  - For improved yield
  - For improved performance.
- Industry needs:
  - Off the shelf manufacturing equipment
  - In-line diagnostics for improved quality control/yield
  - Validation of long-term reliability
  - Bridge the performance gap between small cells and commercial modules.

3.1.1.8 Manufacturing of cadmium telluride thin film solar cells

Three key process categorizations are utilized in CdTe PV module manufacturing. The first one resembles semiconductor deposition, where the semiconductor material, accountable for the sunlight conversion to electricity, is deposited; second, PV cell formation; and third, the ultimate module assembly and test are completed. First Solar's CdTe process technology is constructed on the sublimation property of CdTe. At specific pressure and temperature conditions, CdTe disintegrates in its parent compounds of Cd and Te. Those species in gaseous phase are deposited onto the substrate surface in the form of a thin CdTe layer. The First Solar method utilizes a high rate vapor deposition technology that can deposit the thin semiconductor layer in less than 40 seconds (First Solar, 2016). The first process sequence begins with the deposition on a glass substrate of a thin tin oxide layer that acts as a transparent conductive oxide (TCO) layer. In fact, the TCO layer is applied to the glass by the glass provider. Afterwards, very thin CdS (window) layers followed by a CdTe (absorber) thin layer are deposited. The CdS and CdTe layers are deposited utilizing powders of the same materials via vapor deposition technique. Next, $CdCl_2$ treatment and a thermal treatment is applied to re-crystallize the structure and enhance the electronic properties of the device. The $CdCl_2$ is washed off the module after the recrystallization process is finished. Lastly, a metal layer is deposited to make the back contact. Subsequently, the individual photovoltaic cells are interconnected in series using a laser scribe technology, followed by the lamination process where a transitional polymeric adhesive and glass plate are positioned and thermally sealed together with the glass substrate. The final module is shaped of a series connected CdTe PV cells with a film thickness under 10 μm and about 7 g/m^2 of cadmium content, encapsulated, insulated with solar edge tape, and sealed between two glass plates of about 3 mm thick each (First Solar, 2016).

Table 3.1.4 shows the list of manufacturers of CdTe technology over the last 10 years. The first name that pops up in CdTe module manufacturing is First Solar Inc. First Solar PV modules are the first thin film modules to reach over 10 GWp of module installations worldwide with manufacturing costs of $\sim\$0.6\ W^{-1}$, where India alone has over 1 GWp (First Solar, 2016). Integrating each production step, First Solar manufactures the modules on high throughput, automated lines from semiconductor deposition to final assembly and test – all in one continuous process. The whole flow, from a piece of glass to a completed solar module, takes less than 2.5 hours, whereas a multicrystalline silicon-based module needs at least one day and crystalline silicon-based modules need

Table 3.1.4 List of manufacturers worked/working on cadmium telluride technology in last 10 years.

Company	Country
First Solar	Ohio, USA and Kulim, Malaysia
Antec Solar	Germany
Calino	Germany
Arendi	Italy
Primestar Solar	Colorado, USA
Ava Solar	Colorado, USA

2.5–3.5 days to complete (First Solar, 2016). Using CdTe as the semiconductor material, the manufacturer makes it affordable to convert solar energy into the type of electricity we use every day with a CdTe cell efficiency of 22.1% (module efficiency of 18.2%), which is better than multicrystalline silicon-based cell efficiency of 20.4% (First Solar, 2016).

As CdTe modules have arrived at a production cost under USD 0.50/Watt with an energy payback time of around 1 year as system along with 25 years warranty on power output, it is currently cost viable with silicon PV technologies. Nowadays, manufacturers are not competing with the module cost as $/Wp, rather Levelized Cost of Electricity (LCOE) from its installed solar farms as a new business strategy. Therefore every candidate has its role based on overall cost for installed PV sites. For CdTe systems, the energy cost was about 7–8 United States cent/kWh few years ago, which reduced to 5–6 United States cent/kWh by June 2015 and reached 4–5 United States cent/kWh (First Solar, 2016). Therefore stringent LCOE competition can be visualized between silicon-based first generation and CdTe-based second generation solar cells, where both claim to be cheaper than any fossil fuel-based LCOE. The mostly appropriate optoelectronic properties of CdTe fascinate both the scientific communities and industrial companies of photovoltaic materials. Although the hidden efficiency is yet to be realized, the inventive strategies as new designs must be scrutinized for photovoltaic potential. Therefore CdTe photovoltaic technology will keenly benefit further from inexpensive manufacturing costs and higher efficiency and should ultimately direct solar electricity that can economically contend with fossil fuels and other energy sources from now on.

3.1.1.9 Summary

Recent studies suggest that thin films will dominate the worldwide solar photovltaic market in the near future, since they should be capable of reaching a lower price figure (in $/Watt) than Si. Among thin films, CdTe is the favorable choice in this regard. Significant research has been performed with a CdTe absorber material and stable, low resistance, nonrectifying back contact, but still there many other avenues to enhance the conversion efficiency by improving V_{oc} and J_{sc}. Insertion of suitable BSF layers for ultra-thin CdTe solar cells might overcome the back-surface recombination loss and form a stable back contact for achieving higher V_{oc}, J_{sc}, FF and finally conversion efficiency. To achieve these goals, innovative design, proper methods, and cost-effective fabrication technology must be developed.

References

Alnajjar, A. A., Alias, M. F. A., Almatuk, R. A., & Al-Douri, A. A. J. (2009). The characteristics of anisotype CdS/CdTe heterojunction. *Renewable Energy*, *34*(10), 2160–2163.

Amin, N., Ahmad Shahahmadi, S., Chelvanathan, P., Rahman, K. S., Istiaque Hossain, M., & Akhtaruzzaman, M. D. (2017). *Solar photovoltaic technologies: From inception toward the most reliable energy resource.*

Amin, N., & Rahman, K. S. (2017). *Close-spaced sublimation (CSS): A low-cost, high-yield deposition system for cadmium telluride (CdTe) thin film solar cells*, Modern *Technologies* for *Creating* the Thin-Film *Systems* and *Coatings* (361).

Arce-Plaza, A., et al. (2018). CdTe thin films: deposition techniques and applications. *Coatings and Thin-Film Technologies*, 131–148.

Aris, K. A., et al. (2018). A comparative study on thermally and laser annealed copper and silver doped CdTe thin film solar cells. *Solar Energy*, *173*, 1–6. Available from https://doi.org/10.1016/j.solener.2018.07.009.

Balasubramanian, U. (2004). Indium oxide as a high resistivity buffer layer for CdTe/CdS thin film solar cells.

Barri, K. et al., (2005). Introduction of Cu in CdS and its effect on CdTe solar cells. In *Conference record of the thirty-first IEEE photovoltaic specialists conference, 2005* (pp. 287–290).

Bätzner, D. L., Romeo, A., Zogg, H., Wendt, R., & Tiwari, A. N. (2001). Development of efficient and stable back contacts on CdTe/CdS solar cells. *Thin Solid Films*, *387* (1–2), 151–154.

Bonnet, D. & Rabenhorst H., New results on the development of a thin-film p-CdTe-n-CdS heterojunction solar cell, In *Photovoltaic specialists conference*, 9th, Silver Spring, MD, 1972, pp. 129–132.

Chopra, K. L., Paulson, P. D., & Dutta, V. (2004). Thin-film solar cells: an overview. *Progress in Photovoltaics: Research and Applications*, *12*(2–3), 69–92.

Enriquez, J. P., & Mathew, X. (2004). XRD study of the grain growth in CdTe films annealed at different temperatures. *Solar Energy Materials and Solar Cells*, *81*(3), 363–369.

Ferekides, C. S., Balasubramanian, U., Mamazza, R., Viswanathan, V., Zhao, H., & Morel, D. L. (2004). CdTe thin film solar cells: device and technology issues. *Solar Energy*, *77*(6), 823–830.

First Solar. (2016). Manufacturing process-advanced and continuous process: Module construction process. <http://firstsolar.com/en/Technologies-and-Capabilities/PV-Modules/First-Solar-Series-4-Modules/Manufacturing-Process>.

McCandless, B. E., & Dobson, K. D. (2004). Processing options for CdTe thin film solar cells. *Solar Energy*, *77*(6), 839–856.

McCandless, B. E., & Sites, J. R. (2003). Cadmium telluride solar cells. In L. S. Hedegus (Eds.), *Handbook of photovoltaic science and engineering* (New York: John Wiley Sons, Ltd.).

Mullins, J. T., & Taguchi, T. (1992). Non-linear optical properties of the Cd1 − xZnxS: Cd1 − yZnyS strained-layer superlattice system. *Journal of Crystal Growth*, *117*(1–4), 501–504.

Nawarange, A. V., & Compaan, A. D. (2011). Optimization of back contacts for CdTe solar cells using sputtered Cu x Te. In *2011 37th IEEE photovoltaic specialists conference* (pp. 1317–1321).

Nowshad, A. (2001). Study of high efficiency CdTe ultra thin film solar cells by close-spaced sublimation.

Potlog, T., Ghimpu, L., & Antoniuc, C. (2007). Comparative study of CdS/CdTe cells fabricated with and without evaporated Te-layer. *Thin Solid Films*, *515*(15), 5824–5827.

Rahman, K. S., et al. (2019). Influence of deposition time in CdTe thin film properties grown by Close-Spaced Sublimation (CSS) for photovoltaic application. *Results Physics*, *14*.

Singh, V. P., & McClure, J. C. (2003). Design issues in the fabrication of CdS–CdTe solar cells on molybdenum foil substrates. *Solar Energy Materials and Solar Cells*, *76*(3), 369–385.

Terheggen, M., et al. (2003). Structural and chemical interface characterization of CdTe solar cells by transmission electron microscopy. *Thin Solid Films*, *431*, 262–266.

Wood, D. A., Lane, D. W., Rogers, K. D., & Coath, J. A. (1999). Optical properties of CdS x Te 1 – x polycrystalline thin films. *Journal of Electronic Materials*, *28*(12), 1403–1408.

SUBCHAPTER

Copper indium gallium selenide solar cells 3.2

Yulisa Binti Mohd. Yusoff
Institute of Sustainable Energy, Universiti Tenaga Nasional (@The National Energy University), Kajang, Malaysia

3.2.1 Introduction to copper indium gallium selenide solar cells

The first thin film solar cell was created in 1883 by Charles E. Fritts. In his invention, a thin sheet of selenium in between two dissimilar sheets of metal—the top layer was made from a very thin and semitransparent sheet of gold and serves as the anode while the base layer, the cathode, was made from various types of metals, such as brass, zinc, iron and copper (Fritts, 1883; Fritts, 1885). While studying his invention, Fritts noticed the photovoltaic effect and became the first person to suggest that the selenium cells could convert energy from light directly to electrical energy—a theory later confirmed by W. Siemens (Fritts, 1885; Siemens, 1885).

The solar cells made by metal-semiconductor junctions allow current to flow in one direction and not the other. As a result, the devices became known as *barrier layer cells* (Archer, 2015). In the early 20th century, two more types of barrier layer cells were invented. The first is the copper oxide cell, first reported in 1916, and the second is thallous sulfide cell in 1920 (Case, 1920; Kennard & Dieterich, 1916). Although these early PV devices had efficiencies of around only 1%, these early researchers proved the ability of using compound semiconducting materials as the basis for PV devices. In 1930, B. Lange suggested the possibility of using certain types of semiconducting materials to create PV devices effective for a specific spectral region (Archer, 2015).

Research on copper-based thin film PV devices began concurrently with that of silicon PV devices, fueled by the prospects of higher output power to device weight ratio for extra-terrestrial applications (Archer, 2015). The first device, created in 1954 by Reynolds and his team, had a p-type layer made from copper sulfide (Cu_2S) matched to a cadmium sulfide (CdS) n-type layer with a photovoltaic conversion efficiency of 6% (Reynolds, Leies, Antes, & Marburger, 1954). Despite its promising outlook, devices with Cu_2S layers were notorious for their instability and research was abandoned in the 1980s (Hamakawa, 2004).

To overcome the stability issues of Cu_2S, scientists at the University of Maine and Boeing Aerospace Company experimented with adding indium (In) and gallium (Ga) to the Cu_2S layer, and replacing the sulfur (S) with selenium (Se)

Comprehensive Guide on Organic and Inorganic Solar Cells. DOI: https://doi.org/10.1016/B978-0-323-85529-7.00001-3

(Kazmerski, White, & Morgan, 1976; Mickelsen & Chen, 1980; Mickelsen, Chen, Hsiao, & Lowe, 1984). At the time of writing, the highest recorded efficiency for laboratory-scale $Cu(In,Ga)(S,Se)_2$ Copper indium gallium selenide (CIGS) thin film solar cells stands at 23.35% (Green et al., 2020).

CIGS solar cells are also known as chalcopyrite-based solar cells. The efficiency and stability of these solar cells fast approach that of conventional crystalline silicon solar cells, but with the added advantages of being lower in cost and requiring less raw materials. These advantages combine to give CIGS solar cells a much shorter energy payback time as compared to conventional crystalline silicon solar cells.

Despite its promising outlook, the full potential of CIGS solar cells has been hampered by slow market growth (see Figs. 3.2.1 and 3.2.2). This is a strong

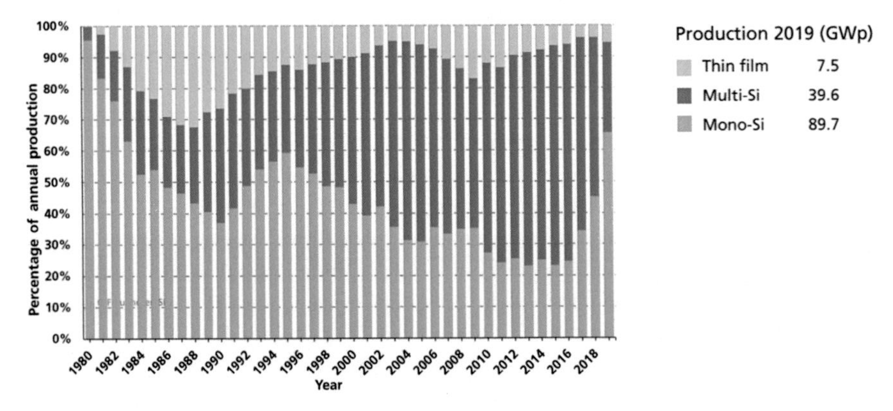

FIGURE 3.2.1

Global annual photovoltaic production by technology.

Ref: ©Fraunhofer ISE: Photovoltaics Report, updated: 16 September 2020.

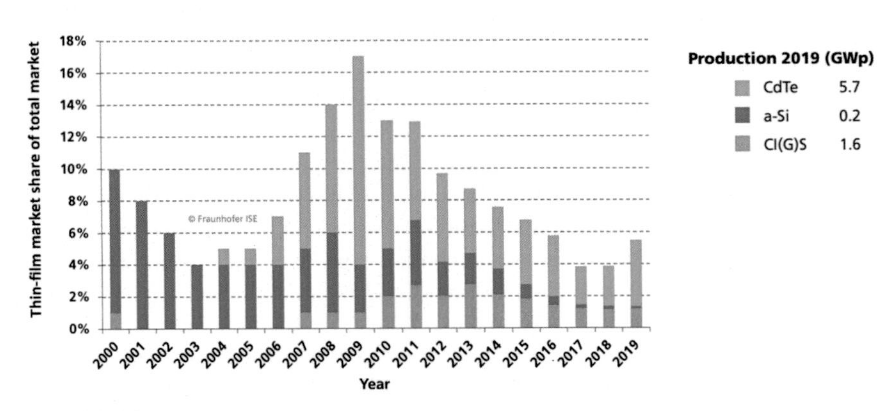

FIGURE 3.2.2

Annual market share of thin film photovoltaic technologies.

Ref: ©Fraunhofer ISE: Photovoltaics Report, updated: 16 September 2020.

indicator that more bridges linking scientific research and commercialization must be built. In the ensuing sections of this chapter, important findings in the research and fabrication of CIGS solar cells, along with the material properties vital for photovoltaic application, are presented in a concise manner.

3.2.2 Copper indium gallium selenide device fabrication

A typical of CIGS solar cell architecture is shown in Fig. 3.2.3. The cross section image of an actual CIGS solar cell taken using a scanning electron microscope is shown in Fig. 3.2.4. As can be seen in both figures, the solar cell is made up of several layers of semiconducting thin films that are epitaxially grown layer-by-layer, from the bottom up. Sunlight enters the solar cell from the window layer; therefore CIGS photovoltaic devices belong in the category of solar cells known as substrate type.

The fabrication methods for CIGS solar cells can be divided into two main categories depending on the growth environment during the fabrication process: with vacuum and vacuum-free. The choice of fabrication method depends on numerous factors such as cost, deposition time, complexity of process, reproducibility of samples, and scaling up capabilities for large area deposition, to name a few. The discussion on this topic will be covered in the latter part of this section. However, the discussion will be limited to the most important fabrication processes for each thin film layer in the CIGS solar cell stack, instead of an all-inclusive and comprehensive discussion. Before the subject of fabrication process can be broached, it is crucial to present a detailed overview of the individual layers comprising the CIGS solar cell stack.

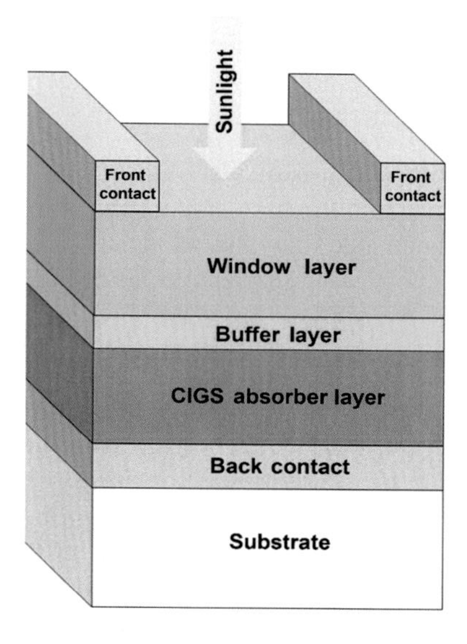

FIGURE 3.2.3

Typical copper indium gallium selenide solar cell architecture.

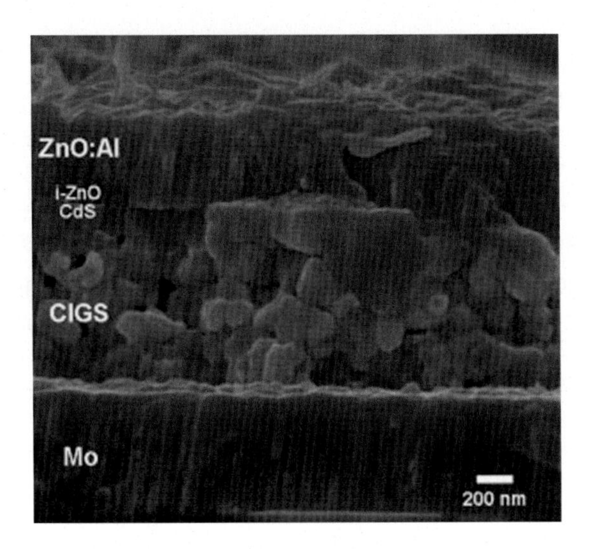

FIGURE 3.2.4

SEM cross-section of CIGS solar cell (Liu et al., 2013).

3.2.2.1 General structure

3.2.2.1.1 Substrates

The choice of substrate material is very important, particularly for thin film solar cells. This is because the total device thickness is typically less than 5 μm. Thus these devices need to be fabricated on sturdy substrates that can serve as the base to hold the whole solar cell together. Ideal substrate materials for CIGS solar cells should meet the following criteria (Paire et al., 2014):

1. Resistant to high temperatures, particularly during absorber layer deposition.
2. Chemically nonreactive with the back contact and absorber layers.
3. Resistant to moisture and able to withstand harsh operating conditions.
4. Has a smooth surface conducive for depositions of thin films.
5. Strong and robust.
6. Has similar thermal expansion coefficient with the metallic back contact and absorber layers.

Soda lime glass (SLG) is often the choice substrate material. Soda lime is not only cheap to procure, but the sodium (Na) content serves as a source of dopant and is an important requirement for producing highly efficient CIGS PV devices. During the deposition process of the CIGS absorber layer, Na from the SLG diffuses through the back contact into the absorber layer being deposited. Numerous studies have shown that Na eliminates Se vacancies and increases the effective p-type doping of the absorber layer. Na addition also passivates the nonradiative recombination centers present at grain boundaries, thereby increasing the conductivity of $Cu(In,Ga)Se_2$ thin films. These resulted in devices with higher open-circuit voltage and enhanced fill-factor (Ishizuka et al., 2009; Kronik, Cahen, & Schock, 1998; Salome, Rodriguez-Alvarez, & Sadewasser, 2015; Shin et al., 2016). Structurally, however, conflicting observations have been made by different groups on the effects of Na on grain size and preferential growth of Cu $(In,Ga)Se_2$ thin films (Salome et al., 2015).

Other types of glass that can withstand high temperatures, such as borosilicate, aluminosilicate, and even gorilla glass can be used as substrates. However, these types of glass is much more expensive than SLG and have varying levels of Na contents in their composition. Too much Na can also degrade the photovoltaic efficiency of CIGS devices due to the creation of deep level traps in the absorber layer (Yoon, Seong, & Jeong, 2012). To gain more control over the amount of Na that diffuses to the absorber layer, some device manufacturers have opted to incorporate a blocking layer between the glass substrate and back contact and provide the necessary Na dopant by other means. The ability to use a more controllable supply of Na also means that other substrates that do not contain Na can also be used to fabricate CIGS solar cells. These include metal foils, polymers, and heat resistant plastic.

3.2.2.1.2 Back contact

The most widely used metallic back contact in the fabrication of CIGS PV devices is molybdenum (Mo). It was established that during the deposition of the $Cu(In,Ga)Se_2$ layer, Mo reacts with Se to form $MoSe_2$ which aids the formation of an ohmic contact at the interface between the absorber and Mo layers (Abou-Ras et al., 2005; Kohara, Nishiwaki, Hashimoto, Negami, & Wada, 2001). Mo back contact also aids in controlling the diffusion rate of Na from the SLG substrate into the absorber layer (Bommersbach et al., 2011; Yoon et al., 2012; Zhu et al., 2012). Mo is also durable enough to withstand the high temperature process used to grow the absorber layer. It has a sheet resistance of approximately 0.2 Ω/square and can bind the solar cell structure to the substrate (Niki et al., 2010).

Several research groups have also investigated the suitability of other metals as alternatives to Mo. Matson et al. looked at gold, nickel, aluminum and silver and found that only gold and nickel could form ohmic contact (Matson Jamjoum, Buonaquisti, Russell, and Ahrenkiel, 1984). Almost two decades later, Orgassa et al. compared tungsten, chromium, tantalum, niobium, vanadium, titanium and manganese to molybdenum. The researchers discovered that only tungsten rivaled Mo as a back contact material for CIGS solar cells (Orgassa, Schock, & Werner, 2003).

Most CIGS device manufacturers employ sputter deposition technique to deposit the Mo layer (Niki et al., 2010). The two sputtering modes that are often used are radio-frequency (RF) mode and DC magnetron mode (DC) mode. The basic principles of sputter deposition in both modes will be covered in the section titled "Fabrication Technology" in the latter part of this chapter.

3.2.2.1.3 Absorber layer

The absorber layer is a semiconducting material often considered the heart of all thin film solar cells. It is aptly named because it is the layer that absorbs the highest number of photons and in response excites electrons into the conduction band to create photocurrent. Due to this, the absorber layers of all thin film solar cells are selected from semiconducting materials with bandgap energies that coincide with the photon-rich region of the solar spectrum.

CIGS solar cells are a family of thin film solar cells where the absorber layer is a constituent of the elements copper (Cu), indium (In), gallium (Ga), selenium (Se), and sulfur (S) with the chemical formula $Cu(In_{1-x}Ga_x)(S_{1-y}Se_y)_2$. The variable x represents the percentage for Ga, in decimal form, over the total content of In and Ga in the film. While variable y represents the decimal form percentage for Se calculated against the total amount of S and Se. Therefore CIGS thin films with varying bandgap energies and optoelectronic properties can be fabricated simply by varying the amounts of Ga versus In and Se versus S in the film, as seen in Fig. 3.2.5 on the following page.

All the copper chalcopyrite alloys in the $Cu(In_{1-x}Ga_x)(S_{1-y}Se_y)_2$ system are direct bandgap semiconducting materials and possess a large tolerance to off-stoichiometry. It is believed that Cu chalcopyrite thin films are ideal for photovoltaic applications due to the high optical absorption coefficient and carrier mobility associated with Cu (Rockett, 2010). There are also no miscibility gaps within the entire system, hence, any desired alloys can be produced. Highly efficient devices usually have $0.15 \leq x \leq 0.35$ which produce $Cu(In,Ga)(Se,S)_2$ films with bandgaps of approximately 1.1–1.2 eV (Jung et al., 2010; Turcu, Pakma, & Rau, 2002).

Bulk $Cu(In,Ga)Se_2$ belong to the $I-III-VI_2$ group of semiconducting materials that crystallize in the tetragonal chalcopyrite structure (space group *I-42d*) and obeys the octet rule (Fan, Wu, & Yu, 2014). Tetragonal chalcopyrite crystal

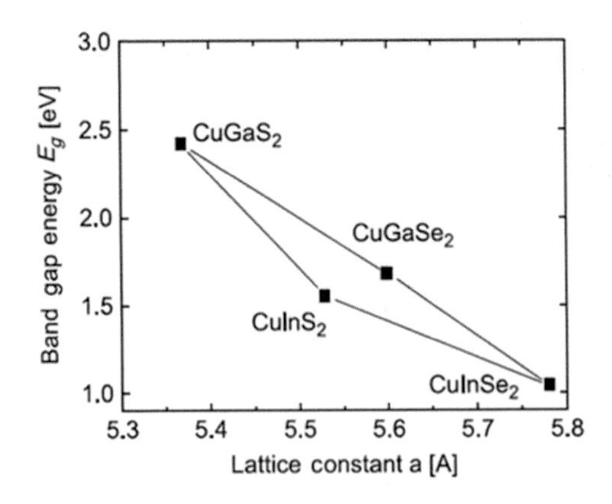

FIGURE 3.2.5

Bandgap energy versus the lattice constants *a* of the $Cu(In,Ga)(Se,S)_2$ alloy system (Rau & Schock, 2014).

structure is a derivative of zincblende, which is, in turn, a derivative of the diamond or face centered cubic structure. Therefore the arrangement of atoms in all three crystal systems is the same, as shown in Fig. 3.2.6. However, because of the difference in the properties of the atom species involved in the covalent bonding, the lengths of the bonds between atoms are different.

In the tetragonal chalcopyrite structure, cations from groups I and III are each tetrahedrally bonded to four anions from group VI. As for the group VI anions, each are tetrahedrally bonded to two cations from group I and two cations from group III (Boyle, McCandless, Shafarman, & Birkmire, 2014). This arrangement is evident in the tetragonal chalcopyrite unit cell illustrated in Fig. 3.2.7. Although it is indiscernible in the figure, the actual bond lengths between I–VI and III–VI atoms are not equal.

The crystal structures of semiconductors play a large role in determining the electronic band structures. It is common for semiconducting compounds with similar crystal structures to exhibit similar band features. In the case of Cu chalcopyrite semiconducting materials, the valence band maxima are the antibonding states of Cu 3d orbital and p orbital of chalcogenide elements, Se and/or S, in group VI. The conduction band minima, on the other hand, are the antibonding states of s orbital of group III elements and p orbital of group VI. First principles calculations have also proven that the conduction band minima of Cu chalcopyrite semiconductors can be raised by reducing the III–VI bond lengths. The bandgap of Cu chalcopyrite materials can be increased by alloying with smaller atoms from either group III or VI (Fan et al., 2014).

Lattice parameters or lattice constants are used to describe the Bravais lattice lengths a, b, and c along with the angles α, β, and γ. For crystals that align in the

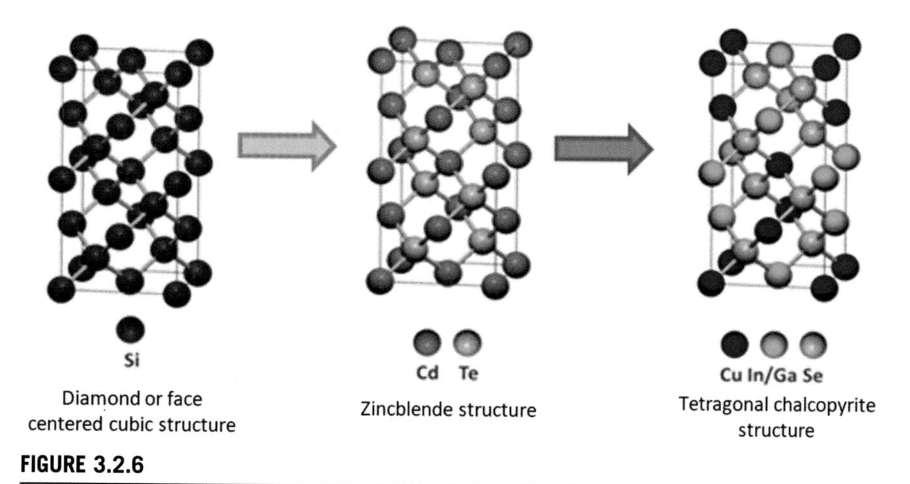

FIGURE 3.2.6

Atomic arrangement in diamond, zincblende, and tetragonal chalcopyrite crystal systems for Si, CdTe, and $Cu(In,Ga)Se_2$.

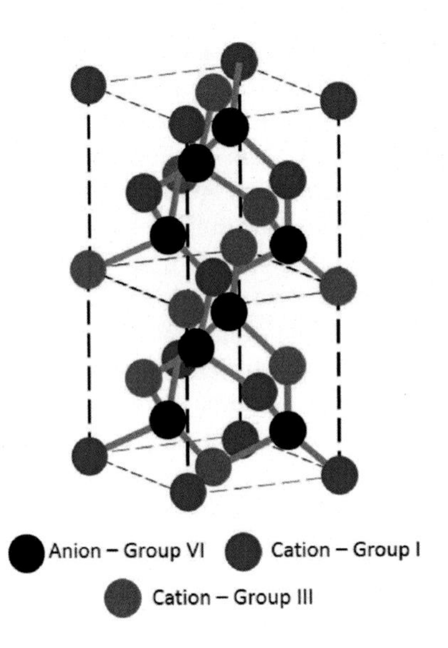

FIGURE 3.2.7

Unit cell of I-III-VI_2 family of semiconducting materials with tetragonal chalcopyrite crystal structure (Boyle et al., 2014).

Table 3.2.1 Lattice parameters for bulk $CuInSe_2$ and $CuGaSe_2$ obtained at room temperature.

Semiconductor type	*a* (Å)	*c* (Å)	Ratio of *c/a*
$CuInSe_2$	5.78	11.55	2.000
$CuGaSe_2$	5.61	11.00	1.960

Source: *Madelung, O. (2004a). I–III–VI2 compounds.pdf. In Semiconductors: Data handbook (pp. 289–328) (3rd ed.). New York: Springer-Verlag Berlin Heidelberg.*

tetragonal system, $a = b \neq c$ and $\alpha = \beta = \gamma = 90$ degrees. Table 3.2.1 lists the lattice parameters a and c for bulk $CuInSe_2$ and $CuGaSe_2$ at room temperature.

3.2.2.1.4 Buffer layer

In CIGS PV devices, the buffer layer assumes the role of the heterojunction counterpart to the p-type CIGS absorber layer. As n-type semiconductors, the buffer layer, along with the subsequent window layers, participates in the formation of the p- or n-junction and generation of the built-in electric field across the depletion region that forms the backbone of thin film solar cell operation.

The first Cu chalcopyrite-based PV device that demonstrated high conversion efficiency was fabricated using single crystal p-type $CuInSe_2$ paired with n-type CdS. The reported efficiency of this device was 12% and since then, CdS has remained the choice n-type heterojunction counterpart for making CIGS PV devices (Shay, Wagner, & Kasper, 1975; Wagner, Shay, Migliorato, & Kasper, 1974).

Over the years, efforts to increase device efficiency led researchers to include wider bandgap n-type semiconductors as part of the device architecture for Cu chalcopyrite solar cells. However, efforts to omit the CdS buffer layer altogether proved detrimental to the device performance (Lauermann et al., 2007; Orgassa, Rau, Nguyen, Werner Schock, & Werner, 2002; Rau & Schmidt, 2001). Likewise, efforts to replace the CdS buffer layer with wider bandgap semiconductors did not produce the expected results and only proved the superiority of CdS over alternative buffer materials (Böer, 2011; Kato, 2017). For this reason, the discussion on buffer layer material will be limited to CdS and ZnS as this is the leading choice of Cd-free buffer material at the time of writing.

It has been theorized that Cd forms a beneficial donor-type defects at the surface of p-type Cu chalcopyrite absorbers. The density of these donor-type defects also happens to be at the optimum level and at just the right position in the forbidden gap (Naghavi, Hildebrandt, Bouttemy, Etcheberry, & Lincot, 2016). Thus replacing the CdS buffer layer requires more than just choosing a wider bandgap n-type semiconducting material with suitable band alignment with the absorber layer.

CdS and ZnS belong to the II–VI group of semiconducting materials that crystallize in two structures: cubic zincblende and hexagonal wurtzite. Hexagonal CdS is more commonly found and easier to synthesize as compared to cubic leading to the assumption that cubic CdS is metastable at temperatures between 20°C and 90°C. The opposite is true for ZnS. At room temperature, cubic ZnS is more common but with inclusions of hexagonal ZnS (Madelung, 2004b).

In cubic zincblende structures, the Bravais lattice lengths $a = b = c$ and angles $\alpha = \beta = \gamma = 90$ degrees. As for hexagonal wurtzite structures, the Bravais lattice lengths $a = b \neq c$ and angles $\alpha = \beta = 90$ degrees, $\gamma = 120$ degrees (Pillai 2010). For CdS, $a = 4.1348$ Å and $c = 6.7490$ Å. For cubic zincblende ZnS, $a = 5.4053$ Å and for hexagonal wurtzite ZnS, $a = 3.820$ Å $c = 6.260$ Å (Madelung, 2004b).

In both CdS and ZnS compound semiconductors, each group II atom is covalently bonded to four S atoms and vice versa forming a tetrahedral bond. Fig. 3.2.8A–C show the illustrations of a tetrahedral bond, cubic zincblende, and hexagonal wurtzite structures, respectively.

In the early days, CIGS PV devices had two layers of CdS. The first layer is the undoped CdS that is the heterojunction counterpart to the Cu(In,Ga)Se_2 layer, while the second layer of CdS was commonly doped with In or Ga and acted as the window layer. However, the thick CdS layers resulted in optical losses in the shorter wavelength region below 520 nm (Shafarman, Siebentritt, & Stolt, 2011). Nowadays, CIGS devices have only one layer of buffer of approximately 50 nm thick. The choice deposition method for buffer layer preparation is chemical bath deposition (CBD).

3.2.2.1.5 Window layer

The final layer to be deposited during the fabrication of CIGS solar cells is the window layer. This is the first layer in the solar cell stack that receives photons.

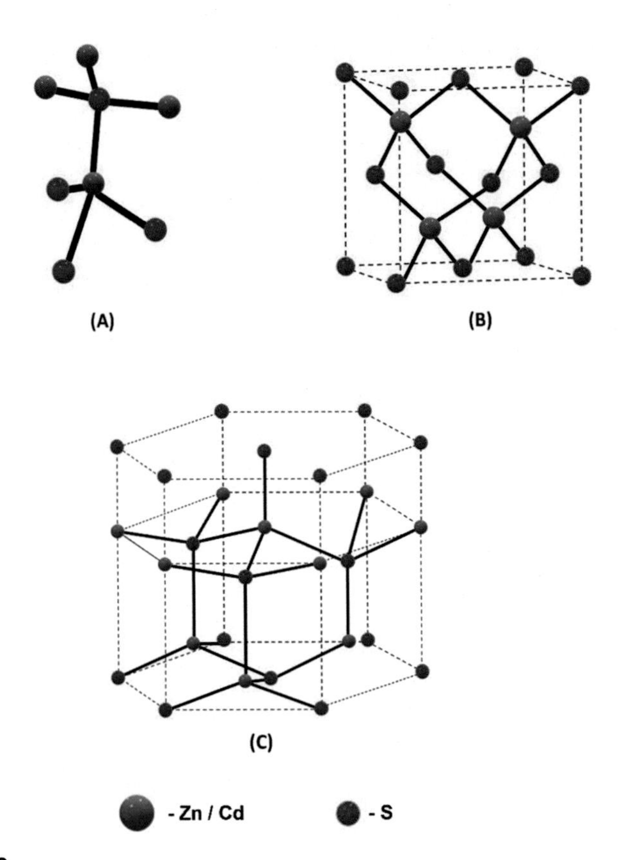

FIGURE 3.2.8

(A) The tetrahedral bond in CdS and ZnS, (B) The cubic zincblende structure for CdS and ZnS, and (C) The hexagonal wurtzite structure for CdS and ZnS.

In modern CIGS solar cells, the window layer comprises two separate thin films, the first window layer to be deposited after the buffer is highly resistive while the second is less resistive. The first window layer on a typical CIGS PV device is a thin layer of highly resistive intrinsic zinc oxide layer (i-ZnO). This layer lessens the effects associated with nonuniform electronic quality in the polycrystalline Cu (In,Ga)Se_2 layer and shunt paths due to pin-holes in the buffer layer (Rau & Schmidt, 2001). The second window layer, usually a type of doped ZnO such as aluminum-doped (ZnO:Al), functions as the front contact of the device, and its properties influence the lateral current transport.

In thin film solar cells, the semiconducting material that form the window layer has the largest bandgap of all the thin films making up the solar cell. The large bandgap ensures that the window layer is transparent to most photons in the visible region of the solar spectrum. Most of the incident photons in this region will be transmitted by the window layer to the underlying layers of the solar cell, which ensures that the most photon absorption happens in the absorber layer. Naghavi et al. (2010) published a review on window layers for CIGS solar cells. The researchers surmised that an effective window layer should have the following criteria (Naghavi et al., 2010):

1. must have high transparency to incoming photons,
2. the interface that the window layer material forms with the absorber must have low recombination losses,
3. able to mitigate the negative impact of physical and optoelectronic defects of the absorber layer, and
4. possess low sheet resistance to ensure maximum carrier collection by the top grid.

The window layers of solar cells are usually made from a family of semiconducting materials known as transparent conducting oxides (TCO). The most common TCOs used for making thin film PVs are derivatives of tin oxide, indium oxide (ITO), zinc oxide, and cadmium oxide (Delahoy & Guo, 2011). Metal oxides are very insulative owing to the high bandgap nature of these semiconducting materials. The conductivity of TCOs are boosted by doping the material with suitable dopants, using off-stoichiometry films or by improving the crystallinity of the deposited films by performing postdeposition treatments such as annealing (Calnan & Tiwari, 2010). The choice of TCO depends on many factors such as the optical and electrical performance of the semiconducting material, its surface morphology, sturdiness during processing, ecofriendliness, flexibility, hardness, work function, and deposition effects on the underlying layers in the solar cell stack. Hence, the choice of deposition method and the corresponding deposition parameters are vital in the preparation of photovoltaic quality TCO thin films (Delahoy & Guo, 2011).

In the production of CIGS solar cells, the two most widely used TCOs are tin-doped ITO and zinc oxide highly doped with gallium, aluminum, or boron. At present, the standard deposition method for TCOs for CIGS solar cells is sputtering at low temperature (Naghavi et al., 2010).

3.2.2.2 Fabrication technology

The discussion of CIGS fabrication technology in this chapter will be limited to the commonly used processes proven to produce stable CIGS solar devices. The techniques mentioned in this section can be applied to both laboratory-scale devices and larger modules.

3.2.2.2.1 Physical vapor deposition

Thin films can be grown using a variety of methods. Physical vapor deposition (PVD) is a category of techniques whereby the thin films are grown on a substrate atom-by-atom or molecule-by-molecule, directly from the source material without involving any chemical reaction processes. In other words, the source material and resultant thin film are composed of the same elements. During PVD, the source material, in the form of a solid, is transformed directly into vapor phase, transported to the substrate where the molecules condense, and thin films are formed. PVDs typically takes place in special chamber placed under low pressure. The two forms of PVD presented in this section are sputter deposition and evaporation.

3.22.2.1.1 Sputter deposition process

Sputter deposition is the process of growing thin films from a source material known as a sputter target. It is a nonthermal vaporization process whereby the particles on the surface of the target are physically ejected by the colliding action of an energetic bombarding particle, usually in the form of ionized sputtering gas. The parametric conditions that must be taken into consideration during sputter deposition are:

1. substrate temperature
2. power
3. gas flow rate
4. chamber pressure
5. sputtering pressure
6. deposition time

Sputter deposition of films composed of multiple elements can be accomplished using two routes: one is cosputtering which uses multiple sputter targets, while the second is single target sputtering which uses a target made of multiple elements that matches the film to be deposited. The first route offers greater control over the deposition of each element thus offering greater control of the composition of the deposited film and this route is generally lower in cost too because elemental sputter targets are less costly. However, this route can be more troublesome especially in uniformity and conformity control over large deposition areas. The uniformity and conformity issues over large deposition areas can be overcome by taking the second route of single target sputtering. However, a compromise in cost is inevitable as the multielement sputter targets are more expensive to procure. Also, the quality of the deposited film depends very much on the quality of the sputter target so the sputter targets must only be sourced from reputable suppliers.

In the sputtering chamber, the substrate is mounted to the anode while the target is mounted on the cathode. Throughout the deposition process, sputtering gas, usually argon, is flowed into the sputtering chamber at a constant pressure. When plasma is induced, the argon gas is ionized into

Ar^+ ions and accelerates toward the target surface where it then collides with the particle on the target surface. The momentum of the collision causes the surface particles to be ejected from the target resulting in the growth of thin film on the substrate. This process is illustrated in Fig. 3.2.9.

Two variants of sputter deposition process are used in the fabrication of CIGS solar cells, viz. DC mode and RF mode. For making CIGS PV devices on SLG substrates, DC mode sputter deposition is used to grow the Mo back contact. The absorber layer

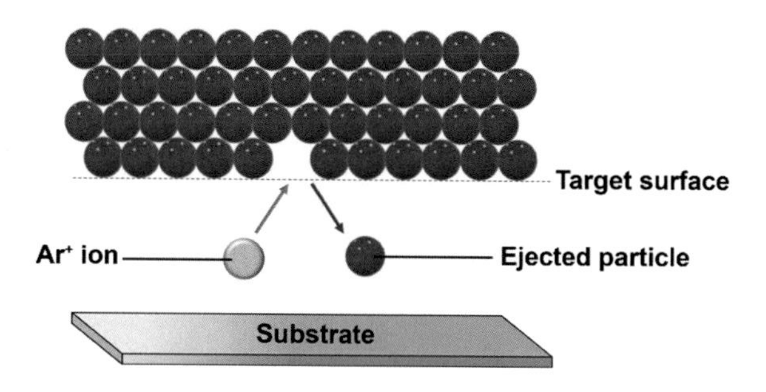

FIGURE 3.2.9

Image representation of sputter deposition process.

can be grown by either RF-mode sputter deposition or evaporation (covered in the proceeding section). The window layers are also commonly grown using sputter deposition.

3.22.2.1.2 Thermal evaporation deposition process

As the name suggests, thermal evaporation deposition is a thin film growth process wherein the solid source material is converted into vapor form using heat. During this process, temperatures exceeding 1000°C are needed to transform solid elemental sources of Cu, In, and Ga into vapors (Shafarman et al., 2011).

This conventional thermal evaporation process was later adapted to make the process more suitable for large scale production. This improved method was dubbed as the coevaporation process because sources of Cu, In, and Ga can be evaporated simultaneously onto the substrate in the presence of Se gas to form the CIGS absorber layer. A schematic of this coevaporation process is shown in Fig. 3.2.10 below.

In this technique, the elemental sources are contained in separate effusion cells with line-of-sight delivery to the substrate. This setup allows the evaporation temperature of each element to be separately controlled. Typical temperature ranges are 1300°C–1400°C for Cu, 1000°C–1100°C for In, 1150°C–1250°C for Ga, and 250°C–350°C for Se (Shafarman et al., 2011). The deposition rate of each elemental source is controlled using a spectrometer with feed-back loop while Se is always evaporated in excess. This setup enables precise control over the deposited rate of each metallic element during the growth process. Hence, a variety of growth profiles can be implemented using the same apparatus. Four growth profiles that produced solar cells with at least 16% efficiency are shown in Fig. 3.2.11A–D below (Rau & Schock, 2014; Shafarman et al., 2011). The film growth rate usually varies between 20 and 200 nm/s. Therefore the total time needed to grow a 2 μm thick film is between 10 and 90 min (Shafarman et al., 2011).

3.22.2.1.3 Chemical bath deposition process

CBD is used to deposit the buffer layer during the generation of CIGS solar cells. In a nutshell, CBD is a thin film epitaxial growth technique that takes place in an aqueous solution. The aqueous solution, known as the chemical bath, contains a

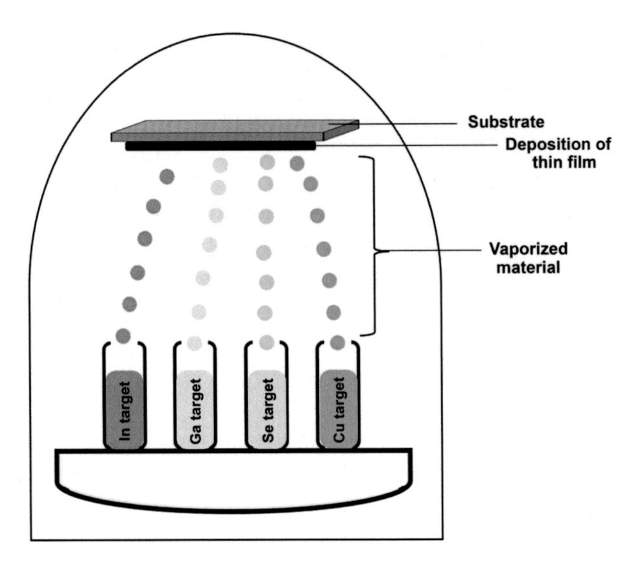

FIGURE 3.2.10

Schematic diagram of the coevaporation growth process for copper indium gallium selenide thin films (Shafarman et al., 2011).

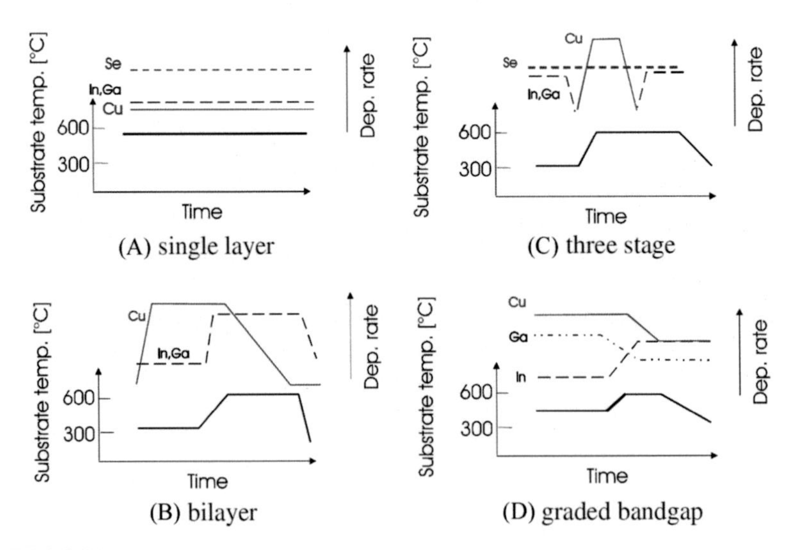

FIGURE 3.2.11

Four deposition profiles used to grow copper indium gallium selenide thin films using coevaporation process (Rau & Schock, 2015).

mixture of a complexing agent, usually ammonium hydroxide (NH_4OH), and deionized water. A sulfur source (typically thiourea) and a Cd source (a form of Cd salt such as cadmium sulfate) are added to the aqueous bath solution and the chemical reaction that takes place in the bath results in a layer of CdS to be deposited on the surfaces of the submerged substrates.

According to Pawar et al. in (Pawar, Pawar, Kim, Joo, & Lokhande, 2011), there are four main sequential events that take place during a CBD process that uses ammonium hydroxide as the complexing agent. These are:

1. Equilibration of the complexating agent and deionized water. Precise pH control throughout the whole deposition process is a prerequisite for epitaxial growth of good quality CdS films.
2. Creation of ionic Cd ligands through the decomposition of Cd salt in the chemical bath. This causes Cd^{2+} ions to be released in the chemical bath.

   The complexating agent, NH_4OH, reacts with the Cd^{2+} and encourages the formation of cadmium tetra-amine complex ion, $Cd(NH_3)_4^{2+}$. The chemical reaction for this process is:

$$Cd^{2+} + 4NH_4OH \Leftrightarrow Cd(NH_3)_4^{2+} + 4H_2O \tag{3.2.1}$$

3. Decomposition of thiourea in the chemical bath through hydrolysis resulting in the release of S^{2-} ions into the chemical bath.

$$CS(NH_2)_2 + OH^- \Leftrightarrow SH^- + CN_2H_2 + H_2O \tag{3.2.2}$$

$$SH^- + OH^- \Leftrightarrow S^{2-} + H_2O \tag{3.2.3}$$

4. Finally, the Cd tetra-amine complex ions react with the sulfur ions to form solid CdS on the submerged substrate.

$$Cd(NH_3)_4^{2+} + S^{2-} \Leftrightarrow CdS + 4NH_3 \tag{3.2.4}$$

The quality of the thin film depends highly on the pH level and the temperature of the chemical bath. For ammonia-based CBD process, the acceptable range of pH levels is between 9 and 11, with 10 being the optimum (Pawar et al., 2011). As for the bath temperature, various groups reported that the lower range of bath temperatures, that is, between 60°C to 70°C, as the most suitable for producing PV quality CdS thin films (Seo et al., 2013; Zhou, Hu, & Wu, 2013).

Fig. 3.2.12 below contains the schematic diagram for a typical CBD process. A magnetic stirrer is used to ensure uniform mixing of all the ionic species in the chemical bath.

Metal sulfide thin films grown using the CBD process often contain traces of hydroxides and/or oxides as well. Hence, it is common to see chemical bath-deposited thin film to be described as $M_x(S,O,OH)_y$ or $M_xS_y(O,OH)$ where M represents an element from group II of the periodic table.

The formation of these hydroxides and/or oxides is competing directly with the formation of metal sulfides, as can be seen by the chemical reaction for the formation of $Cd(OH)_2$ below. Thus if the formation of $Cd(OH)_2$ is high, this will

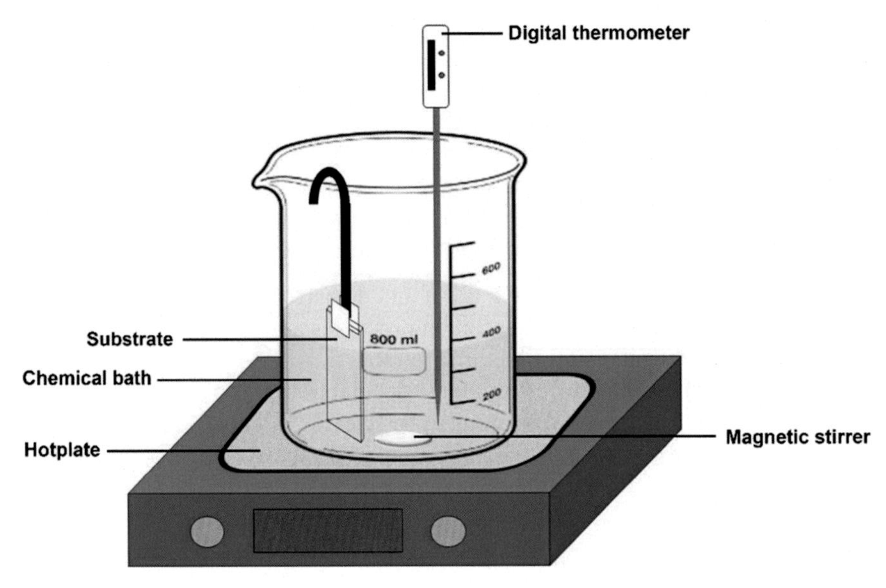

FIGURE 3.2.12

Schematic diagram of a typical chemical bath deposition process.

hinder the formation of desired CdS (Pawar et al., 2011). In addition, $Cd(OH)_2$ is a precipitate in the chemical bath and may adhere to the surface of the substrate causing large colloids to form.

$$Cd^{2+} + 2OH^{-} \Leftrightarrow Cd(OH)_2 \quad (3.2.5)$$

The inability to control the generation of these metal hydroxides and/or oxides alongside the desired metal sulfide is considered by many as a major disadvantage of the CBD process. The second disadvantage is the large amount of toxic chemical waste that requires additional expenses to recycle and treat.

3.2.2.3 Material properties

In semiconductor device fabrication, five main groups of material properties are crucial to the device operations. These are: physical, chemical, electrical, optical, and mechanical properties of the thin films making up the device. Physical properties, that is, the crystallographic properties, for the CIGS absorber and CdS buffer layers have been covered in the previous section. In the section below, only the optical and electrical properties for the absorber and buffer layer will be discussed.

Material properties of the thin films formed are dependent the process parameters and geometry of the deposition system which include, among others, the specific type of deposition process used, process temperature, gas flow rate, pressure, contamination, nucleation rate of the resultant adatoms, and in the case of sputter deposition the properties of the bombarding ionized gas itself (flux, particle mass, and energy).

3.2.2.3.1 Optical properties

3.22.3.1.4 Copper indium gallium selenide absorber layer

Optical properties describe the interaction between semiconducting materials and light. Not surprisingly, these are important parameters in determining the suitability of a particular semiconductor for photovoltaic application. Since $Cu(In,Ga)Se_2$ is alloyed from $CuInSe_2$ and $CuGaSe_2$, the optical properties of these two semiconductors form the lower and upper limits for $Cu(In,Ga)Se_2$ semiconducting material.

Undeniably, the most important attribute of a semiconductor is its bandgap energy, especially when a semiconducting material is intended for PV applications. Semiconducting materials with small bandgap energies are closer in properties to conductors under illumination, whereas semiconducting materials with large bandgap energies are highly resistive and possess properties close to that of insulating materials when illuminated. This ability to directly manipulate the conductivity of the material under illumination by changing and tuning the bandgap of the material is a highly valuable feature of CIGS semiconducting materials.

The chemical formula for the alloy CIGS is $Cu(In_{1-x},Ga_x)Se_2$ where x is the concentration of gallium in the film. When $x = 0$, the bandgap was 1.04 eV and when $x = 1$, the bandgap was 1.68 eV (Madelung, 2004a). Thus it is theoretically possible to deposit $Cu(In_{1-x},Ga_x)Se_2$ films with bandgap values ranging from 1.04 to 1.68 eV simply by varying the concentration of Ga in the film. The best performing devices, however, have Ga concentrations between 0.15 and 0.35 which corresponds to $Cu(In,Ga)Se_2$ films with bandgaps of approximately 1.1−1.2 eV (Jung et al., 2010; Turcu et al., 2002).

Jaffe and Zunger discussed the bandgap theory of Cu chalcopyrite materials in reference (Jaffe & Zunger, 1983). Fig. 3.2.12A and B show the band structure for $CuInSe_2$ and $CuGaSe_2$ modeled after the calculations performed by Jaffe and Zunger (1983). The shaded area in both figures indicates the respective forbidden gaps. The contribution of the antibonding states of Cu $3d$ and Se $4p$ orbitals in forming the valence band maxima is also pictured in the figures. Also evident is the direct bandgap characteristic common for all $Cu(In_{1-x},Ga_x)(Se_{1-y},S_y)_2$ family of semiconducting materials.

Another property crucial for PV application is optical absorbance of solar irradiation. Unlike the bandgap, which is a derived parameter, absorbance by a semiconductor can be directly measured and examined using various spectrometers. A commonly used method of bandgap calculation uses the measured absorbance (or the transmittance) to approximate E_g with a Tauc plot.

Fig. 3.2.14 shows the measured absorbance for $CuInSe_2$ and CdTe contrasted against AM 1.5 irradiance spectra. $CuInSe_2$ has a smaller bandgap ($E_g = 1.04$ eV) than CdTe ($E_g = 1.45$ eV), therefore $CuInSe_2$ can absorb more low energy photons compared to CdTe.

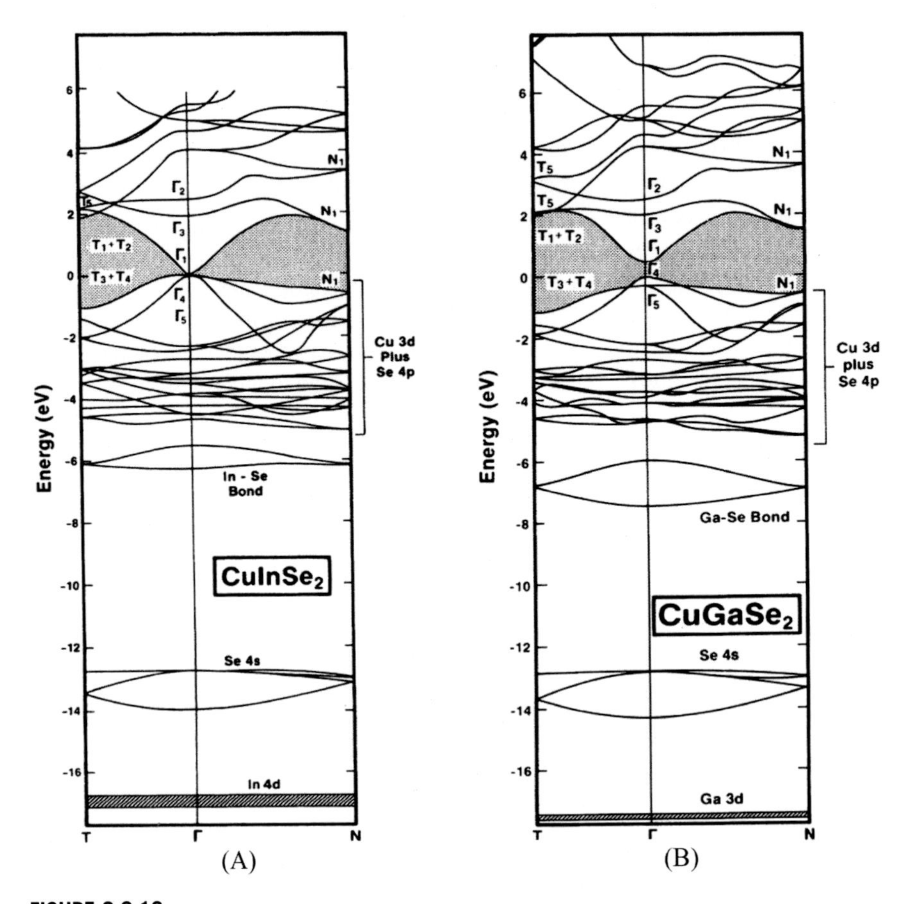

FIGURE 3.2.13

The energy versus momentum plots for (A) $CuInSe_2$ and (B) $CuGaSe_2$. The *y*-axis represents energy while the *x*-axis represents the crystal momentum (Jaffe & Zunger, 1983).

The ability to control the composition of In and Ga, which is equivalent to direct manipulation of the bandgap, is a very useful and interesting feature of Cu $(In_{1-x},Ga_x)Se_2$. This ability opens the possibility to engineer the bandgap of Cu $(In_{1-x},Ga_x)Se_2$ thin films to create absorber layers with graded bandgaps, a concept discernible in Fig. 3.2.15A–C. Absorbers with graded bandgaps can channel the flow of electrons in the absorber layer toward the depletion region where the carrier collection probability is highest. This translates to higher J_{SC} and increased photovoltaic efficiency (Nakada, 2012).

This graded bandgap feature is the cornerstone for highly efficient CIGS PV devices. Most seasoned researchers believe that, during fabrication, attaining an absorber layer with a double graded bandgap is more important than having an

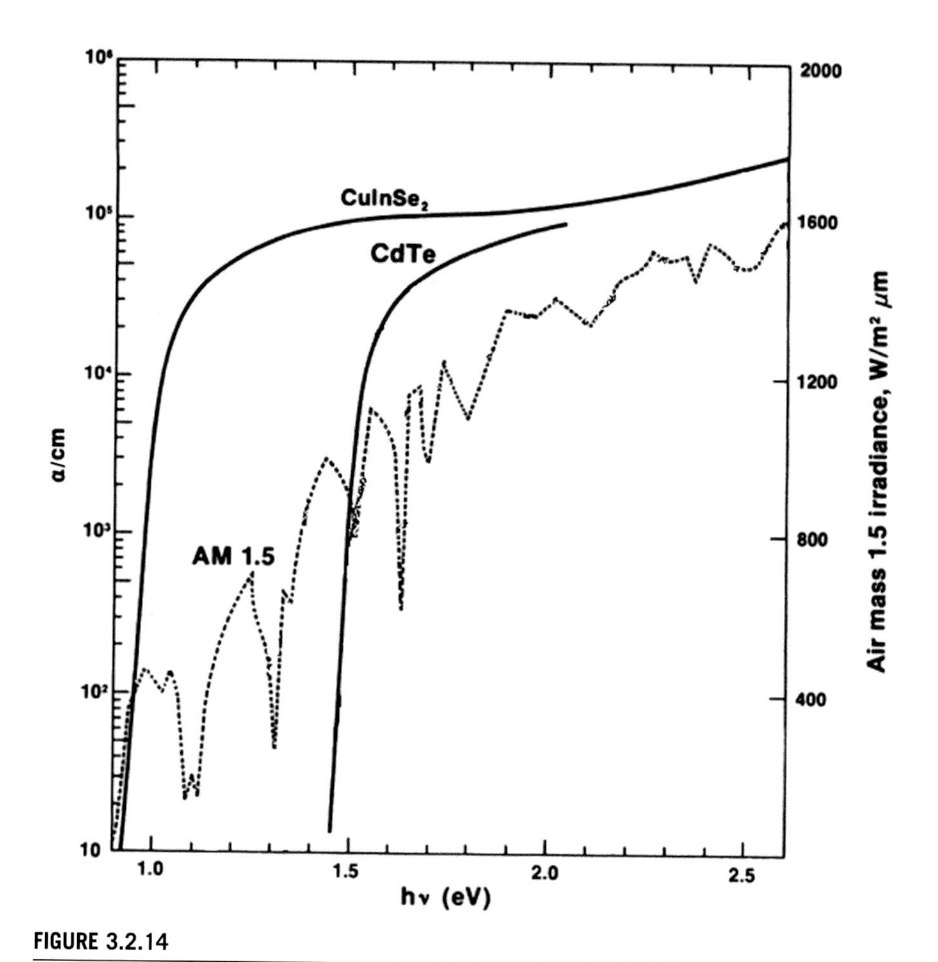

FIGURE 3.2.14

Absorbance spectra for CuInSe2 and CdTe contrasted against Air Mass 1.5 irradiance spectrum (Jaffe & Zunger, 1983).

absorber layer with large grains (Contreras et al., 1996; Nakada, 2012; Ramanathan et al., 2003; Ramanathan, Teeter, Keane, & Noufi, 2005; Repins et al., 2009).

3.22.3.1.5 Cadmium sulfide buffer layer

The energy-momentum (E-k) diagrams for CdS is shown in Fig. 3.2.16. Only the E-k diagram for hexagonal wurtzite CdS is shown below because the diagram for cubic zincblende CdS is not provided in the cited reference material. Some say that cubic zincblende is metastable at temperatures ranging from 20°C to 90°C, and this variation of CdS can only be found at room temperature in epitaxial

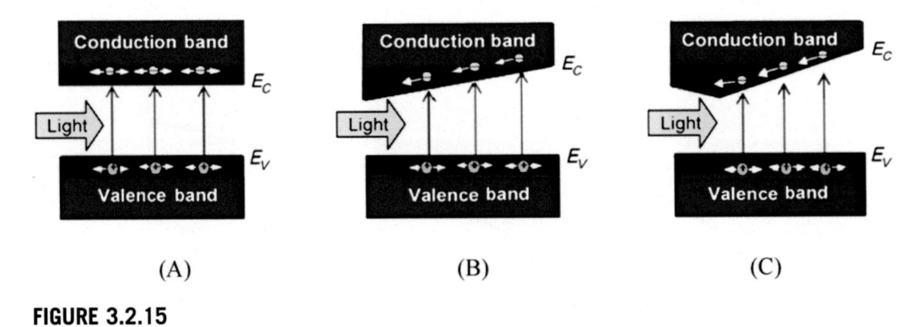

FIGURE 3.2.15

Three types of bandgap profiles in copper indium gallium selenide absorber layer (A) flat bandgap, (B) single graded bandgap, and (C) double graded bandgap (Nakada, 2012).

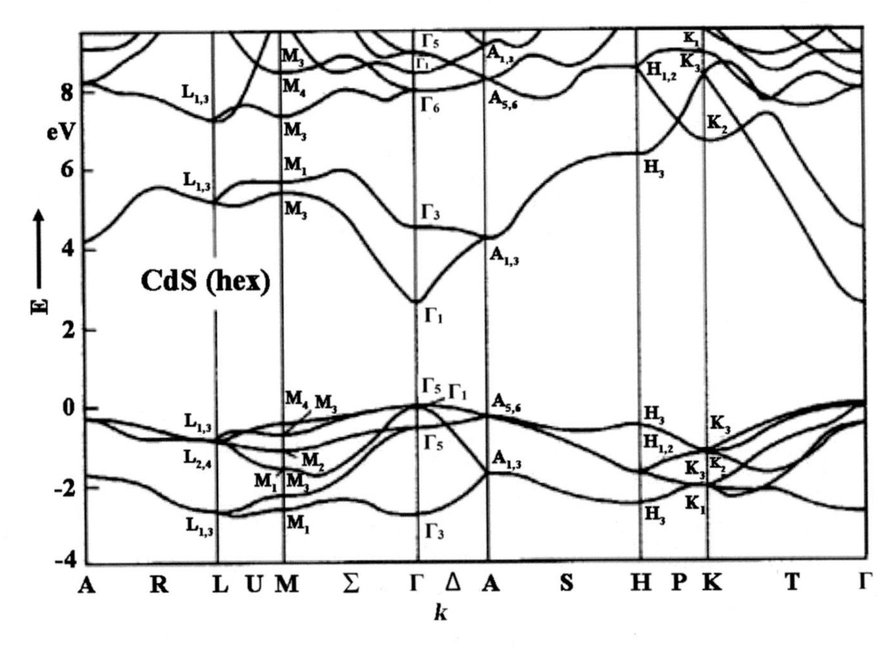

FIGURE 3.2.16

Energy-momentum diagram for hexagonal wurtzite cadmium sulfide (Madelung, 2004b).

layers (Madelung, 2004b). In the figure, the y-axis represents the bonding and antibonding energy while the x-axis represents the crystal momentum. Since the valence band maxima occurred at the same momentum value as the conduction band minima in all the figures, CdS is classified as direct bandgap semiconducting material. According to Madelung also, the bandgap value for single-crystal CdS is 2.482 eV (Madelung, 2004b).

3.2.2.3.2 Electrical properties

3.22.3.2.6 Copper indium gallium selenide absorber layer

Electrical properties express the behavior of charge carriers inside a semiconducting material. The commonly reported parameters for thin films used in PV applications are the conductivity, the carrier concentration, mobility, and lifetime. All the aforementioned parameters can be directly measured except for the conductivity.

Intrinsic semiconductors such as silicon need to be doped with acceptors or donors to infuse excess holes (p-type) or electrons (n-type). Ternary chalcopyrite semiconductors automatically become p-type or n-type depending on the growth process and condition. In the case of $Cu(In,Ga)Se_2$ thin films, p-type conductivity is observed if the film is Cu-poor and annealed under high Se pressure, while n-type conductivity is observed in films that are Cu-rich and Se-poor (Rau & Schock, 2015). This self-doping capability results from intrinsic defects present in the semiconductor.

In their published work, Rincon and Marquez identified twelve intrinsic defects expected to exist in undoped $CuInSe_2$ samples (Rincon & Marquez, 1999). These defects can be broken down as shown below.

1. Three types of donor point defects: V_{Se}, Cu_i, and In_i.
2. Three types of acceptor point defects: V_{Cu}, V_{In}, and Se_i.
3. One donor antisite defect: In_{Cu}.
4. One acceptor antisite defect: Cu_{In}.
5. Two donor antisite defects arising from anion-cation disorder: Se_{Cu} and In_{Se}.
6. Two acceptor antisite defects arising from anion-cation disorder: Se_{In} and Cu_{Se}.

Similar intrinsic defects are expected to exist in undoped $CuGaSe_2$ samples, with Ga in the place of In (Wei & Zhang, 2005). Since $Cu(In,Ga)Se_2$ is an alloy of $CuInSe_2$ and $CuGaSe_2$, the combined expected defects present in both parent semiconducting materials can be expected in $Cu(In,Ga)Se_2$ samples.

Although there are minor discrepancies in the actual reported values for the defect formation energy of copper vacancy by several research groups, these groups are in consensus that V_{Cu} requires the lowest formation energy among all the intrinsic defects in both $CuInSe_2$ and $CuGaSe_2$ semiconductors (Rincon & Marquez, 1999; Wei & Zhang, 2005; Zhang, Wei, & Zunger, 1998; Zunger & Zhang, 1997). So V_{Cu} can be found in vast quantities in $Cu(In,Ga)Se_2$ samples and is the most dominant contributor of the p-type conductivity in the semiconductor.

The electrical properties of a semiconducting thin film can reveal its quality. Large deviations from standard or commonly reported values often indicate a film of poor quality with an abundance of traps and recombination centers that impedes the movement of charge carriers thus resulting in PV devices with low efficiency. Repins et al. analyzed the electrical properties of the $Cu(In,Ga)Se_2$ absorber layer used in the solar cell that was verified to be

20.0% efficient. Based on the analysis, several recommendations were made pertaining to the minimum requirements for absorber layer electrical properties to attain high photovoltaic conversion efficiency. These recommendations are listed below (Repins et al., 2009).

1. Carrier concentration, p, should be at least $2 \times 10^{16}/cm^3$.
2. Carrier mobility, μ, should be at least 100 cm^2/V-sec.
3. Carrier lifetime, τ, should be at least 50 ns.

3.22.3.2.7 Cadmium sulfide buffer layer

Similar to the CIGS absorber layer, the intrinsic defects contained within the physical structure of CdS thin films are the factors that influence its electrical properties. Two categories of defects can be expected in CdS—point defects and antisite defects. These defect types were identified by Varley and Lordi and are listed below (Varley & Lordi, 2013):

1. Donor point defects: V_S, Cd_i and Zn_i.
2. Acceptor point defects: V_{Cd} and V_{Zn}.
3. Donor antisite defects: Cd_S and Zn_S.
4. Acceptor antisite defects: S_{Cd} and S_{Zn}.

The formation energies of the intrinsic defects in CdS are shown in Fig. 3.2.17. In the figure, all the antisite defects are indicated as having high formation energies, therefore the antisite defects have a low level of occurrence in

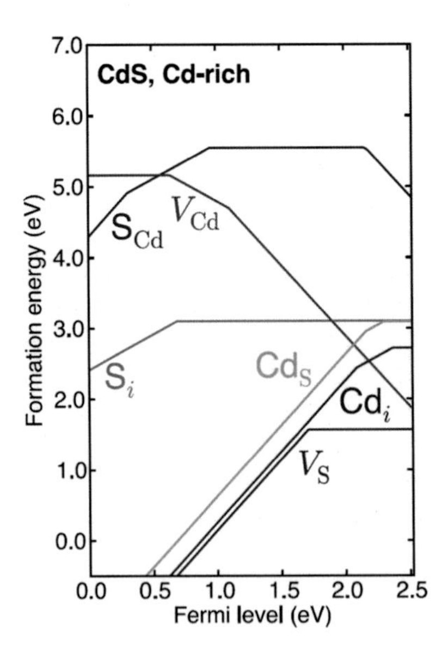

FIGURE 3.2.17

Formation energies for intrinsic point defects in cadmium sulfide (Varley & Lordi, 2013).

CdS films. However, the donor defect V_S possesses the lowest formation energy and is therefore the most prominent defect type in CdS. V_S is the main contributor for n-type conductivity in CdS.

In an article highlighting the device parameters for NREL's 20.0% efficient CIGS solar cell, the only parameter listed for the CdS buffer layer was the thickness (50 nm) (Repins et al., 2009). The article did not include the electrical properties for the buffer layer, unlike the other layers. A plausible explanation could be that researchers at NREL believed that the electrical properties, such as carrier concentration, mobility and lifetime of the buffer layer are of lesser importance than its role in aligning the bandgap at the heterojunction interface. A brief discussion on heterojunction formation in CIGS solar cells with CdS buffers is presented in the next section.

3.2.2.4 Heterojunction formation in copper indium gallium selenide solar cells

When the n-type buffer layer is epitaxially joined to the p-type absorber, an electrical imbalance occurs at the interface because of the charge distributions in the two dissimilar semiconductors. To correct this electrical imbalance and return to a thermal equilibrium state, excess electrons in the n-type semiconductor diffuse through the pn-junction to the p-type semiconductor. This carrier diffusion process results in the shifting of the conduction and valence bands in both semiconducting materials at the vicinity of the pn-junction. The shift is a phenomenon known as band bending and the degree of shifting that takes place depends on the number of excess carriers in each semiconductor before being epitaxially joined.

In this thesis, the n-type buffer layer is the heterojunction partner to the p-type absorber layer. In CIGS solar cells, the term "heterojunction partner" implicitly implies that the n-type buffer layer must fulfill certain requirements before it can be incorporated into the solar cell device structure. These requirements are listed below (Naghavi et al., 2010).

1. The bandgap must be wide enough to limit optical absorption by the buffer.
2. The semiconducting material must be able to create a favorable line-up with the conduction bands of both the absorber and ZnO window layers.
3. The buffer material must be able to create an interface with low defect density with the absorber layer.
4. It must be able to shift the Fermi level of the absorber layer near the interface as close toward the conduction band as possible to reduce the effects of interface defects.

The first two requirements are easily fulfilled by many semiconducting materials with bandgap energies between the bandgap of the absorber and intrinsic ZnO layers. The latter two requirements, however, have proven to be notoriously difficult to fulfill such that only CBD grown CdS buffer layers seem to be suitable for this role. Therefore to propose alternative Cd-free buffer layers or an alternative deposition process to CBD with greater compatibility with the vacuum-based deposition processes of the other layers, an in-depth comprehension of the contributions of CBD grown CdS on the photovoltaic performance of CIGS solar cells is a necessary first task.

An investigation on the beneficial effects of chemical bath-deposited CdS buffer layer on heterojunction formation with $CuInSe_2$ absorbers was carried out by a group of researchers and their findings were reported in (Hunger et al., 2007). The group chose evaporation as the "dry" deposition method for CdS. Prior to that, some of the absorber samples were treated with cadmium partial electrolyte in a solution of cadmium sulfate and aqueous ammonia. This "wet" treatment with cadmium partial electrolyte mimics the treatment received by absorber samples submerged in the chemical bath during CBD before adding thiourea. Finally, the results of surface treated and nonsurface treated CdS/$CuInSe_2$ samples were compared using high-resolution synchrotron X-ray photoelectron spectroscopy. The conclusions drawn from this research are:

1. The type inversion at the surface of the $CuInSe_2$ absorbers shifts the valence band maximum to 0.95 eV below the Fermi level.
2. The cadmium partial electrolyte treatment did not cause further shifting of the band alignments but induces a thin layer of Cd(Se,OH) to form on the surface of the absorber. This thin layer (according to the researchers, a few monolayers in thickness) of Cd(Se,OH) plays a very significant role on the growth process of CdS.
3. The researchers observed a reduction in the degree of band bending at the interface between the absorber and buffer layers of nonsurface treated samples. This indicates that the beneficial type inversion at the absorber layer surface has been reversed and therefore higher interface recombination rates can be expected from nonsurface treated samples.
4. The researchers also carried out surface photovoltage by photoelectron spectroscopy (SPV) measurements on the surface of the samples after the CdS layer is deposited. In the samples given the "wet" treatment process, SPV of 150–250 mV was observed. However, SPV was not measured on samples not given the "wet" treatment. This led the researchers to conclude that the surface recombination velocity was so fast in nontreated samples that the SPV measurements were actually shunted. This shunting effect was attributed to large surface defect in nontreated samples.
5. The immersion of Cu chalcopyrite absorber layers during cadmium partial electrolyte treatment reduces the surface recombination velocity while removing surface impurities simultaneously.

Another interesting conclusion that can be drawn from the study is that a short treatment in an aqueous ammonia solution mixed with cadmium salt before CdS buffer layer growth by "dry" deposition processes may produce the same positive effects as chemical bath-deposited CdS buffer layers.

3.2.3 Conclusion

CIGS solar cells have had the good fortune of outperforming other polycrystalline thin film photovoltaic devices despite the limited, but growing, understanding of the material properties and electronic behavior of the device. Researchers have long observed many favorable outcomes at the interfaces and bulk materials of CIGS devices that are naturally occurring during device fabrication process. Thus further exploration in empirical science is needed to explain these phenomena. A sound knowledge base will also benefit the commercialization efforts for CIGS solar cells while also pushing the efficiencies beyond the present 23% maximum.

References

Abou-Ras, D., Kostorz, G., Bremaud, D., Kälin, M., Kurdesau, F. V., Tiwari, A. N., & Döbeli, M. (2005). Formation and characterisation of $MoSe_2$ for $Cu(In,Ga)Se_2$ based solar cells. *Thin Solid Films*, *480–481*, 433–438. Available from https://doi.org/10.1016/j.tsf.2004.11.098.

Archer, M. D. (2015). The past and present. In M. D. Archer, & M. A. Green (Eds.), *Clean energy from photovoltaics* (2nd ed., pp. 1–39). London: Imperial College Press.

Böer, K. W. (2011). Cadmium sulfide enhances solar cell efficiency. *Energy Conversion and Management*, *52*(1), 426–430. Available from https://doi.org/10.1016/j.enconman.2010.07.017.

Bommersbach, P., Arzel, L., Tomassini, M., Gautron, E., Leyder, C., Urien, M., et al. (2011). Influence of Mo back contact porosity on co-evaporated $Cu(In,Ga)Se_2$ thin film properties and related solar cell. *Progress in Photovoltaics: Research and Applications*. Available from https://doi.org/10.1002/pip.1193.

Boyle, J. H., McCandless, B. E., Shafarman, W. N., & Birkmire, R. W. (2014). Structural and optical properties of $(Ag,Cu)(In,Ga)Se_2$ polycrystalline thin film alloys. *Journal of Applied Physics*, *115*(22). Available from https://doi.org/10.1063/1.4880243.

Calnan, S., & Tiwari, A. N. (2010). High mobility transparent conducting oxides for thin film solar cells. *Thin Solid Films*, *518*(7), 1839–1849. Available from https://doi.org/10.1016/j.tsf.2009.09.044.

Case, T. W. (1920). Thalofide Cell"—A new photo-electric. *Physical Review*, *15*, 0–3.

Contreras, M., et al. (1996). High efficiency graded bandgap thin-film polycrystalline Cu(In,Ga) Se_2-based solar cells. *Solar Energy Materials and Solar Cells*, *41–42*, 231–246. Available from https://doi.org/10.1016/0927-0248(95)00145-X.

Delahoy, A. E., & Guo, S. (2011). Transparent conducting oxides for photovoltaics. In A. Luque, & S. S. Hegedus (Eds.), *Handbook of photovoltaic science and engineering* (2nd ed., pp. 716–796). Chennai: John Wiley & Sons.

Fan, F., Wu, L., & Yu, S. (2014). Energetic I-III-VI2 and I2-II-IV-VI4 nanocrystals: Synthesis, photovoltaic and thermoelectric applications. *Energy & Environmental Science*, *7*, 190–208. Available from https://doi.org/10.1039/C3EE41437J.

Fritts, C. E. (1883). On a new form of selenium cell, and some electrical discoveries made by its use. *American Journal of Science*, *s3–26*(156), 465–472. Available from https://doi.org/10.2475/ajs.s3-26.156.465.

Fritts, C. E. (1885). On the fritts selenium cells and batteries. *Journal of the Franklin Institute*, *119*, 221–232.

Green, M., Dunlop, E., Hohl-Ebinger, J., Yoshita, M., Kopidakis, N., & Hao, X. (2020). Solar cell efficiency tables (version 57). *Progress in Photovoltaics: Research and Applications*, 1–13. Available from https://doi.org/10.1002/pip.3371, no. November.

Hamakawa, Y. (Ed.), (2004). *Thin-film solar cells next generation photovoltaics and its applications*. New York: Springer-Verlag Berlin Heidelberg.

Hunger, R., Lebedev, M., Sakurai, K., Schulmeyer, T., Mayer, T., Klein, A., et al. (2007). Junction formation of $CuInSe_2$ with CdS: A comparative study of 'dry' and 'wet' interfaces. *Thin Solid Films*, *515*(15), 6112–6118. Available from https://doi.org/10.1016/j.tsf.2006.12.120.

Ishizuka, S., Yamada, A., Islam, M. M., Shibata, H., Fons, P., Sakurai, T., et al. (2009). Na-induced variations in the structural, optical, and electrical properties of Cu (In,Ga) Se_2 thin films. *Journal of Applied Physics*, *106*(3). Available from https://doi.org/10.1063/1.3190528.

Jaffe, J. E., & Zunger, A. (1983). Electronic structure of the ternary chalcopyrite semiconductors $CuAlS_2$, $CuGaS_2$, $CuInS_2$, $CuAlSe_2$, $CuGaSe_2$, and $CuInSe_2$. *Physical Review B*, *28*(10), 5822–5847. Available from https://doi.org/10.1103/PhysRevB.28.5822.

Jung, S., Ahn, S., Yun, J. H., Gwak, J., Kim, D., & Yoon, K. (2010). Effects of Ga contents on properties of CIGS thin films and solar cells fabricated by co-evaporation technique. *Current Applied Physics*, *10*(4), 990–996. Available from https://doi.org/10.1016/j.cap.2009.11.082.

Kato, T. (2017). Cu (In, Ga)$(Se,S)_2$ solar cell research in solar frontier: Progress and current status temperatures. *Japanese Journal of Applied Physics*, *56*. Available from https://doi.org/10.7567/JJAP.56.04CA02.

Kazmerski, L. L., White, F. R., & Morgan, G. K. (1976). Thinfilm $CuInSe_2$/CdS heterojunction solar cells. *Applied Physics Letters*, *29*, 268–270. Available from https://doi.org/10.1063/1.89041.

Kennard, E. H., & Dieterich, E. O. (1916). An effect of light upon the contact potential of selenium and cuprous oxide. *Physical Review*, *9*, 58–63.

Kohara,N.,Nishiwaki,S.,Hashimoto,Y.,Negami,T.,&Wada,T.(2001).Electricalpropertiesofthe Cu(In,Ga)Se_2/$MoSe_2$/Mo structure. *Solar Energy Materials and Solar Cells*, *67*(1–4), 209–215.Availablefromhttps://doi.org/10.1016/S0927-0248(00)00283-X.

Kronik, B. L., Cahen, D., & Schock, H. W. (1998). Effects of sodium on polycrystalline Cu(In, Ga)Se_2 and its solar cell performance. *Advanced Materials*, *10*(1), 31–36. Available from https://doi.org/10.1002/(SICI)1521-4095(199801)10:1 < 31::AID-ADMA31 > 3.0.CO;2-3.

Lauermann, I., Loreck, C., Grimm, A., Klenk, R., Mönig, H., Lux-Steiner, M., et al. (2007). Cu-accumulation at the interface between sputter-(Zn,Mg)O and Cu(In,Ga)(S, Se)2—A key to understanding the need for buffer layers? *Thin Solid Films*, *515*(15), 6015–6019. Available from https://doi.org/10.1016/j.tsf.2006.12.172.

Liu, J., Zhuang, D., Luan, H., Cao, M., Xie, M., & Li, X. (2013). Preparation of Cu(In,Ga) Se2 thin film by sputtering from Cu(In,Ga)Se2 quaternary target. *Progress in Natural Science: Materials International*, *23*(2), 133–138. Available from https://doi.org/10.1016/j.pnsc.2013.02.006.

Madelung, O. (2004a). *I–III–VI2 compounds.pdf*. *Semiconductors: Data handbook* (3rd ed., pp. 289–328). New York: Springer-Verlag Berlin Heidelberg.

Madelung, O. (2004b). *II-VI compounds.pdf* (pp. 173–244). New Delhi: Springer-Verlag Berlin Heidelberg.

Matson, R. J., Jamjoum, O., Buonaquisti, A. D., Russell, P. E., & Ahrenkiel., R. K. (1984). Metal contacts to $CuInSe_2$. *Solar Cells, 11*(3), 301–305. Available from https://doi.org/10.1016/0379-6787(84)90019-X.

Mickelsen, R. A., & Chen, W. S. (1980). High photocurrent polycrystalline thin-film CdS/$CuInSe_2$ solar cella. *Applied Physics Letters, 36*(5), 371–373. Available from https://doi.org/10.1063/1.91491.

Mickelsen, R. A., Chen, W. S., Hsiao, Y. R., & Lowe, V. E. (1984). Polycrystalline Thin-Film $CuInSe_2$/CdZnS Solar Cells. *IEEE Transactions on Electron Devices, 31*(5), 542–546. Available from https://doi.org/10.1109/T-ED.1984.21566.

Naghavi, N., Abou-Ras, D., Allsop, N., Barreau, N., Bücheler, S., Ennaoui, C., et al. (2010). Buffer layers and transparent conducting oxides for chalcopyrite Cu $(In,Ga)(S,Se)_2$ based thin film photovoltaics: present status and current developments. *Progress in Photovoltaics: Research and Applications, 18*, 411–433. Available from https://doi.org/10.1002/pip.955, no. April.

Naghavi, N., Hildebrandt, T., Bouttemy, M., Etcheberry, A., & Lincot, D. (2016). Impact of the deposition conditions of buffer and windows layers on lowering the metastability effects in $Cu(In,Ga)Se_2$ /Zn(S,O)-based solar cell. *Proceedings of SPIE, 9749*, 97491I. Available from https://doi.org/10.1117/12.2223151.

Nakada, T. (2012). Invited paper: CIGS-based thin film solar cells and modules: Unique material properties. *Electronic Materials Letters., 8*(2), 179–185. Available from https://doi.org/10.1007/s13391-012-2034-x.

Niki, S., Contreras, M., Repins, I., Powalla, M., Kushiya, K., Ishizuka, S., & Matsubara, K. (2010). CIGS absorbers and processes. *Progress in Photovoltaics: Research and Applications., 18*(6), 453–466. Available from https://doi.org/10.1002/pip.969.

Orgassa, K., Rau, U., Nguyen, Q., Werner Schock, H., & Werner, J. H. (2002). Role of the CdS buffer layer as an active optical element in $Cu(In,Ga)Se_2$ thin-film solar cells. *Progress in Photovoltaics: Research and Applications, 10*(7), 457–463. Available from https://doi.org/10.1002/pip.438.

Orgassa, K., Schock, H. W., & Werner, J. H. (2003). Alternative back contact materials for thin film $Cu(In,Ga)Se_2$ solar cells. *Thin Solid Films, 431–432*, 387–391. Available from https://doi.org/10.1016/S0040-6090(03)00257-8.

Paire, M., Delbos, S., Vidal, J., Naghavi, N., & Guillemoles, J. F. (2014). 7 Chalcogenide thin-film solar cells. In G. Conibeer, & A. Willoughby (Eds.), *Solar cell materials: Developing technologies* (1st ed., pp. 145–215).

Pawar, S. M., Pawar, B. S., Kim, J. H., Joo, O.-S., & Lokhande, C. D. (2011). Recent status of chemical bath deposited metal chalcogenide and metal oxide thin films. *Current Applied Physics., 11*(2), 117–161. Available from https://doi.org/10.1016/j.cap.2010.07.007.

Ramanathan, K., Contreras, M. A., Perkins, C. L., Asher, S., Hasoon, F. S., Keane, J., et al. (2003). Properties of 19.2% efficiency ZnO/CdS/$CuInGaSe_2$ thin-film solar cells. *Progress in Photovoltaics: Research and Applications, 11*(4), 225–230. Available from https://doi.org/10.1002/pip.494.

Ramanathan, K., Teeter, G., Keane, J. C., & Noufi, R. (2005). Properties of high-efficiency CuInGaSe2 thin film solar cells. *Thin Solid Films, 480–481*, 499–502. Available from https://doi.org/10.1016/j.tsf.2004.11.050.

Rau, U., & Schock, H. W. (2014). Cigs solar modules. In A. McEvoy, T. Markvart, & L. Castaner (Eds.), *Solar cells: Material, manufacture and operations* (2nd ed., pp. 261–302). Amsterdam: Academic Press.

Rau, U., & Schmidt, M. (2001). Electronic properties of ZnO/CdS/Cu(In,Ga)Se_2 solar cells—Aspects of heterojunction formation. *Thin Solid Films*, *387*(1–2), 141–146. Available from https://doi.org/10.1016/S0040-6090(00)01737-5.

Rau, U., & Schock, H. W. (2015). Cu (In,Ga)Se_2 and related solar cells. In M. D. Archer, & M. Green (Eds.), *Clean energy from photovoltaics* (2nd ed., pp. 245–305). London: Imperial College Press.

Repins, I., Glynn, S., Duenow, J., Coutts, T.J., Metzger, W.K., & Contreras, M.A. (2009). *Required material properties for high-efficiency CIGS modules*, doi: 10.1117/12.828365.

Reynolds, D. C., Leies, L. W., Antes, L. L., & Marburger, R. E. (1954). Photovoltaic effect in cadmium sulfide. *Physical Review*, *96*, 533.

Rincon, C., & Marquez, R. (1999). Defect physics of the $CuInSe_2$ chalcopyrite semiconductor. *The Journal of Physics and Chemistry of Solids*, *60*, 1865–1873.

Rockett, A. A. (2010). Current status and opportunities in chalcopyrite solar cells. *Current Opinion in Solid State & Materials Science*, *14*(6), 143–148. Available from https://doi.org/10.1016/j.cossms.2010.08.001.

Salome, P. M. P., Rodriguez-Alvarez, H., & Sadewasser, S. (2015). Incorporation of alkali metals in chalcogenide solar cells. *Solar Energy Materials and Solar Cells*, *143*, 9–20. Available from https://doi.org/10.1016/j.solmat.2015.06.011.

Seo, H.-K., Ok, E. A., Kim, W. M., Park, J. K., Seong, T. Y., Lee, D. W., et al. (2013). Electrical and optical characterization of the influence of chemical bath deposition time and temperature on CdS/Cu(In,Ga)Se_2 junction properties in Cu(In,Ga)Se_2 solar cells. *Thin Solid Films*, 0–4. Available from https://doi.org/10.1016/j.tsf.2013.05.024.

Shafarman, W. N., Siebentritt, S., & Stolt, L. (2011). Cu(InGa)Se_2 solar cells. In A. Luque, & S. Hegedus (Eds.), *Handbook of photovoltaic science and engineering* (2nd ed., pp. 546–599). John Wiley & Sons.

Shay, J. L., Wagner, S., & Kasper, H. M. (1975). Efficient $CuInSe_2$/CdS solar cells. *Applied Physics Letters*, *27*(2), 89–90. Available from https://doi.org/10.1063/1.88372.

Shin, D., Kim, J., Gershon, T., Mankad, R., Hopstaken, M., Guha, S., et al. (2016). Effects of the incorporation of alkali elements on Cu(In,Ga)Se_2 thin film solar cells. *Solar Energy Materials and Solar Cells*, *157*, 695–702. Available from https://doi.org/10.1016/j.solmat.2016.07.015.

Siemens, W. (1885). On the electro motive action of illuminated selenium, discovered by Mr. Fritts, of New York. *Journal of the Franklin Institute*, *119*(6), 453–456. Available from https://doi.org/10.1016/0016-0032(85)90176-0.

Turcu, M., Pakma, O., & Rau, U. (2002). Interdependence of absorber composition and recombination mechanism in Cu(In,Ga)$(Se,S)_2$ heterojunction solar cells. *Applied Physics Letters*, *80*(14), 2598–2600. Available from https://doi.org/10.1063/1.1467621.

Varley, J. B., & Lordi, V. (2013). Electrical properties of point defects in CdS and ZnS. *Applied Physics Letters*, *103*(10). Available from https://doi.org/10.1063/1.4819492.

Wagner, S., Shay, J. L., Migliorato, P., & Kasper, H. M. (1974). $CuInSe_2$ /CdS heterojunction photovoltaic detectors. *Applied Physics Letters*, *25*(8), 434. Available from https://doi.org/10.1063/1.1655537.

Wei, S. H., & Zhang, S. B. (2005). Defect properties of $CuInSe_2$ and $CuGaSe_2$. *The Journal of Physics and Chemistry of Solids*, *66*(11), 1994–1999. Available from https://doi.org/10.1016/j.jpcs.2005.10.003.

Yoon, J.-H., Seong, T.-Y., & Jeong, J. (2012). Effect of a Mo back contact on Na diffusion in CIGS thin film solar cells. *Progress in Photovoltaics: Research and Applications*, *21*(1), 58–63. Available from https://doi.org/10.1002/pip.2193.

Zhang, S. B., Wei, S., & Zunger, A. (1998). Defect physics of the $CuInSe_2$ chalcopyrite semiconductor. *Physical Review B*, *57*(16), 9642–9656.

Zhou, L., Hu, X., & Wu, S. (2013). Effects of deposition temperature on the performance of CdS films with chemical bath deposition. *Surface and Coatings Technology*, *228*, S171–S174. Available from https://doi.org/10.1016/j.surfcoat.2012.06.047.

Zhu, X., Zhou, Z., Wang, Y., Zhang, L., Li, A., & Huang, F. (2012). Determining factor of $MoSe_2$ formation in Cu(In,Ga)Se_2 solar cells. *Solar Energy Materials and Solar Cells*, *101*, 57–61. Available from https://doi.org/10.1016/j.solmat.2012.02.015.

Zunger, A., & Zhang, S. B. (1997). Revisiting the defect physics in $CuInSe_2$/and $CuGaSe_2$/. In Proceedings of the *twenty sixth IEEE photovoltaic specialists conference* (pp. 313–318). Available from https://doi.org/10.1109/PVSC.1997.654091.

SUBCHAPTER

CZTS solar cells 3.3

Puvaneswaran Chelvanathan[1,2] and Nadrah Azmi[3,4]

[1]*Solar Energy Research Institute (SERI), The National University of Malaysia, Bangi, Malaysia*
[2]*Graphene & Advanced 2D Materials Research Group (GAMRG), School of Engineering and Technology, Sunway University, Jalan Universiti, Bandar Sunway, Malaysia*
[3]*College of Engineering, Universiti Tenaga Nasional, Jalan Ikram-UNITEN, Kajang, Malaysia*
[4]*Institute of Sustainable Energy (ISE), Universiti Tenaga Nasional, Jalan Ikram-UNITEN, Kajang, Malaysia*

3.3.1 Introduction to CZTS thin film solar cells

Thin film photovoltaic (PV) technologies mainly have a direct bandgap material, and some of these thin film PV technologies, such as CdTe and CIGS, have gone through the commercialization phase with power conversion efficiencies up to 19% in large-scale production. However, considering that some of the materials have issues with toxicity (Cd and Te) and availability (Te and In), new emerging solar absorber materials with non-toxic, abundant, and cost-effective materials are highly sought. Thus, compared to other materials, Cu_2ZnSnS_4 (CZTS), which is composed of Cu, Zn, Sn, and S, with crustal abundance of 68, 79, 2.2, and 420 ppm, respectively, is recognized as the next generation low cost thin film PV technology (Ravindiran & Praveenkumar, 2018; Suryawanshi et al., 2013). The arrangement of atoms and bandgap of the CZTS crystal structure practically equals a certain chalcopyrite crystal structure called copper indium disulfide (CIS). The CZTS crystal structure is inferred when two indium cations in the form of a tetragonal unit cell, also one type of chalcopyrite lattice, are filled in for two tin cations, whereas two indium cations replace two zinc cations. In terms of crystal structure, CZTS usually appears in either structure as a stannite or kesterite type, depending on the methods used to grow the materials. These two crystallographic structures are identical yet have contrasts in the arrangement of copper and zinc atoms, lattice parameter, and total energy value (Saha, 2020).

The Shinshu University laboratory pioneered the first development of CZTS thin film in 1988 and consequently demonstrated that CZTS thin film possesses an ideal direct bandgap of 1.45 eV together with a high absorption coefficient value of 10^4 cm^{-1} in visible light, which is an optimal characteristic for PV applications (Ito & Nakazawa, 1988). The aforementioned effort led by K. Ito and T. Nakazawa utilized atom beam sputtering technique to deposit the CZTS from a

Comprehensive Guide on Organic and Inorganic Solar Cells. DOI: https://doi.org/10.1016/B978-0-323-85529-7.00002-5

sputtering target made up of stoichiometric quaternary CZTS powder. It was shown that a heterojunction diode comprised of cadmium tin oxide transparent conducting film and CZTS film exhibited an open circuit voltage of 0.165 V under AM1.5 illumination. Significant breakthroughs in the performance of CZTS based thin film device were achieved two decades later, stemming from systematic investigation led by H. Katagiri from Nagaoka National College of Technology, which culminated in the realization of a 6.7% efficient CZTS device in 2008 by using two-step technique consisting of co-sputtering of precursors and the sulfurization process (Katagiri et al., 2009). In 2011, IBM (T. J. Watson Research Center, USA) successfully fabricated a CZTS thin film PV device with a power conversion efficiency of 8.4% by using a vacuum thermal evaporation technique to deposit CZTS thin film on the glass substrates, which then underwent thermal annealing in a sulfur-based environment (Shin et al., 2013). In 2016, Toyota Center R&D Labs, Japan improved the efficiency of the CZTS device 9.4% by optimizing the thickness of the two-layered CZTS precursors and post-annealing temperature of the CdS buffer layer (Tajima, Umehara, Hasegawa, Mise, & Itoh, 2017). In 2018, a research group from the University of New South Wales, Australia led by Xiaojing Hao successfully breached the 10% efficiency mark for CZTS solar cells through heterojunction heat treatment (Yan et al., 2018). Detailed information on these state-of-the-art, record-breaking CZTS devices are discussed progressively at the end of this chapter.

CZTS is largely considered to have lower costs and reduced environmental impact due to its use of more abundant elements. Over the years, CZTS thin film solar cells have shown advantages on the materials side as they have comparative structures to the earlier large scale commercialized CIGS thin film solar cells, which make it possible to elevate the power conversion efficiency of CZTS by adjusting the ideas and expertise from CIGS. Hence, this material showed availability issues and will have a longer payback time than CZTS material. Further research on various markets (see Fig. 3.3.1) were achieved to better suit the needs (Wang et al., 2021). Besides, potential limitations and promising pathways towards the commercialization of CZTS needs more research until it can be fully accomplished.

3.3.2 CZTS device architecture and fabrication techniques

The conventional structure of a CZTS solar cell is shown in Fig. 3.3.2. The generic structure of the CTZS solar cell is adapted from the CIGS solar cell.

Soda lime glass (SLG) is used as the preferred substrate mainly due to its cheap cost and most importantly, it acts as a source of sodium which diffuses into the CZTS absorber layer and improves the structural and opto-electrical properties through a doping mechanism (Gershon et al., 2015). Refractory metal, molybdenum (Mo) acts as a stable back contact due to its excellent electrical and chemical properties. Mo can withstand high processing temperatures of 500°C–600°C

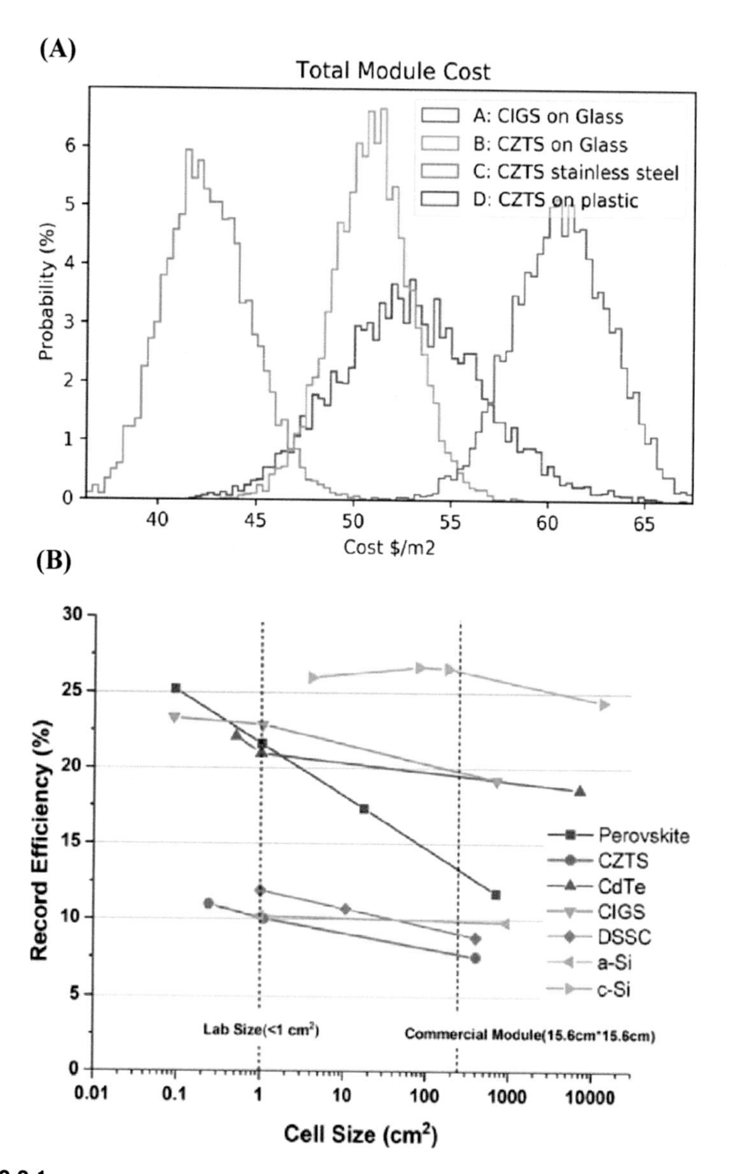

FIGURE 3.3.1

(A) Total module cost for different product types based on the assumption of 1 GW per year manufacturing volume and (B) Record efficiencies of each technology at different cells per module area.

Reproduced with permission Wang, A., Chang, N. L., Sun, K., Xue, C., Egan, R. J., Li, J. . . . Hao, X. (2021). Analysis of manufacturing cost and market niches for Cu2ZnSnS4 (CZTS) solar cells. Sustainable Energy & Fuels, 5, *1044–1058. Copy right 2021, Royal Society of Chemistry, Sustainable Energy & Fuels.*

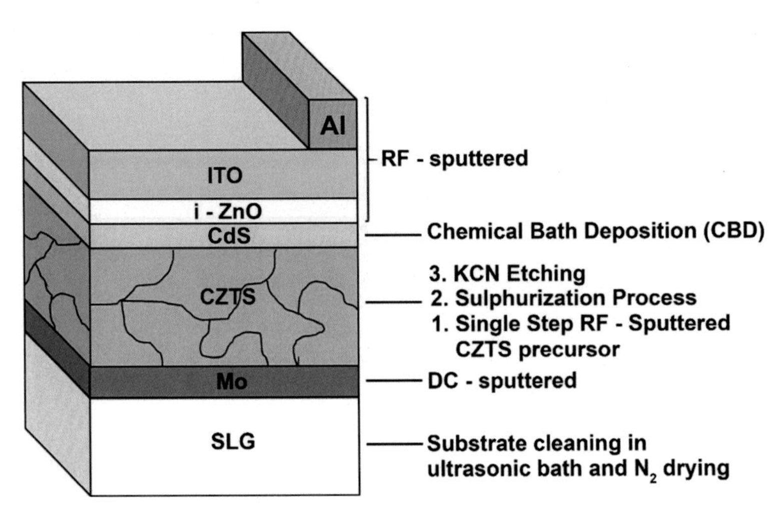

FIGURE 3.3.2

Device architecture and complete fabrication process of CZTS solar cells.

during CIGSe absorber layer formation. Since the processing temperature of CZTS is in a similar range to CIGSe, an Mo back contact has been a common choice for CZTS solar cell. CdS acts as a n-type heterojunction partner for p-type CZTS absorber layer which, together, form the p-n heterojunction which plays a crucial role in the charge carrier separation process. The transparent conducting oxide (TCO) window layer has to perform a dual function of transmitting photons into the solar cell and conducting electron current from the absorber layer to be collected by the front contact (FC) grid.

3.3.2.1 Substrate

SLG has been used as the substrate for the fabrication of CZTS solar cells. The primary motivation of using SLG-based substrates is due to the spontaneous sodium (Na) diffusion into the absorber layer, which has been proven to enhance the material properties of the CZTS layer (Bansal, Chandra Mohanty, & Singh, 2020). Intentional sodium doping through ionic treatment involving Na-containing salt has also been shown to improve the optoelectronic properties of the CZTS absorber layer (Azmi et al., 2021). SLG is cut into small pieces measuring 1 cm × 1 cm for material characterization and 1.25 cm × 2.5 cm for CZTS solar cell device fabrication. The thickness of the SLG is typically around 1.1 mm to 2 mm. Prior to the thin film deposition step, the SLG samples were ultrasonically cleaned in methanol for 15 minutes followed by acetone for 15 minutes, then again in methanol for 15 minutes, and lastly in de-ionized water for 30 minutes. The cleaned substrates were then dried by dry N_2 stream under the fume hood to ensure no liquid residue (watermarks) was left on top of the SLG surface. The setup of the cleaning system

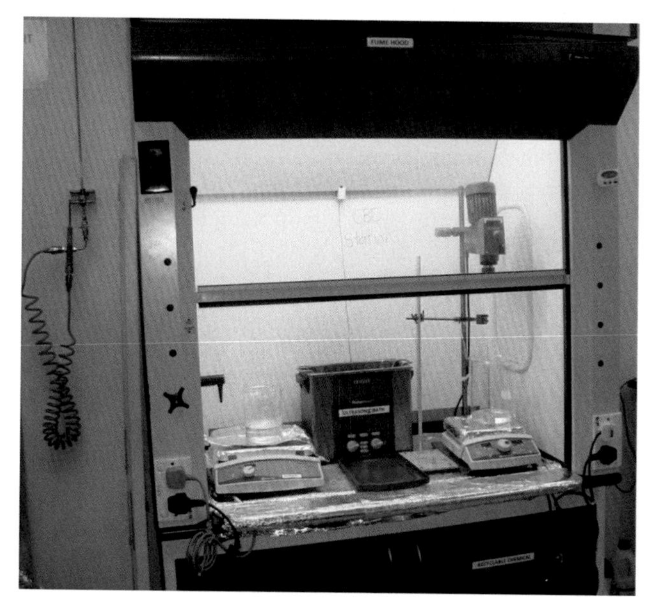

FIGURE 3.3.3

Cleaning system used to clean SLG substrates.

is shown in Fig. 3.3.3. The cleaning process is crucial to minimizing the foreign particle contamination which can degrade the quality of the thin film and solar cell device. After completing the cleaning and drying process, the samples were immediately loaded into a deposition chamber.

3.3.2.2 Molybdenum back contact

Physical vapor deposition (PVD) was used to deposit Mo, CZTS, i-ZnO, ITO, and Al thin films. The magnetron sputtering technique is a PVD technique that is particularly attractive for the deposition of thin films because sputter deposited films typically have better adhesion on the substrate compared to other techniques (Kelly & Arnell, 2000). Sputter-deposited films have a composition close to that of the source material. Sputtering is based on the theory that particle to particle collisions will involve an elastic transfer of momentum, which can apply a thin film to the substrate. In this technique, ions are accelerated towards a target by utilizing electric and magnetic fields. These ions are usually derived by exciting a neutral gas into ion plasma. As the ions are accelerated and bombard the target surface, they dislodge the target atoms. The ejected atoms attach themselves to the substrate and a thin film of target material is produced. To promote a uniform film growth, the substrate is rotated with a rotatable substrate holder where the rotational speed is variable. The substrate can also be heated to enhance the nucleation process during the film growth. The sputtering process offers a flexible platform for thin film deposition research as numerous variables can be manipulated and varied to observe the effects on the film growth condition. Table 3.3.1 shows the sputtering process parameters employed for Mo thin film deposition.

The typical Mo thin film thickness, which is conventionally adopted as a back contact for thin film solar cells, is in the range of 0.8–1.2 μm (Chelvanathan et al., 2017). Fig. 3.3.4 shows the surface morphology and cross-sectional view of the Mo back contact on SLG.

The electrical resistivity of a DC-sputtered Mo back contact is typically in the range of 3×10^{-5} to 6×10^{-5} Ω.cm (Chelvanathan et al., 2015). The conventional structure of the CZTS solar cell device is adapted from the successfully implemented CIGSe solar cell structure. Hence, most key device level development and findings in CZTS have taken place in substrate type structure. One of the key research findings in CIGSe thin film solar cell is the ability of p-type CIGSe to contact the Mo back contact (Wada, Kohara, Negami, & Nishitani, 1996). During the CIGSe absorber layer deposition and selenization processes, spontaneous or unintentional formation of $MoSe_2$ layer on the order of few tens of nm is found to facilitate the formation of an ohmic contact. Formation of an ohmic back contact is very important to reduce the series resistance of any solar cells, which ultimately enhances the device performance. In CZTS solar cells, instead of a p-$MoSe_2$ interfacial layer, an MoS_2 layer is formed between the interface of the CZTS and Mo layers. The formation of a MoS_2 layer has been observed and reported in literature review by various research CZTS groups through numerous material characterization techniques such as X-ray diffraction (XRD), scanning electron microscope (SEM), transmission electron microscopy, and Raman spectroscopy (Biccari et al., 2011; Chelvanathan et al., 2018; Liu et al., 2017; Scragg et al., 2013). Influences of MoS_2 interfacial layer have also been studied numerically, with the majority of the simulation outcomes highlighting the detrimental effects of this layer to the overall performance of CZTS device, particularly when MoS_2 possesses n-type electrical conductivity (Çetinkaya, 2019; Chelvanathan et al., 2012; Ferdaous et al., 2019). For comprehensive insights on the back contact interface related issues in kesterite-based devices, readers are encouraged to peruse a review article by V. Karade et al. (2019).

Table 3.3.1 DC magnetron sputtering parameters for Mo back contact.

Sputtering parameter	Mo
Substrate temperature	200°C
DC power	100 W
Argon gas flow	2 SCCM
Base pressure	10^{-6} Torr
Operating pressure	12 mTorr
Deposition time	120 minutes
Film thickness range	1.0 μm

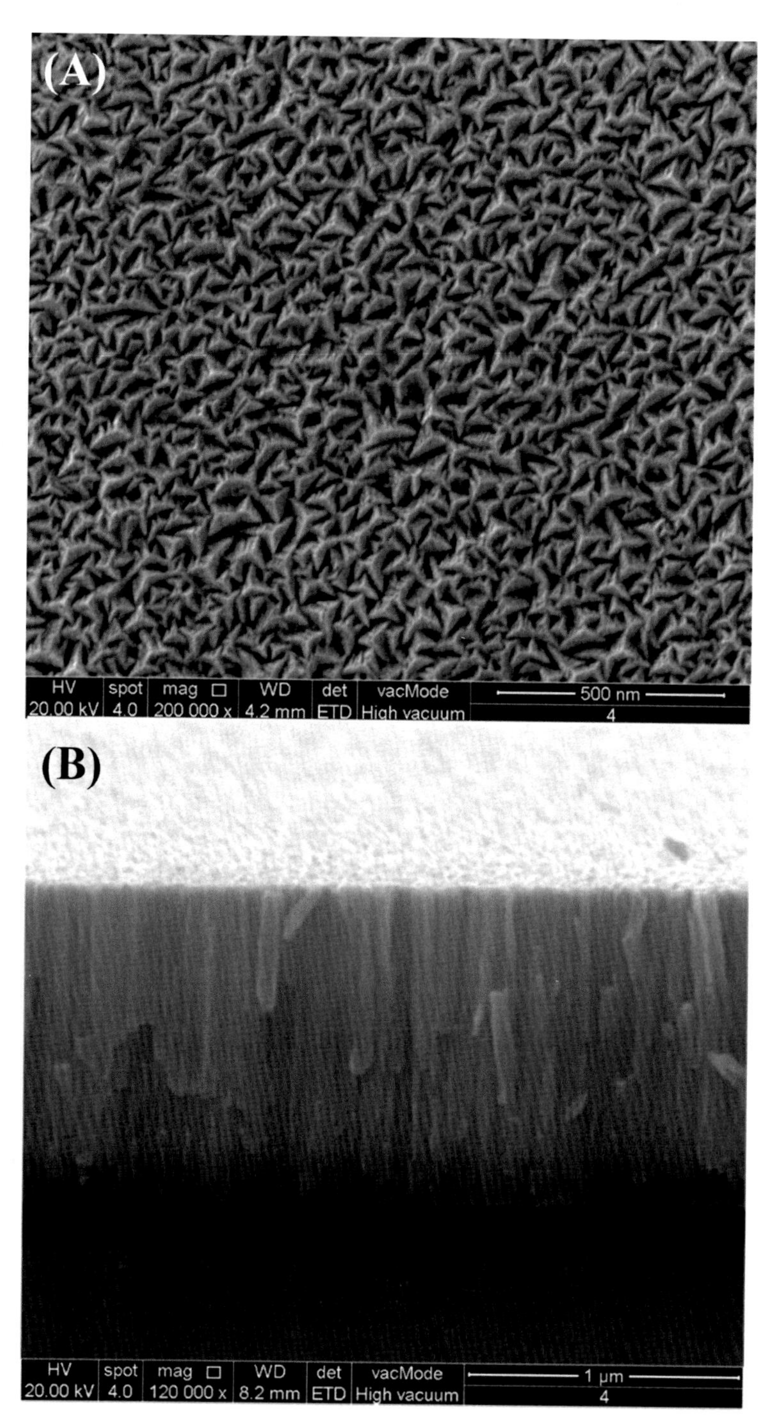

FIGURE 3.3.4

(A) Surface morphology and (B) cross-section of a Mo thin film.

3.3.2.3 CZTS absorber layer

The techniques to fabricate a CZTS absorber layer can be generally classified into two main categories: (1) nonvacuum based process and (2) vacuum based process. Table 3.3.2 shows the variants for each category. The fabrication methods as shown in Table 3.3.2, primarily serve as a physical route to deposit elemental or compound precursors onto targeted substrates, with a desired composition. Hence, in this stage of deposition, control of precise elemental composition and layer uniformity are given major emphasis. Then, precursors undergo a heat treatment in a sulfur environment termed as sulfurization process, which is a compulsory step to achieve micrometer-sized grains through the coalescence process.

The sulfurization process plays a pivotal role in improving the microstructural and optoelectronic properties of the CZTS absorber layer. Despite various deposition processes, most CZTS thin films exhibit common general material properties regardless of the fabrication technique such as p-type conductivity, band gap 1.4–1.5 eV, and absorption coefficient in the range 10^4 cm^{-1}. The highest NREL-certified efficiency of pure sulfide and quaternary-based CZTS thin film solar cell devices is 11.0%, as achieved by UNSW (Yan et al., 2018). The fabrication route of the CZTS absorber layer in this champion cell is through co-sputtering of Cu/ZnS/SnS by using metal (Cu) and (ZnS and SnS) ceramic targets in a multi-target sputtering system followed by sulfurization process. From an industrial point of view, a multi-target sputtering system generally has higher capital and operating cost due to the additional redundant components such as deposition chamber, vacuum pumping system, and sputtering power supplies compared to a single target deposition system (He et al., 2014; Wang et al., 2013). Although multi-target precursor sputtering method has produced the highest efficiency of CZTS solar cells, the inherent process disadvantages, such as higher cost and longer process time, could retard the commercialization effort.

Table 3.3.2 Classification of CZTS fabrication techniques.

Process	Technique
Nonvacuum	• Spin coating • Spray pyrolysis • Electrodeposition • Successive ionic layer adsorption and reaction • Inkjet printing • Chemical bath deposition • Screen printing
Vacuum	• Sputtering • Thermal evaporation • Pulsed laser deposition • Molecular beam epitaxy • Electron beam deposition

Hence, an alternative CZTS deposition process with lower capital and operating cost and shorter process time to increase the throughput is necessary for successful realization of CZTS thin film solar cells as a source of clean energy. One of the most promising techniques that fulfills the aforesaid criteria is single quaternary CZTS compound target sputtering. However, this technique has not been investigated with the same rigor as the multi-target sputtering target method, hence leaving a substantially gray area in terms of the resulting structural, chemical, compositional, and optoelectronic properties of the CZTS precursors and sulfurized films. A lack of in-depth understanding of this vital information has led to insignificant progress of CZTS devices fabricated through single quaternary compound target sputtering techniques. Hence, in the subsequent sections of this chapter, the feasibility aspects of single quaternary CZTS compound target sputtering method to produce a PV grade CZTS absorber layer has been identified as an important and crucial challenge that must systematically addressed.

3.3.2.4 CdS buffer layer

In this research, the CdS buffer layer is fabricated using a chemical bath deposition process, which is relatively cheap compared to other deposition methods. The principal difference between the CBD process and sputtering lies on the method used to produce source material for deposition. In sputtering, the atoms of the target material are produced via physical bombardment of energetic particles, whereas, in CBD, ions of the elements that make up the material are prepared chemically and a specified chemical reaction is induced in a buffer solution in a controlled environment so that the desired material can be deposited as a thin film on the substrate. The CBD process to deposit CdS needs a Cd source, S source, and a buffer solution. Table 3.3.3 shows the chemicals used with the corresponding molarity, and Fig. 3.3.5 shows the apparatus used to perform the CBD process. Before CdS deposition by CBD method, the sulfurized CZTS films were etched with 10 wt.% KCN solution for 2.5 minutes to remove the semi-metallic-like Cu_xS secondary phase. This step is crucial to reduce the possible shunting mechanism caused by the highly conductive secondary phase compound.

During deposition, the temperature inside the deposition beaker is kept between 65°C to 70°C. The total dip time of the full process is around 20–30 minutes, and the whole process is done under the fume hood. During the deposition process, the CdS precipitates on top of the substrate and on the bottom

Table 3.3.3 Chemical reagents used in the CdS-CBD process.

Purpose	Chemical	Molarity
Cadmium source	Cadmium Sulfate, $CdSO_4$	0.002 M
Sulfur source	Thiourea, $SC(NH_2)_2$	0.05 M
Buffer solution	Ammonia, NH_4OH	3.5 M

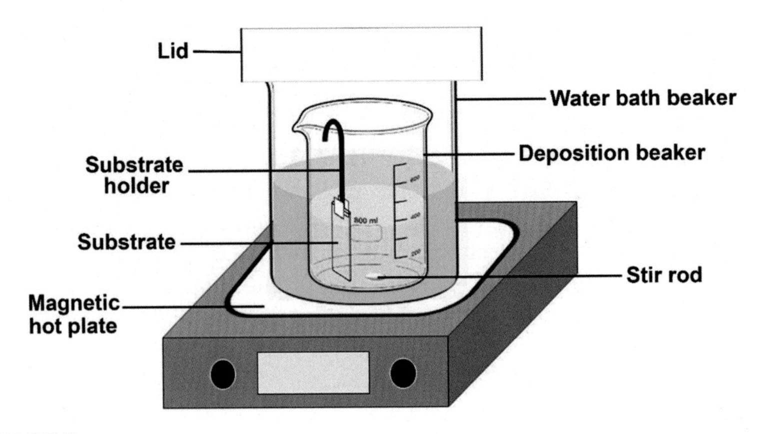

FIGURE 3.3.5

CBD setup for CdS deposition.

of the deposition chamber. To promote the precipitation of a CdS layer on top of the substrate, a stir rod is rotated at 100–200 revolutions per minute.

The key chemical reactions are given in Eqs. (3.3.1) and (3.3.2)

$$CdSO_4 + NH_4OH \rightarrow \left[Cd(NH_3)_4\right]^{2+} + SO_4^{2-} + OH^- \quad (3.3.1)$$

$$\left[Cd(NH_3)_4\right]^{2+} + SC(NH_2)_2 + 2OH^- + NH_4OH \rightarrow \underline{\mathbf{CdS}} + (NH_2)CO + H_2O + 4NH_3 \quad (3.3.2)$$

3.3.2.5 i-ZnO (high resistive transparent layer), ITO (transparent conducting oxide) and Al (front contact)

After the deposition of the chalcogenide buffer layer, the subsequent deposited layers are i-ZnO and ITO. The i-ZnO layer acts as an additional buffer layer on top of the chalcogenide buffer layer. Meanwhile, ITO is a TCO layer that collects the photo-generated electrons before they are passed to the Al top metal grid. Both i-ZnO and ITO layers are deposited using RF sputtering in this study, and Table 3.3.4 shows the sputtering deposition parameters that have been used.

The estimated thickness for the i-ZnO layer is around 80 to 100 nm, and ITO layer is estimated around 700 to 800 nm. Last, the metal front grid Al is deposited by RF sputtering. The RF sputtering deposition parameters for the Al top metal grid are given in Table 3.3.5.

Conventionally, the top metal grid is deposited through a thermal evaporation process. In this study, it is discovered that RF sputtering is a suitable process for Al top metal grid deposition. The estimated thickness for the Al top metal grid is around 300–500 nm.

Table 3.3.4 Sputtering deposition parameters for i-ZnO and ITO layers.

Process parameters	i-ZnO	ITO
Substrate temperature	Room temperature	200°C
RF power	50 W	125 W
Argon gas flow	4 SCCM	8 SCCM
Base pressure	0.1 mTorr	0.025 mTorr
Operating pressure	7.5 mTorr	7 mTorr
Deposition time	20 minutes	120 minutes
Film thickness	50–80 nm	300–400 nm

Table 3.3.5 Sputtering deposition parameters for the Al top metal grid.

Process parameters	Al top metal grid
Substrate temperature	Room temperature
DC power	100 W
Argon gas flow	5 SCCM
Base pressure	0.025 mTorr
Operating pressure	7.6 mTorr
Deposition time	80 minutes
Film thickness	300–500 nm

3.3.3 CZTS thin films by RF-sputtering from single quaternary compound targets

In this section, the resulting material properties of the CZTS precursor deposited by RF magnetron sputtering technique are discussed. The baseline sputtering conditions are listed in Table 3.3.6.

Material properties such as structural, compositional, electrical, and optoelectronic were obtained using appropriate techniques, and the resulting outcomes are correlated.

3.3.3.1 Structural and surface morphological properties

Fig. 3.3.6 shows the XRD pattern for as-sputtered CZTS thin film deposited on SLG/Mo substrate from 10–80 degrees. As-sputtered film exhibits three diffraction peaks belonging to the CZTS crystal structure with the dominant peak corresponding to (112) crystallographic plane orientation. The other two peaks of (220) and (312), which are shown in the inset of Fig. 3.3.6, have relatively low intensity compared to the diffraction peak of (112) preferred orientation. All the peaks can be indexed to the kesterite CZTS phase according to JCPDS card # 26–0575 (Xie et al., 2013). Hence, as-sputtered CZTS thin films sputtered on SLG/Mo substrates

Table 3.3.6 Sputtering process variables for CZTS precursor thin films.

Parameters	Condition
Target	Cu_2ZnSnS_4 (99.99% purity)
Substrate	SLG, Mo-coated SLG
RF power	50 W
Sputtering gas	Pure argon (4 SCCM)
Base pressure	5×10^{-6} Torr
Working pressure	4.0×10^{-2} Torr
Working distance	11 cm
Deposition time	360 minutes
Substrate temperature	Room temperature (RT)

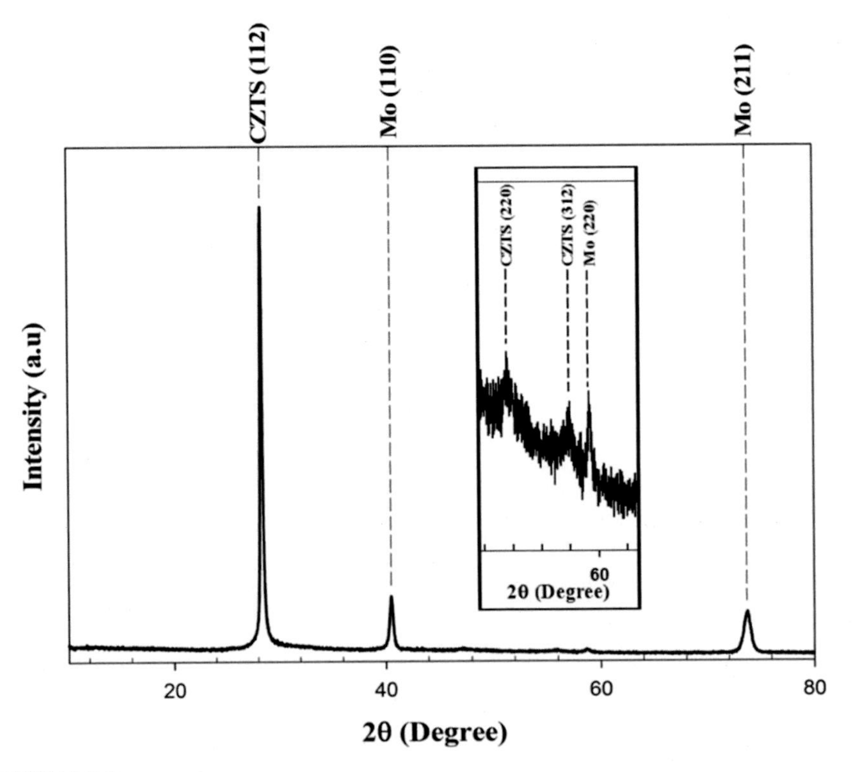

FIGURE 3.3.6

XRD pattern of as-sputtered CZTS thin film on Mo-coated SLG.

in this study possess polycrystalline structure with high texture towards the (112) crystallographic plane. Spatial configuration of substrates on the substrate holder during the sputtering process is predicted to have a significant impact on the crystal structure of the deposited films. Due to the confocal sputter gun configuration (target and sputter gun is at 45 degrees) used in this system, the energy level and trajectory of the incoming sputtered particles hurling towards different areas of the substrate holder can vary significantly. The energy of incoming sputtered particles during the sputtering process influences the nucleation process during thin film growth, which determines the microstructural properties.

Fig. 3.3.7 shows the substrate-sample holder configuration investigated in this study. Eight substrates measuring 1 by 1 cm have been placed as in the configuration shown above, and CZTS sputter deposition was carried out using the same baseline recipe in Table 3.3.2. The XRD patterns and peak analysis are shown in the subsequent figures.

Fig. 3.3.8A until H show the XRD patterns and XRD peak analysis for both CZTS films on bare SLG and on SLG/Mo substrates as a function of substrate position as described in Fig. 3.3.7. All sputtered films, regardless of the substrate type and position on the substrate holder, exhibit the same pattern of diffraction peaks described earlier with (112) preferred orientation with trivial (220) and (312) peaks. However, substrate positions near the center of the substrate holder (position 1 and 2) result in equally intense (112) peaks for both types of substrates. The intensity gradually decreases for films sputtered further away from the center position. Closer scrutiny at the (220) peaks reveal that slight increases

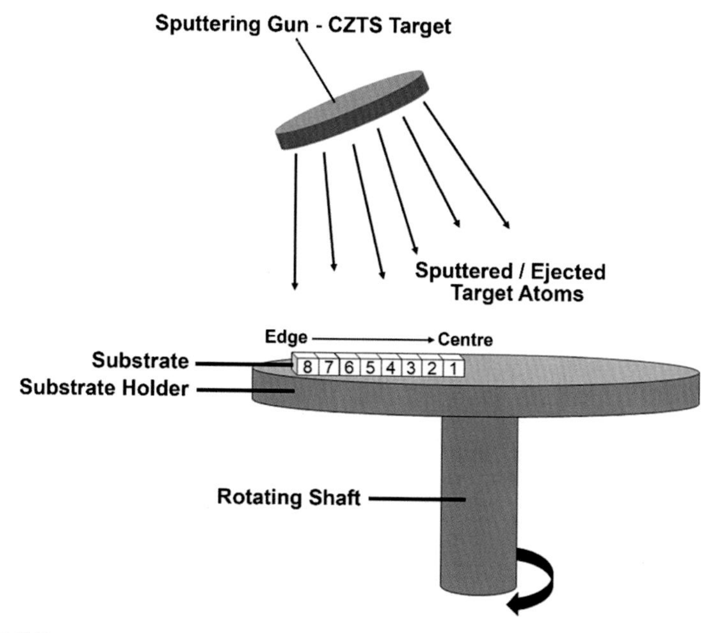

FIGURE 3.3.7

Spatial configuration of substrates during the RF-sputtering process.

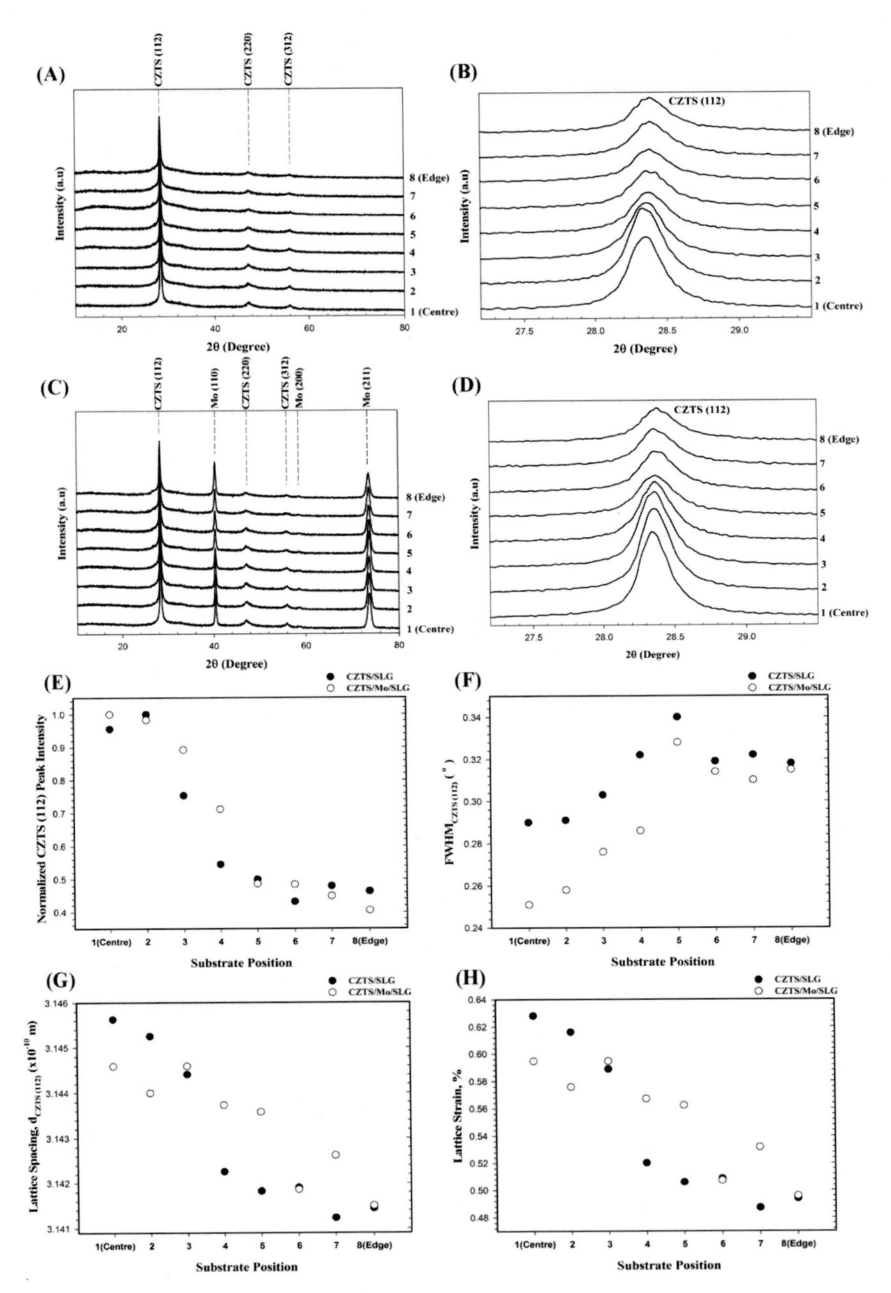

FIGURE 3.3.8

(A) XRD patterns of as-sputtered SLG/CZTS thin films, (B) CZTS (112) peak profiles of as-sputtered SLG/CZTS thin films, (C) XRD patterns of as-sputtered SLG/Mo/CZTS thin films, (D) CZTS (112) peak profiles of as-sputtered SLG/Mo/CZTS thin films, (E) Normalized CZTS (112) peak intensity for various substrate positions, (F) FWHM of CZTS (112) peak for various substrate positions, (G) Lattice spacing, d of CZTS (112) plane for various substrate positions, and (H) Lattice strain of CZTS thin films for various substrate positions.

in the peak intensity are also observed. Fig. 3.3.8B and Fig. 3.3.8D show the CZTS (112) peak profiles in which changes in peak shift and peak intensity are observable. Normalized CZTS (112) peak intensity is the highest for CZTS film deposited at the center position and gradually decreases as the deposition occurs away from the center as shown in Fig. 3.3.8E. On the other hand, full width half maximum (FWHM) values of (112) peaks also indicate the lowest values for substrates at the center position and increase as the deposition occurs away from the center as shown in Fig. 3.3.8F. Lower FWHM values corresponds to larger grain sizes. Hence, by combining both findings (normalized peak intensity and FWHM), it can be concluded that better crystallization of CZTS crystal is induced at the center of the substrate holder. Sputtered particles that arrive at the center of substrate holder are postulated to have higher kinetic energy.

This kinetic energy is converted into thermal energy upon impact of collision on the substrate. Additional thermal energy facilitates the grain nucleation and coalescence process, which can be a contributing factor to the observed increase in crystallinity. On the other hand, lattice strain decreases progressively as films deposited further away from the center of the substrate holder as shown in Fig. 3.3.8H. Films sputtered at the edge (position 8) have the lowest values of strain for both types of substrates used in this study. Films sputtered at the center have higher strain which could be due to the rapid crystallization by the energetic sputtered particles. Rapid crystallization inhibits lattice relaxation and manifests physically through higher strain. Higher strained precursor film is less desirable as the subsequent mandatory sulfurization process will cause the CZTS precursors to peel of due to excessive tensile force. Although the film sputtered on the center position has higher crystallinity, the pre-requisite criterion of low strained precursor is much more crucial. The difference in the degree of crystallinity between films deposited at the center and the edge is quantitatively diminutive. Furthermore, rigorous crystallization of CZTS film through a grain coalescence mechanism actually takes place during the sulfurization process. Hence, CZTS films with the lowest strain (position 8) were chosen as the preferred films for subsequent experimental investigation throughout this study. For all experiments related to CZTS sputter deposition, the substrates were placed at the edge of the substrate holder (position 8).

Fig. 3.3.9 shows the surface morphology (A,C and D) with different magnification power and cross-sectional (B) images of as-sputtered CZTS film deposited on top SLG/Mo substrate. Images of (A) and (C) shows uniform deposition by RF-sputter deposition which results in homogeneous microstructure and surface morphology. No micro-cracks, dome formation, or film peeling are observed from the SEM images. Higher magnification images, (D) using FESEM, reveal that as-deposited CZTS grains are on the order of 50–200 nm. Fine vertical structures of columnar grains with no delamination from the Mo layer are observed from the cross-sectional view. This thickness of sputtered film is around 2 μm. Scanning electron microscopy images presented herein suggest RF-sputtering

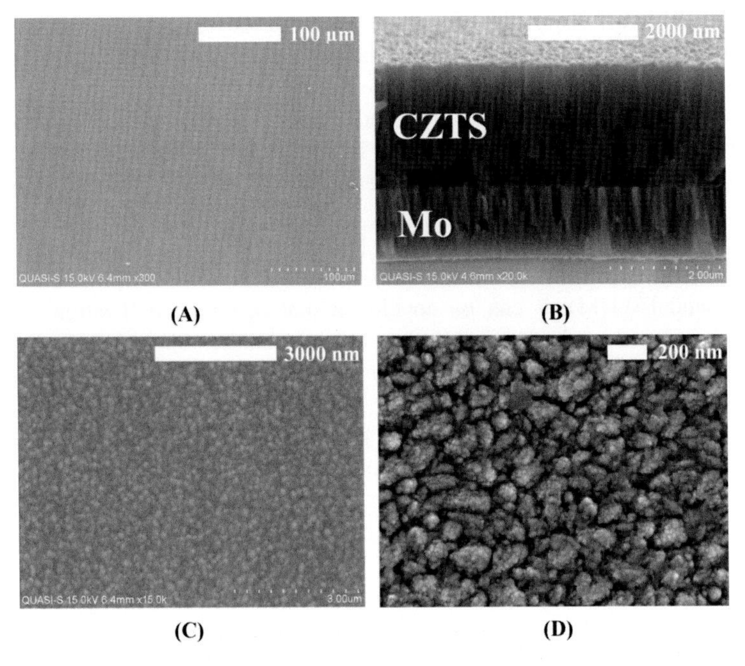

FIGURE 3.3.9

Surface morphology and cross-sectional images of as-sputtered SLG/Mo/CZTS film.

method is a suitable method to deposit CZTS precursors at least in terms thin film uniformity and physically free from major macrostructure defects.

3.3.3.2 Compositional and phase analysis

The CZTS quaternary target used in this study possesses the following atomic concentration shown in Table 3.3.7. Atomic concentration for a hypothetical stoichiometric CZTS compound is also shown for comparison.

Zn-rich and Cu-poor targets have been chosen as the source to deposit the precursors in this study, in accordance with the well-established fact that the highest performing CZTS devices are usually comprised of Zn-rich and Cu-poor composition (Schorr et al., 2019; Yu & Carter, 2016). EDX measurements on the surface as well on the freshly cleaved cross-sectional area were performed to understand the composition of the as-sputtered film and the resulting deviation from the initial composition of the target. Fig. 3.3.10A and B show the EDX probed region for the surface and cross-sectional area of the as-sputtered CZTS film, respectively.

Quantitative values of the atomic concentrations obtained from EDX measurements are given in Fig. 3.3.11.

Table 3.3.7 Atomic concentration of expected stoichiometric CZTS film and CZTS quaternary target used in this study.

Element	Atomic concentration % (stoichiometric)	Atomic concentration % (target)
Cu	25	20.3
Zn	12.5	16.2
Sn	12.5	12.6
S	50	50.9

The dotted line serves as a guide, which coincides with the normalized atomic concentrations of the target. Cu and Zn record higher concentrations while Sn and S possess lower concentrations relative to the target for both the surface and cross-sectional probed area. During sputter deposition, both mass loss and mass gain of any particular element is due to the following mechanism. Mass gain could only occur if there is a possibility of selective physical etching of Cu and Zn atoms compared to Sn and S atoms from the target and vice versa for mass loss. Another possible mechanism that could explain the difference in atomic concentration is different mass transport during sputter deposition. Assuming no selective physical etching mechanism is in play, preferential mass transport of certain species of atoms more than other species could be readily accounted for the observed phenomenon. No further experimental attempts were taken to investigate this matter due to the scope of this study. The most crucial finding from this section of experiments and the corresponding outcome through EDX measurement is that the target's elemental composition could not be translated to the as-sputtered CZTS films. This means that we are working with a sputtered precursor with a slightly altered composition from the initially targeted Zn-rich, Cu-poor elemental ratio. Fig. 3.3.12 shows the elemental ratio in a specific format widely used in the CZTS research community.

By looking at the Fig. 3.3.12, the Zn-rich, Cu-poor target used as source results in precursor with higher Cu content which makes it Cu-"less poor" or relatively Cu-rich as the Cu/Zn+Sn ratio increases close to unity which is the stoichiometric value. Similarly, even higher Zn/Sn and Cu/Sn were calculated due to severe loss of Sn, which renders the as-sputtered films to be even higher with Zn content comparatively or even Zn-'more rich' than it is intended to be. Spatial homogeneity of all the elements on the surface is given in the Fig. 3.3.13, while Fig. 3.3.14 shows the EDX line scan across the cross-sectional area.

Sn and S are fairly homogeneous throughout the probed area. Cu shows slightly non-uniform coverage and Zn exhibits pronounced segregation on top of surface evident from the EDX mapping (emergence of spots of black background). Since Zn content on the surface is equal to the Zn content of the target as shown in Fig. 3.3.11, the notion of segregation induced by mass gain or loss could be ruled out effectively.

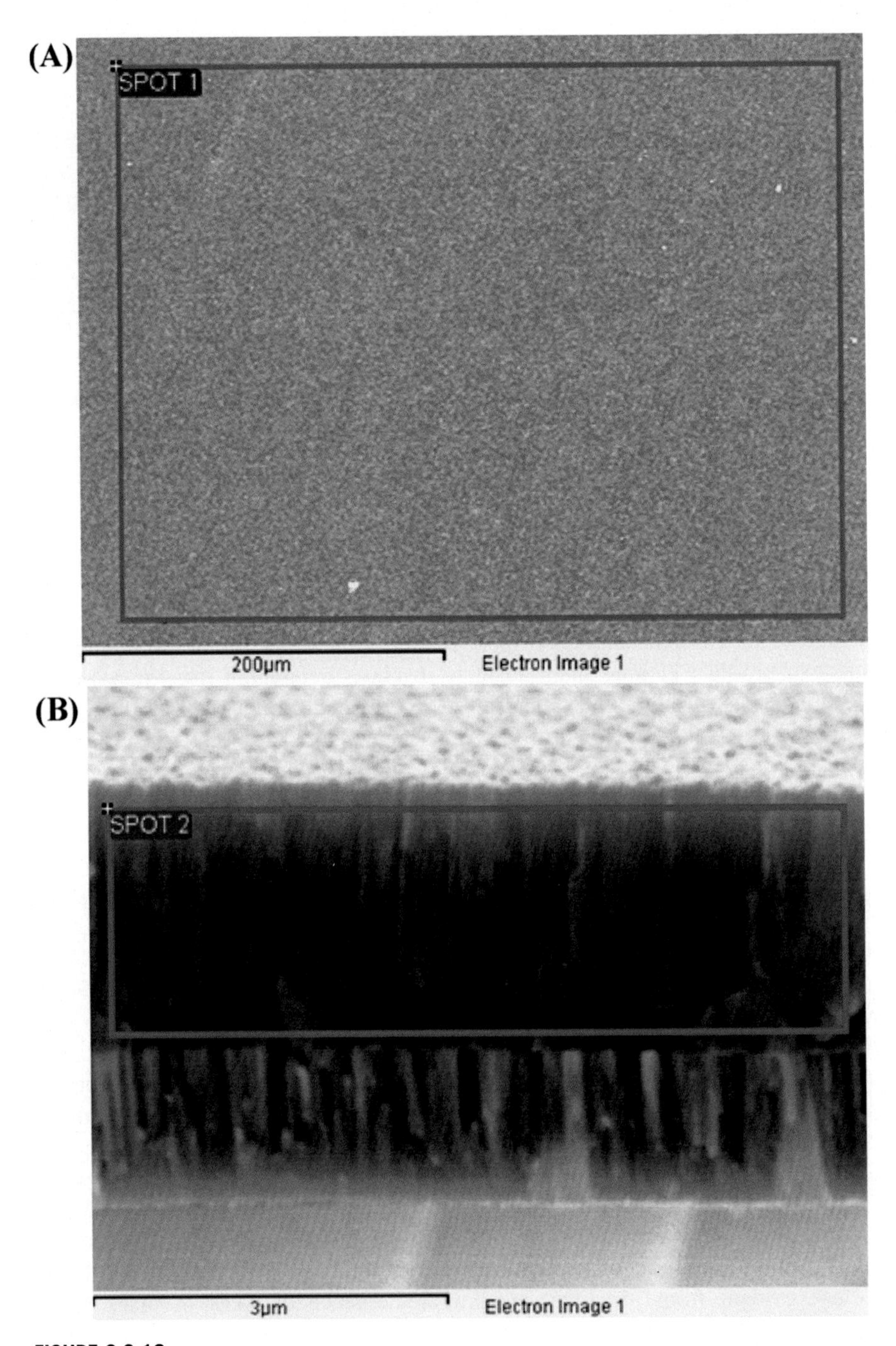

FIGURE 3.3.10

(A) EDX probed spot on the as-sputtered CZTS film surface and (B) EDX probed spot on as-sputtered CZTS film cross section.

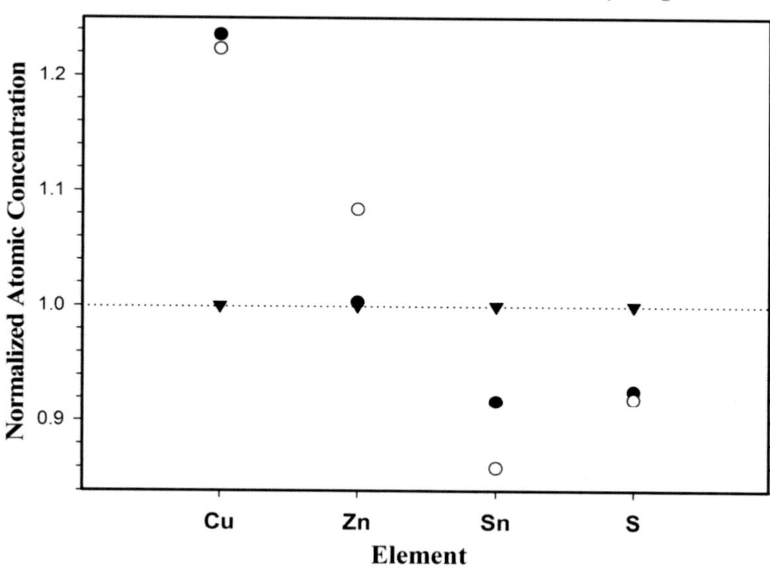

FIGURE 3.3.11

Normalized atomic concentration of CZTS quaternary target and as-sputtered film.

Fig. 3.3.14 reveals that Cu plot shows some minor segregation on the top and on the bottom of the cross-sectional layer. On other hand, Sn has a slightly increasing concentration when approaching the middle of the layer and gradually decreases thereafter towards the back contact. Zn has a slightly increased concentration at the bottom of the layer, while the S signal is amplified in the Mo region due to the identical Ka1 energy of Mo and S elements. Fig. 3.3.15(A) and (B) (zoomed plot for clarity) show the Raman spectra for as-sputtered CZTS film. As anticipated, due to the Zn rich composition of the target and resulting as-sputtered film, weak ZnS signal centered on 642 cm^{-1} which corresponds to + LA vibration mode is observed indicating existence of ZnS compound in the CZTS film (Xiong, Wang, Reese, Lew Yan Voon, & Eklund, 2004). Hence, it is confirmed that the ZnS phase co-exists with the CZTS phase and characterized by Raman peak centered on 332 cm^{-1}, which is known as the disordered kesterite phase. The presence this phase that appears as in this study is due to the existence of highly disordered distribution of Zn and Cu atoms in the cation sublattice. Hence, high concentrations of Cu_{Zn} and Zn_{Cu} antisite defects due to the similar size of the Cu and Zn cations exist and this phase is referred to in the literature as the disordered kesterite phase and is characterized by a dominant A_1 symmetry Raman peak at 332 cm^{-1} as observed in this study (Dimitrievska, Fairbrother, Pérez-Rodríguez, Saucedo, & Izquierdo-Roca, 2014). Due to the absence of supply of intentional thermal

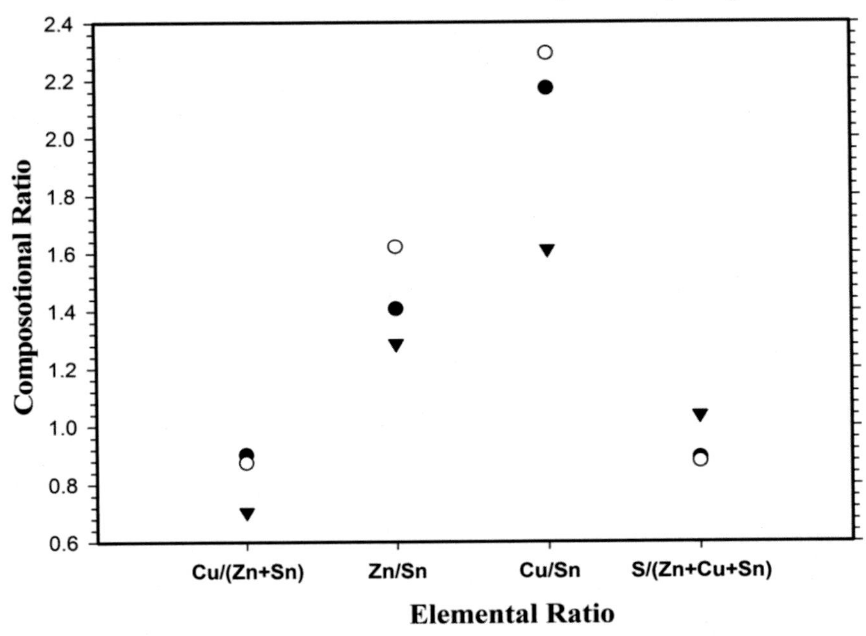

FIGURE 3.3.12

Elemental compositional ratios of CZTS quaternary target and as-sputtered film.

energy during sputter deposition, cation re-arrangement in the lattice will likely be in a minimum state explaining the non-existence of kesterite phase, which, had it existed, would cause a Raman signal at 338 cm^{-1} (Azmi et al., 2021). All the other possible peaks which could arise from CZTS decomposition are marked for reference. No distinct and credible peaks can be associated to any of the binary and ternary sulfide compounds. However, further peak fitting and thorough analyses are proposed to completely deconvolute the Raman spectrum to its individual contributing signal.

3.3.3.3 Electrical properties

The average electrical properties are shown in Table 3.3.8. Consistent with the defect physics predictions, Zn-rich and Cu poor composed CZTS film results in p-type electrical conductivity mechanism as confirmed by Hall effect measurement studies (Chen, Yang, Gong, Walsh, & Wei, 2010). The high carrier concentration around 10^{20} cm^{-3} is probably due to the highly compensating and heavily doped nature of CZTS semiconductor (Grossberg, Raadik, Raudoja, & Krustok, 2014).

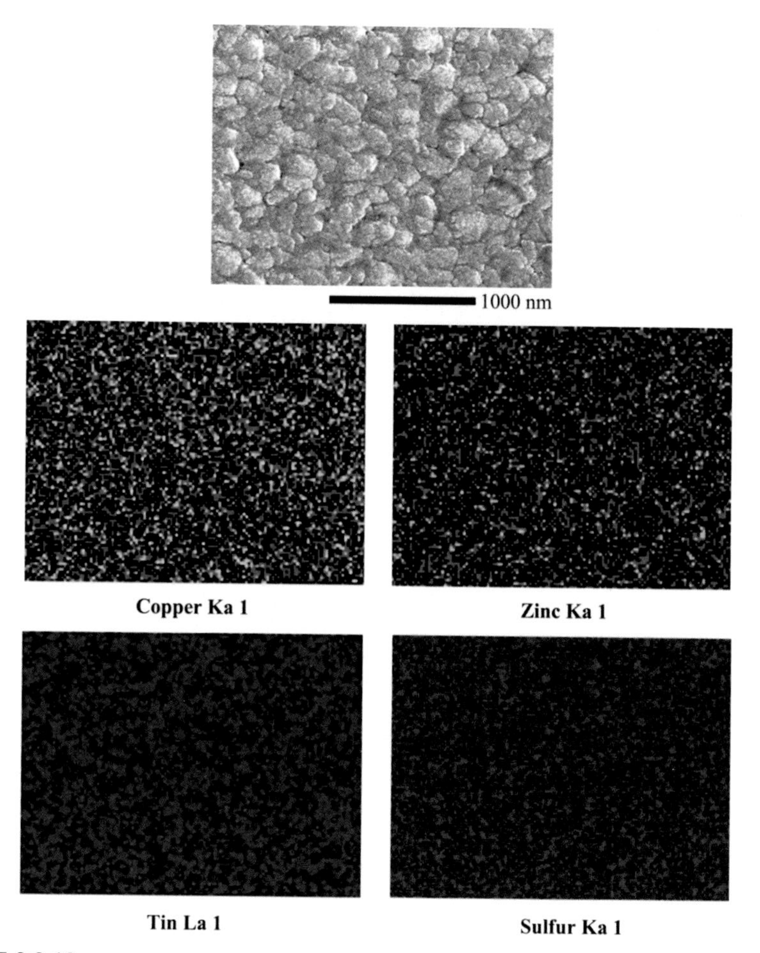

FIGURE 3.3.13

Surface EDX mapping plot of as-sputtered CZTS film.

3.3.3.4 Effects of natural oxidation on As-sputtered CZTS films

This study on the natural oxidation of CZTS sputtered films was investigated by visual observation of the films. A subtle color change was observed on CZTS film surfaces which were unintentionally left exposed to atmospheric ambient. Fig. 3.3.16 shows two surface morphologies of CZTS films maintained in a vacuum or ambient environment.

As-sputtered film kept in vacuum desiccator possesses smooth and uniform morphology. On the other hand, film left in atmospheric ambient exhibits 'freckled' morphology localized non-uniformly throughout the surface. Fig. 3.3.17 shows the closer physical surface morphology images of as-sputtered CZTS films

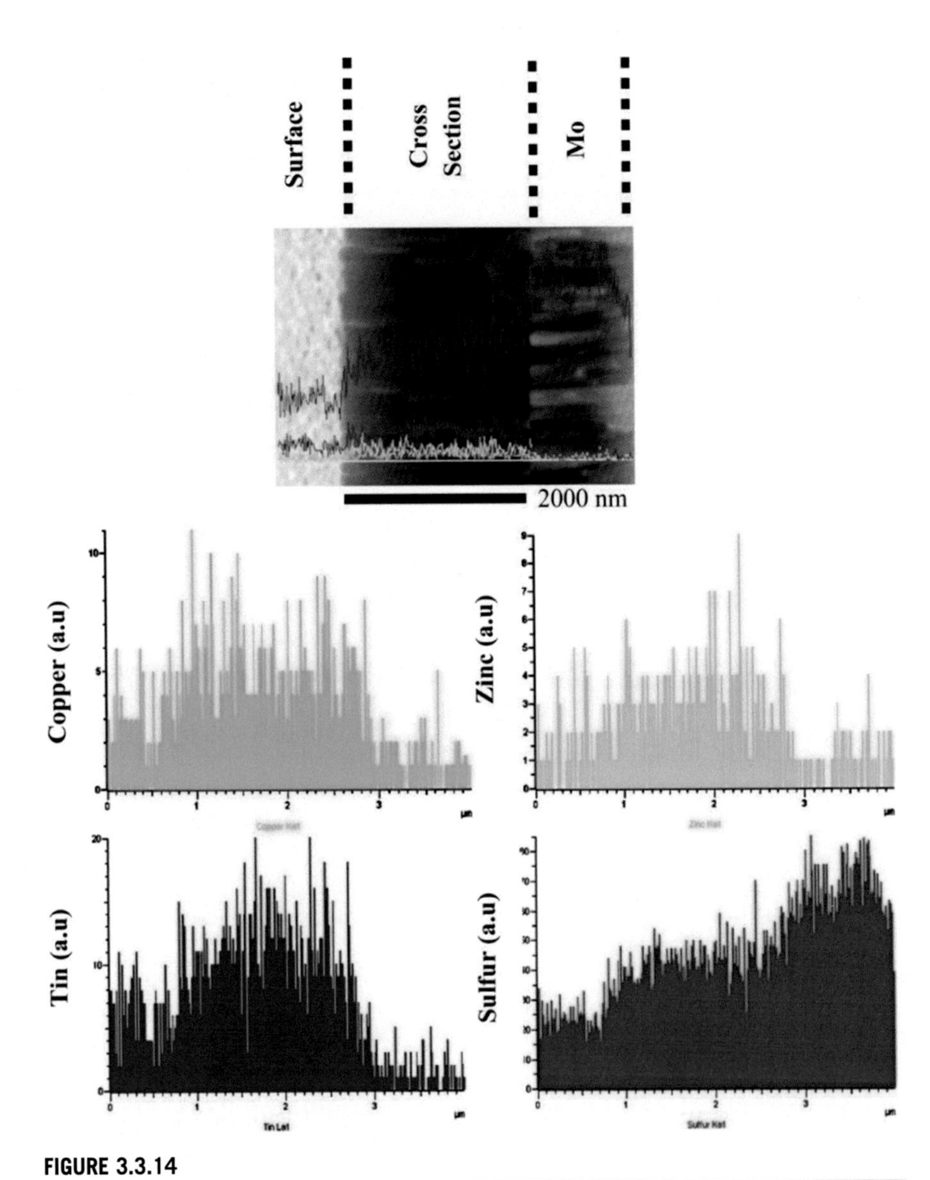

FIGURE 3.3.14

Cross sectional EDX line scan of as-sputtered CZTS film.

suspected to be oxidized naturally due to atmospheric ambient and moisture exposure. EDX measurements at two spots, as shown below, were carried out to determine the corresponding elemental composition.

Table 3.3.9 shows the atomic concentration of the EDX probed CZTS surface morphologies as shown in Fig. 3.3.16.

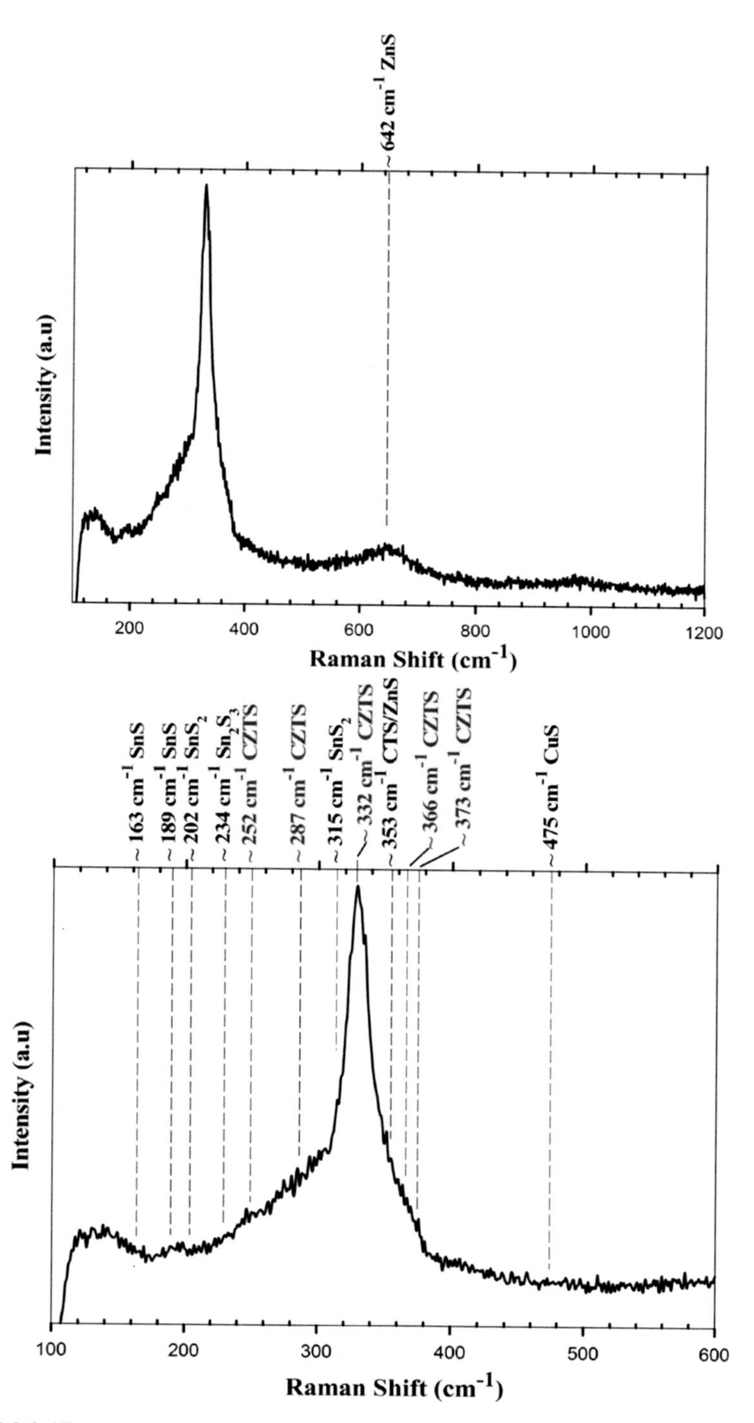

FIGURE 3.3.15

(A) Raman spectrum of as-sputtered CZTS film, and (B) Detailed Raman spectrum of as-sputtered CZTS film.

Table 3.3.8 Type of conductivity, carrier concentration, mobility, and resistivity of as-sputtered CZTS film on bare SLG substrate.

Electrical properties	Value
Conductivity	p
Carrier concentration [cm^{-3}]	2.420×10^{20}
Mobility [cm^2/Vs]	2.568×10^{-01}
Resistivity [Ω.cm]	1.004×10^{-1}

The oxygen concentration in the 'freckled' surface is almost thrice as compared to the smooth surfaced area. Hence from this quantitative elemental analysis, the oxygen penetration play a major role in the observed surface modification of CZTS films. Table 3.3.10 shows a comparison of the normalized Cu, Zn, Sn and S atomic concentration of the CZTS films. Two distinct monotonic trends could be observed by correlating the oxygen content, which are an increase of Zn content and decrease of Sn as the oxygen concentration becomes higher. Zn concentration records increase more than 50% whereas Sn decreases by 30% compared to the initial atomic concentration of the as-sputtered CZTS film stored in a vacuum desiccator.

EDX measurement in linescan mode was performed to understand the elemental distribution on the various surface microstructures which formed on the oxidized CZTS film. The obtained elemental distribution is plotted as shown in Fig. 3.3.17. This linescan covers three distinct surface morphologies which can be categorized by the varying image contrast and structural features. The linescan plot shown in Fig. 3.3.18 confirms that the oxygen content is higher in the darker region of the 'freckled' surface morphology. Step-wise increases and decreases of oxygen content, as pointed out by the four vertical arrows in the EDX linescan, suggest that the induced surface modification is highly oxygen dependent. Existence of varying oxygen content in the three distinct non-uniform surface morphologies is postulated to affect the corresponding material properties after the sulfurization process.

Fig. 3.3.19 shows the surface of sulfurized of CZTS films of both stored in vacuum desiccator and kept in atmospheric ambient all the time prior to sulfurization process. After sulfurization, the CZTS film not exposed to ambient conditions exhibits grains of micrometer size with uniform shape and features. As expected, sulfurized CZTS film initially kept in the atmospheric ambient also exhibits non-uniform surface in Fig. 3.3.19B and in a larger surface area as shown in Fig. 3.3.20 as shown below.

The numbered spots of 1, 2, and 3 on the image depict regions with notable surface variations. Spot 3 consists of severely cracked films which has the brightest contrast among the labeled areas. The resulting contrast level of the probed area during SEM imaging process serves as a quantitative measure of electrical resistivity of the film. Higher contrast (bright image) indicates electrons

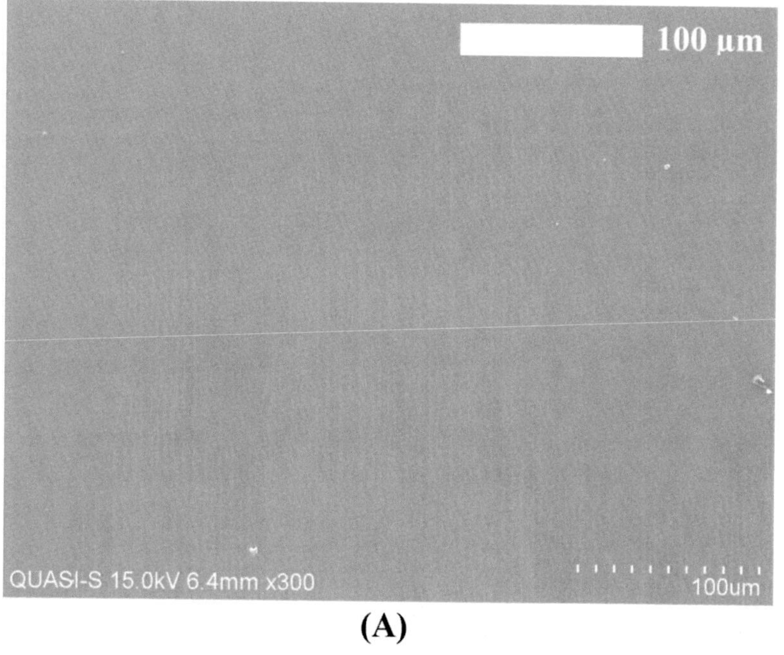

(A)

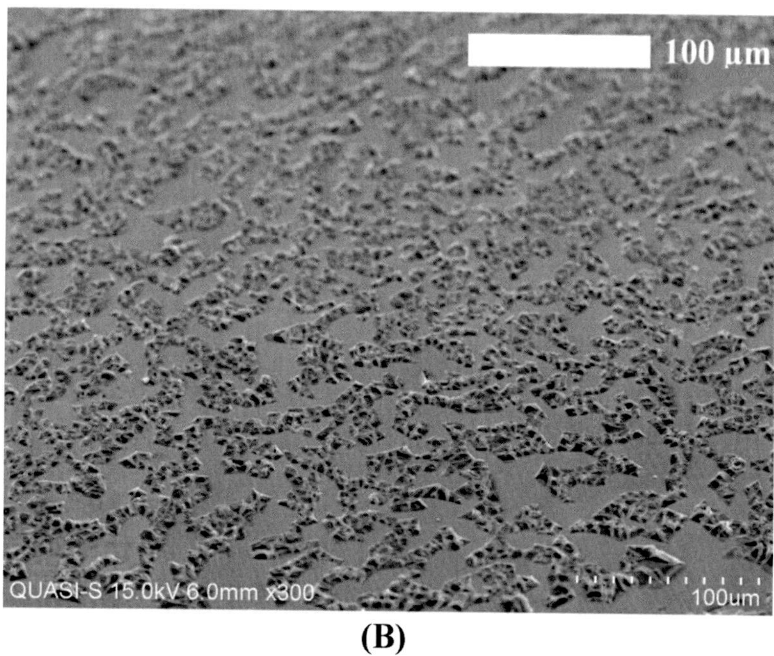

(B)

FIGURE 3.3.16

Surface morphology of as-sputtered CTZS films. (A) Stored in vacuum desiccator. (B) Kept in atmospheric ambient.

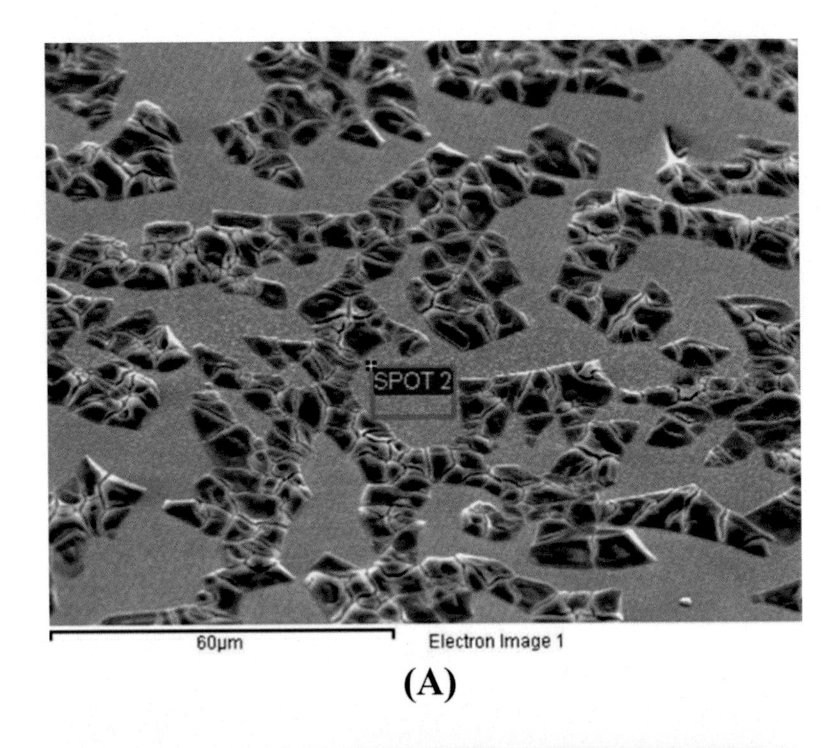

(A)

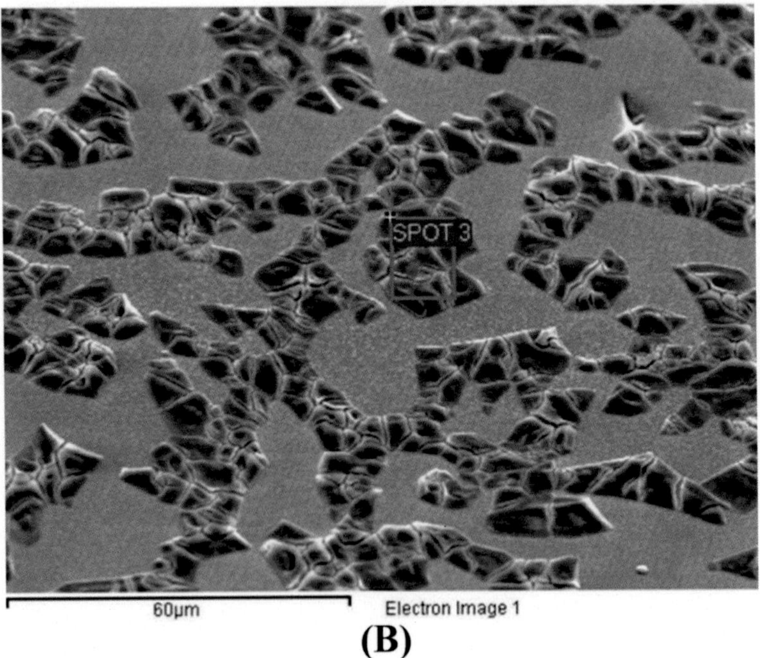

(B)

FIGURE 3.3.17

Probed EDX spots on oxidized CZTS films. (A) On smooth surface morphology and (B) on 'freckled' morphology.

Table 3.3.9 Atomic concentrations at various surface morphologies as shown in Figure 4.62A and Figure 4.43.

Element	Smooth, SPOT 2 (%)	'Freckled', SPOT 3 (%)
Cu	23.04	9.78
Zn	16.52	9.41
Sn	7.97	2.65
S	33.90	15.22
O	**18.57**	**62.93**

Table 3.3.10 Normalized Cu, Zn, Sn, and S atomic concentrations at various surface morphologies as shown in Fig. 3.3.16A and Fig. 3.3.17.

Element	Fig. 3.3.16A (%)	Fig. 3.3.17A smooth SPOT 2 (%)	Fig. 3.3.17B: 'Freckled' SPOT 3 (%)
Cu	25.08	27.72	23.10
Zn	**16.25**	**20.40**	**25.65**
Sn	**11.25**	**9.85**	**7.48**
S	47.12	42.03	43.76

are accumulated in that area duc to the inability of electrons to flow through the film. In another words, the film has a relatively higher resistivity. The phenomenon of electron accumulation which renders the obtained image to be bright in contrast is called the charging effect. Spot 2 consist of powder like microstructure while spot 1 exhibits uniform granular like grains. The 45 degrees tilt angle of the SEM images reveals that all three layers possess different spatial distribution not only on the x and y-axes but also in the z-axis. Layer of Spot 3 exists on top of layer of Spot 2 and finally these two layers are piled on layer of Spot 1. Table 3.3.11 shows the atomic concentration of the EDX probed CZTS surface morphologies as shown in Fig. 3.3.19A and Fig. 3.3.20.

Similar trends of variation in Zn and Sn contents are observed for sulfurized CZTS film, whereby Zn increases while Sn decreases mimicking the monotonic trend observed for as-sputtered CZTS film (see Table 3.3.10). The mechanism of oxygen penetration in terms of oxidation of Cu, Zn, Sn or S elements is beyond this scope of this study. However, the most crucial information able to be inferred from this experimental study is that to fabricate uniform CZTS thin film from sulfurization process, the oxygen content of the as-sputtered CZTS film must be strictly controlled. Segregation of Zn onto to the top surface is observed in oxygen rich regions of CZTS films. Higher oxygen level induces higher Zn segregation which becomes highly resistive upon sulfurization process. Zn-rich layer on the surface appears to be bright in contrast and possesses poor surface properties such as

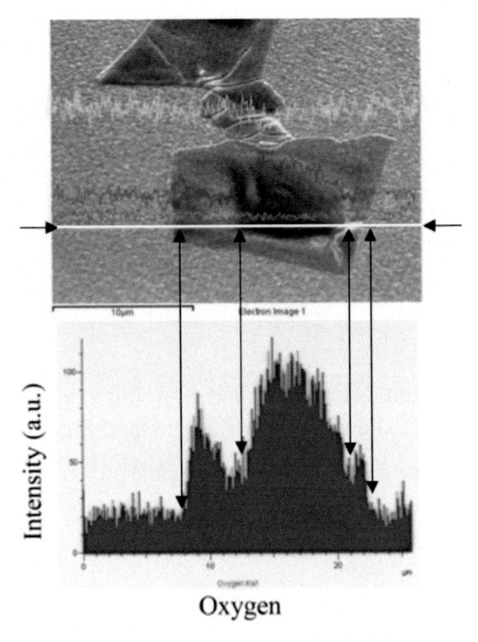

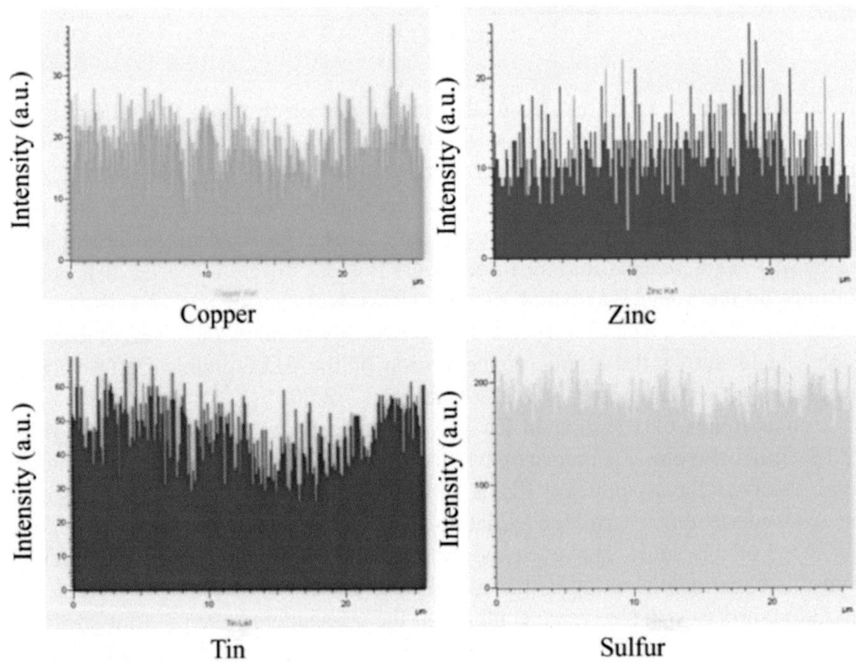

FIGURE 3.3.18

EDX linescan measurement on oxidized CZTS film surface.

(A)

(B)

FIGURE 3.3.19

Surface morphology of sulfurized CTZS films. (A) stored in vacuum desiccator prior to sulfurization (B) kept in atmospheric ambient prior to sulfurization.

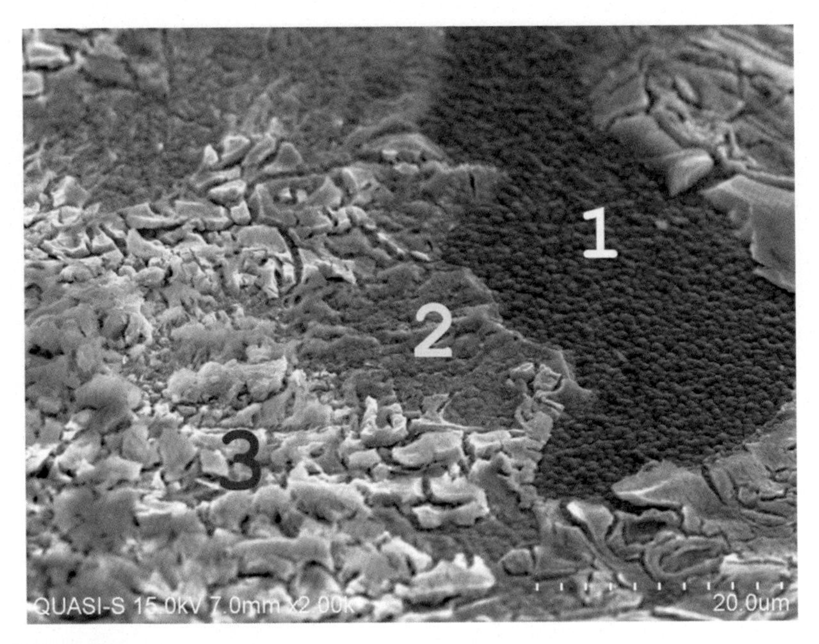

FIGURE 3.3.20

Surface morphology of sulfurized CZTS film initially kept in the atmospheric ambient prior to sulfurization. (Tilted: 45 degrees).

Table 3.3.11 Atomic concentrations of Cu, Zn, Sn and S at various surface morphologies as shown in Fig. 3.3.19A and 20.

Element	Fig. 3.3.19A (%)	Fig. 3.3.20 SPOT 1 (%)	Fig. 3.3.20 SPOT 2 (%)	Fig. 3.3.20 SPOT 3 (%)
Cu	24.21	27.28	29.26	21.87
Zn	**15.76**	**19.64**	**22.07**	**27.29**
Sn	**11.74**	**9.46**	**9.11**	**8.01**
S	48.29	43.62	43.34	42.83

cracks and pinholes. Non-homogeneous CZTS films, in terms of distribution of elemental composition and surface morphological properties, are anticipated to result in non-homogenous structural, compositional, and optoelectronic properties. These undesired effects should be mitigated to carry out an effective experimental study of optimization of CZTS absorber through variation in sulfurization process parameters. Hence, all the as-sputtered CZTS films were stored in vacuum desiccator all the time prior to any subsequent material processing step or thin film characterization process throughout this study unless stated otherwise.

3.3.3.5 Sulfurization of CZTS precursor thin film

As-sputtered CZTS films were subjected to the sulfurization process with various hold times. The effective hold time was varied from 15 to 75 minutes. Table 3.3.12 summarizes the process parameters used during the sulfurization process.

The main objective of this study is to investigate the effects of sulfurization holding time on the resulting CZTS thin film material properties. Table 3.3.13 shows the physical condition of CZTS thin film 5 days after the sulfurization process visually observed. No peeling off failures and macrostructure defects were observed for all films.

Table 3.3.14 shows a summary of the various aspects in terms of sulfurization process time used in this study.

Fig. 3.3.21 shows the XRD patterns of sulfurized CZTS films from 10 to 25 degrees. The SnS_2 (001) phase is eliminated for the CZTS films sulfurized for 75 minutes, indicating better or complete CZTS crystallization process. A shoulder

Table 3.3.12 Sulfurization process parameters for CZTS thin films used in this sulfurization time study.

Parameters	Condition
Stack configuration	CZTS/Mo/SLG, CZTS/SLG
Heating ramp-rate	10°C/minute
Cooling rate	Natural cooling
Base pressure	90 mTorr (0.0001 atm)
Working pressure	760 Torr (1 atm)
Background gas	Purified N_2 (99.99 %)
Sulfur content	250 mg
Hold time	1. 15 minutes
	2. 30 minutes
	3. 45 minutes
	4. 60 minutes
	5. 75 minutes
Sulfurization temperature	580°C

Table 3.3.13 Physical condition of CZTS thin films after sulfurization process.

Hold time (minutes)	Film condition
15	Intact with no cracking/peeling off
30	Intact with no cracking/peeling off
45	Intact with no cracking/peeling off
60	Intact with no cracking/peeling off
75	Intact with no cracking/peeling off

Table 3.3.14 Sulfurization process total time for various hold times at 580°C used in this study (unit of time: minute).

Hold time @ 580°C	Dwell time (25°C<x<580°C)	Total time
15	55.5	70.5
30	55.5	85.5
45	55.5	100.5
60	55.5	115.5
75	55.5	130.5

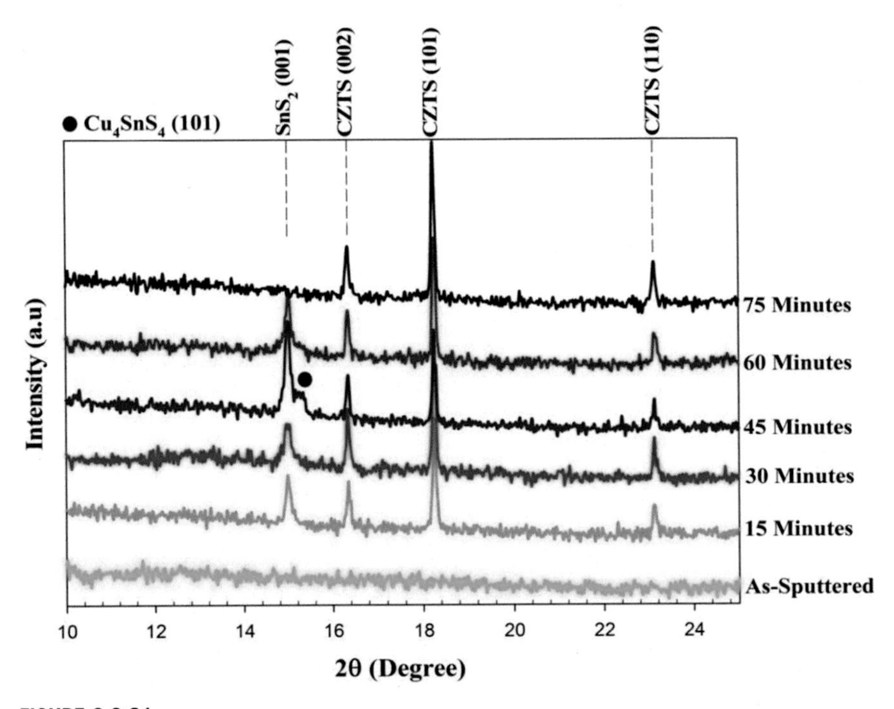

FIGURE 3.3.21

XRD patterns from 10 to 25 degrees of CZTS films sulfurized with different sulfurization times.

peak at the SnS_2 (001) compound for CZTS film sulfurized for 45 minutes can be identified as a Cu_4SnS_4 ternary compound with (101) preferred orientation. The existence of both SnS_2 and Cu_4SnS_4 compounds implies that two Cu_2SnS_3 decomposition mechanisms are simultaneously at play. By comparing the peak intensities of both compounds, it can be deduced that the decomposition mechanism that produces SnS_2 compound is more pronounced than the other mechanism that results in Cu_4SnS_4 compound. Fig. 3.3.22 shows the normalized intensity of CZTS (112)

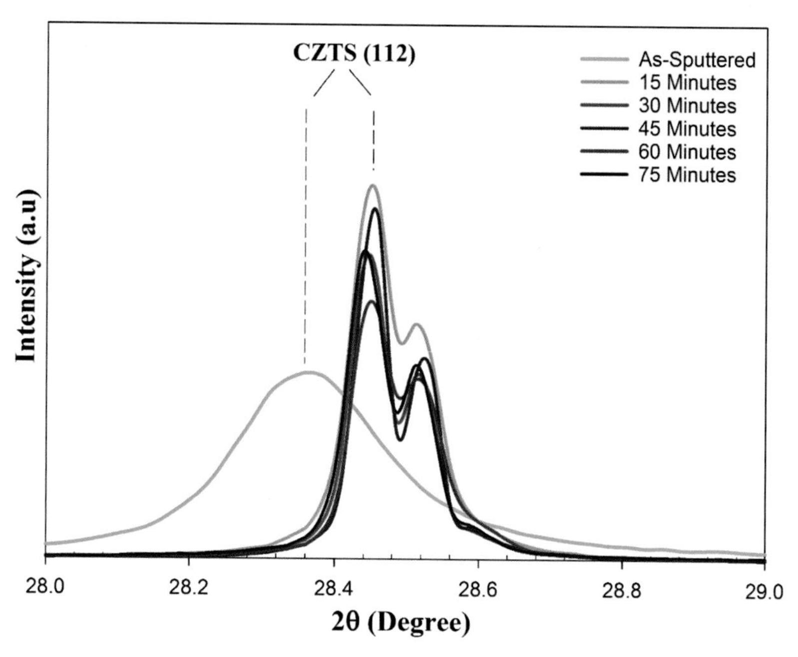

FIGURE 3.3.22

Normalized XRD patterns from 28 to 29 degrees of CZTS films sulfurized with different sulfurization times.

peak and Fig. 3.3.23 shows the surface morphology and cross section of CZTS films sulfurized with different hold times.

No particular trend could be observed in the CZTS (112) peak evolution regarding the corresponding hold time. On the other hand, CZTS grain size increases with an increase in hold time as evidenced from the cross-sectional SEM images. This enlargement is due to the grain coalescence process in which neighboring grains physically in contact with one and another merge together through grain boundary migration. Micrometer-sized grains form at an onset of sulfurization hold time of 30 minutes and becomes larger up to more than 2 μm when the hold time is 75 minutes. However, the interfacial MoS_2 layer (layer sandwiched between CZTS and Mo in the CZTS film sulfurized at 75 minutes) also becomes thicker (≈500 nm) as the time is increased due to prolonged reaction time between Mo and S at an elevated temperature of 580°C. Segregation of Zn-rich compound on the CZTS surface is minimal for CZTS film sulfurized for 45 minutes. Fig. 3.3.24A and B show the relevant elemental ratios of CZTS films probed by EDX on the surface and cross section meanwhile Figs. 3.3.25 and 3.3.26 show the Raman spectra and its corresponding analysis. The least deviations in terms of elemental ratios between surface and cross section are observed for CZTS film sulfurized for 45 minutes.

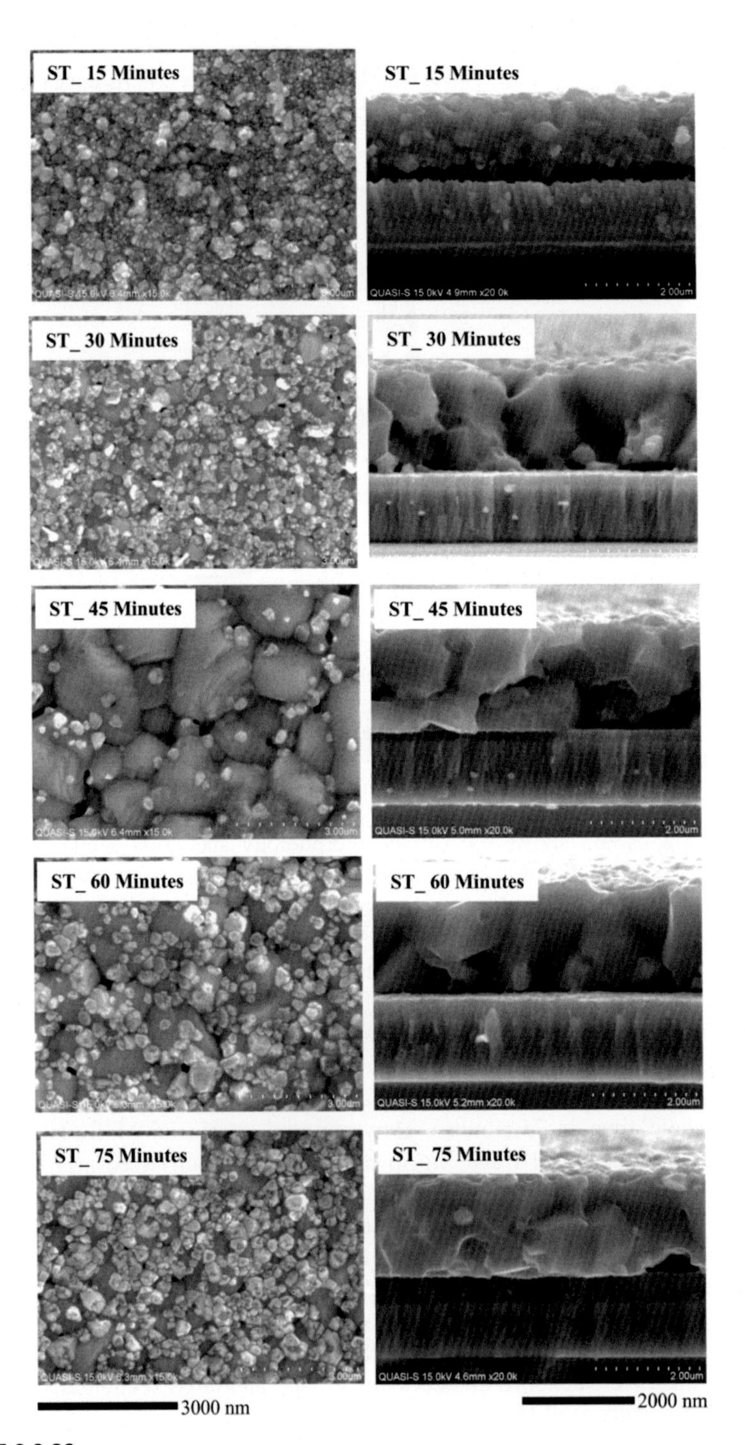

FIGURE 3.3.23

SEM images of CZTS films sulfurized with different hold times.

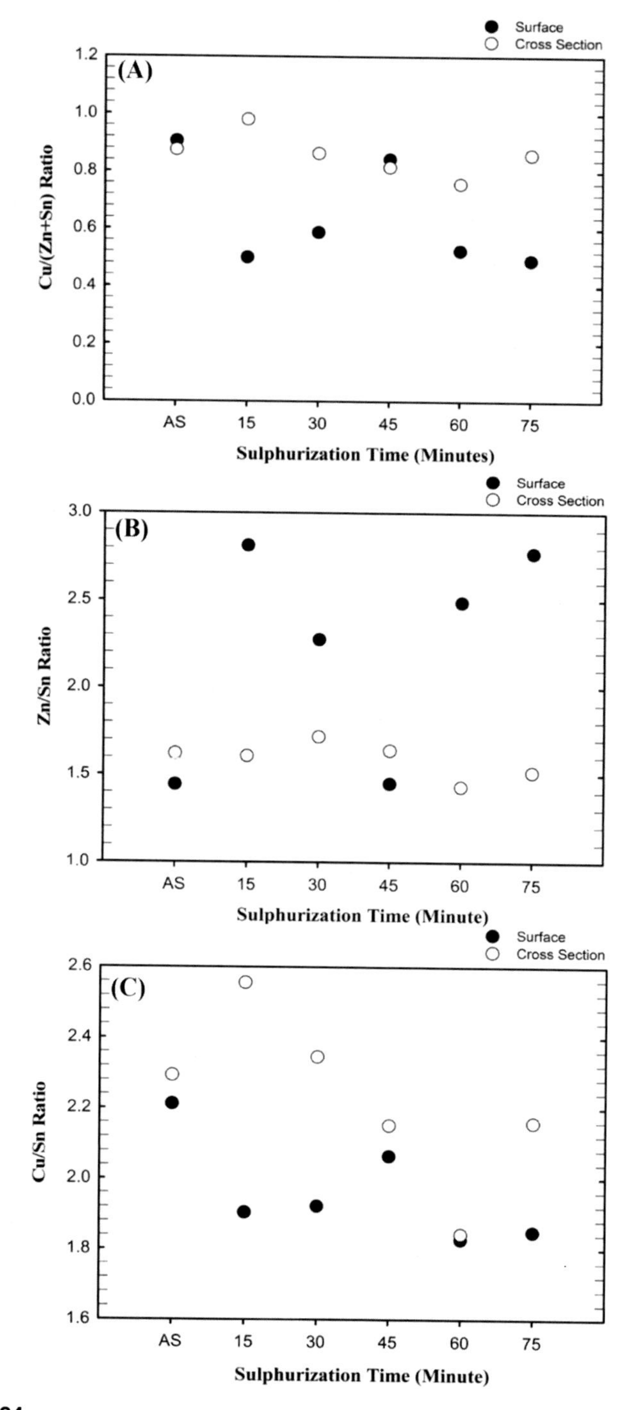

FIGURE 3.3.24

(A) Elemental ratios of CZTS films sulfurized with different hold times. (A) Cu/(Zn+Sn) ratio (B) Zn/Sn ratio and (C) Cu/Sn ratio.

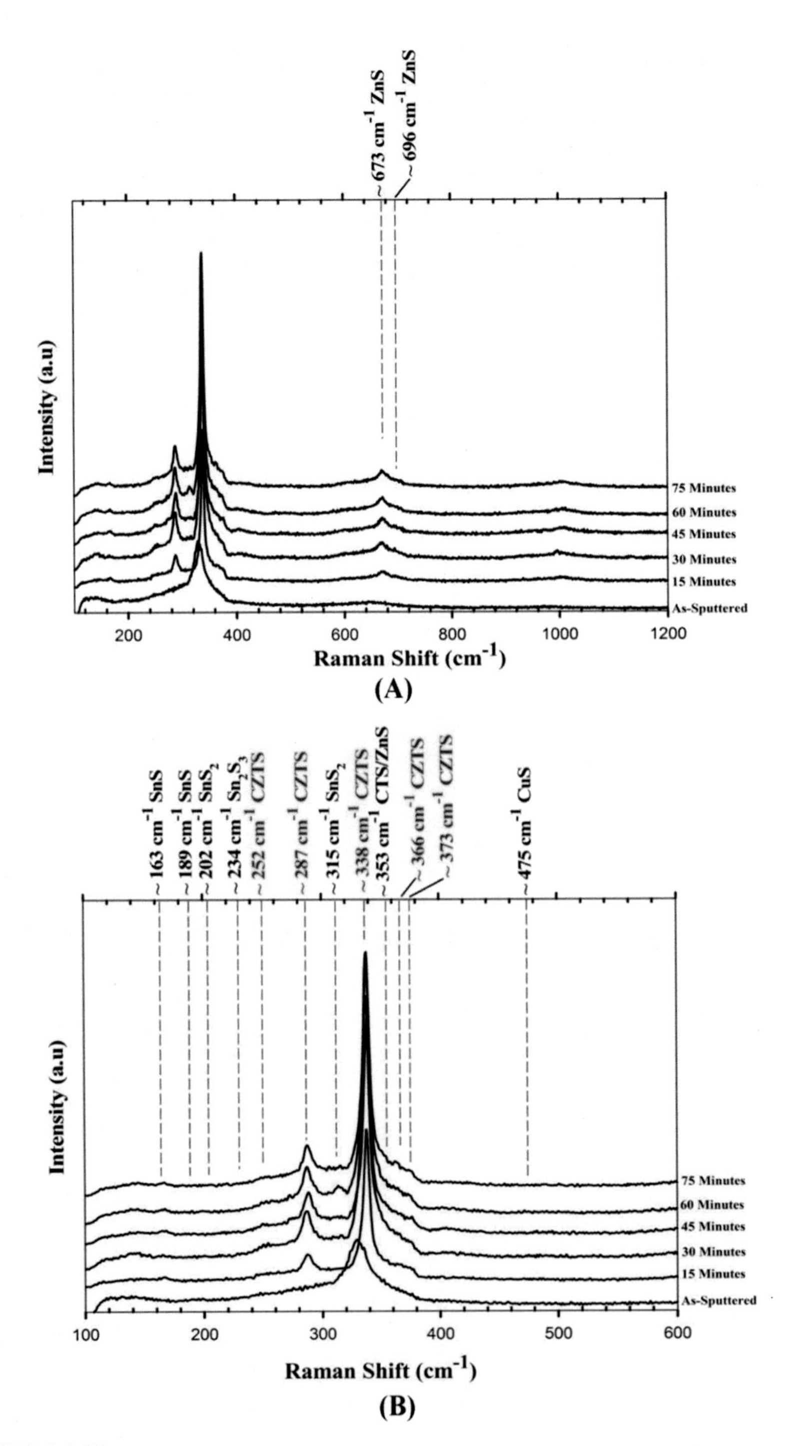

FIGURE 3.3.25

Raman spectra for CZTS films sulfurized with respect to hold time (A) 100 to 1200 cm^{-1} (B) 100 to 600 cm^{-1}.

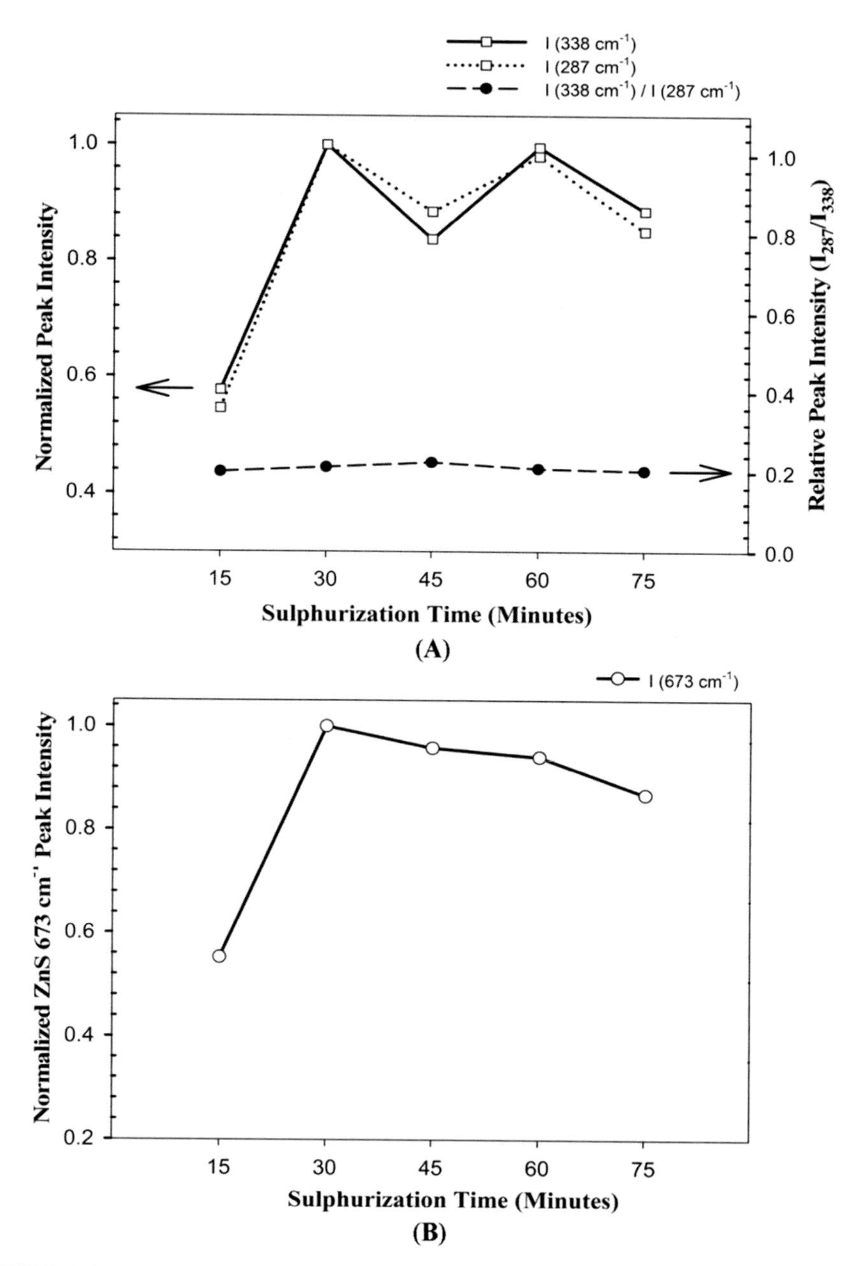

FIGURE 3.3.26

Raman peak analysis for CZTS films sulfurized with different hold times (A) Normalized CZTS peaks (B) Normalized ZnS peak.

Elemental ratios of Cu/(Zn+Sn) and Zn/Sn for CZTS film sulfurized 45 minutes are almost identical with ratios obtained for the as-sputtered film. Hence, uniform distribution of elements and absence of preferential segregation on the surface and the bulk layer can be interlinked as contributing factors to the minimal formation of Zn-rich compound on the surface as observed in the surface SEM image of ST_45 Minutes. This is further supported by the fact that CZTS films with large deviations, especially in Zn/Sn ratio (see Fig. 3.3.24B) between the surface and cross section possess higher Zn-rich compound segregation as evidenced in sulfurized films with holds time of 15, 30, 60, and 75 minutes. Fig. 3.3.26A reveals that the largest increase in peak intensity for CZTS Raman peaks occurs when the hold time increased from 15 to 30 minutes. This is concomitant with the increase in the size of CZTS grains from nanometer to micrometer range as the hold time is increased from 15 to 30 minutes (see Fig. 3.3.23; cross-section images of ST_15 Minutes and ST_30 Minutes). Inducement of larger grain size of CZTS compound implies that higher degree of structural crystallinity and phase homogeneity are present in the film.

3.3.3.6 Optical and optoelectronic properties

Fig. 3.3.27A and B and Fig. 3.3.28A and B show the transmission, reflection, absorption coefficient, and optical band gap of CZTS sulfurized with different hold times.

CZTS film sulfurized for 15 minutes exhibits highest average transmission and lowest reflection spectrum. On the other hand, CZTS films sulfurized for 45, 60, and 75 minutes exhibit the lowest average transmission, and the reflection spectrum slightly higher. The absorption coefficients presented are calculated from the obtained transmission and reflection measurements. Using the calculated absorption coefficient values, an estimate of optical band gap for CZTS film is obtained by linear extrapolation method in the $(\alpha h\nu)^2$ versus $h\nu$ graph in Fig. 3.3.28B. Highest absorption coefficient values for the photon energy ranging from 1.0 to 1.7 eV are obtained for CZTS films sulfurized for 45, 60, and 75 minutes. The absorption coefficient values of all three films are almost identical. This is due the similarities in the obtained transmission and reflection spectra. Hence, the optical properties of CZTS films do not change drastically with sulfurization hold time in the range from 45 to 75 minutes in this study.

Fig. 3.3.29 shows the PL spectrum of all the CZTS films sulfurized with different hold times. CZTS film sulfurized for 75 minutes possesses the most intense PL peak obtained in this entire study. However, CZTS film sulfurized for 45 minutes recorded the lowest PL peak intensity despite the minimal Zn-rich compound segregation on the CZTS surface. This indicates that homogeneous distribution of elements in CZTS film is not the sole variable in determining the optoelectronic properties. Stepwise increment or decrement in the normalized Raman peak plot in Fig. 3.3.26 as the hold time is varied from 15 to 60 minutes is mimicked in the corresponding PL peaks depicted in Fig. 3.3.29. However, sulfurization hold time

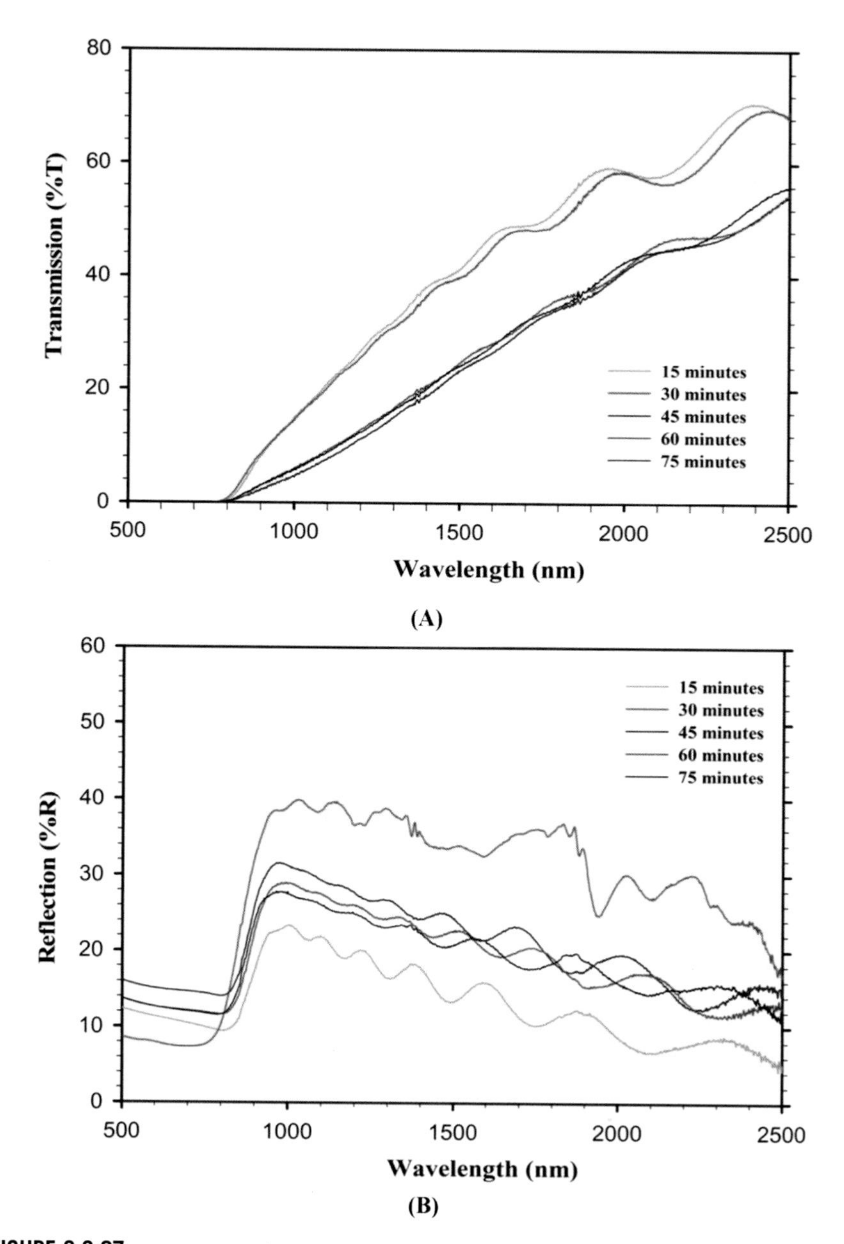

FIGURE 3.3.27

Optical properties for CZTS films sulfurized with different hold times (A) transmission (B) reflection.

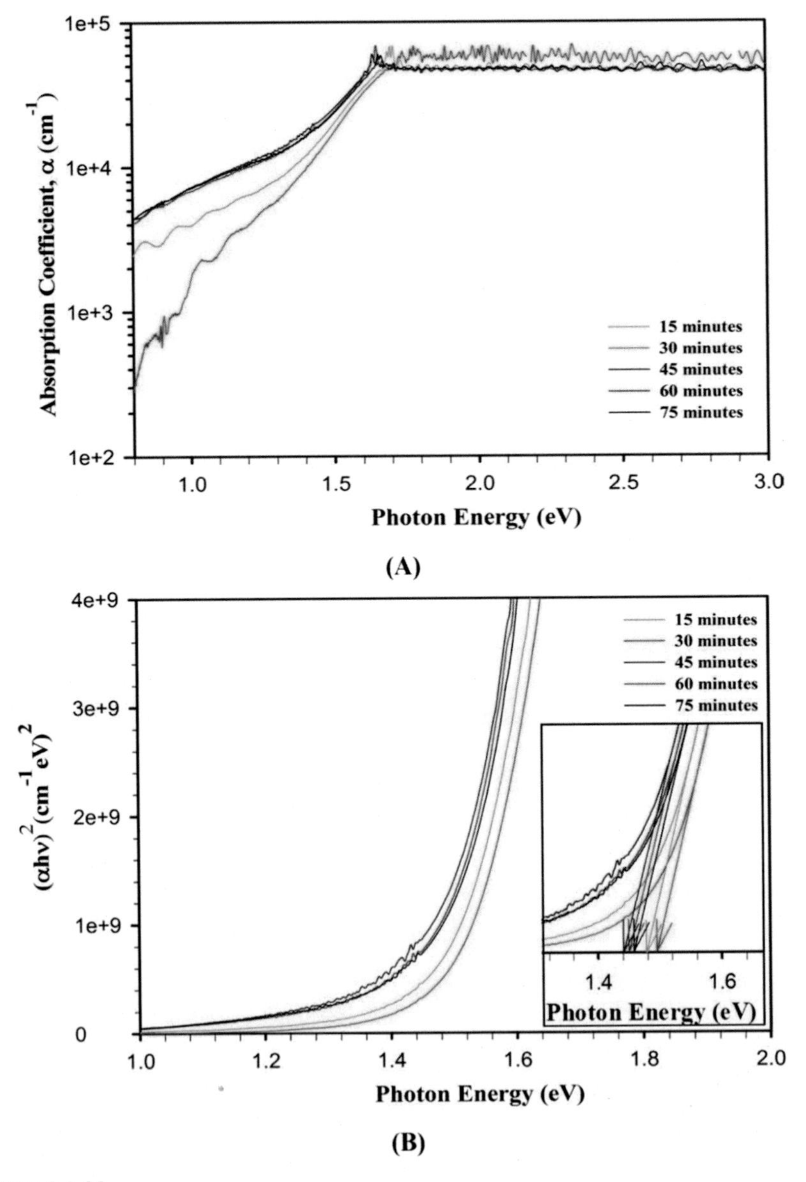

FIGURE 3.3.28

Optical properties for CZTS films sulfurized with different ramp rates (A) absorption coefficient (B) optical band gap.

of 75 minutes which results in a decrease in normalized Raman peak is not replicated in the PL spectra. In contrary, PL peak shows drastic increase in its intensity. The improvement in optoelectronic property of CZTS films sulfurized for

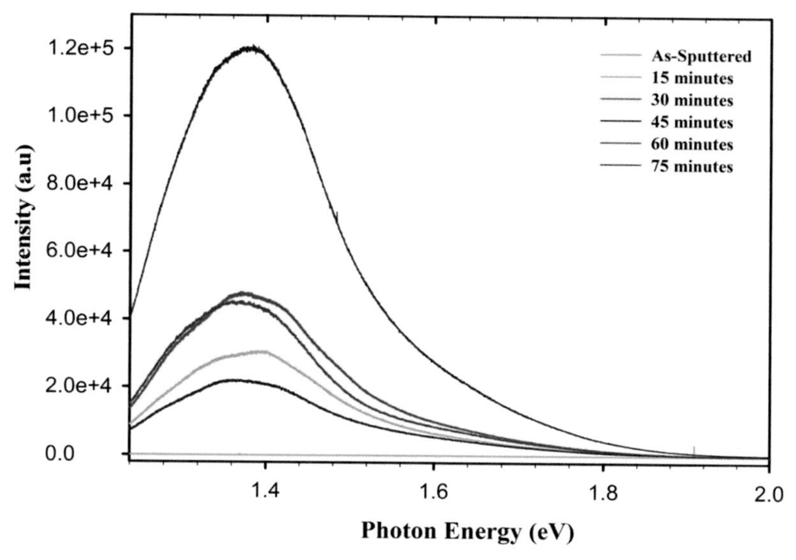

FIGURE 3.3.29

PL spectra for CZTS films sulfurized at different hold times.

Table 3.3.15 Comparison of optical band gap value obtained from UV-Vis-IR analysis with the emission peak(s) obtained from PL spectroscopy for CZTS films sulfurized at different hold times.

Hold time (minutes)	Optical band gap (UV-Vis-IR) (eV)	Emission peak (PL) (eV)	Energy difference, ΔE (meV)
15	1.48	1.4011	78.9
30	1.50	1.3706	129.4
45	1.44	1.3614	78.6
60	1.45	1.3609	89.1
75	1.46	1.3831	76.9

75 minutes is further exemplified in the Table 3.3.15. Table 3.3.15 shows the optical band gap and emission peak values for CZTS films with different hold times.

CZTS film sulfurized for 30 minutes records the highest optical band gap value of 1.50 eV and the highest energy difference between optical band gap and emission peak energy. On the other hand, CZTS film sulfurized for 75 minutes has the lowest energy difference. Coupled with it also possessing the highest PL intensity, it has the best optoelectronic properties among all films fabricated. One of the possible crucial factors which could have resulted in the observed trend is the improved crystallization through grain coalescence process induced by longer sulfurization hold times. This particular CZTS film possesses grains that are almost 3 μm in size and is one of the films with largest grains obtained from this study. In terms of surface morphological properties, CZTS films sulfurized for 45 minutes yields minimal segregation of Zn-rich compound but have the lowest PL peak. On the other hand, CZTS films sulfurized for 75 minutes exhibit the highest PL peaks but with severe surface segregation. Hence, sulfurization hold time 45 to 75 minutes can be considered as a refined window for device validation and further optimization.

3.3.3.6.1 Performance analysis of CZTS thin film solar cells

Post fabrication air annealing of the complete CZTS thin film devices was carried in order to improve the cell performance particularly the open circuit voltage, V_{oc}. Post fabrication air annealing process is originally applied for CIGS devices in order to increase the open circuit voltage through suppression of surface recombination mechanism at the CdS/CIGS interface (Rau et al., 1999). In this study, four temperature-time profiles have been chosen and executed. The open circuit voltage V_{oc}, is given by Eq. (3.3.3),

$$V_{oc} = (nkT/q) \times \ln([J_{ph}/J_0] + 1) \quad (3.3.3)$$

whereby n is the diode quality factor, k is the Boltzmann constant, T is the temperature, q is the value of an electrical point charge, J_{ph} is the photo-generated current density, and J_0 is the reverse saturation current density. Pronounced carrier recombination mechanism is manifested by a large J_0 value which eventually decreases the V_{oc}. Post fabrication air annealing in CIGS device passivates the defects originating from Se deficiency in interfacial region of CIGS/CdS boundary by interdiffusion of Na and O (Kronik et al., 2000). Table 3.3.16 shows the air annealing process parameters used in this study.

An as-fabricated CZTS device not annealed is used as a reference device to observe the effect of different temperature-time profiles on the overall cell

Table 3.3.16 Post fabrication air annealing process parameters for CZTS devices used in this study.

Parameters	Condition
Stack configuration	SLG/Mo/CZTS/CdS/i-ZnO/ITO/Al
Heating ramp-rate	10°C/sec (Hot plate)
Cooling rate	Natural cooling
Ambient	Atmosphere
Working pressure	1 ATM (760 Torr)
Temperature-timeprofile	1. 300°C—2 minutes
	2. 300°C—4 minutes
	3. 400°C—2 minutes
	4. 400°C—4 minutes

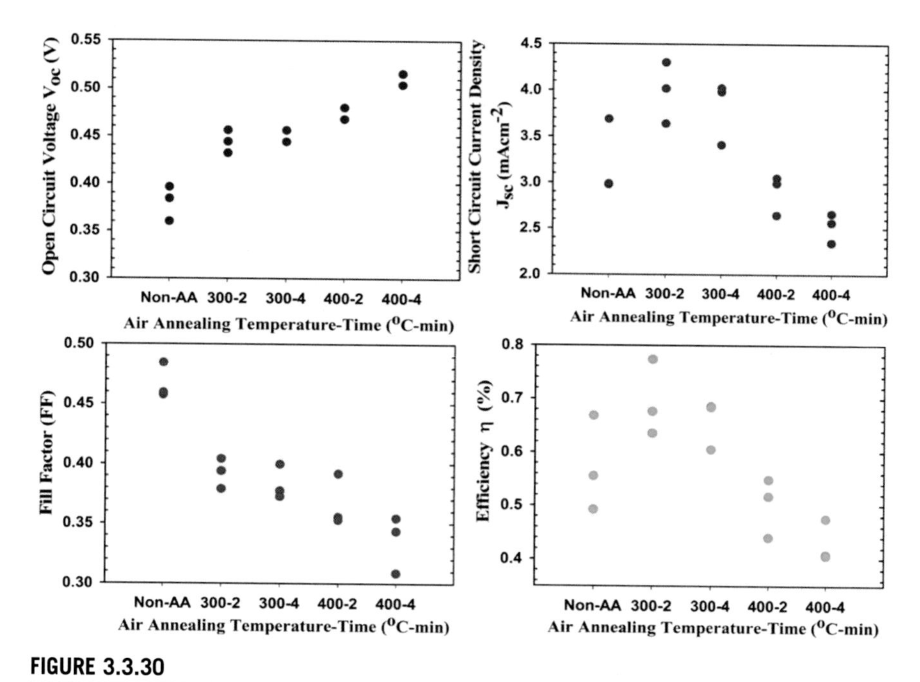

FIGURE 3.3.30

Performance analysis of CZTS devices with different post fabrication air annealing profiles.

performance. Fig. 3.3.30 shows the performance parameters of non-annealed and annealed CZTS devices.

A positive incremental trend is observed for the V_{oc} parameter as the temperature and the time of air annealing is increased from 300°C to 400°C and 2 to 4 minutes, respectively. However, the J_{sc} parameter initially increases for the 2 minute, 300°C profile but gradually decreases thereafter for the other three profiles. On the other hand, FF values of all the air annealed devices are lower than the non-air annealed devices. The highest average efficiency is obtained for device which was annealed for 2 minutes at 300°C which is a slight improvement compared to the non-annealed device. The improvement in V_{oc} could not be exploited to the fullest extent due to the deterioration primarily in the FF values and also the J_{sc}. This could be due to the window layer/TCO thin film stack. Hence, to explain the loss in FF and initial increment in J_{sc} optical and electrical optical properties of i-ZnO/ITO thin film stack were investigated. Fig. 3.3.31A and B show the transmission spectra, and the electrical properties of i-ZnO/ITO thin film stack air annealed at different temperatures.

Air annealing at 300°C increases the optical transmission of i-ZnO/ITO stack as evident from the transmission spectra shown in Fig. 3.3.31A. This improvement is reflected in the initial increase in J_{sc}. However, the carrier mobility of the

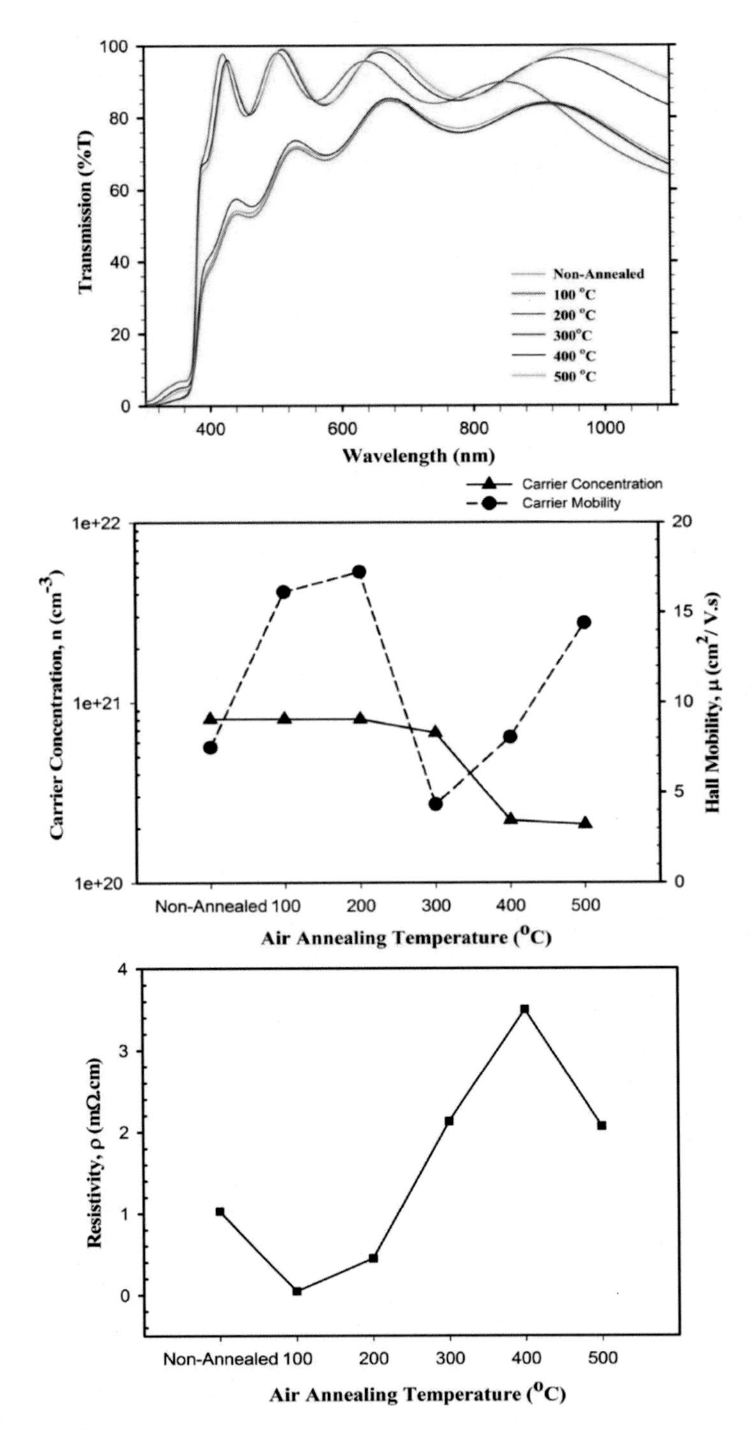

FIGURE 3.3.31

(A) Transmission spectra of i-ZnO/ITO thin film stack air annealed at different temperatures, (B) Carrier concentration and Hall mobility of i-ZnO/ITO thin film stack air annealed at different temperatures, and (C) Resistivity of i-ZnO/ITO thin film stack air annealed at different temperatures.

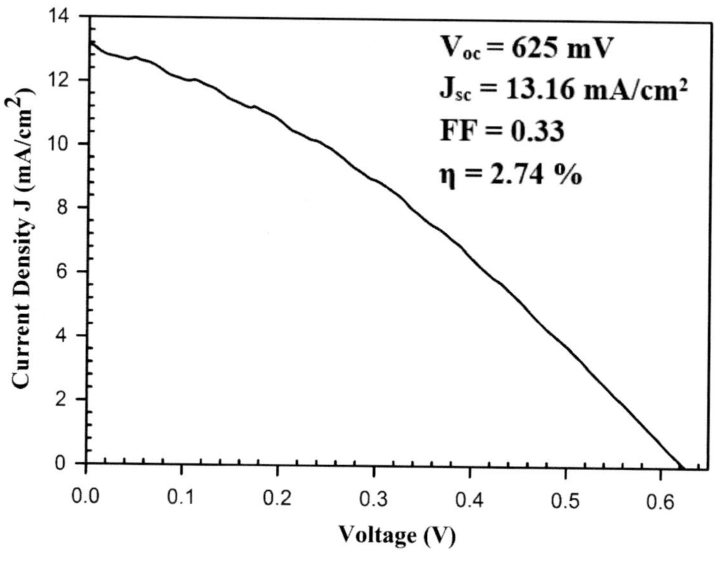

FIGURE 3.3.32

I-V curve of the best CZTS device fabricated in this study.

stack decreases, which in turn induces high resistivity in the film as shown in Fig. 3.3.31C. This increases the series increase of the device and results in decrease in FF value. Hence, post fabrication air annealing process at temperatures of 300°C (4 minutes) and 400°C induces detrimental electrical properties of i-ZnO/ITO stack which overshadows the beneficial optical enhancement. The optimum air annealing profile from this study is 300°C-2 minutes. Fig. 3.3.32 show the I-V curve of the best CZTS device fabricated in this study. The highest conversion efficiency obtained CZTS thin film solar cell from this study for is 2.74%. The active area of the best device is 0.205 cm^2.

3.3.4 Performance of state-of-the-art CZTS thin film solar cells

Until now, it has been important to improve the manufacturing process of CZTS thin film to use CZTS as an absorber layer of the thin film solar cell. Many articles have been published that define and discuss the techniques for CZTS thin film fabrication. There is an example of a PVD technique used for the deposition CZTS thin film which precursors sulfurization, co-evaporation further with thermal annealing, also a single step co-evaporation, pulsed laser deposition, and other more. Every technique has been demonstrated to have its particular characteristics. Keeping this in mind, the wide-scale operations of commercial solar cells are involved in a two-stage process of physical vapor-deposited sulfurization of precursor layers, which will have a tremendous benefit. The whole procedure involves fabrication and sulfurization of precursors.

Methods for the fabrication of CZTS layer stacks precursors and PV characteristics were documented as reported from H. Katagiri et al., which increased the power conversion efficiency from 0.66% up to 6.7% (Katagiri et al., 2008). As it grows, the CZTS precursors using RF magnetron co-sputtering method have shown a thickness of ~1.3 μm using Cu, SnS, and ZnS in a 4-inch target. The precursor parameter is 50 sccm for Ar gas flow rate, 20 rpm for substrate rotation, the substrate temperature in room temperature, and RF powers for Cu, ZnS, and SnS are at 80, 155, and 100 W, respectively. Then, the as-sputtered stack precursors encounter sulfurization to become a CZTS absorber layer. The parameters for sulfurizing are 580°C for operating temperature and 3 hours for time hold in the 20 vol% H_2S with N_2 environment. The CZTS absorber layer thickness after sulfurization is ~2.2 μm. The SEM images portray the cross-sectional view of the CZTS absorber layer after sulfurization, as shown in Fig. 3.3.33.

Nevertheless, the CZTS absorber layer does have a compact structure, yet it clearly can be observed that plenty of voids were formed after the sulfurization

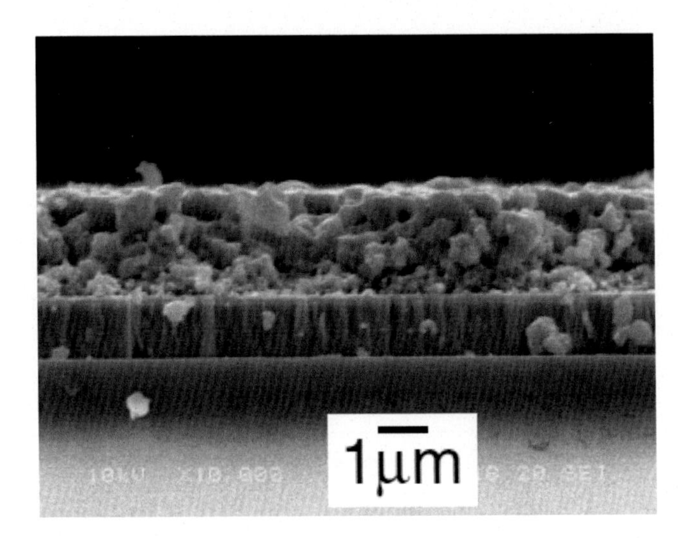

FIGURE 3.3.33

SEM image of the cross-sectional view of the CZTS absorber layer on the Mo electrode layer.

Reproduced with permission. Katagiri, H., Jimbo, K., Yamada, S., Kamimura, T., Maw, W. S., Fukano, T., ... Motohiro, T. (2008). Enhanced conversion efficiencies of Cu2ZnSnS4-based thin film solar cells by using preferential etching technique. Applied Physics Express, 1, *041201, 2008/04/04 Copy right 2021, IOP Science, Applied Physics Express.*

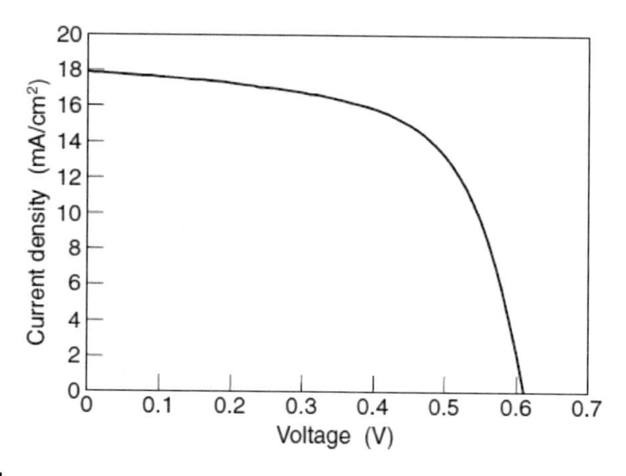

FIGURE 3.3.34

J–V characteristic of the best performance CZTS-based thin film solar cell under AM 1.5 and 100 mW/cm^2 illumination after light soaking for 5 min. $\eta = 6.77\%$, Voc= 610 mV, Jsc =17.9 mA/cm^2, FF = 0:62, Rs =4:25 Ω, and Rsh =370 Ω.

Reproduced with permission. Katagiri, H., Jimbo, K., Yamada, S., Kamimura, T., Maw, W. S., Fukano, T., ... Motohiro, T. (2008). Enhanced conversion efficiencies of Cu2ZnSnS4-based thin film solar cells by using preferential etching technique. Applied Physics Express, 1, *041201, 2008/04/04 Copy right 2021, IOP Science, Applied Physics Express.*

process. After absorber layer deposition, CdS buffer layers were grown using chemical bath deposition (CBD) with a thickness of 70 nm for 20 minute deposition time. Then a ZnO:Al window layer was deposited using sputtering while an Al FC was grown using vacuum evaporation methods. From each fabrication process, Fig. 3.3.34 shows the current density-voltage (J-V) characteristics for CZTS thin film after undergoing deionized water soaking.

As reported by K. J. Yang et al. a power conversion efficiency up to 7.5% was achieved with samples that underwent sulfurization at 570°C for 30 minutes (Yang et al., 2015). The CZTS absorber layer was deposited on Mo using the sputtering technique with a stack precursor of Cu/SnS/ZnS/Mo. The sputtering parameter for each precursor is a Cu target with 150 W using DC power, SnS, and ZnS target with both 200 W using RF power, and the total thickness as sputtered was 330 nm for the CZTS absorber layer. Then, a buffer layer was deposited using CdS through chemical bath deposition with a thickness of 50 nm and a window layer intrinsic ZnO and transparent conductive oxide Al-doped ZnO was deposited using sputtering with the thickness of 50 nm and 300 nm, respectively. The upper layer is the FC Al grid, which was deposited using thermal evaporation with a 500 nm thickness. Fig. 3.3.35 shows SEM images which include surface and cross-section for a sample that undergoes sulfurization with temperature hold for 570°C. From previous work (Katagiri et al., 2008), we can see a clear

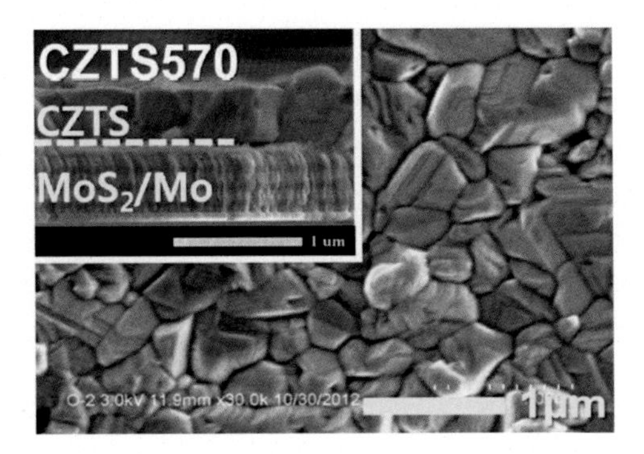

FIGURE 3.3.35

The surface and cross-sectional scanning electron microscopy (SEM) images after sulfurization.

Reproduced with permission. Yang, K.-J., Sim, J.-H., Son, D.-H., Kim, D.-H., Kim, G.Y., Jo, W. . . . Kang, J.-K. (2015). Effects of the compositional ratio distribution with sulfurization temperatures in the absorber layer on the defect and surface electrical characteristics of Cu2ZnSnS4 solar cells. Progress in Photovoltaics: Research and Applications, 23, *1771–1784, 2015/12/01 Copy right 2021, John Wiley & Sons, Inc, Progress in Photovoltaics: Research and Applications.*

difference as the grain size grows and becomes more compact, which is another discussion in this paper.

The Toyota R&D Labs fabricated the next CZTS absorber layer structure, which has increased the power conversion efficiency up to 9.4% (Tajima et al., 2017). Along with Mo being sputtered on soda-lime glass, a double layer of CZTS was grown with an electron beam and RF magnetron sputtering with ~1200 nm total thickness. The first layer of the CZTS absorber layer was sputtered using Cu, Sn, and ZnS targets with a total thickness of ~400 nm; the sample then underwent sulfurization at 580°C for 20 minutes under conditions of 20 vol% H_2S and 80 vol % N_2 in atmospheric pressure. As for the second CZTS absorber layer, the precursor used is the same as the first CZTS, which becomes a ZnS/Sn/Cu/CZTS(1)/Mo/SLG structure with a total thickness of ~800 nm, and then the samples again undergo sulfurization at 500°C for 60 minutes. For the buffer layer, CdS was deposited using chemical bath deposition (CBD) with a varied deposition time in the range of 3 to 12 minutes, and right after CBD, the samples were put through heat treatment at a range of 200°C until 360°C for 20 minutes, along with a window layer of Ga-doped ZnO deposit using a sputtering technique. The final layer is FC Al deposit using electron beam deposition. These studies emphasized the variation of CdS buffer layer thickness, post-annealing, and also doubled the CZTS absorber layer structure. Fig. 3.3.36 shows the J-V curve for sample No. 1 is the baseline and the highest efficiency sample No. 9, respectively.

The next development structure of the CZTS absorber layer showed an increase in power conversion efficiency (PCE) from 8.57% to 11% and originated

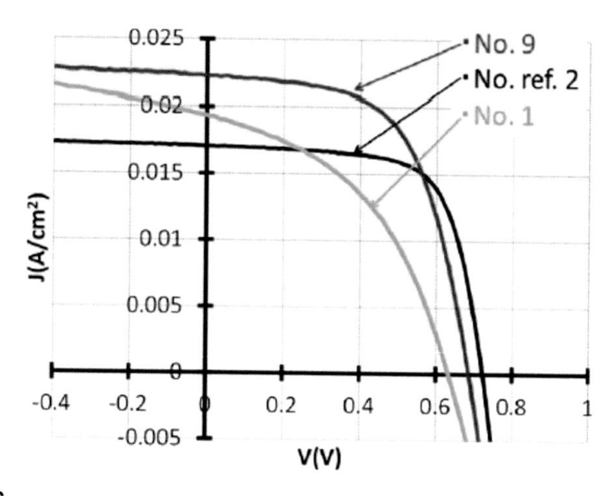

FIGURE 3.3.36

J–V curves of CZTS cells (Nos. ref. (Suryawanshi et al., 2013; Tajima et al., 2017), 1, and 9) under 1.5 AM (100 mW/cm^2) illumination at 25°C.

Reproduced with permission. Tajima, S., Umehara, M., Hasegawa, M., Mise, T., & Itoh, T. (2017). Cu2ZnSnS4 photovoltaic cell with improved efficiency fabricated by high-temperature annealing after CdS buffer-layer deposition. Progress in Photovoltaics: Research and Applications, 25, *14–22, 2017/01/01 Copy right 2021, John Wiley & Sons, Inc, Progress in Photovoltaics: Research and Applications.*

from a research group in University of New South Wales (UNSW). Starting with 8.57% PCE (Liu et al., 2017), they implemented an Al_2O_3 layer on Mo and ultra-thin CZTS absorber layer, which gave a total thickness of ~400 nm after the sulfurization process. The magnetron co-sputtering deposition technique was used to deposit the Cu/SnS/ZnS precursor, and the sulfurization process was performed at 560°C for 5 min where the sulfur and SnS are combined. The next layer is a buffer layer which was CdS grown using CBD with the total thickness of 50 nm. As for the window and TCO layer, they deposited intrinsic ZnO and indium tin oxide (ITO) using RF sputtering with a total thickness of 50 and 300 nm, respectively. FC Al was deposited using thermal evaporation and an anti-reflection layer with Mg_2F with a thickness of 110 nm. By having this structure of adding Al_2O_3 layer at the back contact with the ultrathin CZTS absorber layer it can eliminate the voids and increase the grain size of the CZTS absorber layer as shown in Fig. 3.3.37.

As for the next PCE from the same research group, they achieved 9.2% efficiency with ZnCdS precursor as a buffer layer as shown in Fig. 3.3.38 (Sun et al., 2016). For this work, the CZTS absorber layer was deposited with the co-sputtering technique and underwent sulfurization in a sulfur environment at 560°C. As for the buffer layer, they changed the conventional CdS buffer layer to the ZnCdS buffer layer by using the Successive ionic layer adsorption and reaction (SILAR)

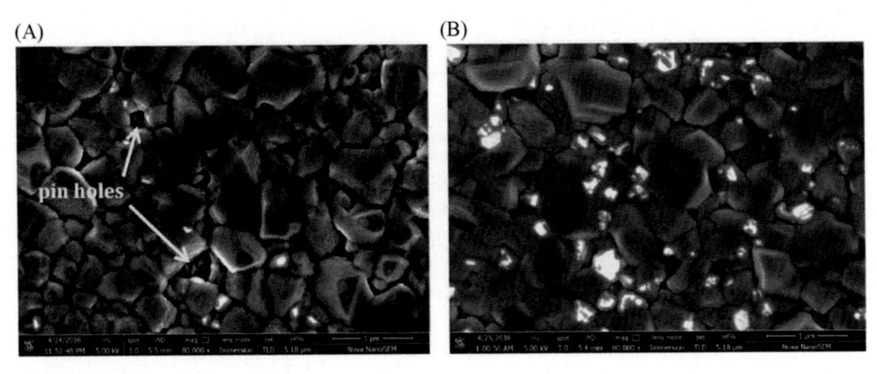

FIGURE 3.3.37

SEM morphologies of ultrathin CZTS absorbers on Mo without (a) and with (b) Al_2O_3 intermediate layer.

Reproduced with permission. Liu, F., Huang, J., Sun, K., Yan, C., Shen, Y., Park, J. ... Hao, X. (2017). Beyond 8% ultrathin kesterite Cu2ZnSnS4 solar cells by interface reaction route controlling and self-organized nanopattern at the back contact. NPG Asia Materials, 9, *e401–e401, 2017/07/01 Copy right 2021, Nature Publishing Group, NPG Asia Materials.*

technique. The total thickness layer grown is 70 nm with 30 times SILAR cycles. After that, i-ZnO and ITO were deposited using the RF magnetron sputtering method with a thickness of 60 and 200 nm, respectively, and the FC was Ag paste. In these studies, they attempted to address the problem of CZTS thin film solar cell issues, which involves the lower value of V_{oc} coming from the unwanted conduction band offset (CBO) at the CZTS/CdS heterojunction.

The next PCE they achieved for CZTS absorber layer solar cells is 11% for small cell area and 10% for the standard cell size, whereby they applied the heat treatment after the deposition of the CdS buffer layer (Yan et al., 2018). For this study, the CZTS absorber layer was deposited by the same parameter from the previous work. In the sulfurization process, they put the samples in sulfur and SnS environment for 3 minutes at 560°C. For the buffer layer, they grew CdS by CBD, and then the samples underwent heat treatment at 270°C, 300°C, and 330°C for 10 minutes. By doing the heat treatment after the heterojunction had been made, it shows an improvement at the elemental interdiffusion of the p-n junction. Fig. 3.3.39 shows the J-V curve for heat treatments at 300°C samples which give the highest PCE compared to other heat treatment temperatures.

The next PCE from this research group is 11.5% by alloying the cadmium (Cd) with the influence of the ratio of Cd/(Zn+Cd) which controlling the CdS thickness and ZnS target power for absorber layer precursor to becoming CZCTS absorber layer (Yan et al., 2017). The CZTS absorber layer was deposited by

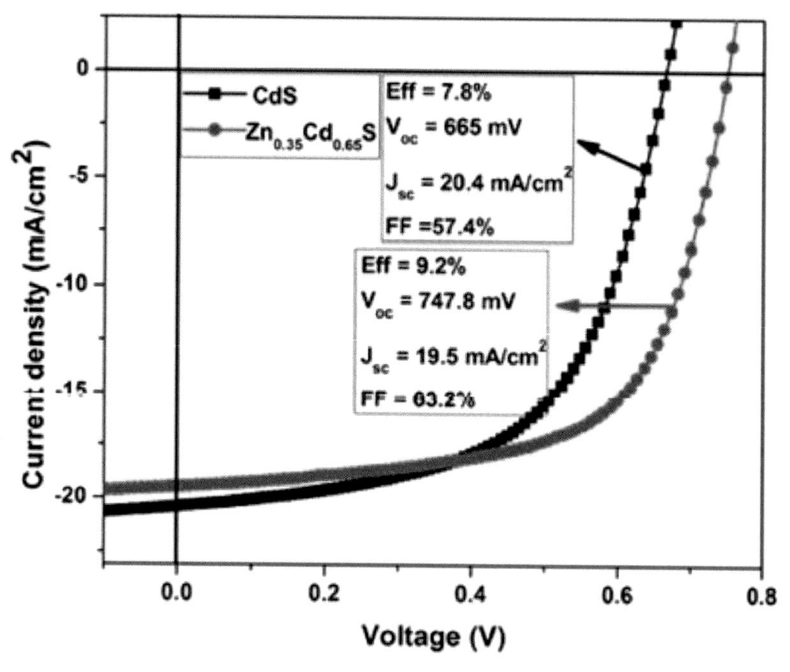

FIGURE 3.3.38

J–V characteristics of CZTS solar cells with CdS and ZnCdS buffer layer.

Reproduced with permission. Sun, K., Yan, C., Liu, F., Huang, J., Zhou, F., Stride, J. ... Hao, X. (2016). Over 9% Efficient kesterite Cu2ZnSnS4 solar cell fabricated by using Zn1-xCdxS buffer layer. Advanced Energy Materials, 6 *John Wiley & Sons, Inc, Advance Energy Materials.*

using the magnetron sputtering technique with a co-sputtering of a precursor of Cu, ZnS, and SnS on Mo back contact. The sulfurization process was done in an environment of sulfur and SnS at 560°C for few minutes. For baseline CZTS precursor, each of the gun powers used were 17 W for Cu, 70 W for ZnS, and 38 W for SnS while for CZCTS, each of the gun powers used were 17 W for Cu, 50 W for ZnS, and 38 W for SnS, with a 3″ target. For the buffer layer, they used two rounds of CBD to produce the CdS buffer layer which has ~120 nm of thickness. For the four layers above, it is the same method as previous work. Table 3.3.17 shows the device characteristics for CZTS and CZCTS solar cells. From this work, they found the addition of Cd did a good impact in terms of the structure crystallinity and the device performance of the solar cells. Until this date, developing kesterite CZTS solar cells with highest efficiency is still held by the UNSW research group.

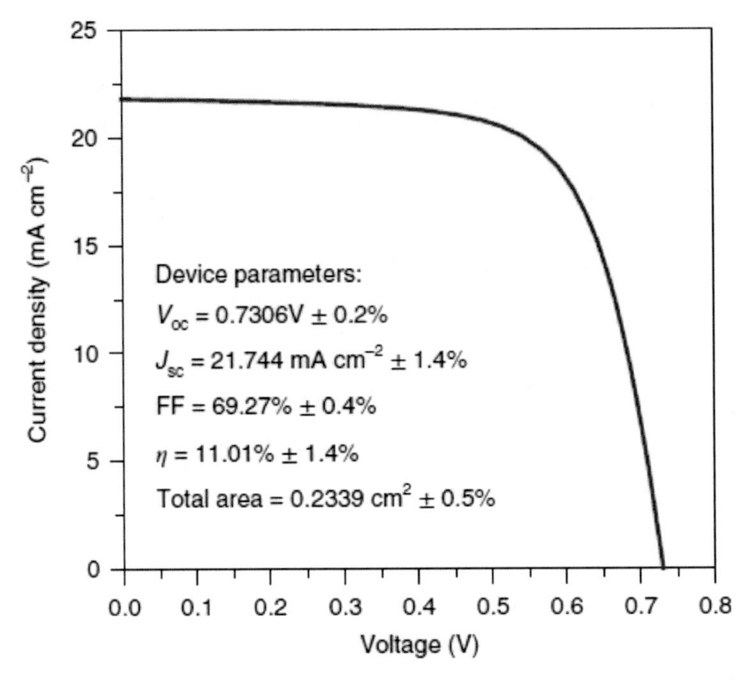

FIGURE 3.3.39

Certified J-V curve.

Reproduced with permission. Yan, C., Huang, J., Sun, K., Johnston, S., Zhang, Y., Sun, H. ... Hao, X. (2018). Cu2ZnSnS4 solar cells with over 10% power conversion efficiency enabled by heterojunction heat treatment. Nature Energy, 3, *764–772, 2018/09/01 Copy right 2021, Nature Publishing Group, Nature Energy.*

Table 3.3.17 Device characteristics of the CZTS and CZCTS solar cells (Yan et al., 2017).

Absorber	V_{oc} (mV)	J_{SC} (mA/cm^2)	FF (%)	Eff (%)	E_g/q.V_{OC} (mV)	$R_{S,L}$ (Ωcm^2)	$G_{s,L}$ (mS cm^{-2})	A	J_0 (A/cm^2)
CZTS	683	20.7	62.5	8.8	847	0.96	0.67	3.3	8.1×10^{-6}
CZCTS	650	26.7	66.1	11.5	730	0.45	1.0	2.5	1.2×10^{-6}

Source*: Reproduced with permission Yan, C., K. Sun, J. Huang, S. Johnston, F. Liu, B.P. Veettil, ... and X. Hao, (2017) "Beyond 11% efficient sulfide kesterite Cu2ZnxCd1−xSnS4 solar cell: Effects of cadmium alloying," ACS Energy Letters, 2, pp. 930-936, 2017/04/14. Copyright 2021, American Chemical Society, ACS Energy Letters.*

3.3.5 Conclusion

The success of second generation solar cells, particularly the CIGS thin film solar cell is exemplified by the emergence of numerous PV companies around globe. This has added another dimension in the silicon-dominated PV market in terms of lower production cost stemming from low material usage as well low energy intensive manufacturing process. However, to achieve terawatt PV, indium-containing CIGS quaternary material is poised to fall short due to the scarcity of In, but on the other hand, CZTS is a relatively new absorber material yet to reach the full commercialization stage. It has already proved itself as a promising material as it contains non-toxic, cheap, earth abundant materials and has shown excellent photo-absorption properties. Currently, it is still in the R&D phase in lab and sub-module scale. More systematic and in-depth studies are urgently required to boost the power conversion efficiency beyond 15%, which will be a timely catalyst for pre-commercialization efforts.

References

Azmi, N., Chelvanathan, P., Yusoff, Y., Ferdaous, M. T., Zuhdi, A. W. M., Tiong, S. K., & Amin, N. (2021). Enhancing microstructural and optoelectronic properties of CZTS thin films by post deposition ionic treatment. *Materials Letters*, *285*, 129117, 2021/02/15/.

Bansal, N., Chandra Mohanty, B., & Singh, K. (2020). Designing composition tuned glasses with enhanced properties for use as substrate in Cu2ZnSnS4 based thin film solar cells. *Journal of Alloys and Compounds*, *819*, 152984, 2020/04/05/.

Biccari, F., Chierchia, R., Valentini, M., Mangiapane, P., Salza, E., Malerba, C., … Mittiga, A. (2011). Fabrication of Cu2ZnSnS4 solar cells by sulfurization of evaporated precursors. *Energy Procedia*, *10*, 187–191, 2011/01/01/.

Çetinkaya, S. (2019). Study of electrical effect of transition-metal dichalcogenide-MoS2 layer on the performance characteristic of Cu2ZnSnS4 based solar cells using wxAMPS. *Optik*, *181*, 627–638, 2019/03/01/.

Chelvanathan, P., Hossain, M. I., Husna, J., Alghoul, M., Sopian, K., & Amin, N. (2012). Effects of transition metal dichalcogenide molybdenum disulfide layer formation in copper–zinc–tin–sulfur solar cells from numerical analysis. *Japanese Journal of Applied Physics*, *51*, 10NC32, 2012/10/22.

Chelvanathan, P., Shahahmadi, S. A., Arith, F., Sobayel, K., Aktharuzzaman, M., Sopian, K., … Amin, N. (2017). Effects of RF magnetron sputtering deposition process parameters on the properties of molybdenum thin films. *Thin Solid Films*, *638*, 213–219, 2017/09/30/.

Chelvanathan, P., Shahahmadi, S. A., Ferdaous, M. T., Sapeli, M. M. I., Sopian, K., & Amin, N. (2018). Controllable formation of MoS2 via preferred crystallographic orientation modulation of DC-sputtered Mo thin film. *Materials Letters*, *219*, 174–177, 2018/05/15/.

Chelvanathan, P., Zakaria, Z., Yusoff, Y., Akhtaruzzaman, M., Alam, M. M., Alghoul, M. A., ... Amin, N. (2015). Annealing effect in structural and electrical properties of sputtered Mo thin film. *Applied Surface Science*, *334*, 129–137, 2015/04/15/.

Chen, S., Yang, J.-H., Gong, X. G., Walsh, A., & Wei, S.-H. (2010). Intrinsic point defects and complexes in the quaternary kesterite semiconductor Cu_2ZnSnS_4. *Physical Review B*, *81*, 245204.

Dimitrievska, M., Fairbrother, A., Pérez-Rodríguez, A., Saucedo, E., & Izquierdo-Roca, V. (2014). Raman scattering crystalline assessment of polycrystalline Cu2ZnSnS4 thin films for sustainable photovoltaic technologies: Phonon confinement model. *Acta Materialia*, *70*, 272–280, 2014/05/15/.

Ferdaous, M. T., Shahahmadi, S. A., Chelvanathan, P., Akhtaruzzaman, M., Alharbi, F. H., Sopian, K., ... Amin, N. (2019). Elucidating the role of interfacial MoS2 layer in Cu2ZnSnS4 thin film solar cells by numerical analysis. *Solar Energy*, *178*, 162–172, 2019/01/15/.

Gershon, T., Shin, B., Bojarczuk, N., Hopstaken, M., Mitzi, D. B., & Guha, S. (2015). The role of sodium as a surfactant and suppressor of non-radiative recombination at internal surfaces in Cu2ZnSnS4. *Advanced Energy Materials*, *5*, 1400849.

Grossberg, M., Raadik, T., Raudoja, J., & Krustok, J. (2014). Photoluminescence study of defect clusters in Cu2ZnSnS4 polycrystals. *Current Applied Physics*, *14*, 447–450, 2014/03/01/.

He, J., Sun, L., Chen, Y., Jiang, J., Yang, P., & Chu, J. (2014). Cu2ZnSnS4 thin film solar cell utilizing rapid thermal process of precursors sputtered from a quaternary target: A promising application in industrial processes. *RSC Advances*, *4*, 43080–43086.

Ito, K., & Nakazawa, T. (1988). Electrical and optical properties of stannite-type quaternary semiconductor thin films. *Japanese Journal of Applied Physics*, *27*, 2094–2097, 1988/11/20.

Karade, V., Lokhande, A., Babar, P., Gang, M. G., Suryawanshi, M., Patil, P., & Kim, J. H. (2019). Insights into kesterite's back contact interface: A status review. *Solar Energy Materials and Solar Cells*, *200*, 109911, 2019/09/15/.

Katagiri, H., Jimbo, K., Yamada, S., Kamimura, T., Maw, W. S., Fukano, T., ... Motohiro, T. (2008). Enhanced conversion efficiencies of Cu2ZnSnS4-based thin film solar cells by using preferential etching technique. *Applied Physics Express*, *1*, 041201, 2008/04/04.

Katagiri, H., Jimbo, K., Maw, W. S., Oishi, K., Yamazaki, M., Araki, H., & Takeuchi, A. (2009). Development of CZTS-based thin film solar cells. *Thin Solid Films*, *517*, 2455–2460, 2009/02/02/.

Kelly, P. J., & Arnell, R. D. (2000). Magnetron sputtering: A review of recent developments and applications. *Vacuum*, *56*, 159–172, 2000/03/01/.

Kronik, L., Rau, U., Guillemoles, J.-F., Braunger, D., Schock, H.-W., & Cahen, D. (2000). Interface redox engineering of Cu(In,Ga)Se2—based solar cells: Oxygen, sodium, and chemical bath effects. *Thin Solid Films*, *361-362*, 353–359, 2000/02/21/.

Liu, F., Huang, J., Sun, K., Yan, C., Shen, Y., Park, J., ... Hao, X. (2017). Beyond 8% ultrathin kesterite Cu2ZnSnS4 solar cells by interface reaction route controlling and self-organized nanopattern at the back contact. *NPG Asia Materials*, *9*, pp. e401-e401, 2017/07/01.

Rau, U., Braunger, D., Herberholz, R., Schock, H. W., Guillemoles, J. F., Kronik, L., & Cahen, D. (1999). Oxygenation and air-annealing effects on the electronic properties of Cu(In,Ga)Se2 films and devices. *Journal of Applied Physics*, *86*, 497–505, 1999/07/01.

Ravindiran, M., & Praveenkumar, C. (2018). Status review and the future prospects of CZTS based solar cell—A novel approach on the device structure and material modeling for CZTS based photovoltaic device. *Renewable and Sustainable Energy Reviews, 94*, 317–329, 2018/10/01/.

Saha, S. (2020). A status review on Cu2ZnSn(S, Se)4-based thin-film solar cells. *International Journal of Photoenergy, 2020*, 3036413, 2020/09/01.

Schorr, S., Gurieva, G., Guc, M., Dimitrievska, M., Pérez-Rodríguez, A., Izquierdo-Roca, V., ... Merino, J. M. (2019). Point defects, compositional fluctuations, and secondary phases in non-stoichiometric kesterites. *Journal of Physics: Energy, 2*, 012002, 2019/12/10.

Scragg, J. J., Kubart, T., Wätjen, J. T., Ericson, T., Linnarsson, M. K., & Platzer-Björkman, C. (2013). Effects of back contact instability on Cu2ZnSnS4 devices and processes. *Chemistry of Materials, 25*, 3162–3171, 2013/08/13.

Shin, B., Gunawan, O., Zhu, Y., Bojarczuk, N. A., Chey, S. J., & Guha, S. (2013). Thin film solar cell with 8.4% power conversion efficiency using an earth-abundant Cu2ZnSnS4 absorber. *Progress in Photovoltaics: Research and Applications, 21*, 72–76, 2013/01/01.

Sun, K., Yan, C., Liu, F., Huang, J., Zhou, F., Stride, J., ... Hao, X. (2016). Over 9% efficient kesterite Cu2ZnSnS4 solar cell fabricated by using Zn1-xCdxS buffer layer. *Advanced Energy Materials, 6*.

Suryawanshi, M. P., Agawane, G. L., Bhosale, S. M., Shin, S. W., Patil, P. S., Kim, J. H., & Moholkar, A. V. (2013). CZTS based thin film solar cells: A status review. *Materials Technology, 28*, 98–109, 2013/03/01.

Tajima, S., Umehara, M., Hasegawa, M., Mise, T., & Itoh, T. (2017). Cu2ZnSnS4 photovoltaic cell with improved efficiency fabricated by high-temperature annealing after CdS buffer-layer deposition. *Progress in Photovoltaics: Research and Applications, 25*, 14–22, 2017/01/01.

Wada, T., Kohara, N., Negami, T., & Nishitani, M. (1996). Chemical and structural characterization of $\bf Cu(In,Ga)Se_{2}/Mo$ interface in $\bf Cu(In,Ga)Se_{2}$ solar cells. *Japanese Journal of Applied Physics, 35*, L1253–L1256, 1996/10/01.

Wang, A., Chang, N. L., Sun, K., Xue, C., Egan, R. J., Li, J., ... Hao, X. (2021). Analysis of manufacturing cost and market niches for Cu2ZnSnS4 (CZTS) solar cells. *Sustainable Energy & Fuels, 5*, 1044–1058.

Wang, J., Li, S., Cai, J., Shen, B., Ren, Y., & Qin, G. (2013). Cu2ZnSnS4 thin films: Facile and cost-effective preparation by RF-magnetron sputtering and texture control. *Journal of Alloys and Compounds, 552*, 418–422, 2013/03/05/.

Xie, M., Zhuang, D., Zhao, M., Zhuang, Z., Ouyang, L., Li, X., & Song, J. (2013). Preparation and characterization of Cu2ZnSnS4 thin films and solar cells fabricated from quaternary Cu-Zn-Sn-S target. *International Journal of Photoenergy, 2013*, 929454, 2013/12/25.

Xiong, Q., Wang, J., Reese, O., Lew Yan Voon, L. C., & Eklund, P. C. (2004). Raman scattering from surface phonons in rectangular cross-sectional w-ZnS nanowires. *Nano Letters, 4*, 1991–1996, 2004/10/01.

Yan, C., Huang, J., Sun, K., Johnston, S., Zhang, Y., Sun, H., ... Hao, X. (2018). Cu2ZnSnS4 solar cells with over 10% power conversion efficiency enabled by heterojunction heat treatment. *Nature Energy, 3*, 764–772, 2018/09/01.

Yan, C., Sun, K., Huang, J., Johnston, S., Liu, F., Veettil, B. P., ... Hao, X. (2017). Beyond 11% efficient sulfide kesterite Cu2ZnxCd1−xSnS4 solar cell: Effects of cadmium alloying. *ACS Energy Letters*, *2*, 930–936, 2017/04/14.

Yang, K.-J., Sim, J.-H., Son, D.-H., Kim, D.-H., Kim, G. Y., Jo, W., ... Kang, J.-K. (2015). Effects of the compositional ratio distribution with sulfurization temperatures in the absorber layer on the defect and surface electrical characteristics of Cu2ZnSnS4 solar cells. *Progress in Photovoltaics: Research and Applications*, *23*, 1771–1784, 2015/12/01.

Yu, K., & Carter, E. A. (2016). Determining and controlling the stoichiometry of Cu2ZnSnS4 photovoltaics: The physics and its implications. *Chemistry of Materials*, *28*, 4415–4420, 2016/06/28.

SUBCHAPTER

Novel chalcogenides and their fabrication techniques 3.4

Md. Khan Sobayel Bin Rafiq and Md. Akhtaruzzaman
Solar Energy Research Institute, Universiti Kebangsaan Malaysia, Bangi, Malaysia

3.4.1 Introduction

If a material's free charges are immobile in one spatial dimension but mobile in the other two, it is defined as an atomically thin two-dimensional (2-D) material. This property, in comparison to bulk materials or thin films, helps 2-D materials to perform better once applied. Molecular beam epitaxial (MBE) growth III−V semiconductors sparked interest in 2-D films in the late 1970s and early 1980s (Choi, Lahiri, Seelaboyina, & Kang, 2010; Geim & Novoselov, 2007). Transition metal dichalcogenides (TMDCs) are a group of hexagonal-structured, van der Waals-bonded, layered materials with the molecular formula MX_2 (where M = transition metals such as Mo, W, and Nb, and X = chalcogens such as S, Se, and Te) and the space group P63/mmc. Since TMDC materials have strong in-plane bonding and weak out-of-plane bonding, cxfoliation of these materials into a single crystal, two-dimensional flake of atomic level thickness is possible. There are over 150 exotic layered materials that can be broken into a single atom thick film, including molybdenum disulfide (MoS_2), (Radisavljevic, Radenovic, Brivio, Giacometti, & Kis, 2011) molybdenum di-selenide ($MoSe_2$), (Mahatha, Patel, & Menon, 2012) silicene, (Aufray et al., 2010; Vogt et al., 2012) boron nitride (BN), (Pakdel, Zhi, Bando, & Golberg, 2012) tungsten disulfide (WS_2), (Hwang et al., 2012) germanene (Ni et al., 2012) and tungsten diselenide (WSe_2). When we look closely at these materials, we notice that the effects of their remarkable properties are more dimension-dependent than size-dependent, necessitating the development of a new research field. Based on their chemical structure and structural configurations, atomically thin 2-D materials may be metallic, semimetallic, or semiconducting. In addition, these materials, such as graphene, have optical, chemical, and mechanical properties that make them suitable for applications in electronics, photonics, composites, and energy storage. Electronic and optoelectronic properties of some TMDC materials and heterostructures have been particularly impressive. For example, monolayer MoS_2 has a direct intrinsic bandgap of 1.8 eV, good transistor mobility (200 cm^2/V per second), and quantum confinement. The inherent semiconducting property of the 2-D MoS_2 structure is advantageous in this situation, since it avoids the need for doping or supernarrow ribbons, which would be needed to achieve a comparable energy bandgap in graphene. As a result, 2-D MoS_2- or WS_2-based field effect transistors have high current on/off ratios of 1:108 and low power

Comprehensive Guide on Organic and Inorganic Solar Cells. DOI: https://doi.org/10.1016/B978-0-323-85529-7.00012-8

dissipation even at room temperature, making them suitable for future low-power electronics (Bao, Cai, Kim, Sridhara, & Fuhrer, 2013; Kim et al., 2012).

3.4.2 History of transition metal dichalcogenides

TMDCs have a long and successful proven history. Linus Pauling was the first to discover their structure in 1923 (Dickinson & Pauling, 1923). Around 60 TMDCs were identified by the late 1960s, with at least 40 of them having a layered structure (Wilson & Yoffe, 1969). Robert Frindt published the first papers on the use of adhesive tapes for processing ultrathin MoS_2 layers in 1963 (Frindt & Yoffe, 1963), and monolayer MoS_2 suspensions were first achieved in 1986 (Joensen, Frindt, & Morrison, 1986) (Fig. 3.4.1).

In the 1990s, Reshef Tenne and others led significant efforts in the field of inorganic fullerenes and nanotubes, beginning with the discovery of WS_2 nanotubes and nested particles (Tenne, Margulis, Genut, & Hodes, 1992), followed by the synthesis of MoS_2 nanotubes and nanoparticles (Feldman, Wasserman, Srolovitz, & Tenne, 1995). The rapid growth of graphene-related research, which began in 2004, sparked the development of techniques well suited for working with layered materials, paving the way for new research into TMDCs and, in particular, ultrathin films.

3.4.3 Crystal structures and physical properties

TMDCs exhibit a variety of structural stages as a function of the transition metal atoms' different coordination spheres. Metal atoms are coordinated in either a

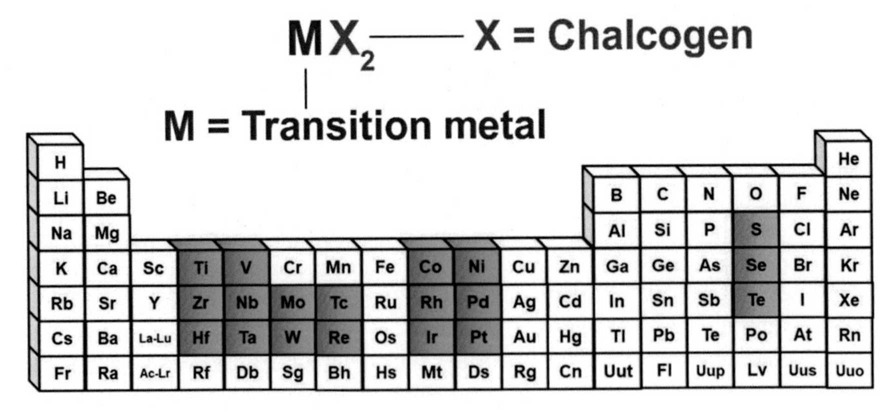

FIGURE 3.4.1

Transition metal dichalcogenides materials in the periodic table.

trigonal prismatic (2H) or an octahedral (1T) manner in these two typical structural phases. The metal atoms are positioned in close packaging and form hexagonal layers during crystallization of TMDCs. On both sides of the metal layers, chalcogen atoms form close hexagonal layers, leading to a trigonal-prismatic unit cell with a= 0.3154 nm and c= 1.236 nm for WS_2 (Mayorov et al., 2011). This means that the TMDC building units are two dimensional, intensely bonded X-M-X layers, weakly binding together with the van der Waals forces along the unit cell axis (Bunch et al., 2008). The weak c-axis bonding also explains the large unit cell dimension. The anisotropic bonding of these materials leads to strong electric and mechanical characteristics, which can be used in a beneficial manner. TMDC single crystals grow preferably along the (100) or (010) axes, resulting in small, plate-like crystals with large planes (001).

3.4.3.1 Lattice structure

The tungsten dichalcogenide lattice structure is similar to its molybdenum counterparts but slightly bigger. Therefore traditional techniques of characterization make it quite difficult to distinguish between WX_2 and MoX_2 (Berkdemir et al., 2013). This is a common behavior for heavy dichalcogenides that tend to have lower anisotropic symmetry. The large size of the transition metal atom makes the structure sensitive to small changes, resulting in massive differences in the overall structure. Since the arrangements of active sites (the basal plane's edges) depend on all 2D TMDC applications, the material properties can change significantly. It is reported that a small modification of the synthesis path leads to significant variations in the electronic properties of the material. Such changes are more severe for exfoliation approaches controlling the formation of active sites (Lorchat et al., 2016).

3.4.3.2 1T, 2H, and 3R phases

A variety of polymorphic structures arise in multilayered TMDs, as each layer can have one or two coordination phases. The three most common polymorphs are 1T, 2H, and 3R, whereby the number of layers in the crystallographic unit indicates the number of cells, and the latter indicates the type of symmetry exhibited; T means tetragonal (Group D3d), H means hexagonal (Group D3h), and R indicates rhombohedral (Group C5 3rd). The trigonal prismic phase is also known as the 2H phase (or 1H for a single layer), and it can be described by means of hexagonal symmetry (d3h group) and is a trigonal-prismic coordination of the metal atoms. That means that the sulfur atoms in single layers are aligned vertically along the z-axis, and the stacking sequence is then AbA, where A and b denote chalcogen and metal atoms, respectively (Fig. 3.4.2).

The octahedral phase has a tetragonal symmetry (D3d) corresponding to an octahedral atom coordination. In the octahedral phase, traditionally called the 1T phase, one of the sulfur layers is moved in comparison with the others, which

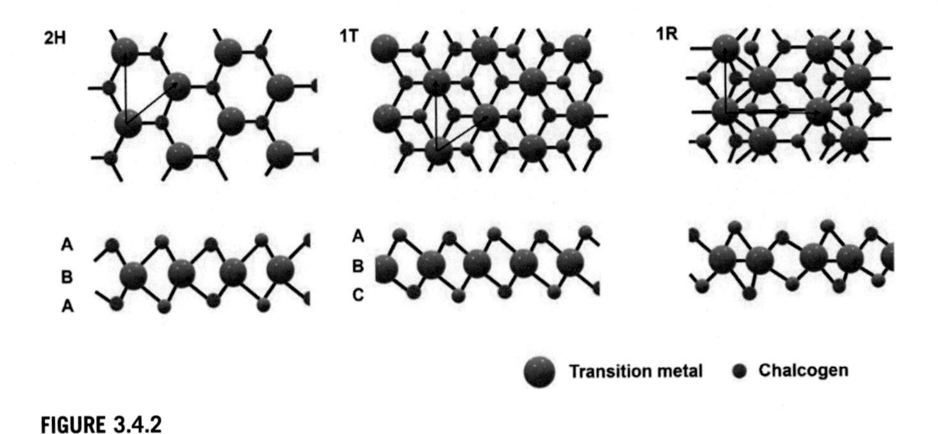

FIGURE 3.4.2

1T, 2H, and 3R phases of WS_2.

results in the AbC stacking sequence. The d-orbital metal filling directly affects the atomic structure of the TMD layers. During the 1H phase, the d orbital is divided into three degenerated states (dz^2, dx^2-y^2, xy, and dxy), yz with an energy gap between dz^2 and dx^2-y^2, and xy orbitals of B 1 eV. For 1T-phase tetragonal symmetry, the d metal orbits degenerate into orbitals dxy, yz, zx (t^2g), and dx^2-y^2, z^2 (for example). The e^2g orbital can be filled with up to six electrons. Since the p orbital of chalcogens is far lower in energy than the Fermi level, the nature of the MX_2 compounds is determined by the filling of d orbitals. Fully filled orbitals lead to semiconduction, while partial filling induces metallic behavior. In addition, two various forms of 1H layer stacking achieve rhombohedral symmetry (3R, symmetry C5 3v) with AbA-CaC-BcB stacking sequence. The 1T shape is metallic, while both 2H and 3R have a semiconducting behavior. In all cases, these phases have been investigated by depositing a well-defined mono- or few-layer TMDC on a substratum because the 1T phase in mass materials is less stable. Moreover, it is reported that chemical and thermal treatments can convert the metastable 3R phase into the most stable 2H phase (Eftekhari, 2017).

3.4.3.3 Band structure

In contrast to metallic graphene, WS_2 has both direct and indirect bandgap semiconductors in bulk form, with energy gaps of 1.3 and 1.5 eV, respectively. Quantum confinement effects cause an indirect-to-direct transition when reduced to a monolayer, resulting in 1.9 and 2.0 eV indirect and direct gaps for single layer WS_2 (Wang, Kalantar-Zadeh, Kis, Coleman, & Strano, 2012). Fig. 3.4.3 depicts the band structures of bulk and monolayer WS_2 determined from first principles. The valence band maximum is at the Γ stage, and the conduction band minimum is nearly halfway along the Γ-K path, forming the indirect bandgap transition (Ganatra & Zhang, 2014).

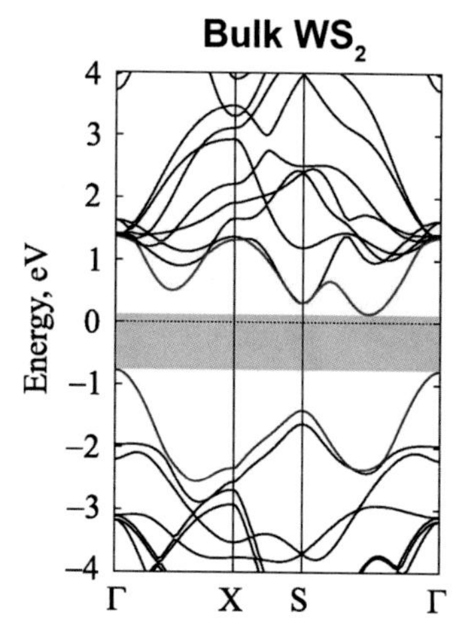

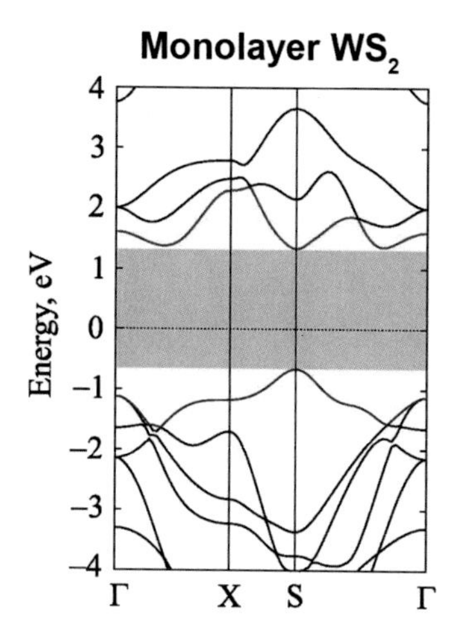

FIGURE 3.4.3

Band structures calculated from first-principles DFT for bulk and monolayer WS_2. The dashed lines indicate the Fermi level. The arrows indicate the fundamental bandgap (direct or indirect). The blue and green curves represent the top of the valence band and bottom of the conduction band, respectively.

Adapted from: Do Muoi, Nguyen N. Hieu, Huong T.T. Phung, Huynh V. Phuc, B. Amin, Bui D. Hoi, ... P.T.T. Le. (2019). Electronic properties of WS2 and WSe2 monolayers with biaxial strain: A first-principles study, Chemical Physics, *519, 69–73, ISSN 0301-0104, https://doi.org/10.1016/j.chemphys.2018.12.004.*

3.4.3.4 Electronic properties

In 2D TMD layers, the mobility of carriers is affected by four factors like (1) acoustic and optical phonon scattering, (2) Coulombic scattering at charged impurities, (3) surface interface phonon scattering, and (4) roughness scattering (Kaasbjerg, Thygesen, & Jacobsen, 2012). With increasing temperature, phonon scattering has a greater impact on carrier mobility. The optical component dominates at higher temperatures, while the acoustic component dominates at lower temperatures. While the electronic properties of WS_2 layers have received less attention, theoretical models suggest that, due to its lower effective mass, WS_2 should have the highest mobility among semiconducting 2D TMDCs (Liu, Kumar, Ouyang, & Guo, 2011). In early reports, the carrier mobility of liquid-gated multilayer and single-layer WS_2 was reported to be 44 cm^2/V per second (Jo, Ubrig, Berger, Kuzmenko, & Morpurgo, 2014). However, Ovchinnikov et al. recently fabricated n-type transistors with a high room-temperature on/off current ratio based on monolayer and bilayer WS_2. At room temperature, the systems showed higher carrier mobilities of up to 60 cm^2/V per second single-layer WS_2, which saturates at 140 cm^2/V per second when the temperature falls to 83K (Ovchinnikov, Allain, Huang, Dumcenco, & Kis, 2014).

3.4.4 Optical properties

The optical properties of WS_2 are directly influenced by the evolution of their electronic architectures as their thickness decreases. Changes in photoconductivity, absorption spectra, and photoluminescence indicate the transition from an indirect to a direct bandgap and an increase in bandgap energy (PL). Single-layer WS_2 has also been shown to have exceptional room temperature PL from flake edges, outperforming all existing 2D layered TMDs. Single-layer MoS_2 and WS_2 sheets are exceptionally thin and exhibit high optical clarity in the visible light spectrum ($>90\%$) (Park et al., 2015). 2D MoS_2 and WS_2 are well suited for optoelectronic applications such as sensors, phototransistors, and organic light emitting devices for next-generation solid-state lighting panels and high-resolution displays due to their combination of high clarity, high conductivity, and bandgap tunability depending on layer thickness (Zhang et al., 2013).

3.4.5 Fabrication of transition metal dichalcogenides

Based on the usage and applications, TMDC materials can be fabricated in many ways, such as sputtering, chemical bath deposition, chemical vapor deposition (CVD), etc. In this section, we will highlight the synthesis and fabrication of MoS_2 and WS_2, which are most promising amongst all TMDCs.

3.4.5.1 Mechanical exfoliation

Micromechanical exfoliation was invented by Novoselov and Geim in 2004 for graphene synthesis and has since been highly used for the synthesis of 2-D materials due to its flexibility and low cost. Mechanical exfoliation is a simple material synthesis process that produces one to many layers of 2-D crystalline flakes while preserving the crystal structure and properties. Easy scotch tape is used to exfoliate an atomic layer-thick TMDC from the bulk sheet/mesas, so those exfoliated layers are deposited on the substrate (Das, Kim, Lee, & Choi, 2014). Mechanical exfoliation is a low-cost way of producing 2-D materials, which is ideal for fundamental analysis. However, the technique is poorly scalable for 2-D flakes. While mechanical exfoliation is capable of fabricating single layer 2-D graphene and TMDC materials, several other factors (stoichiometry and stacking orders) are also important in the efficient fabrication of monolayer MX_2 nanostructures.

3.4.5.2 Chemical synthesis

Chemical synthesis is a low-cost, room-temperature method for producing a variety of carbonaceous and noncarbonaceous 2-D nanosheets. The chemical synthesis method is a bottomsup technique in which atoms/molecules are deposited on a substrate in a step-by-step deposition process to create a 2-D layered structure. Chemical synthesis methods are also being aggressively studied to achieve more optimizations, while solvent exfoliation is now well optimized and widely used to synthesize 2-D nanostructures. These are essentially wet chemical methods that produce 2-D nanostructures by chemical reaction and deposition onto substrates. For 2-D TMDC components, chemical synthesis is a low-cost, scalable synthesis method. Chemical synthesis processes are very useful because they have the potential to generate gram-scale 2-D nanomaterials with less defects and crystalline disorder than other approaches. However, when a TMDC is synthesized chemically, obtaining the optimal stoichiometry in the final product is more complex. Many optimizations would be needed to use such materials in consumer products due to their nonstoichiometric composition and wide variety of physical properties.

3.4.5.3 Chemical vapor deposition

CVD is a high-temperature chemical synthesis technique for depositing a target substance on substrates. Metals, semiconductors, and insulators have all been researched extensively using CVD processes to create thin film coatings. Large-scale homogeneous 2-D TMDC and BN nanosheets have also been synthesized using the CVD technique. By reacting sulfur vapor with metal thin films at high temperatures in an inert environment, CVD is a very simple way of synthesizing metal sulfides. Using a metal thin film and sulfur powder as starting ingredients, the CVD process produces the following reaction [Eq. (3.4.1)]:

$$M(s) + 2S(g) \rightarrow \mathrm{MS2}(s) \tag{3.4.1}$$

where "M" signifies any transition metal thin film, such as Mo, Nb, W, etc. Under a steady flow of N_2, the reaction temperature is maintained at 750°C. CVD methods have been used to demonstrate wide area synthesis of 2-D MoS_2 among the MX_2 community of TMDCs (Lee et al., 2012; Wang, Feng, Wu, & Jiao, 2013; Zhan, Liu, Najmaei, Ajayan, & Lou, 2012). A molybdenum (Mo) thin film is exposed to a high-temperature flow of sulfur (S) vapor in an inert atmosphere, transforming the Mo thin film into a 2-D MoS_2 thin sheet. A metal oxide thin film can be exposed to sulfur vapor in a similar way to reduce the oxide to a 2-D MoS_2 sheet. This reaction, also known as sulfurization, takes place at high temperatures with a steady supply of inert gas (N_2). The starting material, molybdenum trioxide (MoO_3), is reduced to metal disulfide in the overall process using the reactions [Eqs. (3.4.2) and (3.4.3)]. As MoO_3 is exposed to a stream of sulfur vapor at high temperatures, it first becomes a sub oxide, MoO_{3-x}, and then reacts to create a 2-D layered MoS_2 nanostructure (as seen in Fig. 3.4.4):

$$\mathrm{MoO_3(S) + S(g) \rightarrow MoO_{3-\mathit{x}}(S) + SO_{\mathit{x}}(g)} \tag{3.4.2}$$

$$\mathrm{MoO_{3-\mathit{x}}(S) + S(g) \rightarrow MoS_2(S)} \tag{3.4.3}$$

The reduction reaction (sulfurization) of MoO_3 takes place at 650°C, which is a lower temperature synthesis method than the direct reaction method (Zhan et al., 2012).

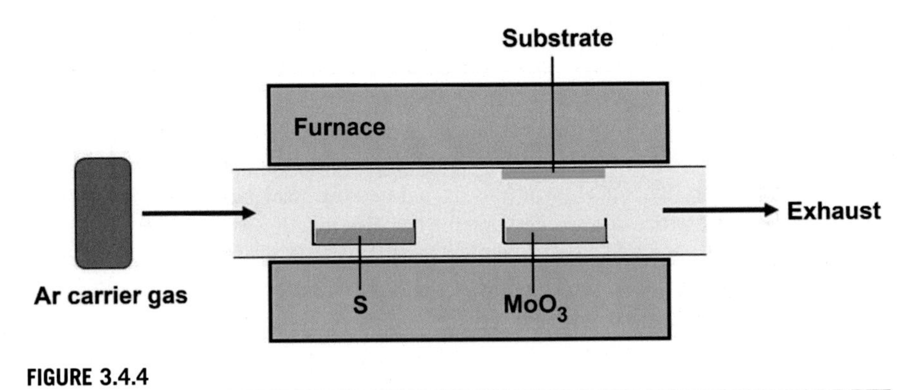

FIGURE 3.4.4

Schematic diagram of the chemical vapor deposition process.

The substrate is placed top down within the CVD chamber for more effective deposition (Lee et al., 2012), as seen in the method diagram in Fig. 3.4.4. This simple process is capable of generating large-scale MoS_2 multilayers; however, it is exceedingly difficult to synthesize a single layer of MoS_2 using this approach.

The CVD synthesis method has the key benefit of producing high-quality, high-purity 2-D nanomaterials with controlled properties. Furthermore, by manipulating various process parameters, the CVD process helps one to monitor the morphology, crystallinity, and defects of 2-D nanostructures. Synthesizing a wide range of new varieties of 2-D nanosheets and their derivatives is also feasible by mixing a variety of rigid, liquid, and gaseous precursor materials. Similarly, nanosheets can be doped and functionalized simply by adding another precursor source to the mix. Customizing atomic-level properties and monitoring stoichiometry and defects in 2-D materials, on the other hand, would necessitate more optimizations and will remain a significant challenge in the CVD growth process.

3.4.5.4 Molecular beam epitaxy

MBE was one of the first scalable approaches for the direct growth of TMDCs (MBE). The growth of GaSe on MoS_2 (Joyce, 1985) is used to demonstrate the theory of action schematically in Fig. 3.4.5. To shape molecular beams, an ultrahigh vacuum chamber (with pressure usually below 10^{10} mbar) requires many sources—in this case, Ga and Se. The thickness and crystallinity of the substrate are controlled in situ by reflection high-energy electron diffraction and low-energy electron diffraction during deposition on a preheated substrate. In the 1980s and 1990s, Atsushi Koma's group published a number of studies on TMDC MBE formation, demonstrating the efficient growth of $MoSe_2$ and GaSe on a variety of substrates, including other layered materials (Koma & Yoshimura, 1986; Koma, 1999; Ohuchi, Parkinson, Ueno, & Koma, 1990; Ohuchi, Shimada, Parkinson, Ueno, & Koma, 1991). MBE-based 2D substance development varies

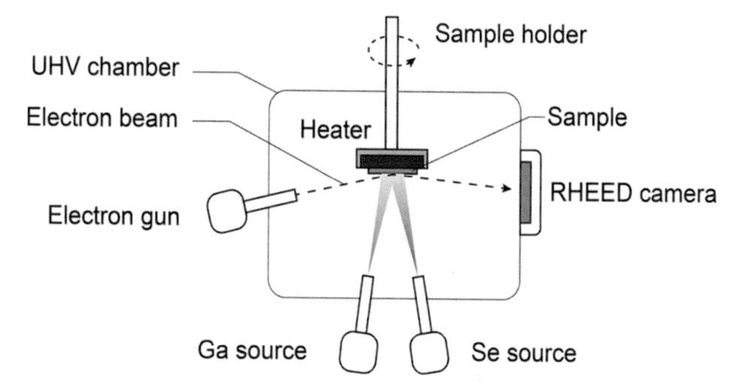

FIGURE 3.4.5

Schematic diagram of the molecular beam epitaxial process.

significantly from standard 3D epitaxy. Since TMDCs do not have hanging bonds, they can be grown on other layered materials and passivated surfaces (such as H-passivated Si) without having to meet lattice matching requirements. The poor van der Waals interactions can be used to achieve epitaxial growth, in which the substrate's lattice structure is used to align the overlayer's orientation (Dumcenco et al., 2015a), which is a prerequisite for producing high-quality films. If the substrate and overlayer are lattice-matched, as in the case of $MoSe_2$ on GaAs, or if a commensuration condition is met, as in the case of $MoSe_2$ on sapphire (Lehtinen et al., 2015), epitaxial growth will occur.

If these conditions are not met, the resulting films, such as MBE-grown $MoSe_2$ on an amorphous SiO_2 substrate (Roy et al., 2016), would be polycrystalline and contain several dislocations. In MBE-grown $MoSe_2$ and $MoTe_2$ on sapphire substrates, certain defects and grain boundaries can cause variable-range hopping via localized states (Ugeda et al., 2014).

MBE paired with scanning tunneling microscopy or angle-resolved photoelectron spectroscopy in ultrahigh vacuum growth chambers given access to the intrinsic properties of materials such as $MoSe_2$ (Barja et al., 2016) and $NbSe_2$. Furthermore, different forms of vertical heterostructures of 2D materials, such as $SnSe_2/WSe_2$ (Aretouli et al., 2016), $MoSe_2/Bi_2Se_3$ (Xenogiannopoulou et al., 2015), and $MoSe_2$/graphite (Vishwanath et al., 2015), may be realized. Thus MBE is a valuable instrument for studying the fundamentals of 2D materials and their variations, though the electrical properties of as-grown films must be improved.

3.4.5.5 Metal organic chemical vapor deposition

Metal–organic CVD (MOCVD) is a CVD-related growth process that uses gas-phase precursors (Chung, Dai, & Ohuchi, 1998; Dumcenco et al., 2015b; Eichfeld

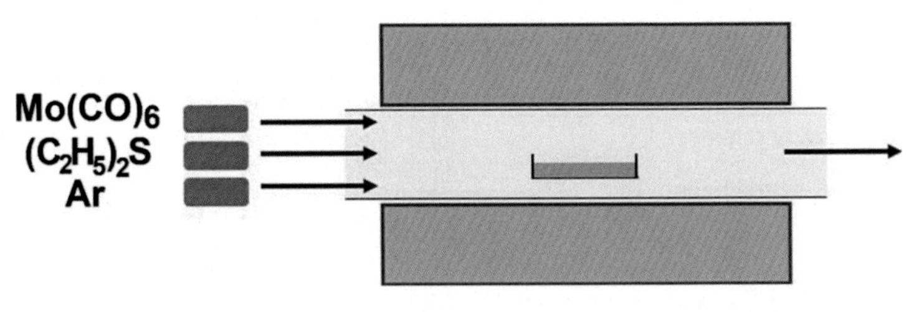

FIGURE 3.4.6

Schematic diagram of metal–organic chemical vapor deposition process.

et al., 2015; Kang et al., 2015) and was first used for TMDC synthesis in the 1990s (Dumcenco et al., 2015b). Gas precursors are fed into the main chamber at a steady rate (Fig. 3.4.6). Wafer-scale synthesis of MoS_2 and WS_2 on a variety of substrates was recently demonstrated, with the resulting samples exhibiting strong electrical properties. The electron mobility of the MoS_2 sample, for example, was 30 cm^2/V per second (Kang et al., 2015). The key focus of the CVD findings is MoS_2 development. Other TMDCs, such as $MoSe_2$, WS_2, WSe_2, ReS_2, $ReSe_2$, $MoTe_2$, or WTe_2, can be grown using both CVD and MOCVD (Chen et al., 2015; Gao et al., 2015; Wang et al., 2014; Zhou et al., 2017). However, the electrical properties of these materials' large-area films are yet to be fully defined, necessitating further research.

3.4.5.6 Sputtering

Sputtering is commonly used to deposit TMDC compounds such as MoS_2 and WS_2 (Fig. 3.4.7). For this reason, magnetrons are also commonly used in thin-film solar cells and tribological applications for film deposition. Reactive magnetron sputtering is a well-known form of large-scale deposition that enables coating substrate sizes of 3 m × 6 m, for example for architectural glass coating (Szczyrbowski, Bräuer, Ruske, Schilling, & Zmelty, 1999). The advantages of plasma-assisted deposition methods like magnetron sputtering are (Ellmer, 2008):

1. low substrate temperature compared to thermally enabled deposition methods,
2. high chemical reactivity by the plasma-assisted separation of species and excitement,
3. suitable for any compound or alloy deposition,
4. possible to produce uniform, compact, dense, and adherent thin films.

However, in the sputtering process, films are grown by the continuous bombardment of energy species (atoms, ions or energy neutrals) through plasma. Hence, any slight deviation in the process parameters (power, gas flow, etc.) can lead to significant changes in the material properties.

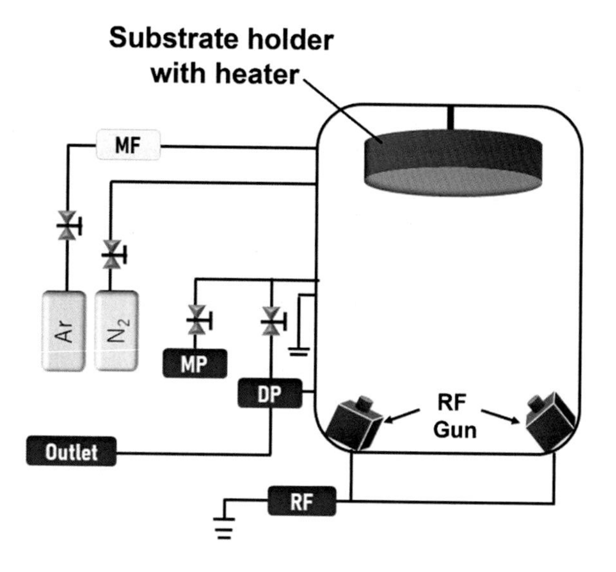

FIGURE 3.4.7

Schematic diagram of the RF magnetron sputtering process.

3.4.6 Applications of transition metal dichalcogenides materials

Because of their robustness and high raw material availability, MoS_2 and WS_2 are the most studied 2D TMDC materials. These materials have interesting properties and promise for a broad variety of applications in low-power electronics, versatile electronics, optoelectronics, straintronics, and spintronics due to their wide range of electrical properties, the atomic-scale thickness of their single layers, and the probability of obtaining a direct bandgap in semiconductor TMDCs. Furthermore, since nanopores based on atomically thin membranes of TMDCs have been shown to provide a rapid, high-resolution tool for DNA sequencing, TMDCs show promise for biophysics applications such as DNA sequencing and personalized medicine.

3.4.7 Conclusions

The vast number of scientists working on TMDCs and the dramatic increase in the number of publications represent the field's rapid advancement. However, the field is still in its preliminary stages in many aspects, with a wide variety of TMDC materials that have yet to be researched or synthesized, implying that there will be many more interesting developments in the near future.

References

Aretouli, K. E., Tsoutsou, D., Tsipas, P., Marquez-Velasco, J., Aminalragia, G. S., Kelaidis, N., ... Dimoulas, A. (2016). Epitaxial 2D $SnSe_2$/ 2D WSe_2 van der Waals heterostructures. *ACS Applied Materials & Interfaces*, *8*, 23222–23229.

Aufray, B., Kara, A., Vizzini, S., Oughaddou, H., Leandri, C., Ealet, B., & Le Lay, G. (2010). Graphene-like silicon nanoribbons on Ag(110): A possible formation of silicene. *Applied Physics Letters*, *96*, 183102–183103.

Bao, W., Cai, X., Kim, D., Sridhara, K., & Fuhrer, M. S. (2013). High mobility ambipolar MoS_2 field effect transistors: Substrate and dielectric effects. *Applied Physics Letters*, *102*, 042104–042104.

Barja, S., Sebastian, W., Zhen-Fei, L., Yi, Z., Hyejin, R., & Miguel, M. U. (2016). Charge density wave order in 1D mirror twin boundaries of single-layer $MoSe_2$. *Nature Physics*, *12*, 751–756.

Berkdemir., Gutiérrez, H. R., Botello-Méndez, A. R., Perea-López, N., Elías, A. L., Chia, C., ... Terrones, M. (2013). Identification of individual & few layers of WS_2 using Raman spectroscopy. *Scientific Reports*, *3*, 01755.

Bunch, J. S., Scott, S. V., Jonathan, S. A., Arend, M. v. d. Z., Jeevak, M. P., Harold, G. C., & Paul, L. M. (2008). Impermeable atomic membranes from graphene sheets. *Nano Letters*, *8*, 2458–2462.

Chen, J., Liu, B., Liu, Y., Tang, W., Nai, C. T., Li, L., ... Loh, K. P. (2015). Chemical vapor deposition of large-sized hexagonal WSe_2 crystals on dielectric substrates. *Advanced Materials*, *27*, 6722–6727.

Choi, W., Lahiri, I., Seelaboyina, R., & Kang, Y. S. (2010). Synthesis of graphene and its applications: A review. *Critical Reviews in Solid State and Materials Sciences*, *35*, 52–71.

Chung, J.-W., Dai, Z. R., & Ohuchi, F. S. (1998). WS_2 thin films by metal organic chemical vapor deposition. *Journal of Crystal Growth*, *186*, 137–150.

Das, S., Kim, M., Lee, J.-W., & Choi, W. (2014). Synthesis, properties, and applications of 2-D materials: A comprehensive review. *Critical Reviews in Solid State and Materials Sciences*, *39*(4), 231–252. Available from https://doi.org/10.1080/10408436.2013.836075.

Dickinson, R. G., & Pauling, L. (1923). The crystal structure of molybdenite. *Journal of the American Chemical Society*, *45*, 1466–1471.

Dumcenco, D., Dmitry, O., Kolyo, M., Predrag, L., Marco, G., & Nicola, M. (2015a). Large-area epitaxial monolayer MoS_2. *ACS Nano*, *9*, 4611–4620.

Dumcenco, D., Dmitry, O., Oriol, L. S., Philippe, G., Duncan, T. L. A., Sorin, L., ... Andras, K. (2015b). Large-area MoS_2 grown using H_2S as the sulfur source. *2D Materials*, *2*, 044005.

Eftekhari, A. (2017). Tungsten dichalcogenides (WS_2, WSe_2, and WTe_2): Materials chemistry and applications. *Journal of Materials Chemistry A*. Available from https://doi.org/10.1039/c7ta04268j.

Eichfeld, S. M., Lorraine, H., Yu-Chuan, L., Aleksander, F. P., Benjamin, K. A., & Glen, B. (2015). Highly scalable, atomically thin WSe_2 grown via metal–organic chemical vapor deposition. *ACS Nano*, *9*, 2080–2087.

Ellmer, K. (2008). *Low temperature plasmas: Fundamentals, technologies & techniques* (Vol. 2, p. 675). *Wiley-VCH*.

Feldman, Y., Wasserman, E., Srolovitz, D. J., & Tenne, R. (1995). High-rate, gas-phase growth of MoS_2 nested inorganic fullerenes and nanotubes. *Science (New York, N.Y.)*, *267*, 222–225.

Frindt, R. F., & Yoffe, A. D. (1963). Physical properties of layer structures: Optical properties and photoconductivity of thin crystals of molybdenum disulfide. *Proceedings of the Royal Society A*, *273*, 69–83.

Ganatra, R., & Zhang, Q. (2014). Few-layer MoS_2: A promising layered semiconductor. *ACS Nano*, *vol 8*, 4074–4099.

Gao, Y., Zhibo, L., Dong-Ming, S., Le, H., Lai-Peng, M., & Li-Chang, Y. (2015). Large-area synthesis of high-quality and uniform monolayer WS_2 on reusable Au foils. *Nature Communications.*, *6*, 8569.

Geim, A. K., & Novoselov, K. S. (2007). The rise of graphene. *Nature Materials*, *6*, 183–191.

Hwang, W. S., Remskar, M., Yan, R., Protasenko, V., Tahy, K., Chae, S. D., ... Jena, D. (2012). Transistors with chemically synthesized layered semiconductor WS_2 exhibiting 10_5 room temperature modulation and ambipolar behavior. *Applied Physics Letters*, *101*, 013107–4.

Jo, S., Ubrig, N., Berger, H., Kuzmenko, A. B., & Morpurgo, A. F. (2014). Mono- and bilayer WS_2 light-emitting transistors. *Nano Letters*, *14*, 2019–2025.

Joensen, P., Frindt, R. F., & Morrison, S. R. (1986). Singlelayer MoS_2. *Materials Research Bulletin*, *21*, 457–461.

Joyce, B. A. (1985). Molecular beam epitaxy. *Reports on Progress in Physics*, *48*, 1637–1697.

Kaasbjerg, K., Thygesen, K. S., & Jacobsen, K. W. (2012). Phonon-limited mobility in n-type single-layer MoS_2 from first principles. *Physical Review B*, *85*, 115317.

Kang, K., Saien, X., Lujie, H., Yimo, H., Pinshane, Y. H., Kin, F. M., ... Jiwoong, P. (2015). High-mobility three-atom-thick semiconducting films with wafer-scale homogeneity. *Nature*, *520*, 656–660.

Kim, S., Konar, A., Hwang, W.-S., Lee, J. H., Lee, J., Yang, J., ... Kim, K. (2012). High-mobility and low-power thin-film transistors based on multilayer MoS_2 crystals. *Nature Communications*, *3*, 1011.

Koma, A., & Yoshimura, K. (1986). Ultrasharp interfaces grown with van der waals epitaxy. *Surface Science*, *174*, 556–560.

Koma, A. (1999). Van der Waals epitaxy for highly latticemismatched systems. *Journal of Crystal Growth*, *201–202*, 236–241.

Lee, Y.-H., Zhang, X.-Q., Zhang, W., Chang, M.-T., Lin, C.-T., Chang, K.-D., ... Lin, T.-W. (2012). Synthesis of large-area MoS_2 atomic layers with chemical vapor deposition. *Advanced Materials*, *24*.

Lehtinen, O., Hannu-Pekka, K., Artem, P., Michael, B. W., Ming-Wei, C., & Tibor, L. (2015). Atomic scale microstructure and properties of Se-deficient two-dimensional $MoSe_2$. *ACS Nano*, *9*, 3274–3283.

Liu, L., Kumar, S. B., Ouyang, Y., & Guo, J. (2011). Performance limits of monolayer transition metal dichalcogenide transistors. *IEEE Transactions on Electron Devices*, *58*, 3042–3047.

Lorchat, E., Lorchat, E., Froehlicher, G., Froehlicher, G., Berciaud, S., & Berciaud, S. (2016). Splitting of interlayer shear modes & photon energy dependent anisotropic Raman response in N-Layer $ReSe_2$ & ReS_2. *ACS Nano*, *10*, 2752–2760. Available from https://doi.org/10.1021/acsnano.5b07844.

Mayorov, A. S., Roman, V. G., Sergey, V. M., Liam, B., Rashid, J., & Leonid, A. P. (2011). Micrometer-scale ballistic transport in encapsulated graphene at room temperature. *Nano Letters*, *11*, 2396–2399.

Mahatha, S. K., Patel, K. D., & Menon, K. S. R. (2012). Electronic structure investigation of MoS_2 and $MoSe_2$ using angle-resolved photoemission spectroscopy and ab initio band structure studies. *Journal of Physics. Condensed Matter: An Institute of Physics Journal*, *24*, 475–504.

Ni, Z., Liu, Q., Tang, K., Zheng, J., Zhou, J., Qin, R., ... Lu, J. (2012). Tunable bandgap in silicene and germanene. *Nano Letters*, *12*.

Ohuchi, F. S., Parkinson, B. A., Ueno, K., & Koma, A. (1990). Van der Waals epitaxial growth and characterization of $MoSe_2$ thin films on SnS_2. *Journal of Applied Physics*, *68*, 2168–2175.

Ohuchi, F. S., Shimada, T., Parkinson, B. A., Ueno, K., & Koma, A. (1991). Growth of $MoSe_2$ thin-films with Van der Waals epitaxy. *Journal of Crystal Growth*, *111*, 1033–1037.

Ovchinnikov, D., Allain, A., Huang, Y.-S., Dumcenco, D., & Kis, A. (2014). Electrical transport properties of single-layer WS_2. *ACS Nano*, *8*, 8174–8181.

Pakdel, A., Zhi, C., Bando, Y., & Golberg, D. (2012). Low-dimensional boron nitride nanomaterials. *Materials Today*, *15*.

Park, J., Choudhary, N., Smith, J., Lee, G., Kim, M., & Choi, W. (2015). Thickness modulated MoS_2 grown by chemical vapor deposition for transparent and flexible electronic devices. *Applied Physics Letters*, *106*, 012104.

Radisavljevic, B., Radenovic, A., Brivio, J., Giacometti, V., & Kis, A. (2011). Single-layer MoS_2 transistors,. *Nature Nanotechnology*, *6*, 147–150.

Roy, A., Hema, C. P. M., Biswarup, S., Kyounghwan, K., Rik, D., & Amritesh, R. (2016). Structural and electrical properties of $MoTe_2$ and $MoSe_2$ grown by molecular beam epitaxy. *ACS Applied Materials & Interfaces*, *8*, 7396–7402.

Szczyrbowski, J., Bräuer, G., Ruske, M., Schilling, H., & Zmelty, A. (1999). New low emissivity coating based on TwinMag® sputtered TiO_2 and Si_3N_4 layers. *Thin Solid Films*, *351*, 254.

Tenne, R., Margulis, L., Genut, M., & Hodes, G. (1992). Polyhedral and cylindrical structures of tungsten disulfide. *Nature*, *360*, 444–446.

Ugeda, M. M., Aaron, J. B., Su-Fei, S., Felipe, H. d. J., Yi, Z., & Diana, Y. Q. (2014). Giant bandgap renormalization and excitonic effects in a monolayer transition metal dichalcogenide semiconductor. *Nature Materials*, *13*, 1091–1095.

Vishwanath, S., Xinyu, L., Sergei, R., Patrick, C. M., Angelica, A., & Stephen, M. (2015). Comprehensive structural and optical characterization of MBE grown $MoSe_2$ on graphite, CaF_2 and graphene. *2D Materials*, *2*, 024007.

Vogt, P., De Padova, P., Quaresima, C., Avila, J., Frantzeskakis, E., Asensio, M. C., ... Le Lay, G. (2012). Silicene: Compelling experimental evidence for graphene like two-dimensional silicon. *Physical Review Letters*, *108*, 155–501.

Wang, Q. H., Kalantar-Zadeh., Kis, K., Coleman, A., & Strano, J. N. (2012). Electronics and optoelectronics of two-dimensional transition metal dichalcogenides. *Nature Nanotechnology*, *7*, 699–712.

Wang, X., Yongji, G., Gang, S., Wai, L. C., Kunttal, K., & Gonglan, Y. (2014). Chemical vapor deposition growth of crystalline monolayer $MoSe_2$. *ACS Nano*, *8*, 5125–5131.

Wang, X., Feng, H., Wu, Y., & Jiao, L. (2013). Controlled synthesis of highly crystalline MoS_2 flakes by chemical vapor deposition. *Journal of the American Chemical Society*, *135*, 5304–5307.

Wilson, J. A., & Yoffe, A. D. (1969). The transition metal dichalcogenides discussion and interpretation of the observed optical, electrical and structural properties. *Advances in Physics*, *18*, 193–335.

Xenogiannopoulou, E., Tsipas, P., Aretouli, K. E., Tsoutsou, D., Giamini, S. A., & Bazioti, C. (2015). High-quality, large-area $MoSe_2$ and $MoSe_2/Bi_2Se_3$ heterostructures on AlN (0001)/Si(111) substrates by molecular beam epitaxy. *Nanoscale*, *7*, 7896–7905.

Zhan, Y., Liu, Z., Najmaei, S., Ajayan, P. M., & Lou, J. (2012). Large-area vapor-phase growth and characterization of MoS_2 atomic layers on a SiO_2 substrate. *Small (Weinheim an der Bergstrasse, Germany)*, *8*.

Zhang, W., Saliba, M., Stranks, S. D., Sun, Y., Shi, X., Wiesner, U., & Snaith, H. J. (2013). Enhancement of perovskite-based solar cells employing core–shell metal nanoparticles. *Nano Letters*, *13*(9), 4505–4510.

Zhou, J., Fucai, L., Junhao, L., Xiangwei, H., Juan, X., Bowei, Z., & Qingsheng, Z. (2017). Large-area and high-quality 2D transition metal telluride. *Advanced Materials*, *29*, 1603471.

CHAPTER

Introduction to organic-inorganic hybrid solar cells

4

Md. Akhtaruzzaman[1], Vidhya Selvanathan[1], Md. Shahiduzzaman[2] and Mohammad Ismail Hossain[3]

[1]*National University of Malaysia, Bangi, Malaysia*

[2]*Nanomaterials Research Institute, Kanazawa University, Kanazawa, Japan*

[3]*Department of Materials Sciences and Engineering, City University of Hong Kong, Kowloon, Hong Kong, P.R. China*

The evolution of photovoltaic technology post crystalline silicon solar cells era led to the exploitation of inorganic thin film devices. To some extent, inorganic solar cells are perceived to be successful due to their ability to attain high efficiencies without high cost fabrication processes compared to silicon solar cells. The higher material and production costs have prevented silicon-based solar cells from being widely adopted as an alternative to fossil fuels. Since the silicon in solar cells must have the features of high purity and quality, it is quite difficult to realize its full potential. Furthermore, in the line of sustainable development, extra care must be taken to avoid environmental contaminations as many harmful solvents like hydrofluoric acid, sulfuric acid, and 1,1,1-trichloroethane have been used in the silicon wafer cleaning process. In addition, silicon dust, using lead (Pb)-containing compounds for soldering, silicon tetrachloride, waste disposal system, land usage, excessive water use for the cooling process are also a major issue of this technology. The heavy weight of the PV panel is also another drawback to rooftop installation. Therefore, silicon is not considered an ideal material for solar cell applications. A potential alternative to silicon solar cells is the advancement in the framework of inorganic thin film solar cells.

To some extent, inorganic solar cells are perceived to be successful due to their ability to attain high efficiencies without high cost fabrication processes compared to silicon solar cells. However, the prerequisite of rare earth elements as part of its composition challenges the status of inorganic solar cells as effective substitutes to first generation silicon photovoltaics. Alongside complex fabrication techniques, inorganic solar cells also suffer in terms of sustainability issues and elevated costs due to material scarcity. In these solar cells, a desirable trade-off can be made by reducing the thickness of the semiconductor layer (reducing cost). In addition, the recycling process of thin film PV also requires attention to prevent environmental contamination of toxic and costly rare earth metals containing compounds such as gallium arsenide, cadmium telluride, etc. The cost of

Comprehensive Guide on Organic and Inorganic Solar Cells. DOI: https://doi.org/10.1016/B978-0-323-85529-7.00004-9

solar cells must drop further and do so quickly to accelerate the transition to a low-carbon economy, which will require new materials that are less expensive, solution processable, lightweight, mechanically flexibility, abundant in nature, tunable optical and electrical properties and environmental friendliness. In these circumstances, researchers introduced organic materials as prospective alternatives to accomplish low cost photovoltaic devices. Inception of organic solar material-based solar cells presented the possibility of photovoltaics with two main advantages. First, the high absorption coefficients of organic molecules translate into the capacity to attain efficient absorption with very thin layers of material. This is beneficial as it limits material usage and allows fabrication of flexible photoactive layers. Second, organic photovoltaics can be fabricated via facile solution-processed techniques without elaborate instrumentation like its predecessor technologies. Unfortunately, the efficiencies and lifetimes of organic solar cells are inferior to their inorganic counterparts. The constant tug-of-war between high efficiency of inorganic photovoltaics and low cost of organic photovoltaics inspired the genesis of the concept of organic-inorganic hybrid solar cells (OIHSC). This category of photovoltaics intelligently assimilates the merits of both organic and inorganic materials to give rise to the fabrication of cost-efficient, highly stable solar cells with high efficiencies as shown in Fig. 4.1. The organic-inorganic hybrid photovoltaic technology also puts a huge emphasis on

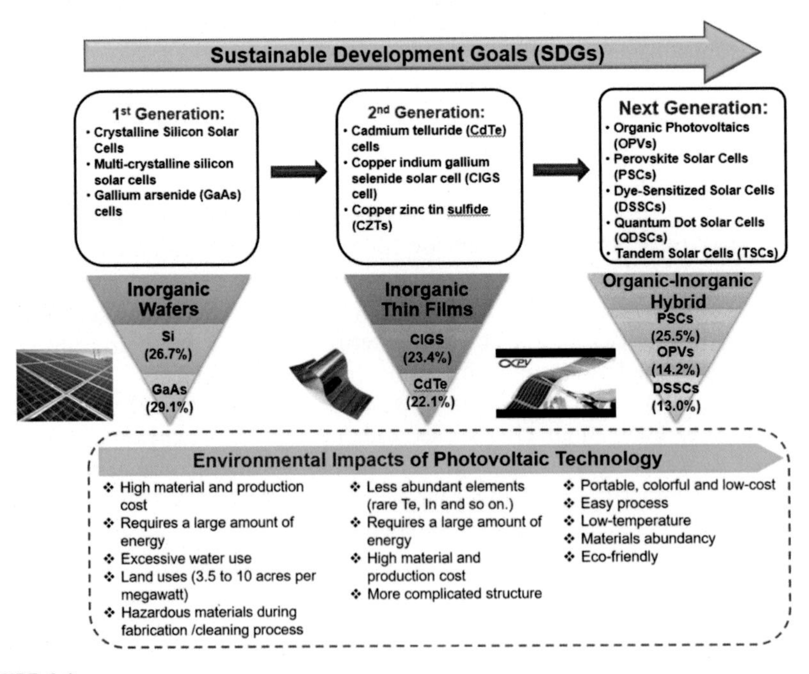

FIGURE 4.1

Evolution of solar conversion devices.

the development of solar cells from abundant and sustainable material resources which is in alignment with the "Affordable and Clean Energy" principle listed in the Sustainable Development Goals.

Given that the idea of OIHSC evolved from improvised organic photovoltaics, the fundamental working principle of organic-inorganic solar cells complies with the excitonic principle of organic solar cells. This is different than the p-n junction principle which forms the basic mechanism of inorganic solar cells in which free charge carriers are generated in the absorber layer with the help of built-in junction potential. In OIHSC, upon photoabsorption in the active layer, an excited state in which the electron-hole pair is still coulombically bound is observed. Such bound electron-hole pairs are termed as excitons. The photogenerated excitons are dissociated at the interface due to the energy level offset of the highest occupied molecular orbital of the organic component (donor) and valence band edge of the inorganic component (acceptor). Upon dissociation, the hole is relocated to the donor and is subsequently transported to the anode, whereas the electron is exported from the acceptor to the cathode. In some organic-inorganic photovoltaic devices, the navigation of charge carriers to their respective electrodes are facilitated by selective materials known as the hole transport layer and electron transport layer. A simple OIHSC device structure is illustrated in Fig. 4.2.

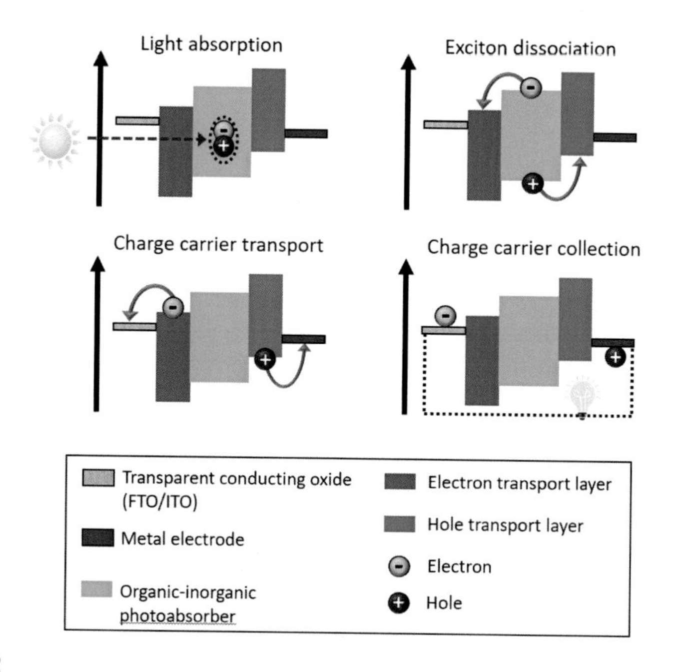

FIGURE 4.2

Device structure and energy band diagram of organic-inorganic hybrid solar cells.

In terms of device architecture, the integration of organic-inorganic material in the photoactive layers can be categorized into three configurations. The first type of interface configuration is the typical organic-inorganic planar structure which mimics conventional inorganic solar cells. In these devices, an organic film is deposited on an inorganic layer, producing a simple OIHSC. The major drawback of such planar configurations is the presence of a single interface which limits excitonic dissociations. Hence, generation of free charge carriers can only occur within the excitonic diffusion length of the single interface. To overcome this issue, an improvised configuration comprised of inorganic compositions blended into organic layers was proposed. This configuration is similar to the bulk heterojunction device structure and has shown to be advantageous in forming extensive interfacial area, thus significantly improving exciton dissociation yield. Nevertheless, the formation of a large dispersion of interface with improved interfacial area can compensate efficient transport of holes and electrons. Therefore, in the latest devices, an ordered nanostructure configuration is adapted to expand the interfacial area and concurrently create a systematic conduction pathway for charge carriers.

Chronological development of OIHSC indicates that this category of photovoltaics originated from novel attempts to incorporate inorganic acceptor materials into the traditional organic photovoltaic devices as a means to improve its stability. In the early developments of OIHSC, conjugated polymeric structures were used as the organic donor material and inorganic semiconductors were introduced as acceptor material in place of conventional fullerene-based molecules. CdSe, CdS, CdTe, Si, PbS, TiO_2, and ZnO are among the inorganic materials incorporated as acceptors in OIHSC.

The revolutionary era in OIHSC technology began with the recognition of the ability to sensitize semiconductors to light with wavelengths longer than that of its bandgap. Inspired by the mechanism adapted by plants in photosynthesis, the first chlorophyll-sensitized zinc oxide photoanode was fabricated in 1972. Despite its failure with respect to device performance, this pioneer attempt paved the way for the ground-breaking work by Brian O'Regan and Michel Gratzel (O'Regan & Grätzel, 1991) who fabricated a photovoltaic device comprising of TiO_2 nanoparticles coated with a monolayer of organic dye. This became the foundation for the invention of contemporary dye-sensitized solar cells (DSSC), also synonymously known as the Gratzel cell. The discovery of DSSC propelled the popularity of OIHSC, eventually creating a new generation of solar cells designated as "third generation photovoltaics." The initiation of the third generation solar cell is an attempt to reconcile the best features from the first two generations, high efficiency and low production costs. Soon, every component of DSSC was scrutinized and improvised to consistently improve the overall performance, stability, cost, and sustainability of the device.

In one attempt, researchers explored the possibility to replace organic dye molecules with semiconducting nanodots as sensitizers, which led to the fabrication of quantum-dot sensitized solar cell. This was further extended via modification of the size of inorganic quantum dots resulting in quantum confinement effects, hence allowing the optimization of the bandgap and light absorption characteristics of the inorganic sensitizer. This class of OIHSC was particularly successful due to the ability of quantum dot to generate multiple excitons.

In 2009, another innovative outcome by T. Miyasaka and co-workers (Kojima, Teshima, Shirai, & Miyasaka, 2009) to employ organic-inorganic hybrid halide-based perovskites as replacements to dye sensitizers in DSSC opened up a new possibility of OIHSC. This device was fabricated with liquid electrolyte as the hole transport material and failed to garner attention due to poor efficiency and device instability. However, the idea of incorporating perovskite-based materials into photovoltaics began to evolve with new material formulation and fabrication techniques. Eventually, in 2012, N. G. Park and colleagues (Kim et al., 2012) reported the milestone study of the first perovskite-based solid state heterojunction solar cell. This marked the arrival of a new class of OIHSC known as perovskite solar cells (PSC). Within a decade, the progress of PSC with a power conversion efficiency elevated from 3.8% to 25.5% in the laboratory scale. This was deemed as a hallmark achievement within the photovoltaic industry, given the fact that other solar cell technologies took more than thirty years to reach similar device performance (Fig. 4.3).

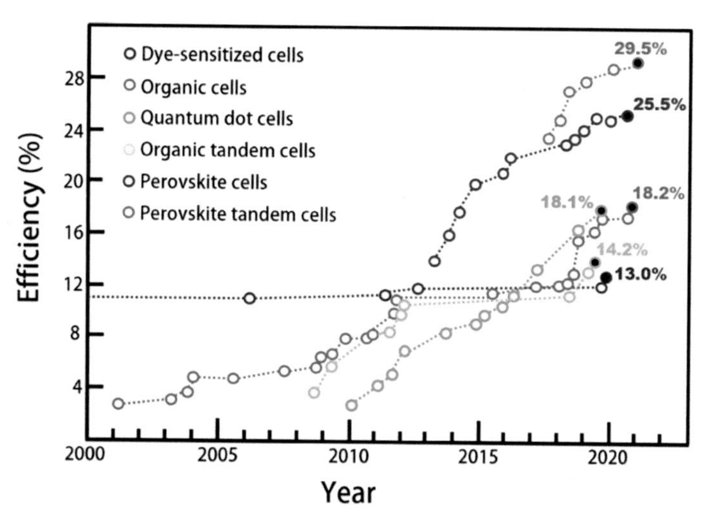

FIGURE 4.3

Record efficiencies achieved for third-generation thin-film cells with the data shown according to the most recent National Renewable Energy Laboratory efficiency report (Green et al., 2021).

Hence, the upcoming chapter is dedicated to explore the fundamentals of various OIHSC technologies. In Subchapter 4.1, readers are introduced to the concept of using dye-sensitized semiconductors for solar cells, detailing the device fabrication and challenges. In Subchapter 4.2, the use of quantum dot sensitized solar cells and their fabrication process are explored. In Subchapter 4.3, the basic and detailed background of organic-inorganic hybrid perovskite materials, and their working principle are discussed emphasizing the role of hybrid materials in the production of highly efficient and stable PSCs. Subchapter 4.4 includes the fundamental understanding of tandem solar cells and their optics and the realization of efficient tandem solar cells (Fig. 4.4).

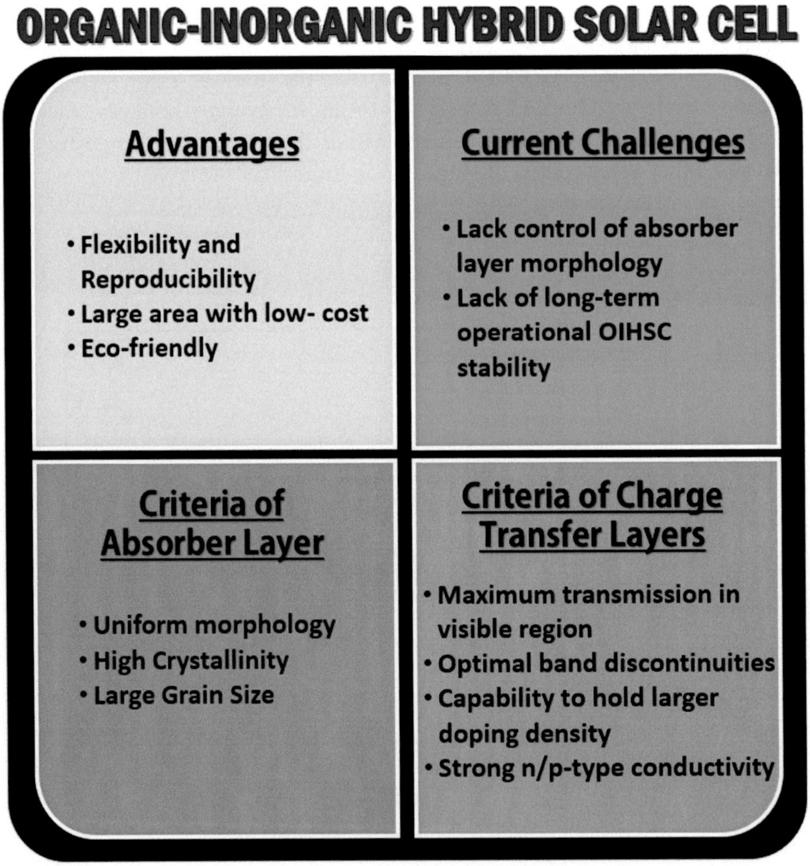

FIGURE 4.4

Key research gaps in the fabrication of OIHSC. *OIHSC*, organic-inorganic hybrid solar cells.

References

Green, M., Dunlop, E., Hohl-Ebinger, J., Yoshita, M., Kopidakis, N., & Hao, X.-j. (2021). Solar cell efficiency tables (version 57). *Progress in Photovoltaics*, *29*, 3–15.

Kim, H.-S., Lee, C.-R., Im, J.-H., Lee, K.-B., Moehl, T., Marchioro, A., ... Park, N.-G. (2012). Lead iodide perovskite sensitized all-solid-state submicron thin film mesoscopic solar cell with efficiency exceeding 9%. *Scientific Reports*, *2*(1), 591.

Kojima, A., Teshima, K., Shirai, Y., & Miyasaka, T. (2009). Organometal halide perovskites as visible-light sensitizers for photovoltaic cells. *Journal of the American Chemical Society*, *131*(17), 6050–6051.

O'Regan, B., & Grätzel, M. (1991). A low-cost, high-efficiency solar cell based on dye-sensitized colloidal TiO_2 films. *Nature*, *353*(6346), 737–740.

SUBCHAPTER

Dye-sensitized solar cells 4.1

Md. Akhtaruzzaman[1], Vidhya Selvanathan[2] and A.K. Mahmud Hassan[2]
[1]Solar Energy Research Institute, Universiti Kebangsaan Malaysia, Bangi, Malaysia
[2]Universiti Kebangsaan Malaysia, Bangi, Malaysia

The dye-sensitized solar cell (DSSC) is a third generation photovoltaic device, developed first by O'Regan and Grätzel in 1991. The conceptualization of DSSC draws inspiration from photosynthesis in which chlorophyll only plays a role in light harvesting but does not participate in charge transfer (CT) (O'Regan, & Grätzel, 1991). Similarly, in DSSCs, charge generation takes place at semiconductor-dye interface while charge transport is performed by the semiconductor and electrolyte. This feature is what differentiates DSSC from conventional photovoltaic systems, where the semiconductor guides both processes (Grätzel, 2003). By assigning the processes to different components, the necessity to use a material with both superior light harvesting and carrier transport properties can be avoided. This greatly favors DSSC in terms of ease and cost of fabrication. The technology also provides areas for improvisation as spectral property optimization can be performed by altering the dye molecule, while charge transport properties can be improved by optimizing the semiconductor and electrolyte composition (Nazeeruddin, Baranoff, & Grätzel, 2011).

4.1.1 Working principle

The architecture of DSSC is comprised of a photoanode made from a semiconductor with a layer of dye molecules adsorbed on its surface, a counter electrode (CE), and a layer of electrolyte sandwiched between the two electrodes (Fig. 4.1.1). The electrolyte consists of a redox couple, traditionally the iodide/triiodide couple.

$$S_{\text{adsorbed}} + h\nu \rightarrow S^{*}{}_{\text{adsorbed}} \tag{4.1.1}$$

$$S^{*}{}_{\text{adsorbed}} \rightarrow S^{+}{}_{\text{adsorbed}} + e^{-}{}_{\text{injected}} \tag{4.1.2}$$

$$I_3^{-} + 2e^{-} \rightarrow 3I^{-} \tag{4.1.3}$$

$$S^{+}{}_{\text{adsorbed}} + \frac{3}{2}I^{-} \rightarrow S_{\text{adsorbed}} + \frac{1}{2}I_3^{-} \tag{4.1.4}$$

Comprehensive Guide on Organic and Inorganic Solar Cells. DOI: https://doi.org/10.1016/B978-0-323-85529-7.00011-6

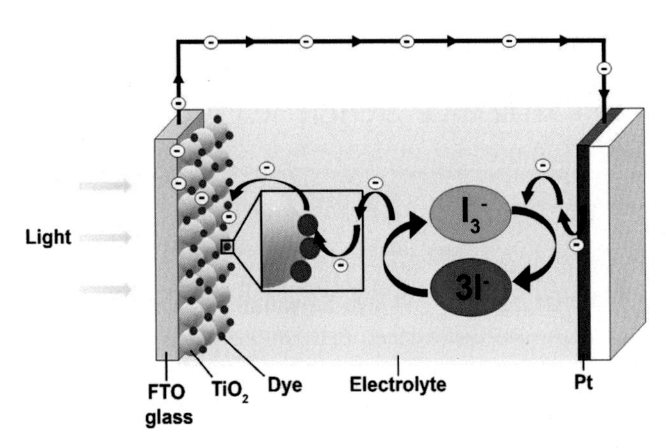

FIGURE 4.1.1

Schematic representation of the dye-sensitized solar cell.

Upon illumination, the dye molecules absorb the incident photons, which promotes the electrons of the dye from the highest occupied molecular orbital (HOMO) to the lowest unoccupied molecular orbital (LUMO) [Eq. (4.1.1)]. The photoexcited electrons then enter the conduction band (CB) of the semiconductor, leaving the dye molecules in an oxidized state [Eq. (4.1.2)]. The injected electrons then move through the TiO_2 layer and enter the external load, subsequently reaching the counter electrode. At the cathode, the electrons reduce triiodide into iodide ions, which then travel toward the photoanode [Eq. (4.1.3)]. At the photoanode-dye interface, the iodide ions are reduced to triiodide species, hence releasing electrons that are returned to the oxidized dye molecule [Eq. (4.1.4)]. With the regeneration of dye molecules, the circuit is completed. Besides the mechanism explained above, there are a number of recombination reactions that may take place within the cell. The photoinjected electrons may recombine to the oxidized dye molecules or the oxidized species of the redox couple.

4.1.2 Photoanode

The photoanode is an essential component of the DSSC. It acts as a ground or baseline for dye adsorption and medium for charge collection and transportation from exited dye molecules to the transferring conducting oxide (TCO) substrate. Ideal dye molecule adsorption and smooth CT deprived of experiencing any recombination are the key aspects of the photoanode. Other features include:

1. maximum dye loading capabilities for effective light absorption and surface area,
2. high transparency to allow maximum incident photons to reach the absorber,

3. high charge carrier mobility to ease electron transportation, and
4. chemically nonreactive toward the redox mediator to diminish recombination.

Semiconductor materials such as monocrystaline-Si, polycrystalline-Si, GaAs, CdS, and InP have been utilized as photoanodes but undergo chemical reactions with the redox shuttle and suffer cell degradation (Ali et al., 2016; Raj & Prasanth, 2016). The photoanode of TiO_2 and ZnO was established as the most favorable in DSSCs. However, the bandgap of ZnO and TiO_2 are ~3:3 eV (Akgul, Akgul, Attenkofer, & Winterer, 2013; Özgür et al., 2005) and ~3:2 eV (Tang, Prasad, Sanjinès, Schmid, & Lévy, 1994), respectively, allowing absorption of high energy parts of the solar spectrum and do not take part in the visible light absorption (Sengupta, Das, Mondal, & Mukherjee, 2016). Hence, it is necessary to modify the ZnO and TiO_2 bandgap by means of doping with different negative or positive ions to enhance the absorption toward the visible-NIR spectra (Dai et al., 2010). Moreover, metal oxides, such as SnO_2, Nb_2O_5 and Zn_2SnO_4, have been studied as photoanode constituents for DSSC applications.

4.1.2.1 Titanium oxide

TiO_2 is the most widespread photoanode for DSSC owing to its high chemical inertness, high surface area, and low toxicity (Das, Sengupta, Mondal, & Mukherjee, 2015; Liu, Fang, Liu, & Lin, 2016). Also, TiO_2 possesses a great CB edge for better CT, electron attraction, and surface area (Raj & Prasanth, 2016). TiO_2 occurs in anatase (tetragonal), brookite (orthorhombic), and rutile (tetragonal) crystalline forms (Dai et al., 2010). Though rutile shows the highest stability, anatase has been more commonly chosen for DSSCs due to its higher power conversion efficiency (PCE) and photocatalytic activity (Serikov, Ibrayev, Smagulov, & Kuterbekov, 2017; Shen et al., 2008). Moreover, anatase is an indirect bandgap semiconductor, which offers (Zhang, Zhou, Liu, & Yu, 2014) enhanced carrier lifetimes than direct bandgap rutiles or brookite polymorphs. Furthermore, electrons generated from the photoexcitation of anatase possess the lowest effective mass amongst the three. Nanostructures like nanosheets, nanotubes, nanowires, nanoparticles, and nanorods of TiO_2 have been synthesized and used as photoanodes in DSSCs (Sengupta et al., 2016). The adapted synthesis approaches of TiO_2 are spray pyrolysis, screen printing, spin coating, sol-gel, hydrothermal, chemical vapor deposition (CVD), and electrochemical routes (Raj & Prasanth, 2016). Also, the possibility of fabricating bendable DSSCs through the engagement of plastic substrates like ITO/poly (ethylene naphthalate) replacing FTO/glass has also been investigated (Chen et al., 2015; Wang, Tang, He, Li, & Yu, 2015).

4.1.2.1.1 TiO_2 doped with metallic cations

A number of studies have been conducted to enhance the optoelectronic properties of TiO_2 by doping different cations and anions as photoanodes in DSSCs.

TiO_2 doped with Zr was proposed by Dürr, Rosselli, Yasuda, and Nelles (2006). It was discovered that increasing the amount of Zr doping led to increased open circuit voltage (V_{oc}), which was correlated to the shifting of the CB edge toward higher energies. However, higher CB was responsible for the reduced driving force of the electron injection resulting in lower device efficiency. The 1% Zr concentration was determined to be the optimal doping amount to deliver a PCE of 8.1%, in comparison with intrinsic TiO_2 (7.0%). Tungsten doping at 0.2% into the TiO_2 photoanode exhibited 9.1% efficiency due to an enlarged J_{sc} resulting from enhanced driving energy and lowered charge recombination (Zhang, Li, et al., 2011; Zhang, Liu, Huang, Zhou, & Wang, 2011). Meanwhile, formation of an intermediate band between the VB and CB of the TiO_2 photoanode rather than CB shifting with W doping was proposed by Tong et al. (2014). The intermediate energy level aided in low energy exited electron migration from VB to CB. So, a synergistic impact on improved short circuit current density (J_{sc}) and PCE was observed due to photoconversion and better carrier lifetime. Photoanode TiO_2 doping with Nb^{5+} was investigated by Chandiran, Sauvage, Etgar, and Graetzel (2011). Nb doping was attributed to bandgap incremental increases and back electron transfer (BET) inhibition to redox couples which, in turn, offered PCE values of 8.1% compared to that of a pure TiO_2 photoanode. Lü et al. (2010) also demonstrated doping of a TiO_2 photoanode with high concentration Nb (2.5–7.5 mol %). The conductivity of TiO_2 was highly improvised and resulted from a flat-band potential's (V_{fb}) positive shift. This phenomenon also promoted charge injection and CT greatly. However, Nikolay, Larina, Shevaleevskiy, and Ahn (2011) opposed the outcomes from the aforementioned investigation. Based on Nikolay et al.'s study, Nb doping concentrations higher than 2.5 mol% would constrict LD, resulting in lower PCE. They projected optimal Nb doping was 1.5–2.5 mol% to generate the highest PCE. TiO_2 doped with Cr (Cr-TiO_2) was studied by Kim, Kim, Kim, and Han (2008) in a Cr–TiO_2/TiO_2/FTO multilayer photoanode, and a bilayer TiO_2 was formed. The double layer photoanode produced 18.3% more PCE than that of pure TiO_2 photoanode due to the presence of lower carrier recombination. Anatase-based TiO_2 was doped with Sc (0.2% concentration) for DSSCs and displayed a high PCE of 9.6% (Latini et al., 2013). Doping of Ce of 0.1% concentration into TiO_2 produced a higher PCE of 7.65% than that of TiO_2 alone (7.2%). Divalent Ce (Ce^{4+}/Ce^{3+}) changed the CB minima to more significantly govern DSSC performance (Zhang, et al., 2012). TiO_2 doped with other metallic dopants, such as Sn^{4+} (Duan et al., 2012), Sb^{3+} (Wang et al., 2012), In^{3+} (Bakhshayesh & Farajisafiloo, 2015), Ga^{3+} and Y^{3+} (Chandiran et al., 2011), were also reported in various studies.

4.1.2.1.2 TiO_2 doped with nonmetallic anions

Moreover, numerous investigations have been performed on TiO_2 photoanodes doped with nonmetal elements like N_2, B, F, C, and S. Ma et al. improved the DSSC performance via a N_2-doped TiO_2 (N-TiO_2) photoanode. N_2 doping was responsible for enhanced light absorption in the 400–700 nm range of the solar spectrum with superior dye adsorption onto the TiO_2 surface (Ma, Akiyama, Abe, & Imai, 2005). Guo, Shen, Wu, Gao, and Ma (2011); Guo, Wu, et al. (2011) also observed similar attributes with a N-TiO_2 photoanode with enhanced photo current, which corresponded to better PCE. Increased dye adsorption was also observed by a N-doped TiO_2 photoanode, leading to higher J_{sc} (19.05 ma/cm^2) and a higher cell efficiency of 10.10%. TiO_2 nanotubes doped with boron (B) (B-TiO_2) were investigated by Subramanian et al. as the photoanode (Subramanian & Wang, 2012). The B-TiO_2 photoanode exhibited a PCE of 3.44% while an intrinsic TiO_2 nanotube produced a PCE of 3.02%. The efficiency enlargement of B-TiO_2 could be attributed to improved carrier lifetime and proper band alignment between the LUMO and CB of dye molecule and the TiO_2 photoanode, respectively, which were advantageous for quicker charge injection. It was also reported that B doping showed better crystalline properties (Tian et al., 2011). Hollow spheres possess more active surface area than nanorods or nanowire. Doping of F with TiO_2 even produced more active surface area. The PCE values of doped and undoped TiO_2 were 6.31% and 5.62%, individually. The enhancements were credited to higher diffusivity of the redox couple, quicker charge carrier transportation, and lesser charge recombination at the F-TiO_2/electrolyte couple interface (Song et al., 2012).

Saadi et al. (Tabari-Saadi & Mohammadi, 2015) further synthesized carbon (C)-doped TiO_2 (C-TiO_2) hollow domains through a hydrothermal route and deployed them in a DSSC that performed as a scattering layer. A maximum PCE of 8.55% was attained for a DSSC with C-TiO_2 as the photoanode. Doping of S within the TiO_2 matrix was capable of reducing the bandgap via a positive shift (Sun et al., 2012). A mechanically ball milled, doped S-TiO_2 anode showed 24% enlargement in the cell PCE mainly because sufficient photo response took place in the 400–700 nm region. Extended absorption of visible light was detected by Hou et al. while TiO_2 nanocrystals were doped with I, which produced 42% more PCE than a photoanode consisting of TiO_2 alone (Hou et al., 2011).

4.1.2.2 Zinc oxide

Zinc oxide (ZnO) possesses wide bandgaps of ~3.3 eV, which is similar to TiO_2 and thus shows almost identical band energy and photophysical characteristics (Zhang, Dandeneau, Zhou, & Cao, 2009). Interestingly, ZnO shows a superior carrier mobility than anatase TiO_2 (Das et al., 2015). Carrier mobility is a vital electronic attribute, which governs the recombination loss by charge transportation (Zhang et al., 2009). The performance of the ZnO photoanode is restricted due to poor electron injection capability to an adverse TCO response toward acidic dye molecules. However, until now, ZnO has been considered as the most suitable alternative due to its anisotropic growth and crystallization ability. As a result, different types of ZnO allotropes, such as nanowires, nanotubes, and nanocrystals with different properties, can be synthesized.

4.1.2.2.1 ZnO doped with metallic cations

ZnO doped with cations of different metals has been described in many works. A Mg-doped ZnO (Mg-ZnO) photoanode was reported by Raj et al. (2013) through a one-step solvent evaporation technique.

An optimized 5% doping concentration of Mg in Mg-TiO_2 photoanode could produce 4.11% PCE compared with 1.97% PCE of a single ZnO photoanode. The formation of a banyan root-like spread stature aided in enhanced electron transfer, better light absorption, and high redox shuttle diffusion. Meanwhile, an enlargement of the bandgap of the Mg-ZnO photoanode was observed by Guo et al. and prepared by a direct precipitation procedure. This extended bandgap was responsible for improving the PCE from 1.72 to 4.19% relative to the higher V_{oc} due to Mg doping (Guo, Dong, Niu, Qiu, & Wang, 2014). The ordered arrangement of ZnO nanorods were doped with Al and could enhance the nanorod's diameter. This was quite helpful in reducing the charge recombination at the photoanode/redox solution contact point (Tao, Tomita, Wong, & Waki, 2012). This outcome of increased diameter of ZnO nanorods due to Al doping was also validated by Zhu et al. (2013). A nearly similar PCE (0.3%) was stated for an Al-ZnO photoanode. A synergistic impact of reduced recombination and long carrier life time were accountable for such DSSC performance. Spherical ZnO particles were also doped with Sn for the formation of Sn-ZnO photoanode by Wang, Bhattacharjee, Hung, Li and Zeng (2013). The particle size decreased upon Sn doping with ZnO and hence, increased the dye adsorption sites. The device showed a PCE of 0.08% with the Sn-ZnO photoanode.

4.1.2.2.2 ZnO doped with nonmetallic anions

A number of studies have been carried out on the doping of ZnO with nonmetallic anions by many research groups. ZnO nanorods were doped with N_2 to prepare N-ZnO photoanodes by Mahmood, Swain, Han, Kim, and Jung (2014). A cumulative increase of stock solution concentration and synthesis temperature had a tremendous positive impact on the optoelectrical and structural properties of the photoanode, leading to improved photovoltaic performance. Actually, N_2 induced a better carrier concentration and elevated Fermi energy position in the energy band diagram. A comparative study of N_2 doping in ZnO by annealing and solution has been performed by Zhang, Peng, Chen, Chen, and Han (2012) showing PCE values of 1.10% and 2.64%, respectively, due to the inhibition in the recombination level while pure ZnO exhibited only 0.76% efficiency. Zheng et al. doped ZnO nanocrystals with I_2 to prepare an I-ZnO photoanode, which exhibited extended carrier lifetime and reduced carrier recombination (Zheng et al., 2011; Zheng, Ding, Tao, & Chen, 2014). A PCE of 3.43% was testified by Luo et al. (2011) on a DSSC based on F-doped ZnO (F-ZnO) prism-arrayed photoanode which was greater than a pure ZnO (η = 1.04%) photoanode. Doping of F into the ZnO matrix enhanced greater surface area for more dye molecule adsorption and a longer carrier lifetime to reduce recombination and better quantum efficiency. Though F-ZnO showed an upper limit of performance, it has not yet been optimized, so it may still provide a higher PCE. A double layer of boron-doped ZnO (DL-BZO) was used as the photoanode and light scattering layer on DSSC by Mahmood and Sung (2014) displaying a PCE of 7.20%. The upper layer BZO spherical organization facilitated CT and light scattering. On the other hand, a nanoporous BZO layer increased dye adsorption and reduced BET to the electrolyte. Therefore, the DL-BZO photoanode unveiled upgraded photovoltaic performance in comparison with a pure double layer-ZnO anode.

4.1.2.3 Tin oxide (SnO_2)

Photoanode and redox electrolyte play important roles in power production. SnO_2 shows high carrier mobility and a more positive CB edge than TiO_2. The high carrier mobility leads to faster CT to TCO. The high bandgap of ~3.6 eV does not allow formation of oxidative holes in the SnO_2 CB in the presence of high energy light. In that way, it diminishes the dye degradation and improves device stability. However, accelerated kinetics of recombination and poor dye adsorption are the main barriers for higher photovoltaic performance with a SnO_2 photoanode compared to TiO_2. Nevertheless, applying a thin layer of high band gap oxide on SnO_2, such as Al_2O_3 or TiO_2, can resolve that issue (Qian et al., 2009). Fig. 4.1.2 shows the high-resolution TEM images of bare meso-SnO_2 and meso-SnO_2/TiO_2 particles (Weinhardt et al., 2008).

4.1.2.4 Tungsten oxide (WO_3)

WO_3 is a transitional metal oxide with a bandgap between 2.6–3.1 eV range. The WO_3 crystals show absorption in the high energy region of solar spectrum. WO_3

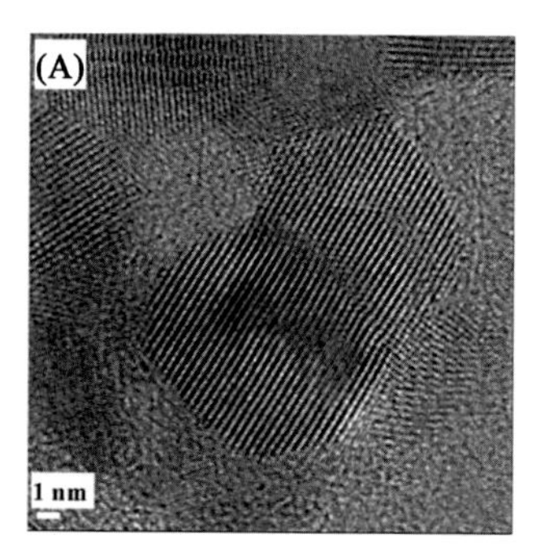

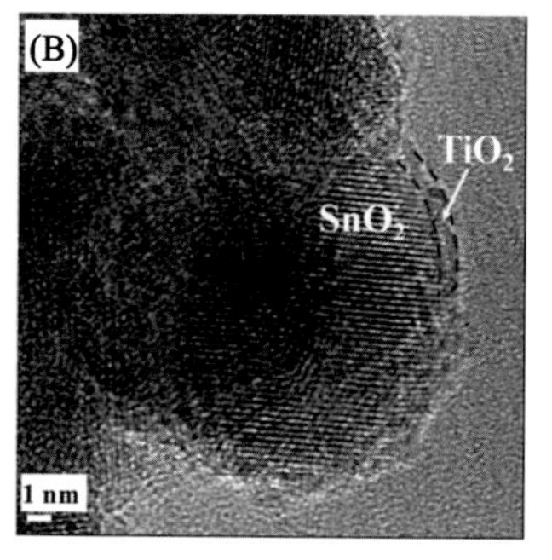

FIGURE 4.1.2

TEM images of meso a) pristine SnO_2 and b) SnO_2/TiO_2 particles

Reproduced with permission from Ramasamy, E., & Lee, J. (2010). Ordered mesoporous SnO_2-based photoanodes for high-performance dye-sensitized solar cells. The Journal of Physical Chemistry C, 114*(50), 22032–22037.*

is a fundamentally and highly robust material. It is even non responsive toward strong acid media that may be encountered in severe atmospheres. WO_3 charge carrier mobility is similar to that of TiO_2 (Weinhardt et al., 2008). WO_3 exists as nanoporous nanoparticles and nanowires, which can be helpful for improving the dye loading. Moreover, WO_3 possesses a 0.5 eV positive CB compared to TiO_2, so it can be expected to be used as a photoanode in DSSC with NIR dyes as the absorber. WO_3 nanotubes present superior response toward redox couples due to higher active surface area compared to orthodox WO_3 atoms (Hara et al., 2011).

4.1.2.5 Fabrication methods

One factor that affects the efficiency of a DSSC is the electron recombination process that occurs between the conducting substrate-electrolyte interfaces. Due to the porous nature of the semiconductor layer, the conducting substrate cannot be entirely insulated from the electrolyte. This issue can be resolved by employing a thin dense blocking layer between the substrate and mesoporous TiO_2 layer. Some of the materials which have been demonstrated to be effective blocking layers are TiO_2 (Manthina & Agrios, 2016), ZnO (Guo, Diao, Wang, & Cai, 2005; Liu et al., 2011), Au (Chang et al., 2011), Nb_2O_5 (Xia, Masaki, Jiang, & Yanagida, 2007), and graphene oxide (Kim, Parvez, & Chhowalla, 2009).

Different deposition approaches of semiconductor film on TCO will be discussed in the next sections. Preparing photoanode film in solution or paste form is preferable than vacuum or gas deposition route since solution phase depositions are easy, time saving, economic, and simple. In solution process, the precursor can exist as paste, solution, emulsion or in colloidal states. The commonly used techniques are spin coating, doctor blade, screen printing, and electrodeposition.

4.1.2.5.1 Spin coating

In spin coating, the precursor chemical solution is dripped on the epicenter of the TCO substrate at a very high spinning speed. The spin coating process is capable of producing a uniform thin film with very little roughness. Due to the high centrifugal force, the solution fully spreads over the substrate. The process is comprised of steps including solution drop casting, spin up, spinoff, and solvent evaporation. Since the method is highly dependent on solvent, a solvent with proper vapor pressure is needed to form a compact and uniform thin film. This is mostly influenced by the boiling temperature and vapor pressure of the solvent as well as ambient conditions where the method is performed. The film thickness could be controlled by solution concentration, spin speed, and time (Hall, Underhill, & Torkelson, 1998). An empirical relationship for determining film thickness is as follows:

$$h = \left(1 - \frac{\rho A}{\rho Ao}\right) \cdot \left(\frac{3\eta . m}{2\rho Ao\omega^2}\right)^{\frac{1}{3}} \tag{4.1.5}$$

where h symbolizes the thickness, ρA and η denote density and viscosity of solution, respectively, ω and m denote angular speed and evaporation rate, correspondingly.

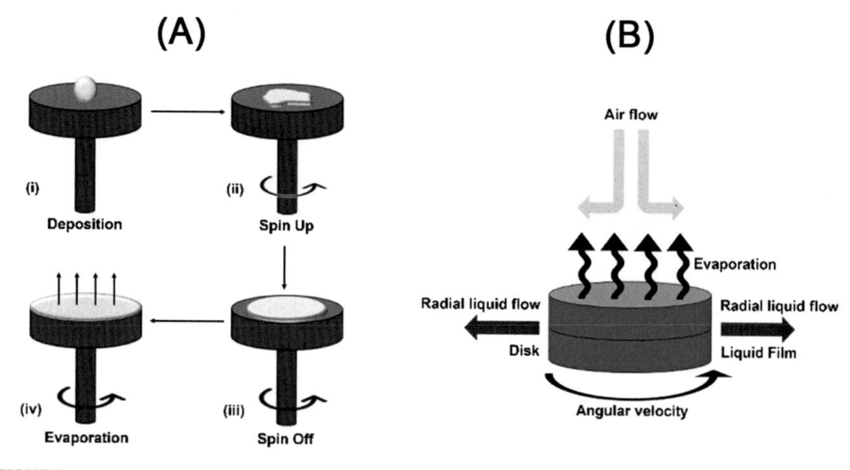

FIGURE 4.1.3

(A) Key stages of the spin coating process and (B) Systematic diagram of the spin coating process.

The great consistency and lower process time make spin coating one of the best functional film preparation techniques especially in microelectronics industries. In spite of its beneficial attributes, there are some drawbacks like high cost, small coating area, requirements of smooth substrates, and defects, such as the circle in the center, streaks, uncoated areas, and pinholes. Fig. 4.1.3A and B illustrate the key stages of the spin coating process and a systematic diagram of the spin coating process, respectively.

4.1.2.5.2 Dip coating

Dip coating is also a liquid-based film preparation technique on a TCO glass substrate. Viscosity, capillary effect, solution concentration, and dipping time are the governing factors of film thickness (Low & Lai, 2018). This kind of preparation requires some heat treatment to achieve a dry film. The retraction speed is designated to maintain the trimming degree inside the Newtonian fluid range. The thickness of film can be measured by the Landua-Levich equation (Landau & Levich, 1988):

$$h = 0.94.\frac{(n-v)^{\left(\frac{2}{3}\right)}}{\gamma_{LV}^{\left(\frac{1}{6}\right)}(\rho.g)^{\left(\frac{1}{2}\right)}} \tag{4.1.6}$$

where viscosity, retraction speed, liquid-vapor surface tension, density, gravitational constant, and film thickness are represented by η, v, γ_{LV}, ρ, g, and h, respectively.

The process is comprised of substrate dipping, startup, film deposition, ditching, and finally solvent removal through evaporation. The diagram of this coating system is presented in Fig. 4.1.4. The substrate is sunken into a solution of coating material for a short time. Then, withdrawing takes place at a continual speed to eliminate any possible waving. The faster the withdrawing speed, the thicker the coating. In the ditching period, additional solution is drained to achieve the expected thin film coating. The spare solution is vaporized during evaporation for drying. This method is well recognized in preparing a TiO_2 photoanode (Ito et al., 2005; Wei, Wan, & Wang, 2006).

The method is ideal for standard range manufacturing. It is capable of producing constant thickness, operation is easy, and many substrates can be coated. Moreover, thin film thickness can be effortlessly tuned without much waste. However, it requires masking for particularly designed device layouts and may require many manifold coating attempts to obtain the proper thickness.

4.1.2.5.3 Doctor blade printing

Doctor blade printing is an alternative method to generate a thin film of a large surface area. In this process, the substrate in glided under the slated blade at a constant velocity with a precise contact angle and altitude (Fig. 4.1.5A). The blade's role is to dispense paste of material on the substrate surface uniformly and equally to obtain the expected film thickness. An alternative way to perform this process is to roll over a glass rod on the substrate containing a precise amount of paste. The rod produces a thin film on the ground of the spacer or around the substrate. This process is economical in the context of spin coating since the material utilization is about 95% once optimized (Krebs, 2009). Fig. 4.1.5B displays a graphic interpretation of tape-casting photoanode for DSSC applications (Dai et al., 2010). Though doctor blading has been considered as the simplest procedure among others, some drawbacks like slow solvent vaporization and particle aggregation is obvious while the solution concentration is high.

4.1.2.5.4 Screen printing

Screen printing is another liquid-based method to produce photoanodes for DSSC. Opening areas of polymer pores or steel mesh (Fig. 4.1.6) are the most important factors in forming a thin film layer (Lee, Kim, Jang, Park, & Choi, 2011). At first, the photoanode paste is placed onto screen and TCO substrate is situated

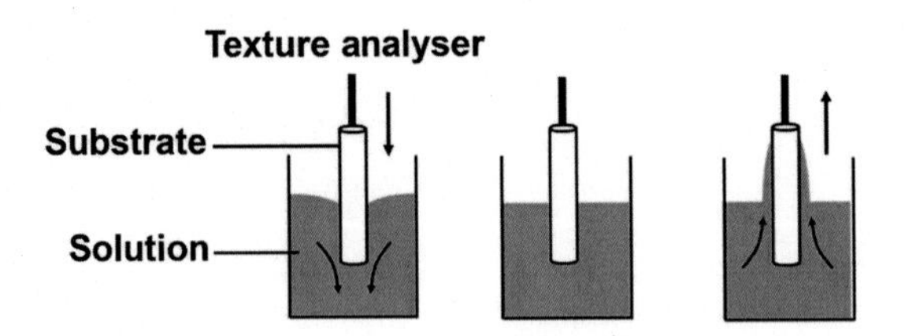

FIGURE 4.1.4

Dipping process (A) immersion, (B) wetting, and (C) withdrawal.

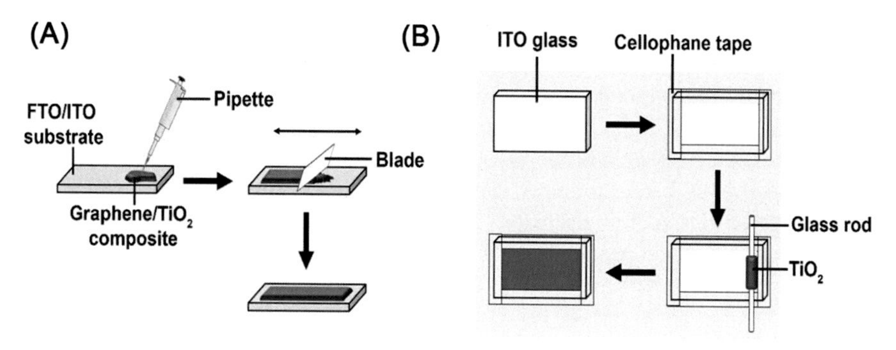

FIGURE 4.1.5

(A) doctor blading and (B) tape-casting procedure to prepare G-TiO_2 photoanodes for dye-sensitized solar cell applications.

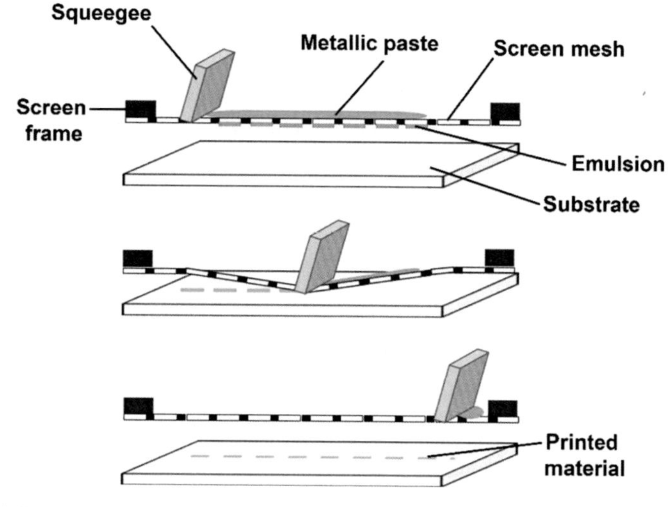

FIGURE 4.1.6

Depiction of screen printing.

underneath. Then this material is conveyed through pressing over the screen with the help of a squeegee so that a definite, ordered, and smooth film can be formed. The transferring amount of material directly relies on the opening of the mesh screen. A number of aspects further affect the process, such as viscosity (η) of paste, speed of spreading, functional gravity, irregularity of the substrate surface, squeegee angle (α), distance (d) between the screen and the substrate (snap off), and its velocity V_d. The relationship between squeezing force (F) and squeezing speed (V_{sq}) is given by the following equation (Liu et al., 2016):

$$F = \eta.V_{sq}\frac{2\alpha.sin\alpha}{\alpha^2 - sin^2\alpha}\cdot\int(Q) \tag{4.1.7}$$

Such a printing method is appropriate for fabricating large area DSSCs modules. Prototype DSSCs having areas of 6000 and 9000 cm^2 (Chen & Chen, 2016; Cui, Zhang, Xing, Feng, & Meng, 2017; Li et al., 2013, 2015; Motlak et al., 2014) have already been fabricated. However, the paste must maintain Newtonian fluid characteristics at low viscosity. The production waste in high volume is also liable toward increased production cost.

4.1.3 Dye

Photosensitizer or dye is a fundamental component of DSSC. It must perform two vital features i.e. absorption of solar spectrum as long as possible and injecting the photogenerated electron into the CB of the semiconductor materials. The ideal photoabsorber ought to hold the features below:

1. The photoabsorber should have a high absorption coefficient specifically in the visible region of the solar spectrum.
2. The photoabsorber or dye should be adsorbed onto the semiconductor surface by the strong anchoring groups attached of the dye molecule.
3. The position of the LUMO level of dye must be higher than the CB of semiconductor for efficient photogenerated CT.
4. The dye should restore its original state through more positive oxidation state than redox shuttle.
5. The photosensitizer should be stable in ambient conditions in the DSSC for as long as 20 years for successful operation.
6. The dye should have better solubility in solvent.

The dye governs the PCE of device mainly in three techniques such as absorbing the incident photons, generating electron–hole pairs effectively upon absorbing incident photon and finally, separating charges and injecting electron to the CB of the semiconductor (Zhang, Yang, Numata, & Han, 2013). Abundant imitation of photoresponding configurations of dyes have been designed to synthesize the features that govern the photon-induced CT mechanism. Over the past decades, abundant steps have been taken to understand and design new absorber dyes for DSSCs. An ideal dye should be mutually electron-rich (donor) as well as electron-poor (acceptor) units connected by a conjugated (π) linker. In general, the electron deprived unit is functionalized by a binder (e.g., -COOH), which is acidic in nature, and attaches the dye molecule onto the semiconductor oxide surface (e.g., TiO_2). Total CT caused by photoexcitation to the acceptor from the donor segments takes place in such a way that the electron wave function aligns to the LUMO of TiO_2. Meanwhile, the hole wave function usually should be confined to the HOMO level of the redox couple but distant from TiO_2 (Lee, et al., 2009).

Commonly, the extended alkyl part is attached at the side of the dye particle. This limits the probable interaction among holes and electrons (TiO_2), consequently inhibiting recombination by creating a blockade in-between said entities. The types of dye employed in DSSC can be dissected into two categories; inorganic dye and organic dye. Inorganic dyes are commonly composed of transition metal complexes such as ruthenium- (Qin & Peng, 2012), copper- (Dragonetti et al., 2019) and zinc- (Daphnomili et al., 2012) based compounds.

4.1.3.1 Inorganic dyes

Ruthenium-complexes exist as the favored absorber for photovoltaic attributes (Kuang et al., 2006). Ru-complexes demonstrate a comprehensive absorption profile of the solar spectrum, appropriate bandgap, and relatively extended excited state lifetime (Haque et al., 2005). The most common Ru-complex dyes are N3 [*cis*-Bis(isothiocyanato) bis(2,2′-bipyridyl-4,4′-dicarboxylato)ruthenium(II)], N719 [*cis*-bis(isothiocyanato)bis(2,2′-bipyridyl-4,4′- dicarboxylato)ruthenium(II)], and black-dye (BD) [2,2″6′,2″-terpyridine]-4,4′,4″-tricarboxylato3N1,N1′,N1″]tris (thiocyanato-N) hydrogen ruthenate(4-) showing a PCE over 10% (Amadelli, Argazzi, Bignozzi, & Scandola, 1990) as shown in the Fig. 4.1.7. Grätzel succeeded to in attaining a PCE of 7.1%–7.9% PCE by using a Ru-complex absorber (Nazeeruddin, Liska, Moser, Vlachopoulos, & Grätzel, 1990; Nazeeruddin, Péchy, & Grätzel, 1997) in 1991. However, Ru-complexes are restricted to absorb a wide range of light spectrum and possess a lower absorption coefficient. The MLCT lets the photogenerated electron to be injected into the CB of the semiconductor TiO_2 proficiently. The CT progression, in conjunction with photon-to-electron transformation, is very effective in Ru-complexes, leaving less need for further development (Nazeeruddin et al., 1993).

4.1.3.2 Organic dyes

DSSC has emerged as a potential candidate for third generation solar cells. Intricate processes to synthesize metal complex sensitizers alongside the environmental burden imposed by these complexes triggered the need to explore a more cost- and nature-friendly alternative. Scheming and synthesizing organic dyes to obtain the expected photochemical properties are naturally possible. Different organic dyes have distinct light harvesting capabilities depending on their molecular conformation. Organic dyes possess a higher absorption coefficient than metal complex dyes (Funabiki et al., 2012). Extensive studies have been conducted on suitable organic dyes with different structures to create long-spectrum absorbing dyes. Efficient dye conformations include donor (D)–conjugation(π)–acceptor (A), D–A, D–D–π–A, (D- π-A)$_2$, D-A-π-A, A-π-D-π-A and so on. Among all the dye structures, the D-π-A conformation has been proved to be the most studied and efficient. The CT mechanism of D-π-A has been illustrated in the Fig. 4.1.8.

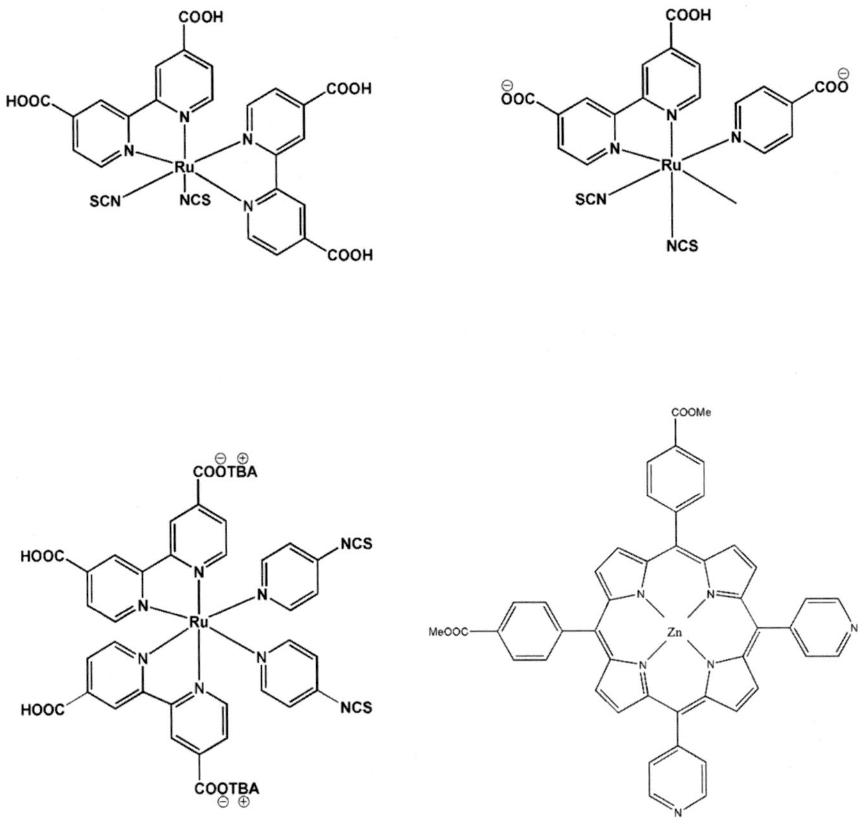

FIGURE 4.1.7

Structure of inorganic dyes.

DSSCs constructed on the Ru-complexes (e.g., BD) absorber showed an efficiency over 10% having bandgap around 1.6 eV (Han et al., 2012). On the other hand, a maximum of the efficient organic dyes has bandgap greater than 2.0 eV. Hence, enough scope is still remaining to achieve the expected HOMO-LUMO gap to improve light harvesting capability and cell performance. The light response attributes of organic dyes can be improved via altering substituents or torsion angles in the D-π-A structure (Hasan et al., 2017). Moreover, the tendency of organic dye aggregation on TiO_2 surface can be controlled by the molecular engineering of dyes (Tong et al., 2014). For instance, K-1 and K-2 are a sequence of novel D-π-A dye, and MK-3 is their carbazole analog (Zhao et al., 2014). DSSCs fabricated on K-dyes exhibited PCE values of 6.62% and 6.73%, respectively, which were greater than MK-3 (4.7%–4.8%), and deprived of substantial reduction of V_{oc} (0.70–0.71 V) yet K-dyes possessed superior bandgaps than MK-3. This consequence could be ascribed to the diverse π-conjugation that was

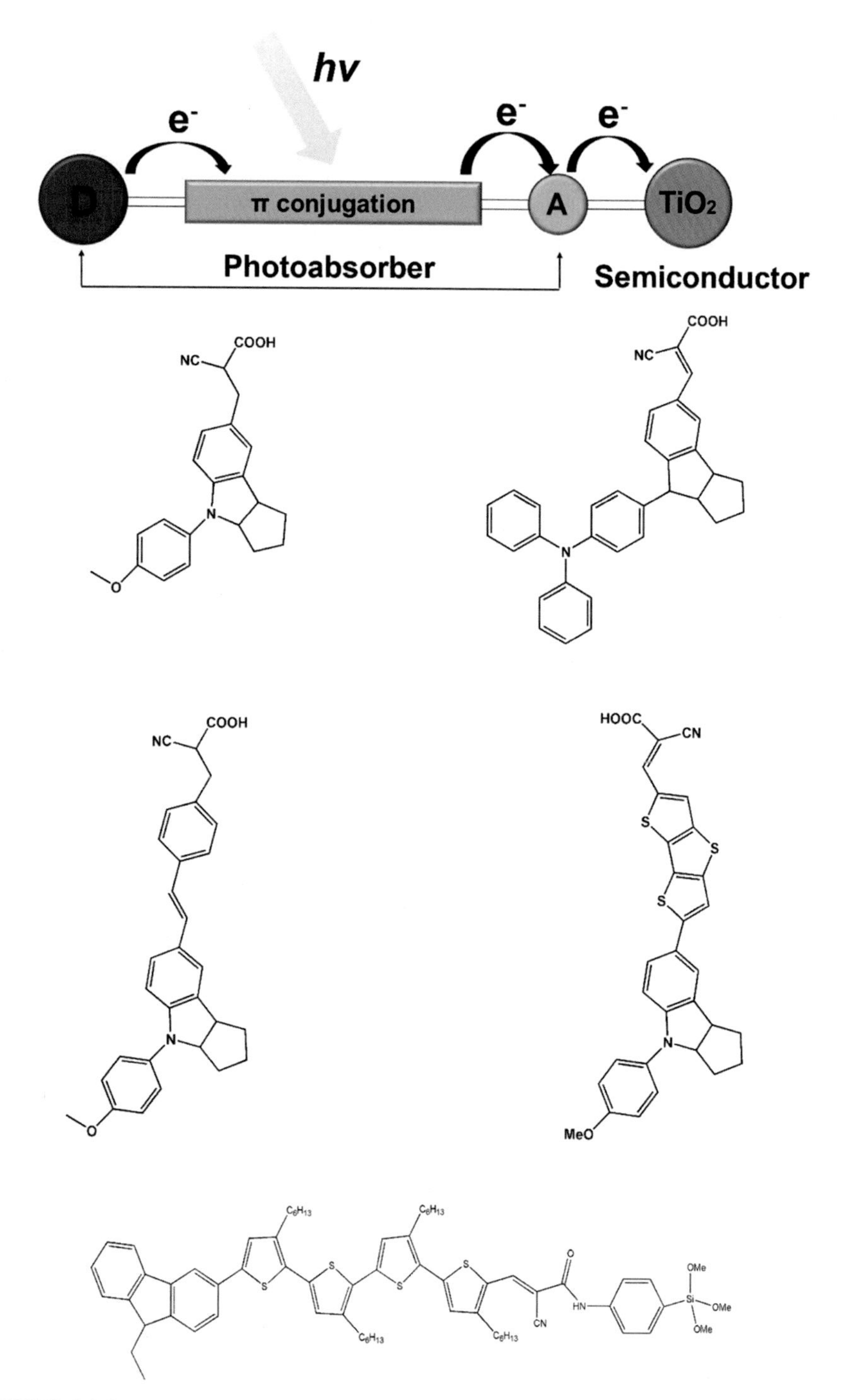

FIGURE 4.1.8

Schematic diagram of D-π-A organic dyes, (B) Molecular structure of organic dyes D-1, U01, AK-01, and Ba-03.

present in the dye structure creating ranges of turning angle due to contain different substituents. Fig. 4.1.8B shows few organic dyes structure. Table 4.1.1 displays photovoltaic performance of DSSCs sensitized with organic dye.

It is important to mention that the PCE of organic dye is governed by the catalytic activity of electrolyte solution. For example, organic dyes show PCE values of 8% and 4% in solution based-redox couples and solid state redox couples, respectively (Mishra, Fischer, & Bäuerle, 2009; Plass, Pelet, Krueger, Grätzel, & Bach, 2002).

4.1.3.3 Natural dyes

Organic dyes from natural sources have been considered as photosensitizers due to the low cost, easy disposability, harmlessness, and ecofriendly nature (Narayan, 2012). Nevertheless, natural photosensitizers aggregate at the surface of photoanodes (Wolfbauer, Bond, Eklund, & MacFarlane, 2001). Some extensively studied pigment groups with examples of their respective plant sources are (Hug, Bader, Mair, & Glatzel, 2014) tabulated in the Table 4.1.2, along with their performance characteristics.

1. Chlorophyll (e.g., arugula, parsley, spinach, henna)
2. Anthocyanin (e.g., rose, lily, pomegranate, red cabbage)
3. Xanthophyll (e.g., marigold, yellow rose)
4. Betacyanin (e.g., cherry, grapes, raspberry, bougainvillea)
5. Carotenoid (e.g., capsicum, walnuts, turmeric).

Table 4.1.1 Current-voltage performance of dye-sensitized solar cells sensitized with organic dye.

Dye	J_{sc} (mA/cm^2)	V_{oc} (mV)	FF	PCE (%)	Reference
D-1	6.42	705	0.750	3.34	Akhtaruzzaman et al. (2013a)
U01	10.70	758	0.740	6.01	Akhtaruzzaman et al. (2013a)
AK-01	15.40	640	0.630	6.20	Akhtaruzzaman et al. (2011)
Ba-03	16.17	595	0.663	6.38	Akhtaruzzaman et al. (2013b)

Table 4.1.2 Natural dyes used in dye-sensitized solar cell.

Structural class	J_{sc} (mA/cm^2)	V_{oc} (mV)	FF	PCE (%)	Reference
Betalains	9.5	425	0.37	1.70	Calogero et al. (2010)
Chlorophyll	11.8	550	0.60	3.90	Wang et al. (2006)
Carotenoid	2.69	686	0.63	1.17	Zhou et al., (2011)
Rutin	2.92	611	0.63	1.12	Zhou et al. (2011)

4.1.4 Counter electrode

The CE is one of the most necessary building materials for DSSC. The CE collects charges from the external circuit and reduces redox species by providing electrons to electrolyte couples, thereby preserving charge neutrality plus contentious dye recover and redox shuttle. TCO (e.g., ITO, FTO) alone cannot reduce the oxidized dye and I_3^- owing to inferior catalytic action and lower charge carrier conductivity (Ameen, Akhtar, Kim, Yang, & Shin, 2010; Bruneaux, Cachet, Froment, Amblard, & Mostafavi, 1989; Ho & Rajeshwar, 1987; Papageorgiou, Maier, & Grätzel, 1997). A perfect CE should achieve small CT resistance (R_{CT}) and great current densities resulting an effective reduction of hole conductor (Ali et al., 2016). Many studies of DSSC address increasing short circuit current density (J_{sc}), open circuit voltage (V_{oc}), and fill factor (FF) for upper efficiency through improved DSSC constituents. In a nutshell, a successful CE should provide the following:

1. greater dynamic presentation to produce great exchange current,
2. unresponsiveness to any potential chemical reaction,
3. very low cost and abundant in supply,
4. mechanically firm and robust, and
5. extraordinary optical transmission and reflection of irradiated light.

The collection of CE is directed by the determination of a specific field of DSSC use. As an example, a building integrating photovoltaic cell ought to build with clear CE, which can integrate the Pt coverage on the TCO. Yet, it is important to use carbon material for a low cost device with the lowest probable sheet resistance aimed at the highest efficiency of device.

4.1.4.1 Bulk Pt as the counter electrode

4.1.4.1.1 Pt nanoparticles

Pt nanoparticles possess great active area, small absorption profile, low CT resistance, and firmness. These stand for some exceptional individualities compared to further noble elements and bulk Pt. The two supreme factors for DSSC efficiency are R_{CT} at CE/redox edge and I_3^- distribution into redox solution (Ferber, Stangl, & Luther, 1998; Hagfeldt et al., 1994). However, cations and diluters affect the aforementioned characteristics, where the Pt breadth and deposition technique equally claim stimulus toward those. A number of Pt deposition methods have been examined to attain optimal conductivity, catalytic action concerning I_3^-, and high reflection effect of Pt (Grätzel, 2000; Kumara, Lim, Lim, Petra, & Ekanayake, 2017). A relative study on properties of Pt–TCO (2–40 nm) as CE formed by electron beam deposition, magnetron sputtering deposition (2, 40, 450 nm), and H_2PtCl_6's thermal decomposition (2–10 nm) specified that the 0.002–0.003 μm range of Pt nanoparticles thickness can deliver great adsorption, ideal permeability, normal CT resistance ($\leq 10\ \Omega\ cm^2$), and great transparency for

DSSC (Fang et al., 2004; Hao, Wu, Lin, & Huang, 2006). The Pt CE preparation such as sputtering (50 nm, 5.17%) (Fang et al., 2004), electrochemical method (40 nm, 5.03%) (Kim, Nah, Noh, Jo, & Kim, 2006), spray deposition method (50 nm, 6.17%) (Iefanova et al., 2014), electrochemical reduction technique (100–300 nm, 9.39%) (Dao & Choi, 2016), thermal decomposition (Tang, Wu, Zheng, Huo, & Lan, 2013) (100–150 nm, 8.15%), bottomsup synthesis method (4.75%) (Faunce et al., 2013), similarly control its properties. The investigation demonstrate that stated routes need an elevated quantity of Pt sources limiting possible usage of rare Pt as the CE. Pt NPs simulate the globular surface of FTO deprived of auxiliary accumulation of Pt NPs due to quicker vanishing of diluents during spray-coat evolution (Chang et al., 2015). Besides, altering the surface of nanomaterials particularly in 3D construction-like nanowires (Piao, Lim, Chang, Lee, & Kim, 2005), nanoflowers (Hsieh et al., 2012), nanotubes (Hino, Ogawa, & Kuramoto, 2006; Sun, Mayers, & Xia, 2003) and nanocups (Jeong et al., 2012) are quite innovative ideas (Fig. 4.1.9). The advantage of this kind association delivers an advanced active exterior area when CE can be formed by uneven ordering of any of the aforementioned nanomaterials.

4.1.4.1.2 Pt composite materials

It has been proposed to focus on the catalytic activity of Pt CE over creating Pt NP-C and PT-polymer compounds as an auxiliary for bulk Pt (Yen et al., 2011). This group prepared Pt-graphene (Pt-G) amalgamations on conducting oxides to form CE by a "water-ethylene glycol method" (Xu, Wang, & Zhu, 2008). It was noticed that the Pt-G complex displayed improved catalytic action due to high CT properties, precise active area with inferior superficial interaction angle, and equivalence to Pt nanoparticle and G, independently. These attribute account for the higher FF of the specific DSSC contributing 6.35% efficacy, compared to the Pt CE (5.47%). Covering of Pt nanoclusters on TCO by poly-N-vinyl-2-pyrrolidone (PVP) was presented by Wei et al. (2006). This two-step plunging is a simple procedure due to its solution's processability. DSSCs with PVP/Pt CE created a reasonable PCE (4.32%) equivalent to sputtered Pt CE (4.55%) demanding only one-tenth of the quantity.

Carbon nanotubes (CNTs) naturally own more electrical conductivity along with exceptional physical durability. Multiwall CNTs (MWCNTs) have established traits and can be a substitute for CNTs. Their amazing specific surface area and faster CT characteristics (Banks & Compton, 2006; Nugent, Santhanam, Rubio, & Ajayan, 2001) are also very encouraging. The (111) orientation of Pt NWs-FTO is a very vital feature to perform as ideal photocatalytic active material. Chiang et al. (2018) prepared PtCoFe, particularly $Pt_{49}Co_{23}Fe_{28}$ NW, which contained higher (111) planes. The CV confirmed lower oxidation-reduction potential energy of 0.53 V for the aforementioned CE certifying enhanced catalytic activities and resulting in the highest J_{sc} of 26.21 $\pm$ 0.21 mA/cm^2. DSSC with Ru(II) dye PRT-22 as an absorber layer and $Pt_{49}Co_{23}Fe_{28}$ as CE produced 12.29% PCE while the Pt CE generated 10.60% PCE.

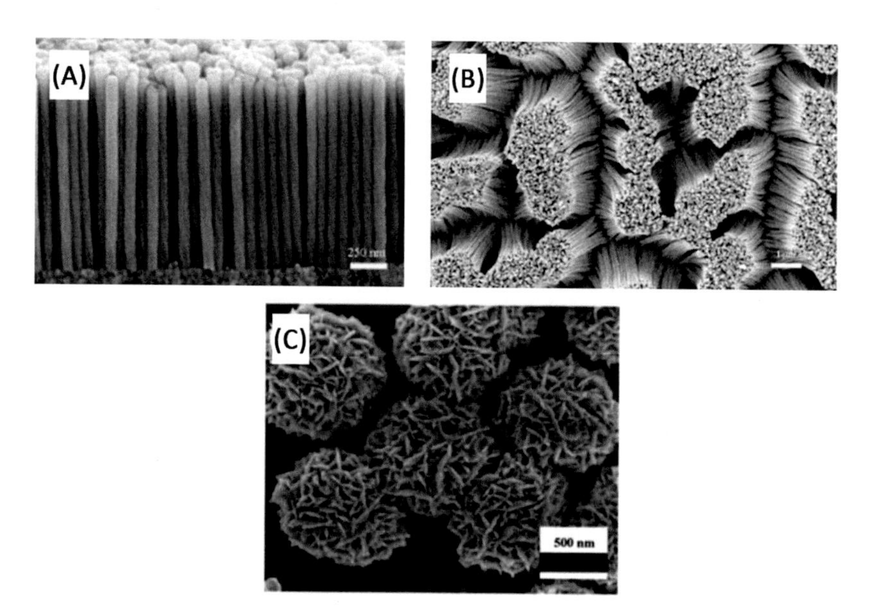

FIGURE 4.1.9

SEM images of (A) side view and (B) top view of Pt nanorods and (C) Pt nanoflowers.

(A and B) reproduced with permission from Piao, Y., Lim, H., Chang, J. Y., Lee, W.-Y., & Kim, H. (2005). Nanostructured materials prepared by use of ordered porous alumina membranes. Electrochimica Acta, 50 *(15), 2997–3013 (C) reproduced with permission from Hsieh, T.-L., Chen, H.-W., Kung, C.-W., Wang, C.-C., Vittal, R., & Ho, K.-C. (2012). A highly efficient dye-sensitized solar cell with a platinum nanoflowers counter electrode.* Journal of Materials Chemistry, *22(12), 5550–5559.*

4.1.4.2 Carbon materials

Carbon (C) is a remarkable reserve material and occurs in different chemical formulas. In exploration of a cheap and plentiful elements to substitute Pt as CE in DSSC, C has been established. Entirely diverse variants of C domestic have revealed reasonable photovoltaic output when integrated as CE in DSSCs (Rhee & Kwon, 2011). By using the CE of C-black dispersed in graphite dust, 6.7% PCE was attained by Kay & Grätzel, (1996).

4.1.4.2.1 Mesoporous carbon

This deviation of C holds large active sites and penetrability that could be twisted toward altered hole radius. Installing C of diverse morphologies can decline CE charge and enhance DSSC performance (Plonska-Brzezinska et al., 2011). Lately, Wu et al. (Wu & Lin, 2019) recommended functionalization of mesoporous carbon by O_2 and S, which led to flaws in the morphological equilibrium through additional disordered C. As a result, substantial quantities of redox dynamic region were amplified, permitting quicker and additional redox response to occur. DSSC with dual O_2 and S-doped C CE unveiled 10.2% PCE, higher than Pt CE (9.4%).

4.1.4.2.2 Graphene

Single graphite layer is called graphene (Gr). Honeycomb-like hexagonal structure of sp^2 hybridized carbon atom forms 2D Gr (Novoselov et al., 2004). Gr possesses amazing physical and electrical characteristics like high carrier mobility, specific surface area, excellent thermal conductivity, high Young's modulus, and very low optical absorbance. All the stated characteristics are extremely expected from an alternative CE for DSSC.

Zhang, Li et al. (2011), Zhang, Liu et al. (2011) investigated the adaptability of graphene nanosheets (GN) CE for DSSC for the first time. GN was produced by comprehensive parting from graphite oxide followed by the reduction. The Raman bands disclose graphitic features of GN with more specific surface sites. GN CE was arranged by screen lithography on FTO and heat treatment was performed at 400°C, that showed 6.8% PCE. It is perceived that forging heat holds a pronounced influence on GN to attain improved glueyness and 3D permeable arrangement. Carbon-based binders might be accountable for flaking off GNs from FTO over 400°C. Different methods to produce GNs are of GO reduction by microwave-assisted chemical method (Hsieh, Yang, & Lin, 2011), electrophoretic deposition method (Choi, Hwang, et al., 2011) and so on. Fig. 4.1.10 symbolizes heteroatom-doped graphene nanoplatelets (GnPs) produced by ball-milling (Kim et al., 2018).

Even though Gr-based DSSC showed lower efficiency compared to FTO-constructed DSSC, it was additionally changed through diverse investigation. Noticeably, it was early stage for expending Gr as CE for DSSC.

4.1.4.3 Gr-based composites

4.1.4.3.1 Graphene-polymer combined material

Kavan et al. (Kavan, Yum, & Grätzel, 2011) reported on Gr-conducting polymer compound as CE in 2006 for the first time. In such preparation, Gr provides reduction of I_3^- by catalytic action whereas conductive polymer provides conducting provision. However, Gr might also be organized as backer for polymer when distributed. Polyaniline (PANI) and poly(3,4-ethylenedioxythiophene) (PEDOT) are frequently positioned in Gr-polymer CE. Gr-PANI was investigated by Wang, Zhuo, and Xing (2012) proposing 6.09% PCE but bulk Pt CE exhibited 6.88%. PCE. Lately Bayram et al. (2020) organized Gr-PANI CE on a single (Gr1) and multilayer (Gr5) G by polymerization through plasma method. Graphene layers' transmittance stretching from 97.6%−86.6% has been observed (Fig. 4.1.11A). The Gr1/PANI and Gr5/PANI CE displayed PCE values of 1.36% and 0.56%, respectively (Fig. 4.1.11B), indicating that Gr thickness influenced DSSC output. The Gr1/PANI CE showed superior PCE compared to pristine Pt CE (Papageorgiou, 2004).

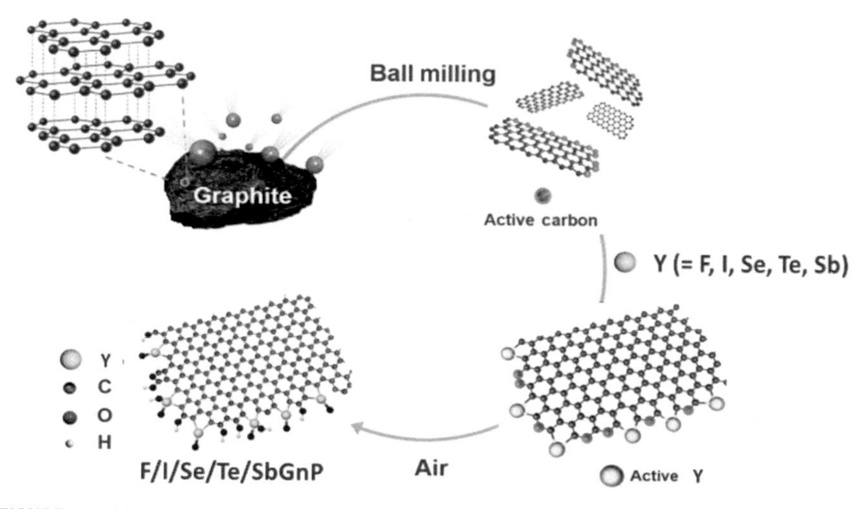

FIGURE 4.1.10

Schematic representation of the preparation of heteroatom-doped edge-functionalized graphene nanoplatelets (GnPs) driven by mechano-chemical ball-milling.

Reproduced with permission from Kim, C.K., et al. (2018). Comparative study of edge-functionalized graphene nanoplatelets as metal-free counter electrodes for highly efficient dye-sensitized solar cells. Materials Today Energy, *9, 67–73.*

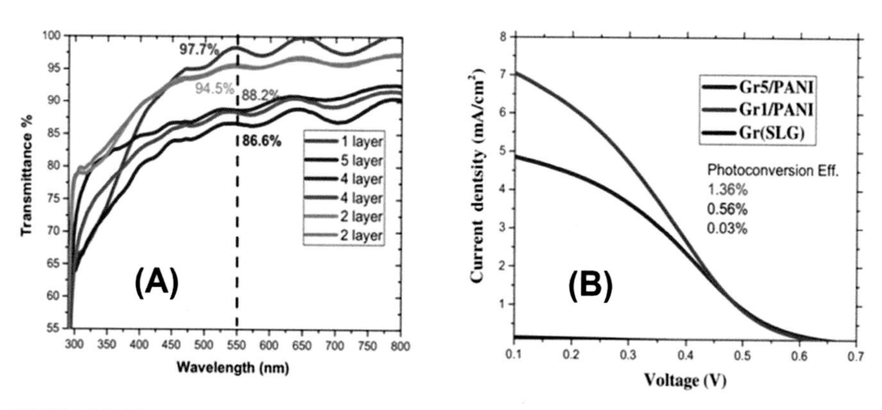

FIGURE 4.1.11

(A) Transmittance spectra of GN of different number of layer; (B) J–V curves of untreated Gr, Gr1/PANI, and Gr5/PANI electrode-based DSSCs.

Reproduced with permission from Bayram, O., et al. (2020). Graphene/polyaniline nanocomposite as platinum-free counter electrode material for dye-sensitized solar cell: its fabrication and photovoltaic performance. Journal of Materials Science: Materials in Electronics, 31*(13), 10288–10297.*

Transition elements, like Ni and Co, might reduce Pt amounts in joining thru Gr to formulate CE. The Gr-metal compound as CE is helpful aimed at amplifying CT amount, predictable reaction concerning the reduction of redox couple over redox activity. The metal occupies sites between Gr sheets easing the dispersion of electrolyte through Gr bridge/backbone, making certain the formation of CE-electrolyte adjacent interaction (Bajpai et al., 2012; Thoi, Sun, Long, & Chang, 2013). A pronounced example was shown by Dou et al. formulating Gr-$Ni1_2P_5$ CE hydrothermally, which showed 5.7% PCE (Dou, Li, Song, & Gao, 2012). Lately, iron carbide (Fe_3C) enfolded up with graphene (Fe_3C@Gr) composite material was projected by Xu et al. (2018) as CE for DSSC.

Reduced graphene oxide (rGO) is prepared from graphite in a honeycomb arrangement due to sp^2 hybridization of C atoms. The main disadvantages of rGO are reduced specific dynamic sites and occurrence of $\pi-\pi$ assembling which increases rGO sheets accumulation resulting to lessening active areas concerning redox shuttle while measured as CE independently (Kavan et al., 2011; Xue et al., 2012). Approaches have been deliberated to control such disadvantages as rGO is inexpensive and abundantly resourced. If those matters might be eradicated effectively, then rGO would take a revolution by substituting expensive Pt. A number of such compounds are FeS_2/rGO doped with Ni (Ni-FeS_2/rGO) (Kumar, Tsai, & Fu, 2020), S-rGO/MoS_2 doped with Ni (Wang et al., 2018), rGO doped with Ag (Ag-rGO) (Mustaffa, Rahman, & Umar, 2020), MoFe amalgam decorated rGO (Shin, Dao, & Choi, 2020) and likewise. The projected Ni-FeS_2/rGO (Kumar et al., 2020) was produced by hydrothermal method. The optimal complex exposed great catalytic action thru high J_{sc}, and 7.60% PCE beating Pt CE (6.69%) ensuing noteworthy backing of inferior R_{CT} in accordance with advanced charge carrier lifetime, independently.

4.1.4.3.2 Graphene–carbon composites

Gr has become a strong nominee as CE in DSSC since it can provide high hole mobility and lower response to humidity and oxygen. Choi, Kim, Hwang, Choi, and Jeon (2011) suggested Gr-MWCNT complexes as CE, where Gr sheet had been enclosed by MWCNT by CVD. The CE unveiled PCE (3%) higher than MWCNT CE. Nevertheless, this process requires a complex fabrication route leading to high cost. A simple route was used for preparing MWNT–Gns composite CE by Battumur et al. (2012). MWNTs and Gr of 60% and 40%, correspondingly, showed 4% PCE. Velten et al. (2012) inspected MWCNT–Gr as a CE in DSSC displaying outstanding PCE of 8.82%, signifying an optimal ratio of MWCNT and Gns.

4.1.4.4 Polymer materials

Theoretically, a single bond alongside a double bond is denoted as conjugation which is accountable for charge carrier transportation attributes present in conductive polymeric materials. The conjugated connection in polymeric structure eases conductivity. Moreover, catalytic action of such kind of polymers mimic Pt to reduce I_3^- depending on polymerization grade and doping states. Polymers like PANI, polypyrrole (PPy), PEDOT:PSS, and corresponding byproducts are broadly renowned as CE for DSSC (Saranya, Rameez, & Subramania, 2015). The characteristic electrocatalytic reaction of polymers is governed by discrete monomer construction. For instance, PProDOT and PEDOT hold alike monomeric arrangement, yet both obviously variated from their monomer unit PPY and PANI. Crosslinked attachments between units throughout polymerization, polymers turn into 3D spongy structures that are beneficial to outstanding film development, promoting required thin films toward active redox motion. Furthermore, the configuration and morphology of such polymers can effortlessly be altered by controlling preparation conditions (Lee et al., 2009). Some contemporary surveys with polymeric compounds, their derivatives, and composites have been mentioned here. Wu et al. (2019) accomplished a sequence Gr/PANI composite. PANI was tamper with Mn^{2+}, Co^{2+}, $Ni^{2+}$and Cu^{2+}. Fig. 4.1.12A–E shows surface topographies of Gr/PANI-Mn^{2+}, Gr/PANI-Co^{2+}, Gr/PANI-Ni^{2+}, Gr/PANI-Cu^{2+}, and Gr/PANI.

It is fairly clear from the pictures that Gr/PANI-Mn^{2+} contains smooth particles and develops an active surface zone compared to any others. Fig. 4.1.12F symbolize current-voltage curve of DSSCs with aforementioned CE. The Gr/PANI-Mn^{2+} as CE presented the utmost PCE posing an encouraging result amongst all the CEs.

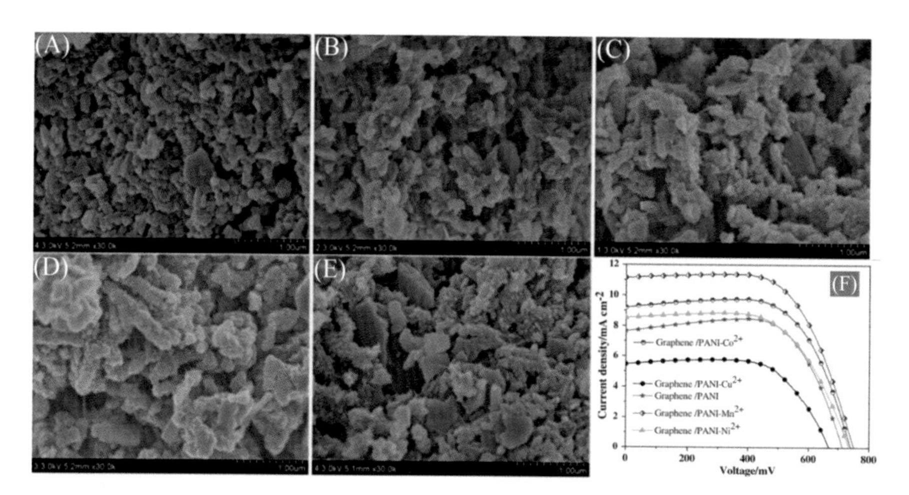

FIGURE 4.1.12

SEM images and J-V curve of the synthesis of the as-prepared Gr/PANI-Mn^{2+}(A), Gr/PANI-Co^{2+}(B), Gr/PANI-Ni^{2+} (C), Gr/PANI-Cu^{2+}(D), Gr/PANI (E) and (F) J-V.

Reproduced with permission from Wu, K., Zhao, Q., Chen, L., Liu, Z., Ruan, B., & Wu, M. (2019). Effect of transition-metal ion doping on electrocatalytic activities of graphene/polyaniline-M^{2+} (Mn^{2+}, Co^{2+}, Ni^{2+}, and Cu^{2+}) composite materials as Pt-free counter electrode in dye-sensitized solar cells. Polymer-Plastics Technology and Materials, 58*(1), 40–46.*

4.1.4.5 Alternative materials

The carrier conductivity of counter electrodes usually follows a sequence as metals and alloys > metal nitrides > carbon materials > multiple metal compounds > conductive polymers, selenides > sulfides > oxides. Greater conductivity guarantees a minor energy cost in CT. However, CEs do not have a straightforward connection with such electrical conductivity. This restriction is favorable since different resources can be positioned aiming at CE preparation. Including C and its derivatives, numerous alternative CE have been proposed. These comprise CoS, NiS, MoS, NiN, MoP, Ni_5P_4, transition metal carbides, nitrides, oxides, and sulfides, etc. (Li, Song, Pan, & Gao, 2011; Sun et al., 2010; Wu et al., 2012). These elements could easily be produced economical route. In 1974, Levy et al. recognized that tungsten carbide (WC) had such an electronic structure that might duplicate the catalytic action of Pt. Diverse devices in energy arenas such as hydrodesulfurization, oxidation hydrogen and methanol have been considered to replace Pt material with transition metals. With the use of CoS and Sn from 2009, the deployment of transition metals sped up (Jiang, Li, & Gao, 2009; Wang et al., 2009).

4.1.5 Electrolyte

As is the case in any electrochemical device, electrolytes are one of the essential components in DSSC. Being mainly responsible for the internal charge transport between the electrodes in order to continually replenish the dye, electrolytes can directly influence photocurrent density (J_{sc}), photovoltage (V_{oc}), and FF of a cell. Some of the crucial prerequisites of an electrolyte in DSSC are (Ardo & Meyer, 2009; Yu, Vlachopoulos, Gorlov, & Kloo, 2011):

1. provide the potential barrier for photovoltaic conversion,
2. rapid transport of charges to prevent the possibility of recombination reactions,
3. good interfacial contact between the mesoporous semiconductor and CE,
4. sustained chemical, thermal, optical, electrochemical, and mechanical stability, and
5. does not exhibit strong absorption in the visible light range to avoid competition with the dye molecule.

Typically, the characterization of the types of electrolytes can be done with respect to their physical states which, as summarized in Fig. 4.1.13, comprises solid, liquid and quasi-solid electrolytes (Wu et al., 2015).

4.1.5.1 Traditional liquid electrolytes

During the inceptive stages of DSSC, organic solvents consisting of iodide/triiodide redox couple were the initial choices of electrolyte used and this liquid electrolyte attained a maximum efficiency of 7.9% (O'Regan, & Grätzel, 1991). Until

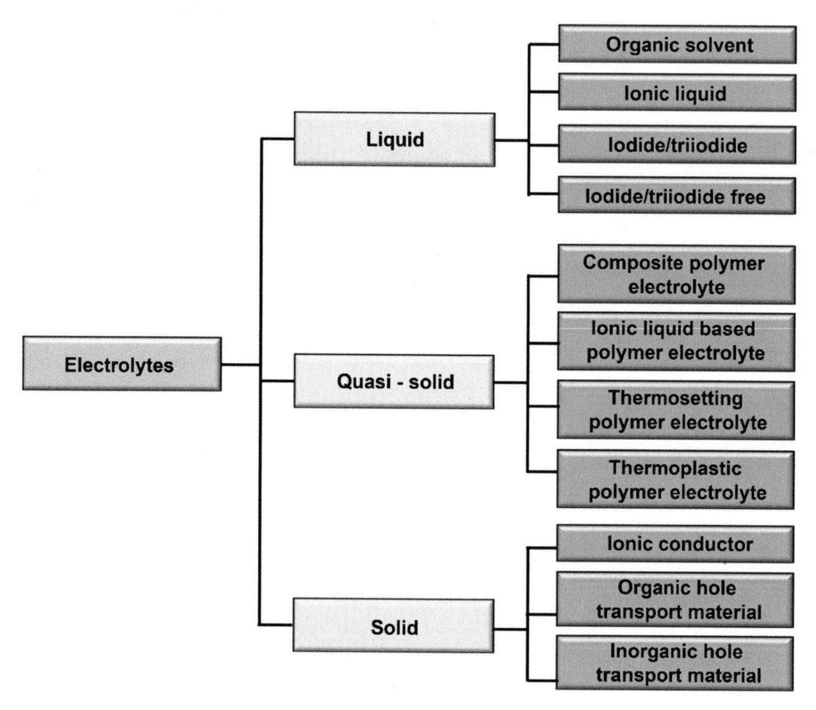

FIGURE 4.1.13

Classification of electrolytes in solar cell.

now, liquid electrolytes remain the most extensive and successful type of electrolytes with a record breaking efficiency of 13% when applied in traditional DSSC (Mathew et al., 2014). The solvents used in liquid electrolytes need to comply with certain conditions such as melting point below −20°C and boiling point above 100°C, high dielectric constant and low viscosity (Yu et al., 2011). So far, there are two categories of solvents which agree with these requirements, namely polar organic solvents and ionic liquids (IL).

Amongst the various polar organic solvents experimented, acetonitrile received particular attention owing to its excellent chemical stability (Hagfeldt & Grätzel, 2000). A study by Hauch et al. pointed out that, upon determination of diffusion constant of triiodide in different solvents and cations, acetonitrile and Li^+ combination achieved the best diffusion constant (Hauch & Georg, 2001). Some of the highest efficiencies in traditional DSSC were also attained with acetonitrile based electrolytes (Joly et al., 2014). However, the low boiling point of acetonitrile (82°C) restrains its usage in commercial solar cells.

Interestingly, water was also proposed as one of the prospective solvents in liquid electrolytes (Desilvestro, Graetzel, Kavan, Moser, & Augustynski, 1985; Kalyanasundaram, Vlachopoulos, Krishnan, Monnier, & Graetzel, 1987; Liska,

Vlachopoulos, Nazeeruddin, Comte, & Graetzel, 1988). For instance, in 2010, Grätzel et al. studied water-based electrolytes using hydrophobic dyes and recorded an efficiency of 2.4%, which remain the highest value obtained for a pure aqueous solvent (Law et al., 2010). The major bottleneck for such aqueous based electrolyte is the oxidation of iodide (I^-) to iodate (IO_3^-) in the presence of water, which inherently reduces the I_3^- ions leading to instability in cell performance (Tributsch, 2004). The selection of water as the electrolyte medium also greatly restricts the options of dyes that can be applied, provided that most of these sensitizers are susceptible to hydrolysis (Hagfeldt, Boschloo, Sun, Kloo, & Pettersson, 2010). Therefore, until these issues remain unsolved, application of aqueous based electrolytes will not be a feasible option.

4.1.5.2 Ionic liquids as electrolytes

IL has been recently established as the new age solvent and highly acclaimed for its superior properties such as multiple solvation interactions with both organic as well as inorganic compounds, high chemical and thermal stability, high ion conductivity and wide electrochemical window (Jayakumar, Mehta, Rao, & Fathima, 2015). This new age solvent has also been nicknamed as the "designer solvent" owing to its ability to alter cation and anion combination adapting to the required application. The role of ILs as electrolytes in DSSC can be in two forms. First, IL can be used as the solvent medium of liquid electrolytes and secondly, it can be a source of charge carriers, just like the conventional inorganic salts, in quasi-solid DSSC (Wu et al., 2015). The latter form will be elaborated in following sections.

One of the preliminary studies involving IL electrolytes is by Papageorgiou et al. who fabricated methyl-hexyl-imidazolium iodide-based DSSC. These cells particularly exhibited outstanding stability with an estimated sensitizer turnover in excess of 50 million (Papageorgiou et al., 1996). The encouraging results have motivated various researchers to explore different types of IL as electrolytes in DSSC and it was found that imidazolium iodide-based IL produced the best performance (Chen et al., 2013). In addition to that, low vapor pressure of these IL is deemed to be an added perk for the solvent as this may translate into lower chances of evaporation and leakages (Bier & Dietrich, 2010). However, despite all the merits, high viscosity and thus low ionic mobility persists to be the major disadvantages of IL based DSSC (Zistler et al., 2006). This is often tackled by introducing a low viscous cosolvent into the system and this auxiliary solvent can either be some organic solvents (Shi et al., 2008) or other low viscosity IL itself (Fan et al., 2010; Ito et al., 2006). With greater understanding of the intrinsic behaviors of IL in the future, more effective means can be introduced to materialize the idea of efficient solvent-free liquid electrolytes for DSSC.

4.1.5.3 Quasi-solid electrolytes

Despite remarkable efficiencies attained using liquid electrolytes, the application of these electrolytes in commercialized DSSCs was deterred by practical flaws such as mechanical instability, leakage, volatilization of solvent, and photodegradation of the dye (Nazeeruddin et al., 2005). On the other hand, the credentials of a completely solid electrolyte do not live up to the required performance often due to poor contact between electrodes (Wu et al., 2008). The trade-off between electrochemical performance and mechanical stability of the electrolyte was achieved by introducing a new class of electrolytes, namely the quasi-solid electrolyte. A quasi-solid state refers to the condition of a material which exists between solid and liquid states. This unique state allows the electrolyte to possess both the cohesive property of solid and diffusive property of liquid (Wang, Zakeeruddin, Moser, et al., 2003). Based on the formation technique and physical features of the material, quasi-solid electrolytes can be dissected into four main categories: composite polymer electrolyte, IL electrolyte, thermosetting electrolyte and thermoplastic electrolyte.

4.1.5.3.1 Composite polymer electrolytes

One of the methods proposed to convert liquid electrolytes into the quasi-solid state is via the addition of inorganic nanoparticles or gelators and the gels produced by this method are named composite polymer electrolytes. This method is notably advantageous due to the formation of an inorganic network within the electrolyte matrix, functioning as a mode of transport for the ions and eventually enhancing charge transport efficiency (Huo et al., 2007).

TiO_2 nanoparticles are one of the widely used organic gelators in quasi-solid electrolytes. For instance, Katsaros et al. fabricated DSSC using polymer electrolytes comprising Polyethylene oxide (PEO), LiI and I_2 incorporated with TiO_2 which displayed an efficiency of 4.2% (Katsaros, Stergiopoulos, Arabatzis, Papadokostaki, & Falaras, 2002). The arrangement of TiO_2 within the polymer chains was as such that a stable three-dimensional network with interstitial gap was created, enabling swift movement of the redox ions. Similarly, PVDF-HFP based electrolyte gelated with TiO_2 was investigated by Huo et al., who apparently observed sixfold improvement in the I_3^- diffusion coefficient (Huo et al., 2007).

Wang et al. attempted to prepare composite electrolyte by including fumed SiO_2 nanoparticles into ILs to produce QSDSSC yielding 7% efficiency (Wang, Zakeeruddin, Comte, et al., 2003). The high performance of the electrolyte inspired subsequent researches with SiO_2 as the inorganic gelator. One such study is by Yoon et al., who highlighted shape dependence of SiO_2 nanomaterials in quasi-solid electrolyte. According to this study, by altering the shape of silica from nanospheres to nanorods, a 38% increase in the overall efficiency was observed (Yoon, Song, Won, & Kang, 2014). Apart from the oxides of silicon and titanium, Al_2O_3 and ZnO nanoparticles based QSDSSC have also been experimented recently (Chi, Roh, Kim, Heo, & Kim, 2013; Singh, Singh, & Kaur, 2016).

Another versatile approach to fabricate composite electrolytes is through the insertion of nanoclay minerals as the gelating agent (Ito, Freitas, Paoli, & Nogueira, 2008; Jin & Chen, 2012; Wang, Deng, et al., 2013). Unique features of nanoclays such as high chemical stability, excellent swelling capability, ion exchange capacity, light scattering property, and rheological property projects them as highly prospective gelators in QSDSSC (Wu et al., 2015). In a recent paper by Mhaisalkar et al., a synthetic nitrate-hydrotalcite nanoclay-based QSDSSC was shown to record efficiencies as high as 9.6%, proving the potential of nanoclays in future works (Wang, Kulkarni, et al., 2013).

4.1.5.3.2 Thermoplastic polymer electrolyte

Typically, thermoplastic polymer electrolyte (TPPE) involves three main components namely polymer, salt and solvent. The primary function of the polymer is to serve as the matrix of the gel in order to bestow mechanical stability. Addition of inorganic salts introduces charge carriers into the matrix which elevates the ionic conductivity of the electrolyte. Most often, the major composition of the electrolyte is made of the solvent component which is regarded as packets of liquid trapped within the polymer chains. The solvent forms a crucial part of the system as it affords the space for charge carrier migration, diminishes crystallinity of the gel, decreases polymer-polymer interaction and increases the free volume and segmental mobility of the system (Wu et al., 2015). The composition of the solvent system can either be any polar organic solvent solely or a mixed system with small organic molecule (called plasticizer) along with the organosolvent (Fig. 4.1.14).

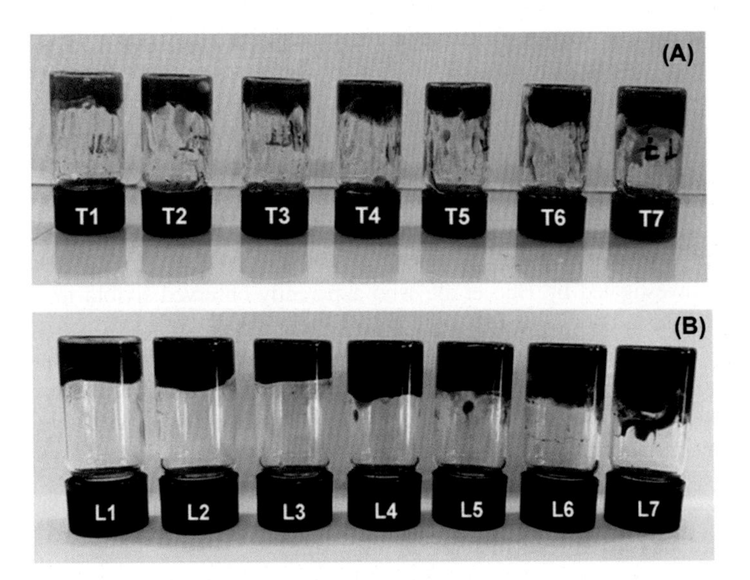

FIGURE 4.1.14

Example of thermoplastic gel electrolytes based on cellulose/starch mixture.

The preparation of the TPPE involves dissolution of the salt in solvent medium in order to form the electrolytic solution and into this, the polymers are added followed by sufficient amount of heating and stirring. During this process, the swelling, adsorption, inflation and entanglement of polymer chains convert the dilute electrolyte system into a gel with viscoelastic properties (Wu et al., 2008). The entire gelation process is due to the establishment of various weak interactions within the electrolyte matrix including hydrogen bonds, van der Waals and electrostatic interaction (Nogueira, Longo, & De, 2004). Since these interactions are highly temperature-dependent, the physical state of the gels (from viscous to dilute condition) can be governed by altering its temperature. Hence, giving rise to the name "thermoplastic gels" which indicates the reversibility of the gels from hard to soft upon heating and vice versa with cooling (Wu, Hao, et al., 2007).

The pioneer work on TPPE began in 1995 by Cao et al., in which Polyacrylonitrile (PAN) was incorporated with ethylene carbonate, propylene carbonate and acetonitrile and sodium iodide (Cao, Oskam, & Searson, 1995). Ever since that, a plethora of studies on synthetic polymers based TPPE in QSDSSC have been carried out. Table 4.1.3 outlines some of these studies.

Amongst the various polymers tested, PEO and their derivatives are the most prevalently used polymer in TPPE (Su'ait, Rahman, & Ahmad, 2015). In fact, some of the highest efficiency for TPPE based QSDSSC were obtained with PEO as the host material (Lee et al., 2018). In 2009, Shi et al. reported QSDSSC fabricated from high molecular weight PEO and the cells boasted a PCE of 10.11% (Shi et al., 2009). It is proposed that the superior performance of PEO-based TPPEs is credited to the presence of multiple ether and polyhydric groups throughout the polymer chains which serves as complexation sites for the cations. As the cations are engaged in an interaction with the polymer chains, the iodide anions are more liberated to move around (Nogueira, Longo, & De Paoli, 2004).

Besides the addition of a single polymer component as the gelating agent, utilization of polymer blends is also a useful procedure adapted to achieve TPPE with desirable properties. By blending Polyvinylidene fluoride (PVDF) into PEO-$LiClO_4$ electrolyte, Jacob et al. had shown that the ionic conductivity can be improved by two orders of magnitude (Jacob, Prabaharan, & Radhakrishna, 1997). In a recent study Hsu et al. demonstrated the difference in photovoltaic performances of DSSC composed of graphene oxide nanosheet-polyaniline (GOS-PANI) nanohybrid/PEO blend gel electrolytes (Hsu, Tseng, & Lee, 2014). In their study, it was evident that due to the improvement in ionic conductivity and electrochemical catalytic activity of the gel electrolyte, higher efficiency was observed for the DSSCs based on the blend in comparison to pristine PEO electrolyte-based DSSC sample.

4.1.5.3.3 Thermosetting polymer electrolyte

Essentially, the preparation of a thermosetting polymer electrolyte (TSPE) is similar to TPPE except for the fact that, gelation process is an effect of crosslinking

Table 4.1.3 Examples of thermoplastic polymer electrolyte based QSDSSC in literature.

Polymer	Electrolyte component	Dye	Efficiency, η (%)	References
Synthetic polymers				
Polyacrylonitrile (PAN)	EC/PC/TPAI/LiI/I_2	N719	6.40	Wanninayake et al. (2016)
Poly(methyl methacrylate)	EC/PC/ TPAI/KI/I_2	N719	3.99	Dissanayake et al. (2014)
Polyvinyl alcohol	EC/PC/DMSO/ TPAI/KI/I_2	N3	4.59	Arof et al. (2014)
PVDF-HFP	MPN/TBP/ KI/I_2	N719	4.75	Huang et al. (2011)
Poly-N-vinyl-2-pyrrolidone	Methanol/KI/ I_2	N719	3.74	Chalkias et al. (2018)
Polyethylene oxide	Acetonitrile/ LiI/KI/ I_2	N719	5.80	Agarwala et al. (2011)
PEG	PC/KI/ I_2	N719	7.22	Wu et al. (2007a)
Natural polymers				
Phthaloyl chitosan	EC/DMF/ TPAI/I_2	N3	6.36	Yusuf et al. (2017)
Hydroxypropyl cellulose	EC/PC/MPII/ NaI/ I_2	N719	5.79	Khanmirzaei, Ramesh, and Ramesh (2015)
Cyanoethylated hydroxypropyl cellulose	MPN/MHII/ TBP/NaI/ I_2	N719	7.55	Huang et al. (2012)
Agarose	NMP/ NaI/ I_2	N/A	4.14	Yang et al. (2011)
Polymer blends				
PEO/PMMA	EC/PC/DMC/ TBP/LiI/ I_2	N719	4.90	Aram, Ehsani, and Khonakdar, (2015)
PVDF-HFP/PAN-VA	Acetonitrile/ LiI/KI/ I_2	N719	6.30	Venkatesan et al. (2014)
Phthaloyl chitosan/PEO	EC/DMF/TPAI/BMII/ I_2	N3	9.61	Buraidah et al. (2017)

reaction with formations of permanent covalent bonds (Li, Wang, Kang, Wang, & Qiu, 2006; Liska et al., 1988). The gels produced by this technique often comprises a three-dimensional network within which the solvent is entrapped. The term "thermosetting" implies the irreversible physical state of the gel with temperature (Wu et al. 2007). In most cases, the visual impression of a TSPE is very similar to solid electrolyte but the internal presence of solvent qualifies them to be quasi-solid electrolytes.

In-situ polymerization by chemical, photochemical, or thermal means is one of the frequent mechanisms employed for the fabrication of TSPEs (Wang, Fang, Lin, Zhou, & Li, 2005). It is believed that by carrying out the polymerization of the electrolytes after assembling between the electrodes, penetration and wetting of the mesoporous layer in photoanode can be improved (Wu et al., 2015). This is also conjectured to reduce unwanted recombination reactions that may diminish cell efficiency. However, the designing of polymerization for QSDSSC can be slightly tricky as there are certain restrictions that need to be imposed. For example, the polymerization should occur without an initiator and complete without a by-product, as both these components may affect photovoltaic properties of the cell (Murai, Mikoshiba, Sumino, & Hayase, 2002). Also, the temperature at which polymerization takes place must not exceed the decomposition temperature of dye molecules.

TSPE fabricated from thermally polymerized poly(ethylene oxide-copropylene oxide) trimethacrylate is one the studies in which enhanced V_{oc} due to suppressed recombination was reported (Komiya, Han, Yamanaka, Islam, & Mitate, 2004). Three polymerizable reactive groups in the oligomer enabled the formation of a three-dimensional polymer network, capable of containing a larger amount of solvent due to larger free volume. Another innovative approach was adapted by Winther-Jensen et al., who combined photocatalysis and photoinduced polymerization (Winther-Jensen, Winther-Jensen, Forsyth, & MacFarlane, 2008). In their work, a novel method of using TiO_2 nanoparticles as both photoinitiator and cogelator in a CT polymerization reaction was presented.

4.1.5.3.4 Ionic liquid polymer electrolyte

As discussed earlier, IL has been used as solvent in liquid electrolyte DSSC. However, the versatile nature of IL also permits it to be used as the charge carrier component in electrolytes. When iodide-based IL is employed in polymer electrolytes as the source of ions, it is called an IL polymer electrolyte (Gorlov & Kloo, 2008).

IL is a widely known ingredient added in TPPE to improve the photovoltaic properties of the gels due to its higher dielectric constant and plasticizing effect. Particularly, imidazolium iodides have been used often in QSDSSC. An important aspect that requires scrutiny when selecting IL as an additive is the alkyl-substituents on the imidazolium cations (Zakeeruddin & Grätzel, 2009). Since these substituents regulate the van der Waals interactions among cations and Coulombic interactions with the anion, the physical properties of the IL are hugely affected (Hwang et al., 2013). As the chain length increases, the room temperature viscosity of the IL increases as an implication of greater van der Waals interactions. Conversely, the shortening of the chain translates into higher lattice energy and increases possibility for solidification of the salt at ambient temperature. In a comparative study of poly (vinylidene fluoride-co-hexafluoropropylene) [PVDF-HFP] containing imidazolium ILs of different carbon chain lengths by Suryanaryanan et al., it was noted that the cell efficiency was mainly controlled by photocurrent density values as photovoltage remained unaffected by different alkyl chains (Suryanarayanan, Lee, Ho, Chen, & Ho, 2007). Although the ionic conductivities decreased, the electron lifetime improved with increasing chain length. Aggregation of imidazolium cations on the photoanode increases in the presence of longer chains, leading to a higher diffusion coefficient of electrons injected in photoanode.

4.1.5.4 Fabrication techniques

Typically, during fabrication of DSSCs, as shown in the Fig. 4.1.15, holes are drilled in the CE layer followed by sandwiching of the CE and photoanode with the help of sealants. The liquid electrolyte is then injected into the predrilled holes, occupying interstitial spaces between the two electrodes. Finally, the holes are sealed with polymer resin.

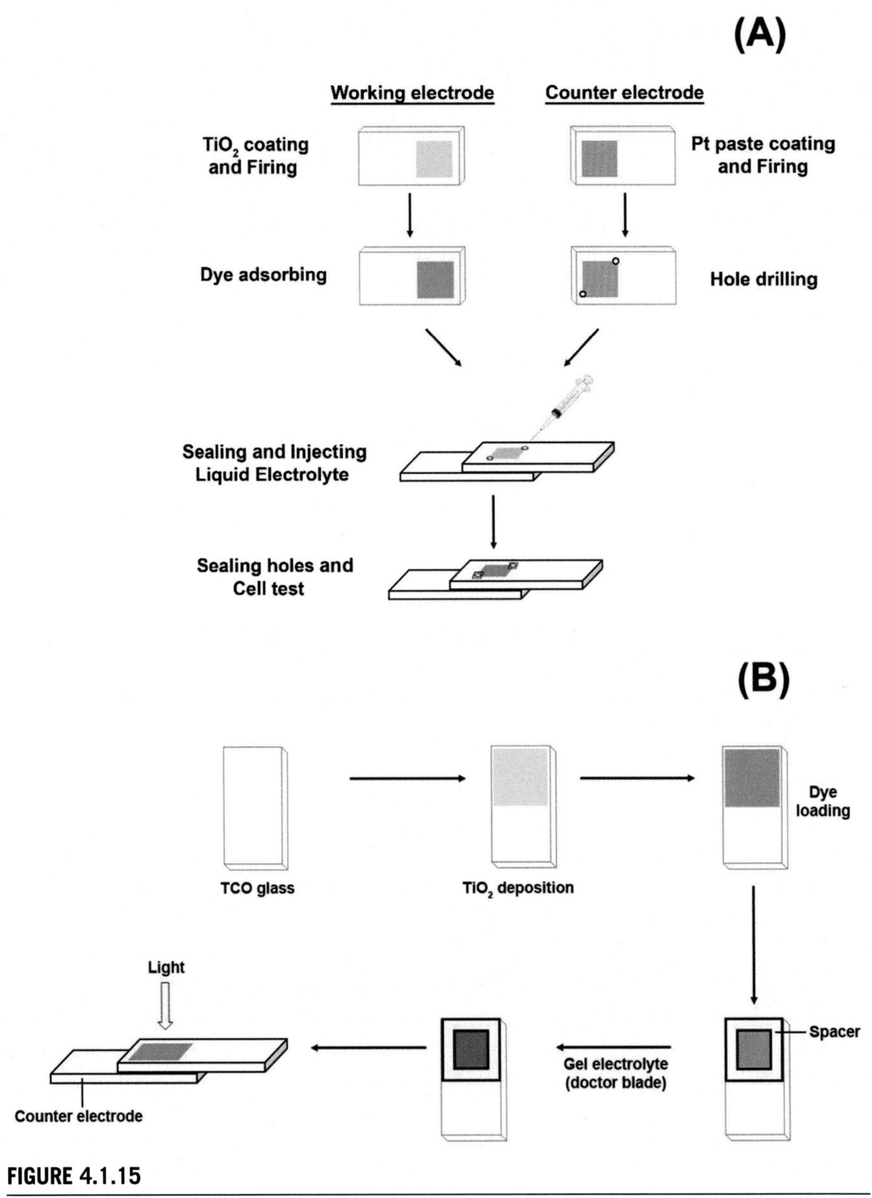

FIGURE 4.1.15

Insertion of (A) liquid electrolytes and (B) quasi-solid electrolyte in dye-sensitized solar cell.

In cases when gel electrolytes are employed in DSSC, the viscous and rigid nature of the gels prevent the application using similar techniques as liquid electrolytes. Hence, thermosetting gels are usually sliced and sandwiched between the electrolytes with subsequent sealing using epoxy glue. On the other hand, thermoplastic gels are smeared on top of the photoanode, which is attached with a polymer film as the spacer to maintain electrolyte thickness. The CE is then clamped on top of the photoanode and upon slight heating the spacer, the polymer film melts to completely seal the cell.

Acknowledgments

The authors would like to acknowledge the support from The National University of Malaysia under Grant LRGS/1/2019/UKM-UKM/6/1. Vidhya Selvanathan wishes to thank Universiti Kebangsaan Malaysia for providing the research grant under the Dana Modal Insan program.

References

Agarwala, S., Thummalakunta, L. N. S. A., Cook, C. A., Peh, C. K. N., Wong, A. S. W., Ke, L., & Ho, G. W. (2011). Co-existence of LiI and KI in filler-free, quasi-solid-state electrolyte for efficient and stable dye-sensitized solar cell. *Journal of Power Sources, 196*(3), 1651–1656. Available from https://doi.org/10.1016/j.jpowsour.2010.08.095.

Akgul, G., Akgul, F. A., Attenkofer, K., & Winterer, M. (2013). Structural properties of zinc oxide and titanium dioxide nanoparticles prepared by chemical vapor synthesis. *Journal of Alloys and Compounds, 554*, 177–181. Available from https://doi.org/10.1016/j.jallcom.2012.11.158.

Akhtaruzzaman, M., Ekramul Mahmud, H. N. M., Islam, A., Ei Shafei, A., Karim, M. R., Sopian, K., ... Yamamoto, Y. (2013). Simple indoline based donor–acceptor dye for high efficiency dye-sensitized solar cells. *Materials Chemistry and Physics, 142*(1), 82–86. Available from https://doi.org/10.1016/j.matchemphys.2013.06.044.

Akhtaruzzaman, M., Islam, A., Yang, F., Asao, N., Kwon, E., Singh, S. P., ... Yamamoto, Y. (2011). A novel metal-free panchromatic TiO_2 sensitizer based on a phenylenevinylene-conjugated unit and an indoline derivative for highly efficient dye-sensitized solar cells. *Chemical Communications, 47*(45), 12400–12402. Available from https://doi.org/10.1039/C1CC15580F.

Akhtaruzzaman, M., Menggenbateer., Islam, A., El-Shafei, A., Asao, N., Jin, T., ... Yamamoto, Y. (2013). Structure–property relationship of different electron donors: Novel organic sensitizers based on fused dithienothiophene π-conjugated linker for high efficiency dye-sensitized solar cells. *Tetrahedron, 69*(16), 3444–3450. Available from https://doi.org/10.1016/j.tet.2013.02.058.

Ali, N., Hussain, A., Ahmed, R., Wang, M. K., Zhao, C., Haq, B. U., & Fu, Y. Q. (2016). Advances in nanostructured thin film materials for solar cell applications. *Renewable and Sustainable Energy Reviews, 59*, 726–737. Available from https://doi.org/10.1016/j.rser.2015.12.268.

Amadelli, R., Argazzi, R., Bignozzi, C. A., & Scandola, F. (1990). Design of antenna-sensitizer polynuclear complexes. Sensitization of titanium dioxide with [Ru(bpy)2

(CN)2]2Ru(bpy(COO)2)22. *Journal of the American Chemical Society*, *112*(20), 7099–7103. Available from https://doi.org/10.1021/ja00176a003.

Ameen, S., Akhtar, M. S., Kim, Y. S., Yang, O. B., & Shin, H.-S. (2010). Sulfamic acid-doped polyaniline nanofibers thin film-based counter electrode: Application in dye-sensitized solar cells. *The Journal of Physical Chemistry C*, *114*(10), 4760–4764. Available from https://doi.org/10.1021/jp912037w.

Aram, E., Ehsani, M., & Khonakdar, H. A. (2015). Improvement of ionic conductivity and performance of quasi-solid-state dye sensitized solar cell using PEO/PMMA gel electrolyte. *Thermochimica Acta*, *615*, 61–67. Available from https://doi.org/10.1016/j.tca.2015.07.006.

Ardo, S., & Meyer, G. J. (2009). Photodriven heterogeneous charge transfer with transition-metal compounds anchored to TiO_2 semiconductor surfaces. *Chemical Society Reviews*, *38*(1), 115–164. Available from https://doi.org/10.1039/B804321N.

Arof, A. K., Naeem, M., Hameed, F., Jayasundara, W. J. M. J. S. R., Careem, M. A., Teo, L. P., & Buraidah, M. H. (2014). Quasi solid state dye-sensitized solar cells based on polyvinyl alcohol (PVA) electrolytes containing I^-/I_3^- redox couple. *Optical and Quantum Electronics*, *46*(1), 143–154. Available from https://doi.org/10.1007/s11082-013-9723-z.

Bajpai, R., Roy, S., kulshrestha, N., Rafiee, J., Koratkar, N., & Misra, D. S. (2012). Graphene supported nickel nanoparticle as a viable replacement for platinum in dye sensitized solar cells. *Nanoscale*, *4*(3), 926–930. Available from https://doi.org/10.1039/C2NR11127F.

Bakhshayesh, A. M., & Farajisafiloo, N. (2015). Efficient dye-sensitised solar cell based on uniform In-doped TiO2 spherical particles. *Applied Physics A*, *120*(1), 199–206. Available from https://doi.org/10.1007/s00339-015-9150-z.

Banks, C. E., & Compton, R. G. (2006). New electrodes for old: From carbon nanotubes to edge plane pyrolytic graphite. *Analyst*, *131*(1), 15–21. Available from https://doi.org/10.1039/B512688F.

Battumur, T., Mujawar, S. H., Truong, Q. T., Ambade, S. B., Lee, D. S., Lee, W., ... Lee, S.-H. (2012). Graphene/carbon nanotubes composites as a counter electrode for dye-sensitized solar cells. *Current Applied Physics*, *12*, e49–e53. Available from https://doi.org/10.1016/j.cap.2011.04.028.

Bayram, O., Igman, E., Guney, H., Demir, Z., Yurtcan, M. T., Cirak, C., ... Simsek, O. (2020). Graphene/polyaniline nanocomposite as platinum-free counter electrode material for dye-sensitized solar cell: Its fabrication and photovoltaic performance. *Journal of Materials Science: Materials in Electronics*, *31*(13), 10288–10297. Available from https://doi.org/10.1007/s10854-020-03575-5.

Bier, M., & Dietrich, S. (2010). Vapour pressure of ionic liquids. *Molecular Physics*, *108*(2), 211–214. Available from https://doi.org/10.1080/00268971003604609.

Bruneaux, J., Cachet, H., Froment, M., Amblard, J., & Mostafavi, M. (1989). Electrochemical behaviour of transparent heavily doped SnO_2 electrodes: Effect of radiolytic grafting of iridium nanoaggregates. *Journal of Electroanalytical Chemistry and Interfacial Electrochemistry*, *269*(2), 375–387. Available from https://doi.org/10.1016/0022-0728(89)85145-9.

Buraidah, M. H., Shah, S., Teo, L. P., Chowdhury, F. I., Careem, M. A., Albinsson, I., ... Arof, A. K. (2017). High efficient dye sensitized solar cells using phthaloylchitosan based gel polymer electrolytes. *Electrochimica Acta*, *245*, 846–853. Available from https://doi.org/10.1016/j.electacta.2017.06.011.

Calogero, G., Di Marco, G., Cazzanti, S., Caramori, S., Argazzi, R., Di Carlo, A., & Bignozzi, C. A. (2010). Efficient dye-sensitized solar cells using red turnip and purple wild sicilian prickly pear fruits. *International journal of molecular sciences*, *11*(1), 254–267. Available from https://doi.org/10.3390/ijms11010254.

Cao, F., Oskam, G., & Searson, P. C. (1995). A solid state, dye sensitized photoelectrochemical cell. *The Journal of Physical Chemistry*, *99*(47), 17071–17073. Available from https://doi.org/10.1021/j100047a003.

Chalkias, D. A., Giannopoulos, D. I., Kollia, E., Petala, A., Kostopoulos, V., & Papanicolaou, G. C. (2018). Preparation of polyvinylpyrrolidone-based polymer electrolytes and their application by in-situ gelation in dye-sensitized solar cells. *Electrochimica Acta*, *271*, 632–640. Available from https://doi.org/10.1016/j.electacta.2018.03.194.

Chandiran, A. K., Sauvage, F., Etgar, L., & Graetzel, M. (2011). Ga^{3+} and Y^{3+} cationic substitution in mesoporous TiO_2 photoanodes for photovoltaic applications. *The Journal of Physical Chemistry C*, *115*(18), 9232–9240. Available from https://doi.org/10.1021/jp1121068.

Chang, H., Cho, K.-C., Kuo, C.-G., Kao, M.-J., Huang, K.-D., Chu, K.-H., & Lin, X.-P. (2011). Application of a Schottky barrier to dye-sensitized solar cells (DSSCs) with multilayer thin films of photoelectrodes. *Journal of Alloys and Compounds*, *509*, S486–S489. Available from https://doi.org/10.1016/j.jallcom.2011.01.162.

Chang, Q., Ma, Z., Lin, Y., Xiao, Y., Huang, L., Xu, S., & Shi, W. (2015). In-situ grown hybrid nanocarbon composite for dye sensitized solar cells. *Electrochimica Acta*, *166*, 134–141. Available from https://doi.org/10.1016/j.electacta.2015.03.092.

Chen, X., Tang, Q., Zhao, Z., Wang, X., He, B., & Yu, L. (2015). One-step growth of well-aligned TiO_2 nanorod arrays for flexible dye-sensitized solar cells. *Chemical Communications*, *51*(10), 1945–1948. Available from https://doi.org/10.1039/C4CC09083G.

Chen, X., Xu, D., Qiu, L., Li, S., Zhang, W., & Yan, F. (2013). Imidazolium functionalized TEMPO/iodide hybrid redox couple for highly efficient dye-sensitized solar cells. *Journal of Materials Chemistry A*, *1*(31), 8759–8765. Available from https://doi.org/10.1039/C3TA11521F.

Chen, Y., & Chen, Z. (2016). Three-dimensional ordered TiO_2 hollow spheres as scattering layer in dye-sensitized solar cells. *Applied Physics A*, *122*(3), 195. Available from https://doi.org/10.1007/s00339-016-9747-x.

Chi, W. S., Roh, D. K., Kim, S. J., Heo, S. Y., & Kim, J. H. (2013). Hybrid electrolytes prepared from ionic liquid-grafted alumina for high-efficiency quasi-solid-state dye-sensitized solar cells. *Nanoscale*, *5*(12), 5341–5348. Available from https://doi.org/10.1039/C3NR00291H.

Chiang, C.-C., Hung, C.-Y., Chou, S.-W., Shyue, J.-J., Cheng, K.-Y., Chang, P.-J., ... Chou, P.-T. (2018). PtCoFe nanowire cathodes boost short-circuit currents of Ru(II)-based dye-sensitized solar cells to a power conversion efficiency of 12.29%. *Advanced Functional Materials*, *28*(3), 1703282. Available from https://doi.org/10.1002/adfm.201703282.

Choi, H., Hwang, S., Bae, H., Kim, S., Kim, H., & Jeon, M. (2011). Electrophoretic graphene for transparent counter electrodes in dye-sensitised solar cells. *Electronics Letters*, *47*(4), 281–283, Retrieved from. Available from https://digital-library.theiet.org/content/journals/10.1049/el.2010.2897.

Choi, H., Kim, H., Hwang, S., Choi, W., & Jeon, M. (2011). Dye-sensitized solar cells using graphene-based carbon nano composite as counter electrode. *Solar Energy Materials and Solar Cells*, *95*(1), 323–325. Available from https://doi.org/10.1016/j.solmat.2010.04.044.

Cui, Z., Zhang, K., Xing, G., Feng, Y., & Meng, S. (2017). Multi-functional 3D *N*-doped TiO_2 microspheres used as scattering layers for dye-sensitized solar cells. *Frontiers of Chemical Science and Engineering*, *11*(3), 395–404. Available from https://doi.org/10.1007/s11705-017-1643-1.

Dai, S., Wu, Y., Sakai, T., Du, Z., Sakai, H., & Abe, M. (2010). Preparation of highly crystalline TiO_2 nanostructures by acid-assisted hydrothermal treatment of hexagonal-structured nanocrystalline titania/cetyltrimethyammonium bromide nanoskeleton. *Nanoscale Research Letters*, *5*(11), 1829. Available from https://doi.org/10.1007/s11671-010-9720-0.

Dao, V.-D., & Choi, H.-S. (2016). Pt nanourchins as efficient and robust counter electrode materials for dye-sensitized solar cells. *ACS Applied Materials & Interfaces*, *8*(1), 1004–1010. Available from https://doi.org/10.1021/acsami.5b11097.

Daphnomili, D., Landrou, G., Prakash Singh, S., Thomas, A., Yesudas, K. K. B., ... Coutsolelos, A. G. (2012). Photophysical, electrochemical and photovoltaic properties of dye sensitized solar cells using a series of pyridyl functionalized porphyrin dyes. *RSC Advances*, *2*(33), 12899–12908. Available from https://doi.org/10.1039/C2RA22129B.

Das, P., Sengupta, D., Mondal, B., & Mukherjee, K. (2015). A review on metallic ion and non-metal doped titania and zinc oxide photo-anodes for dye sensitized solar cells. *Reviews in Advanced Sciences and Engineering*, *4*(4), 271–290. Available from https://doi.org/10.1166/rase.2015.1103.

Desilvestro, J., Graetzel, M., Kavan, L., Moser, J., & Augustynski, J. (1985). Highly efficient sensitization of titanium dioxide. *Journal of the American Chemical Society*, *107*(10), 2988–2990. Available from https://doi.org/10.1021/ja00296a035.

Dissanayake, M. A. K. L., Jayathissa, R., Seneviratne, V. A., Thotawatthage, C. A., Senadeera, G. K. R., & Mellander, B. E. (2014). Polymethylmethacrylate (PMMA) based quasi-solid electrolyte with binary iodide salt for efficiency enhancement in TiO_2 based dye sensitized solar cells. *Solid State Ionics*, *265*, 85–91. Available from https://doi.org/10.1016/j.ssi.2014.07.019.

Dou, Y. Y., Li, G. R., Song, J., & Gao, X. P. (2012). Nickel phosphide-embedded graphene as counter electrode for dye-sensitized solar cells. *Physical Chemistry Chemical Physics*, *14*(4), 1339–1342. Available from https://doi.org/10.1039/C2CP23775J.

Dragonetti, C., Magni, M., Colombo, A., Fagnani, F., Roberto, D., Melchiorre, F., ... Fantacci, S. (2019). Towards efficient sustainable full-copper dye-sensitized solar cells. *Dalton Transactions*, *48*(26), 9703–9711. Available from https://doi.org/10.1039/C9DT00790C.

Duan, Y., Fu, N., Liu, Q., Fang, Y., Zhou, X., Zhang, J., & Lin, Y. (2012). Sn-doped TiO_2 photoanode for dye-sensitized solar cells. *The Journal of Physical Chemistry C*, *116*(16), 8888–8893. Available from https://doi.org/10.1021/jp212517k.

Dürr, M., Rosselli, S., Yasuda, A., & Nelles, G. (2006). Band-gap engineering of metal oxides for dye-sensitized solar cells. *The Journal of Physical Chemistry B*, *110*(43), 21899–21902. Available from https://doi.org/10.1021/jp063857c.

Fan, L., Kang, S., Wu, J., Hao, S., Lan, Z., & Lin, J. (2010). Quasi-solid state dye-sensitized solar cells based on polyvinylpyrrolidone with ionic liquid. *Energy Sources,*

Part A: Recovery, Utilization, and Environmental Effects, *32*(16), 1559–1568. Available from https://doi.org/10.1080/15567030902804764.

Fang, X., Ma, T., Guan, G., Akiyama, M., Kida, T., & Abe, E. (2004). Effect of the thickness of the Pt film coated on a counter electrode on the performance of a dye-sensitized solar cell. *Journal of Electroanalytical Chemistry*, *570*, 257–263.

Faunce, T. A., Lubitz, W., Rutherford, A. W., MacFarlane, D., Moore, G. F., Yang, P., ... Styring, S. (2013). Energy and environment policy case for a global project on artificial photosynthesis. *Energy & Environmental Science*, *6*(3), 695–698. Available from https://doi.org/10.1039/C3EE00063J.

Ferber, J., Stangl, R., & Luther, J. (1998). An electrical model of the dye-sensitized solar cell. *Solar Energy Materials and Solar Cells*, *53*(1), 29–54. Available from https://doi.org/10.1016/S0927-0248(98)00005-1.

Funabiki, K., Mase, H., Saito, Y., Otsuka, A., Hibino, A., Tanaka, N., ... Matsui, M. (2012). Design of NIR-absorbing simple asymmetric squaraine dyes carrying indoline moieties for use in dye-sensitized solar cells with Pt-free electrodes. *Organic Letters*, *14*(5), 1246–1249. Available from https://doi.org/10.1021/ol300054a.

Gorlov, M., & Kloo, L. (2008). Ionic liquid electrolytes for dye-sensitized solar cells. *Dalton Transactions*, *20*, 2655–2666. Available from https://doi.org/10.1039/B716419J.

Grätzel, M. (2000). Perspectives for dye-sensitized nanocrystalline solar cells. *Progress in Photovoltaics: Research and Applications*, *8*(1), 171–185. Available from https://doi.org/10.1002/(SICI)1099-159X(200001/02)8:1<171::AID-PIP300>3.0.CO;2-U.

Grätzel, M. (2003). Dye-sensitized solar cells. *Journal of Photochemistry and Photobiology C: Photochemistry Reviews*, *4*(2), 145–153. Available from https://doi.org/10.1016/S1389-5567(03)00026-1.

Guo, M., Diao, P., Wang, X., & Cai, S. (2005). The effect of hydrothermal growth temperature on preparation and photoelectrochemical performance of ZnO nanorod array films. *Journal of Solid State Chemistry*, *178*(10), 3210–3215. Available from https://doi.org/10.1016/j.jssc.2005.07.013.

Guo, W., Shen, Y., Wu, L., Gao, Y., & Ma, T. (2011). Effect of N dopant amount on the performance of dye-sensitized solar cells based on N-doped TiO_2 electrodes. *The Journal of Physical Chemistry C*, *115*(43), 21494–21499. Available from https://doi.org/10.1021/jp2057496.

Guo, W., Wu, L., Chen, Z., Boschloo, G., Hagfeldt, A., & Ma, T. (2011). Highly efficient dye-sensitized solar cells based on nitrogen-doped titania with excellent stability. *Journal of Photochemistry and Photobiology A: Chemistry*, *219*(2), 180–187. Available from https://doi.org/10.1016/j.jphotochem.2011.01.004.

Guo, X., Dong, H., Niu, G., Qiu, Y., & Wang, L. (2014). Mg doping in nanosheet-based spherical structured ZnO photoanode for quasi-solid dye-sensitized solar cells. *RSC Advances*, *4*(41), 21294–21300. Available from https://doi.org/10.1039/C4RA03188A.

Hagfeldt, A., & Grätzel, M. (2000). Molecular photovoltaics. *Accounts of Chemical Research*, *33*(5), 269–277. Available from https://doi.org/10.1021/ar980112j.

Hagfeldt, A., Boschloo, G., Sun, L., Kloo, L., & Pettersson, H. (2010). Dye-sensitized solar cells. *Chemical Reviews*, *110*(11), 6595–6663. Available from https://doi.org/10.1021/cr900356p.

Hagfeldt, A., Didriksson, B., Palmqvist, T., Lindström, H., Södergren, S., Rensmo, H., & Lindquist, S.-E. (1994). Verification of high efficiencies for the Grätzel-cell. A 7%

efficient solar cell based on dye-sensitized colloidal TiO_2 films. *Solar Energy Materials and Solar Cells*, *31*(4), 481–488. Available from https://doi.org/10.1016/0927-0248(94)90190-2.

Hall, D. B., Underhill, P., & Torkelson, J. M. (1998). Spin coating of thin and ultrathin polymer films. *Polymer Engineering & Science*, *38*(12), 2039–2045. Available from https://doi.org/10.1002/pen.10373.

Han, L., Islam, A., Chen, H., Malapaka, C., Chiranjeevi, B., Zhang, S., ... Yanagida, M. (2012). High-efficiency dye-sensitized solar cell with a novel co-adsorbent. *Energy & Environmental Science*, *5*(3), 6057–6060. Available from https://doi.org/10.1039/C2EE03418B.

Hao, S., Wu, J., Lin, J., & Huang, Y. (2006). Modification of photocathode of dye-sensitized nanocrystalline solar cell with platinum by vacuum coating, thermal decomposition and electroplating. *Composite Interfaces*, *13*(8-9), 899–909. Available from https://doi.org/10.1163/156855406779366732.

Haque, S. A., Palomares, E., Cho, B. M., Green, A. N. M., Hirata, N., Klug, D. R., & Durrant, J. R. (2005). Charge separation versus recombination in dye-sensitized nanocrystalline solar cells: the minimization of kinetic redundancy. *Journal of the American Chemical Society*, *127*(10), 3456–3462. Available from https://doi.org/10.1021/ja0460357.

Hara, K., Zhao, Z.-G., Cui, Y., Miyauchi, M., Miyashita, M., & Mori, S. (2011). Nanocrystalline electrodes based on nanoporous-walled WO_3 nanotubes for organic-dye-sensitized solar cells. *Langmuir*, *27*(20), 12730–12736. Available from https://doi.org/10.1021/la201639f.

Hasan, A. K., Chowdhury, T., Islam, A., Jamal, M., Wadi, A. A., Zin, M. I. M., ... Abdul Wadi, M. (2017). Effect of π-electron system in organic dyes for dye-sensitized solar cells. *Journal of Sustainability Science and Management*, *12*, 103–111.

Hauch, A., & Georg, A. (2001). Diffusion in the electrolyte and charge-transfer reaction at the platinum electrode in dye-sensitized solar cells. *Electrochimica Acta*, *46*(22), 3457–3466. Available from https://doi.org/10.1016/S0013-4686(01)00540-0.

Hino, T., Ogawa, Y., & Kuramoto, N. (2006). Dye-sensitized solar cell with single-walled carbon nanotube thin film prepared by an electrolytic micelle disruption method as the counterelectrode. *Fullerenes, Nanotubes and Carbon Nanostructures*, *14*(4), 607–619. Available from https://doi.org/10.1080/15363830600812183.

Ho, S., & Rajeshwar, K. (1987). Catalytic modification of indium tin oxide electrode surfaces. *Journal of The Electrochemical Society*, *134*(3), 768–769. Available from https://doi.org/10.1149/1.2100555.

Hou, Q., Zheng, Y., Chen, J.-F., Zhou, W., Deng, J., & Tao, X. (2011). Visible-light-response iodine-doped titanium dioxide nanocrystals for dye-sensitized solar cells. *Journal of Materials Chemistry*, *21*(11), 3877–3883. Available from https://doi.org/10.1039/C0JM03327H.

Hsieh, C.-T., Yang, B.-H., & Lin, J.-Y. (2011). One- and two-dimensional carbon nanomaterials as counter electrodes for dye-sensitized solar cells. *Carbon*, *49*(9), 3092–3097. Available from https://doi.org/10.1016/j.carbon.2011.03.031.

Hsieh, T.-L., Chen, H.-W., Kung, C.-W., Wang, C.-C., Vittal, R., & Ho, K.-C. (2012). A highly efficient dye-sensitized solar cell with a platinum nanoflowers counter electrode. *Journal of Materials Chemistry*, *22*(12), 5550–5559. Available from https://doi.org/10.1039/C2JM14623A.

Hsu, Y. C., Tseng, L. C., & Lee, R. H. (2014). Graphene oxide sheet–polyaniline nanohybrids for enhanced photovoltaic performance of dye-sensitized solar cells. *Journal of Polymer Science Part B: Polymer Physics*, *52*(4), 321–332. Available from https://doi.org/10.1002/polb.23416.

Huang, K.-C., Chen, P.-Y., Vittal, R., & Ho, K.-C. (2011). Enhanced performance of a quasi-solid-state dye-sensitized solar cell with aluminum nitride in its gel polymer electrolyte. *Solar Energy Materials and Solar Cells*, *95*(8), 1990–1995. Available from https://doi.org/10.1016/j.solmat.2010.02.047.

Huang, X., Liu, Y., Deng, J., Yi, B., Yu, X., Shen, P., & Tan, S. (2012). A novel polymer gel electrolyte based on cyanoethylated cellulose for dye-sensitized solar cells. *Electrochimica Acta*, *80*, 219–226. Available from https://doi.org/10.1016/j.electacta.2012.07.014.

Hug, H., Bader, M., Mair, P., & Glatzel, T. (2014). Biophotovoltaics: Natural pigments in dye-sensitized solar cells. *Applied Energy*, *115*, 216–225. Available from https://doi.org/10.1016/j.apenergy.2013.10.055.

Huo, Z., Dai, S., Wang, K., Kong, F., Zhang, C., Pan, X., & Fang, X. (2007). Nanocomposite gel electrolyte with large enhanced charge transport properties of an I_3^-/I^- redox couple for quasi-solid-state dye-sensitized solar cells. *Solar Energy Materials and Solar Cells*, *91*(20), 1959–1965. Available from https://doi.org/10.1016/j.solmat.2007.08.003.

Hwang, D., Kim, D. Y., Jo, S. M., Armel, V., MacFarlane, D. R., Kim, D., & Jang, S.-Y. (2013). Highly efficient plastic crystal ionic conductors for solid-state dye-sensitized solar cells. *Scientific Reports*, *3*, 3520. Available from https://doi.org/10.1038/srep03520. Available from https://www.nature.com/articles/srep03520#supplementary-information.

Iefanova, A., Nepal, J., Poudel, P., Davoux, D., Gautam, U., Mallam, V., … Baroughi, M. F. (2014). Transparent platinum counter electrode for efficient semi-transparent dye-sensitized solar cells. *Thin Solid Films*, *562*, 578–584. Available from https://doi.org/10.1016/j.tsf.2014.03.075.

Ito, B. I., Freitas, J. N. d, Paoli, M.-A. D., & Nogueira, A. F. (2008). Application of a composite polymer electrolyte based on montmorillonite in dye-sensitized solar cells. *Journal of the Brazilian Chemical Society*, *19*, 688–696.

Ito, S., Liska, P., Comte, P., Charvet, R., Péchy, P., Bach, U., … Grätzel, M. (2005). Control of dark current in photoelectrochemical ($TiO_2/I^-–I^{3-}$) and dye-sensitized solar cells. *Chemical Communications* (34), 4351–4353. Available from https://doi.org/10.1039/B505718C.

Ito, S., Zakeeruddin, S. M., Humphry-Baker, R., Liska, P., Charvet, R., Comte, P., … Grätzel, M. (2006). High-efficiency organic-dye-sensitized solar cells controlled by nanocrystalline-TiO_2 electrode thickness. *Advanced Materials*, *18*(9), 1202–1205. Available from https://doi.org/10.1002/adma.200502540.

Jacob, M. M. E., Prabaharan, S. R. S., & Radhakrishna, S. (1997). Effect of PEO addition on the electrolytic and thermal properties of PVDF-$LiClO_4$ polymer electrolytes. *Solid State Ionics*, *104*(3), 267–276. Available from https://doi.org/10.1016/S0167-2738(97)00422-0.

Jayakumar, G. C., Mehta, A., Rao, J. R., & Fathima, N. N. (2015). Ionic liquids: New age materials for eco-friendly leather processing. *RSC Advances*, *5*(40), 31998–32005. Available from https://doi.org/10.1039/C5RA02167G.

Jeong, H., Pak, Y., Hwang, Y., Song, H., Lee, K. H., Ko, H. C., & Jung, G. Y. (2012). Enhancing the charge transfer of the counter electrode in dye-sensitized solar cells using periodically aligned platinum nanocups. *Small*, *8*(24), 3757–3761. Available from https://doi.org/10.1002/smll.201201214.

Jiang, Q. W., Li, G. R., & Gao, X. P. (2009). Highly ordered TiN nanotube arrays as counter electrodes for dye-sensitized solar cells. *Chemical Communications*, *44*, 6720–6722. Available from https://doi.org/10.1039/B912776C.

Jin, L., & Chen, D. (2012). Enhancement in photovoltaic performance of phthalocyanine-sensitized solar cells by attapulgite nanoparticles. *Electrochimica Acta*, *72*, 40–45. Available from https://doi.org/10.1016/j.electacta.2012.03.167.

Joly, D., Pellejà, L., Narbey, S., Oswald, F., Chiron, J., Clifford, J. N., ... Demadrille, R. (2014). A robust organic dye for dye sensitized solar cells based on iodine/iodide electrolytes combining high efficiency and outstanding stability. *Scientific Reports*, *4*, 4033. Available from https://doi.org/10.1038/srep04033.

Kalyanasundaram, K., Vlachopoulos, N., Krishnan, V., Monnier, A., & Graetzel, M. (1987). Sensitization of titanium dioxide in the visible light region using zinc porphyrins. *The Journal of Physical Chemistry*, *91*(9), 2342–2347. Available from https://doi.org/10.1021/j100293a027.

Katsaros, G., Stergiopoulos, T., Arabatzis, I. M., Papadokostaki, K. G., & Falaras, P. (2002). A solvent-free composite polymer/inorganic oxide electrolyte for high efficiency solid-state dye-sensitized solar cells. *Journal of Photochemistry and Photobiology A: Chemistry*, *149* (1), 191–198. Available from https://doi.org/10.1016/S1010-6030(02)00027-8.

Kavan, L., Yum, J. H., & Grätzel, M. (2011). Optically transparent cathode for dye-sensitized solar cells based on graphene nanoplatelets. *ACS Nano*, *5*(1), 165–172. Available from https://doi.org/10.1021/nn102353h.

Kay, A., & Grätzel, M. (1996). Low cost photovoltaic modules based on dye sensitized nanocrystalline titanium dioxide and carbon powder. *Solar Energy Materials and Solar Cells*, *44*(1), 99–117. Available from https://doi.org/10.1016/0927-0248(96)00063-3.

Khanmirzaei, M. H., Ramesh, S., & Ramesh, K. (2015). Hydroxypropyl cellulose based non-volatile gel polymer electrolytes for dye-sensitized solar cell applications using 1-methyl-3-propylimidazolium iodide ionic liquid. *Scientific Reports*, *5*, 18056. Available from https://doi.org/10.1038/srep18056.

Kim, C., Kim, K.-S., Kim, H. Y., & Han, Y. S. (2008). Modification of a TiO_2 photoanode by using Cr-doped TiO_2 with an influence on the photovoltaic efficiency of a dye-sensitized solar cell. *Journal of Materials Chemistry*, *18*(47), 5809–5814. Available from https://doi.org/10.1039/B805091K.

Kim, C. K., Kim, H. M., Aftabuzzaman, M., Jeon, I.-Y., Kang, S. H., Eom, Y. K., ... Kim, H. K. (2018). Comparative study of edge-functionalized graphene nanoplatelets as metal-free counter electrodes for highly efficient dye-sensitized solar cells. *Materials Today Energy*, *9*, 67–73. Available from https://doi.org/10.1016/j.mtener.2018.05.003.

Kim, S. R., Parvez, M. K., & Chhowalla, M. (2009). UV-reduction of graphene oxide and its application as an interfacial layer to reduce the back-transport reactions in dye-sensitized solar cells. *Chemical Physics Letters*, *483*(1), 124–127. Available from https://doi.org/10.1016/j.cplett.2009.10.066.

Kim, S.-S., Nah, Y.-C., Noh, Y.-Y., Jo, J., & Kim, D.-Y. (2006). Electrodeposited Pt for cost-efficient and flexible dye-sensitized solar cells. *Electrochimica Acta*, *51*(18), 3814–3819. Available from https://doi.org/10.1016/j.electacta.2005.10.047.

Komiya, R., Han, L., Yamanaka, R., Islam, A., & Mitate, T. (2004). Highly efficient quasi-solid state dye-sensitized solar cell with ion conducting polymer electrolyte. *Journal of Photochemistry and Photobiology A: Chemistry*, *164*(1), 123–127. Available from https://doi.org/10.1016/j.jphotochem.2003.11.015.

Krebs, F. C. (2009). Fabrication and processing of polymer solar cells: A review of printing and coating techniques. *Solar Energy Materials and Solar Cells*, *93*(4), 394–412. Available from https://doi.org/10.1016/j.solmat.2008.10.004.

Kuang, D., Ito, S., Wenger, B., Klein, C., Moser, J.-E., Humphry-Baker, R., ... Grätzel, M. (2006). High molar extinction coefficient heteroleptic ruthenium complexes for thin film dye-sensitized solar cells. *Journal of the American Chemical Society*, *128*(12), 4146–4154. Available from https://doi.org/10.1021/ja058540p.

Kumar, S., Tsai, C.-H., & Fu, Y.-P. (2020). A multifunctional Ni-doped iron pyrite/reduced graphene oxide composite as an efficient counter electrode for DSSCs and as a non-enzymatic hydrogen peroxide electrochemical sensor. *Dalton Transactions*, *49*(25), 8516–8527. Available from https://doi.org/10.1039/D0DT01231A.

Kumara, N. T. R. N., Lim, A., Lim, C. M., Petra, M. I., & Ekanayake, P. (2017). Recent progress and utilization of natural pigments in dye sensitized solar cells: A review. *Renewable and Sustainable Energy Reviews*, *78*, 301–317. Available from https://doi.org/10.1016/j.rser.2017.04.075.

Landau, L., & Levich, B. (1988). Dragging of a liquid by a moving plate. In P. Pelcé (Ed.), *Dynamics of curved fronts* (pp. 141–153). San Diego: Academic Press.

Latini, A., Cavallo, C., Aldibaja, F. K., Gozzi, D., Carta, D., Corrias, A., ... Salviati, G. (2013). Efficiency improvement of DSSC photoanode by scandium doping of mesoporous titania beads. *The Journal of Physical Chemistry C*, *117*(48), 25276–25289. Available from https://doi.org/10.1021/jp409813c.

Law, C., Pathirana, S. C., Li, X., Anderson, A. Y., Barnes, P. R. F., Listorti, A., ... O'Regan, B. C. (2010). Water-based electrolytes for dye-sensitized solar cells. *Advanced Materials*, *22*(40), 4505–4509. Available from https://doi.org/10.1002/adma.201001703.

Lee, C. H., Kim, K. H., Jang, K. U., Park, S. J., & Choi, H. W. (2011). Synthesis of TiO_2 nanotube by hydrothermal method and application for dye-sensitized solar cell. *Molecular Crystals and Liquid Crystals*, *539*(1). Available from https://doi.org/10.1080/15421406.2011.566078, 125/[465]-132/[472].

Lee, C.-W., Lu, H.-P., Lan, C.-M., Huang, Y.-L., Liang, Y.-R., Yen, W.-N., ... Yeh, C.-Y. (2009). Novel zinc porphyrin sensitizers for dye-sensitized solar cells: Synthesis and spectral, electrochemical, and photovoltaic properties. *Chemistry – A European Journal*, *15*(6), 1403–1412. Available from https://doi.org/10.1002/chem.200801572.

Lee, K.-M., Chen, P.-Y., Hsu, C.-Y., Huang, J.-H., Ho, W.-H., Chen, H.-C., & Ho, K.-C. (2009). A high-performance counter electrode based on poly(3,4-alkylenedioxythiophene) for dye-sensitized solar cells. *Journal of Power Sources*, *188*(1), 313–318. Available from https://doi.org/10.1016/j.jpowsour.2008.11.075.

Lee, Y.-L., Shanmugam, V., Tsai, M.-H., Lin, J.-C., Teng, H., & Liu, I. P. (2018). Highly efficient quasi-solid-state dye-sensitized solar cell using polyethylene oxide (PEO) and poly(methyl methacrylate) (PMMA)-based printable electrolytes. *Journal of Materials Chemistry A*. Available from https://doi.org/10.1039/C8TA01729H.

Li, B., Wang, L., Kang, B., Wang, P., & Qiu, Y. (2006). Review of recent progress in solid-state dye-sensitized solar cells. *Solar Energy Materials and Solar Cells*, *90*(5), 549–573. Available from https://doi.org/10.1016/j.solmat.2005.04.039.

Li, G. R., Song, J., Pan, G. L., & Gao, X. P. (2011). Highly Pt-like electrocatalytic activity of transition metal nitrides for dye-sensitized solar cells. *Energy & Environmental Science*, *4*(5), 1680–1683. Available from https://doi.org/10.1039/C1EE01105G.

Li, Y., Ding, J. N., Yuan, N. Y., Bai, L., Hu, H. W., & Wang, X. Q. (2013). The influence of surface treatment on dye-sensitized solar cells based on TiO_2 nanofibers. *Materials Letters*, *97*, 74–77. Available from https://doi.org/10.1016/j.matlet.2013.01.106.

Li, Z.-Q., Chen, W.-C., Guo, F.-L., Mo, L.-E., Hu, L.-H., & Dai, S.-Y. (2015). Mesoporous TiO_2 yolk-shell microspheres for dye-sensitized solar cells with a high efficiency exceeding 11%. *Scientific Reports*, *5*(1), 14178. Available from https://doi.org/10.1038/srep14178.

Liska, P., Vlachopoulos, N., Nazeeruddin, M. K., Comte, P., & Graetzel, M. (1988). cis-Diaquabis(2,2′-bipyridyl-4,4′-dicarboxylate)ruthenium(II) sensitizes wide band gap oxide semiconductors very efficiently over a broad spectral range in the visible. *Journal of the American Chemical Society*, *110*(11), 3686–3687. Available from https://doi.org/10.1021/ja00219a068.

Liu, X., Fang, J., Liu, Y., & Lin, T. (2016). Progress in nanostructured photoanodes for dye-sensitized solar cells. *Frontiers of Materials Science*, *10*(3), 225–237. Available from https://doi.org/10.1007/s11706-016-0341-0.

Liu, Y., Sun, X., Tai, Q., Hu, H., Chen, B., Huang, N., ... Zhao, X.-Z. (2011). Efficiency enhancement in dye-sensitized solar cells by interfacial modification of conducting glass/mesoporous TiO_2 using a novel ZnO compact blocking film. *Journal of Power Sources*, *196*(1), 475–481. Available from https://doi.org/10.1016/j.jpowsour.2010.07.031.

Low, F. W., & Lai, C. W. (2018). Recent developments of graphene-TiO_2 composite nanomaterials as efficient photoelectrodes in dye-sensitized solar cells: A review. *Renewable and Sustainable Energy Reviews*, *82*, 103–125. Available from https://doi.org/10.1016/j.rser.2017.09.024.

Lü, X., Mou, X., Wu, J., Zhang, D., Zhang, L., Huang, F., ... Huang, S. (2010). Improved-performance dye-sensitized solar cells using Nb-doped TiO_2 electrodes: Efficient electron injection and transfer. *Advanced Functional Materials*, *20*(3), 509–515. Available from https://doi.org/10.1002/adfm.200901292.

Luo, L., Tao, W., Hu, X., Xiao, T., Heng, B., Huang, W., ... Tang, Y. (2011). Mesoporous F-doped ZnO prism arrays with significantly enhanced photovoltaic performance for dye-sensitized solar cells. *Journal of Power Sources*, *196*(23), 10518–10525. Available from https://doi.org/10.1016/j.jpowsour.2011.08.011.

Ma, T., Akiyama, M., Abe, E., & Imai, I. (2005). High-efficiency dye-sensitized solar cell based on a nitrogen-doped nanostructured titania electrode. *Nano Letters*, *5*(12), 2543–2547. Available from https://doi.org/10.1021/nl051885l.

Mahmood, K., & Sung, H. J. (2014). A dye-sensitized solar cell based on a boron-doped ZnO (BZO) film with double light-scattering-layers structured photoanode. *Journal of Materials Chemistry A*, *2*(15), 5408–5417. Available from https://doi.org/10.1039/C3TA14305H.

Mahmood, K., Swain, B. S., Han, G.-S., Kim, B.-J., & Jung, H. S. (2014). Polyethylenimine-assisted growth of high-aspect-ratio nitrogen-doped ZnO (NZO) nanorod arrays and their effect on performance of dye-sensitized solar cells. *ACS Applied Materials & Interfaces*, *6*(13), 10028–10043. Available from https://doi.org/10.1021/am500105x.

Manthina, V., & Agrios, A. G. (2016). Blocking layers for nanocomposite photoanodes in dye sensitized solar cells: Comparison of atomic layer deposition and $TiCl_4$ treatment. *Thin Solid Films*, *598*, 54–59. Available from https://doi.org/10.1016/j.tsf.2015.11.054.

Mathew, S., Yella, A., Gao, P., Humphry-Baker, R., Curchod, B. F. E., Ashari-Astani, N., ... Grätzel, M. (2014). Dye-sensitized solar cells with 13% efficiency achieved through the molecular engineering of porphyrin sensitizers. *Nature Chemistry*, *6*, 242. Available from https://doi.org/10.1038/nchem.1861, https://www.nature.com/articles/nchem.1861 #supplementary-information.

Mishra, A., Fischer, M. K. R., & Bäuerle, P. (2009). Metal-free organic dyes for dye-sensitized solar cells: From structure: Property relationships to design rules. *Angewandte Chemie International Edition*, *48*(14), 2474–2499. Available from https://doi.org/10.1002/anie.200804709.

Motlak, M., Akhtar, M. S., Barakat, N. A. M., Hamza, A. M., Yang, O. B., & Kim, H. Y. (2014). High-efficiency electrode based on nitrogen-doped TiO_2 nanofibers for dye-sensitized solar cells. *Electrochimica Acta*, *115*, 493–498. Available from https://doi.org/10.1016/j.electacta.2013.10.212.

Murai, S., Mikoshiba, S., Sumino, H., & Hayase, S. (2002). Quasi-solid dye-sensitized solar cells containing chemically cross-linked gel: How to make gels with a small amount of gelator. *Journal of Photochemistry and Photobiology A: Chemistry*, *148*(1), 33–39. Available from https://doi.org/10.1016/S1010-6030(02)00046-1.

Mustaffa, N., Rahman, M. Y. A., & Umar, A. A. (2020). Dye-sensitized solar cell utilizing silver doped reduced graphene oxide films counter electrode: Influence of annealing temperature on its performance. *Arabian Journal of Chemistry*, *13*(1), 3383–3390. Available from https://doi.org/10.1016/j.arabjc.2018.11.012.

Narayan, M. R. (2012). Review: Dye sensitized solar cells based on natural photosensitizers. *Renewable and Sustainable Energy Reviews*, *16*(1), 208–215. Available from https://doi.org/10.1016/j.rser.2011.07.148.

Nazeeruddin, M. K., Baranoff, E., & Grätzel, M. (2011). Dye-sensitized solar cells: A brief overview. *Solar Energy*, *85*(6), 1172–1178. Available from https://doi.org/10.1016/j.solener.2011.01.018.

Nazeeruddin, M. K., De Angelis, F., Fantacci, S., Selloni, A., Viscardi, G., Liska, P., ... Grätzel, M. (2005). Combined experimental and DFT-TDDFT computational study of photoelectrochemical cell ruthenium sensitizers. *Journal of the American Chemical Society*, *127*(48), 16835–16847. Available from https://doi.org/10.1021/ja052467l.

Nazeeruddin, M. K., Kay, A., Rodicio, I., Humphry-Baker, R., Mueller, E., Liska, P., ... Graetzel, M. (1993). Conversion of light to electricity by cis-X2bis(2,2′-bipyridyl-4,4′-dicarboxylate)ruthenium(II) charge-transfer sensitizers (X = Cl-, Br-, I-, CN-, and SCN-) on nanocrystalline titanium dioxide electrodes. *Journal of the American Chemical Society*, *115*(14), 6382–6390. Available from https://doi.org/10.1021/ja00067a063.

Nazeeruddin, M. K., Liska, P., Moser, J., Vlachopoulos, N., & Grätzel, M. (1990). Conversion of light into electricity with trinuclear ruthenium complexes adsorbed on textured TiO_2 films. *Helvetica Chimica Acta*, *73*(6), 1788–1803. Available from https://doi.org/10.1002/hlca.19900730624.

Nazeeruddin, M. K., Péchy, P., & Grätzel, M. (1997). Efficient panchromatic sensitization of nanocrystalline TiO_2 films by a black dye based on a trithiocyanato–ruthenium complex. *Chemical Communications* (18), 1705–1706. Available from https://doi.org/10.1039/A703277C.

Nikolay, T., Larina, L., Shevaleevskiy, O., & Ahn, B. T. (2011). Electronic structure study of lightly Nb-doped TiO_2 electrode for dye-sensitized solar cells. *Energy & Environmental Science*, *4*(4), 1480–1486. Available from https://doi.org/10.1039/C0EE00678E.

Nogueira, A. F., Longo, C., & De Paoli, M. A. (2004). Polymers in dye sensitized solar cells: Overview and perspectives. *Coordination Chemistry Reviews*, *248*(13), 1455–1468. Available from https://doi.org/10.1016/j.ccr.2004.05.018.

Novoselov, K. S., Geim, A. K., Morozov, S. V., Jiang, D., Zhang, Y., Dubonos, S. V., ... Firsov, A. A. (2004). Electric field effect in atomically thin carbon films. *Science*, *306* (5696), 666–669. Available from https://doi.org/10.1126/science.1102896.

Nugent, J. M., Santhanam, K. S. V., Rubio, A., & Ajayan, P. M. (2001). Fast electron transfer kinetics on multiwalled carbon nanotube microbundle electrodes. *Nano Letters*, *1*(2), 87–91. Available from https://doi.org/10.1021/nl005521z.

O'Regan, B., & Grätzel, M. (1991). A low-cost, high-efficiency solar cell based on dye-sensitized colloidal TiO_2 films. *Nature*, *353*, 737. Available from https://doi.org/10.1038/353737a0.

Özgür, Ü., Alivov, Y. I., Liu, C., Teke, A., Reshchikov, M. A., Doğan, S., ... Morkoç, H. (2005). A comprehensive review of ZnO materials and devices. *Journal of Applied Physics*, *98*(4), 041301. Available from https://doi.org/10.1063/1.1992666.

Papageorgiou, N. (2004). Counter-electrode function in nanocrystalline photoelectrochemical cell configurations. *Coordination Chemistry Reviews*, *248*(13), 1421–1446. Available from https://doi.org/10.1016/j.ccr.2004.03.028.

Papageorgiou, N., Athanassov, Y., Armand, M., Bonhoχte, P., Pettersson, H., Azam, A., & Grätzel, M. (1996). The performance and stability of ambient temperature molten salts for solar cell applications. *Journal of The Electrochemical Society*, *143*(10), 3099–3108. Available from https://doi.org/10.1149/1.1837171.

Papageorgiou, N., Maier, W. F., & Grätzel, M. (1997). An iodine/triiodide reduction electrocatalyst for aqueous and organic media. *Journal of The Electrochemical Society*, *144* (3), 876–884. Available from https://doi.org/10.1149/1.1837502.

Piao, Y., Lim, H., Chang, J. Y., Lee, W.-Y., & Kim, H. (2005). Nanostructured materials prepared by use of ordered porous alumina membranes. *Electrochimica Acta*, *50*(15), 2997–3013. Available from https://doi.org/10.1016/j.electacta.2004.12.043.

Plass, R., Pelet, S., Krueger, J., Grätzel, M., & Bach, U. (2002). Quantum dot sensitization of organic – inorganic hybrid solar cells. *The Journal of Physical Chemistry B*, *106* (31), 7578–7580. Available from https://doi.org/10.1021/jp020453l.

Plonska-Brzezinska, M. E., Lapinski, A., Wilczewska, A. Z., Dubis, A. T., Villalta-Cerdas, A., Winkler, K., & Echegoyen, L. (2011). The synthesis and characterization of carbon nano-onions produced by solution ozonolysis. *Carbon*, *49*(15), 5079–5089. Available from https://doi.org/10.1016/j.carbon.2011.07.027.

Qian, J., Liu, P., Xiao, Y., Jiang, Y., Cao, Y., Ai, X., & Yang, H. (2009). TiO_2-coated multilayered SnO_2 hollow microspheres for dye-sensitized solar cells. *Advanced Materials*, *21*(36), 3663–3667. Available from https://doi.org/10.1002/adma.200900525.

Qin, Y., & Peng, Q. (2012). Ruthenium sensitizers and their applications in dye-sensitized solar cells. *International Journal of Photoenergy*, *2012*, 291579. Available from https://doi.org/10.1155/2012/291579.

Raj, C. C., & Prasanth, R. (2016). A critical review of recent developments in nanomaterials for photoelectrodes in dye sensitized solar cells. *Journal of Power Sources*, *317*, 120–132. Available from https://doi.org/10.1016/j.jpowsour.2016.03.016.

Raj, C. J., Prabakar, K., Karthick, S. N., Hemalatha, K. V., Son, M.-K., & Kim, H.-J. (2013). Banyan root structured Mg-doped ZnO photoanode dye-sensitized solar cells. *The Journal of Physical Chemistry C*, *117*(6), 2600–2607. Available from https://doi.org/10.1021/jp308847g.

Ramasamy, E., & Lee, J. (2010). Ordered mesoporous SnO_2-based photoanodes for high-performance dye-sensitized solar cells. *The Journal of Physical Chemistry C, 114*(50), 22032–22037.

Rhee, S.-W., & Kwon, W. (2011). Key technological elements in dye-sensitized solar cells (DSC). *Korean Journal of Chemical Engineering, 28*(7), 1481–1494. Available from https://doi.org/10.1007/s11814-011-0148-8.

Saranya, K., Rameez, M., & Subramania, A. (2015). Developments in conducting polymer based counter electrodes for dye-sensitized solar cells – An overview. *European Polymer Journal, 66*, 207–227. Available from https://doi.org/10.1016/j.eurpolymj.2015.01.049.

Sengupta, D., Das, P., Mondal, B., & Mukherjee, K. (2016). Effects of doping, morphology and film-thickness of photo-anode materials for dye sensitized solar cell application – A review. *Renewable and Sustainable Energy Reviews, 60*, 356–376. Available from https://doi.org/10.1016/j.rser.2016.01.104.

Serikov, T. M., Ibrayev, N. K., Smagulov, Z. K., & Kuterbekov, K. A. (2017). Influence of annealing on optical and photovoltaic properties of nanostructured TiO_2 films. *IOP Conference Series: Materials Science and Engineering, 168*, 012054. Available from https://doi.org/10.1088/1757-899x/168/1/012054.

Shen, L., Bao, N., Zheng, Y., Gupta, A., An, T., & Yanagisawa, K. (2008). Hydrothermal splitting of titanate fibers to single-crystalline TiO_2 nanostructures with controllable crystalline phase, morphology, microstructure, and photocatalytic activity. *The Journal of Physical Chemistry C, 112*(24), 8809–8818. Available from https://doi.org/10.1021/jp711369e.

Shi, D., Pootrakulchote, N., Li, R., Guo, J., Wang, Y., Zakeeruddin, S. M., ... Wang, P. (2008). New efficiency records for stable dye-sensitized solar cells with low-volatility and ionic liquid electrolytes. *The Journal of Physical Chemistry C, 112*(44), 17046–17050. Available from https://doi.org/10.1021/jp808018h.

Shi, Y., Zhan, C., Wang, L., Ma, B., Gao, R., Zhu, Y., & Qiu, Y. (2009). The electrically conductive function of high-molecular weight poly(ethylene oxide) in polymer gel electrolytes used for dye-sensitized solar cells. *Physical Chemistry Chemical Physics, 11*(21), 4230–4235. Available from https://doi.org/10.1039/B901003C.

Shin, S., Dao, V.-D., & Choi, H.-S. (2020). Incorporating MoFe alloys into reduced graphene oxide as counter electrode catalysts for dye-sensitized solar cells. *Arabian Journal of Chemistry, 13*(1), 2414–2424. Available from https://doi.org/10.1016/j.arabjc.2018.05.006.

Singh, S., Singh, A., & Kaur, N. (2016). Efficiency investigations of organic/inorganic hybrid ZnO nanoparticles based dye-sensitized solar cells. *Journal of Materials, 2016*, 11. Available from https://doi.org/10.1155/2016/9081346.

Song, J., Yang, H. B., Wang, X., Khoo, S. Y., Wong, C. C., Liu, X.-W., & Li, C. M. (2012). Improved utilization of photogenerated charge using fluorine-doped TiO_2 hollow spheres scattering layer in dye-sensitized solar cells. *ACS Applied Materials & Interfaces, 4*(7), 3712–3717. Available from https://doi.org/10.1021/am300801f.

Su'ait, M. S., Rahman, M. Y. A., & Ahmad, A. (2015). Review on polymer electrolyte in dye-sensitized solar cells (DSSCs). *Solar Energy, 115*, 452–470. Available from https://doi.org/10.1016/j.solener.2015.02.043.

Subramanian, A., & Wang, H.-W. (2012). Effects of boron doping in TiO_2 nanotubes and the performance of dye-sensitized solar cells. *Applied Surface Science, 258*(17), 6479–6484. Available from https://doi.org/10.1016/j.apsusc.2012.03.064.

Sun, H., Luo, Y., Zhang, Y., Li, D., Yu, Z., Li, K., & Meng, Q. (2010). In situ preparation of a flexible polyaniline/carbon composite counter electrode and its application in dye-sensitized solar cells. *The Journal of Physical Chemistry C, 114*(26), 11673–11679. Available from https://doi.org/10.1021/jp1030015.

Sun, Q., Zhang, J., Wang, P., Zheng, J., Zhang, X., Cui, Y., ... Zhu, Y. (2012). Sulfur-doped TiO_2 nanocrystalline photoanodes for dye-sensitized solar cells. *Journal of Renewable and Sustainable Energy, 4*(2), 023104. Available from https://doi.org/10.1063/1.3694121.

Sun, Y., Mayers, B., & Xia, Y. (2003). Metal nanostructures with hollow interiors. *Advanced Materials, 15*(7-8), 641–646. Available from https://doi.org/10.1002/adma.200301639.

Suryanarayanan, V., Lee, K.-M., Ho, W.-H., Chen, H.-C., & Ho, K.-C. (2007). A comparative study of gel polymer electrolytes based on PVDF-HFP and liquid electrolytes, containing imidazolinium ionic liquids of different carbon chain lengths in DSSCs. *Solar Energy Materials and Solar Cells, 91*(15), 1467–1471. Available from https://doi.org/10.1016/j.solmat.2007.03.008.

Tabari-Saadi, Y., & Mohammadi, M. R. (2015). Efficient dye-sensitized solar cells based on carbon-doped TiO_2 hollow spheres and nanoparticles. *Journal of Materials Science: Materials in Electronics, 26*(11), 8863–8876. Available from https://doi.org/10.1007/s10854-015-3567-1.

Tang, H., Prasad, K., Sanjinès, R., Schmid, P. E., & Lévy, F. (1994). Electrical and optical properties of TiO_2 anatase thin films. *Journal of Applied Physics, 75*(4), 2042–2047. Available from https://doi.org/10.1063/1.356306.

Tang, Z., Wu, J., Zheng, M., Huo, J., & Lan, Z. (2013). A microporous platinum counter electrode used in dye-sensitized solar cells. *Nano Energy, 2*(5), 622–627. Available from https://doi.org/10.1016/j.nanoen.2013.07.014.

Tao, R., Tomita, T., Wong, R. A., & Waki, K. (2012). Electrochemical and structural analysis of Al-doped ZnO nanorod arrays in dye-sensitized solar cells. *Journal of Power Sources, 214*, 159–165. Available from https://doi.org/10.1016/j.jpowsour.2012.04.071.

Thoi, V. S., Sun, Y., Long, J. R., & Chang, C. J. (2013). Complexes of earth-abundant metals for catalytic electrochemical hydrogen generation under aqueous conditions. *Chemical Society Reviews, 42*(6), 2388–2400. Available from https://doi.org/10.1039/C2CS35272A.

Tian, H., Hu, L., Zhang, C., Chen, S., Sheng, J., Mo, L., ... Dai, S. (2011). Enhanced photovoltaic performance of dye-sensitized solar cells using a highly crystallized mesoporous TiO_2 electrode modified by boron doping. *Journal of Materials Chemistry, 21*(3), 863–868. Available from https://doi.org/10.1039/C0JM02941F.

Tong, Z., Peng, T., Sun, W., Liu, W., Guo, S., & Zhao, X.-Z. (2014). Introducing an intermediate band into dye-sensitized solar cells by W6 + doping into TiO_2 nanocrystalline photoanodes. *The Journal of Physical Chemistry C, 118*(30), 16892–16895. Available from https://doi.org/10.1021/jp500412e.

Tributsch, H. (2004). Dye sensitization solar cells: A critical assessment of the learning curve. *Coordination Chemistry Reviews, 248*(13), 1511–1530. Available from https://doi.org/10.1016/j.ccr.2004.05.030.

Velten, J., Mozer, A. J., Li, D., Officer, D., Wallace, G., Baughman, R., & Zakhidov, A. (2012). Carbon nanotube/graphene nanocomposite as efficient counter electrodes in dye-sensitized solar cells. *Nanotechnology, 23*(8), 085201. Available from https://doi.org/10.1088/0957-4484/23/8/085201.

Venkatesan, S., Obadja, N., Chang, T.-W., Chen, L.-T., & Lee, Y.-L. (2014). Performance improvement of gel- and solid-state dye-sensitized solar cells by utilization the blending effect of poly (vinylidene fluoride-co-hexafluropropylene) and poly (acrylonitrile-*co*-vinyl acetate) co-polymers. *Journal of Power Sources*, *268*, 77–81. Available from https://doi.org/10.1016/j.jpowsour.2014.06.016.

Wang, G., Zhuo, S., & Xing, W. (2012). Graphene/polyaniline nanocomposite as counter electrode of dye-sensitized solar cells. *Materials Letters*, *69*, 27–29. Available from https://doi.org/10.1016/j.matlet.2011.11.086.

Wang, H., Bhattacharjee, R., Hung, I. M., Li, L., & Zeng, R. (2013). Material characteristics and electrochemical performance of Sn-doped ZnO spherical-particle photoanode for dye-sensitized solar cells. *Electrochimica Acta*, *111*, 797–801. Available from https://doi.org/10.1016/j.electacta.2013.07.199.

Wang, L., Fang, S., Lin, Y., Zhou, X., & Li, M. (2005). A 7.72% efficient dye sensitized solar cell based on novel necklace-like polymer gel electrolyte containing latent chemically cross-linked gel electrolyte precursors. *Chemical Communications* (45), 5687–5689. Available from https://doi.org/10.1039/B510335E.

Wang, M., Anghel, A. M., Marsan, B., Cevey Ha, N.-L., Pootrakulchote, N., Zakeeruddin, S. M., & Grätzel, M. (2009). CoS supersedes Pt as efficient electrocatalyst for triiodide reduction in dye-sensitized solar cells. *Journal of the American Chemical Society*, *131* (44), 15976–15977. Available from https://doi.org/10.1021/ja905970y.

Wang, M., Bai, S., Chen, A., Duan, Y., Liu, Q., Li, D., & Lin, Y. (2012). Improved photovoltaic performance of dye-sensitized solar cells by Sb-doped TiO_2 photoanode. *Electrochimica Acta*, *77*, 54–59. Available from https://doi.org/10.1016/j.electacta.2012.05.050.

Wang, P., Zakeeruddin, S. M., Comte, P., Exnar, I., & Grätzel, M. (2003). Gelation of ionic liquid-based electrolytes with silica nanoparticles for quasi-solid-state dye-sensitized solar cells. *Journal of the American Chemical Society*, *125*(5), 1166–1167. Available from https://doi.org/10.1021/ja029294.

Wang, P., Zakeeruddin, S. M., Moser, J. E., Nazeeruddin, M. K., Sekiguchi, T., & Grätzel, M. (2003). A stable quasi-solid-state dye-sensitized solar cell with an amphiphilic ruthenium sensitizer and polymer gel electrolyte. *Nature Materials*, *2*, 402. Available from https://doi.org/10.1038/nmat904.

Wang, X., Deng, R., Kulkarni, S. A., Wang, X., Pramana, S. S., Wong, C. C., ... Mhaisalkar, S. G. (2013). Investigation of the role of anions in hydrotalcite for quasi-solid state dye-sensitized solar cells application. *Journal of Materials Chemistry A*, *1*(13), 4345–4351. Available from https://doi.org/10.1039/C3TA01581E.

Wang, X., Kulkarni, S. A., Ito, B. I., Batabyal, S. K., Nonomura, K., Wong, C. C., ... Uchida, S. (2013). Nanoclay gelation approach toward improved dye-sensitized solar cell efficiencies: An investigation of charge transport and shift in the TiO_2 conduction band. *ACS Applied Materials & Interfaces*, *5*(2), 444–450. Available from https://doi.org/10.1021/am3025454.

Wang, X., Tang, Q., He, B., Li, R., & Yu, L. (2015). 7.35% Efficiency rear-irradiated flexible dye-sensitized solar cells by sealing liquid electrolyte in a groove. *Chemical Communications*, *51*(3), 491–494. Available from https://doi.org/10.1039/C4CC07549H.

Wang, X.-F., Matsuda, A., Koyama, Y., Nagae, H., Sasaki, S.-I., Tamiaki, H., & Wada, Y. (2006). Effects of plant carotenoid spacers on the performance of a dye-sensitized solar cell using a chlorophyll derivative: Enhancement of photocurrent determined by one electron-oxidation potential of each carotenoid. *Chemical Physics Letters*, *423*(4), 470–475. Available from https://doi.org/10.1016/j.cplett.2006.04.008.

Wang, Y., Guo, Y., Chen, W., Luo, Q., Lu, W., Xu, P., ... He, M. (2018). Sulfur-doped reduced graphene oxide/MoS_2 composite with exposed active sites as efficient Pt-free counter electrode for dye-sensitized solar cell. *Applied Surface Science*, *452*, 232–238. Available from https://doi.org/10.1016/j.apsusc.2018.04.276.

Wanninayake, W. M. N. M. B., Premaratne, K., Kumara, G. R. A., & Rajapakse, R. M. G. (2016). Use of lithium iodide and tetrapropylammonium iodide in gel electrolytes for improved performance of quasi-solid-state dye-sensitized solar cells: Recording an efficiency of 6.40%. *Electrochimica Acta*, *191*, 1037–1043. Available from https://doi.org/10.1016/j.electacta.2016.01.108.

Wei, T. C., Wan, C. C., & Wang, Y. Y. (2006). Poly(*N*-vinyl-2-pyrrolidone)-capped platinum nanoclusters on indium-tin oxide glass as counterelectrode for dye-sensitized solar cells. *Applied Physics Letters*, *88*(10), 103122. Available from https://doi.org/10.1063/1.2186069.

Weinhardt, L., Blum, M., Bär, M., Heske, C., Cole, B., Marsen, B., & Miller, E. L. (2008). Electronic surface level positions of WO_3 thin films for photoelectrochemical hydrogen production. *The Journal of Physical Chemistry C*, *112*(8), 3078–3082. Available from https://doi.org/10.1021/jp7100286.

Winther-Jensen, B., Winther-Jensen, O., Forsyth, M., & MacFarlane, D. R. (2008). High rates of oxygen reduction over a vapor phase–polymerized PEDOT electrode. *Science*, *321*(5889), 671–674. Available from https://doi.org/10.1126/science.1159267.

Wolfbauer, G., Bond, A. M., Eklund, J. C., & MacFarlane, D. R. (2001). A channel flow cell system specifically designed to test the efficiency of redox shuttles in dye sensitized solar cells. *Solar Energy Materials and Solar Cells*, *70*(1), 85–101. Available from https://doi.org/10.1016/S0927-0248(00)00413-X.

Wu, J., Lan, Z., Hao, S., Li, P., Lin, J., Huang, M., ... Huang, Y. (2008). Progress on the electrolytes for dye-sensitized solar cells. *Pure and Applied Chemistry*, *Vol. 80*, 2241.

Wu, J., Lan, Z., Lin, J., Huang, M., Huang, Y., Fan, L., & Luo, G. (2015). Electrolytes in dye-sensitized solar cells. *Chemical Reviews*, *115*(5), 2136–2173. Available from https://doi.org/10.1021/cr400675m.

Wu, J. H., Hao, S. C., Lan, Z., Lin, J. M., Huang, M. L., Huang, Y. F., ... Sato, T. (2007). A thermoplastic gel electrolyte for stable quasi-solid-state dye-sensitized solar cells. *Advanced Functional Materials*, *17*(15), 2645–2652. Available from https://doi.org/10.1002/adfm.200600621.

Wu, J. H., Lan, Z., Lin, J. M., Huang, M. L., Hao, S. C., Sato, T., & Yin, S. (2007). A novel thermosetting gel electrolyte for stable quasi-solid-state dye-sensitized solar cells. *Advanced Materials*, *19*(22), 4006–4011. Available from https://doi.org/10.1002/adma.200602886.

Wu, M., Lin, X., Wang, Y., Wang, L., Guo, W., Qi, D., ... Ma, T. (2012). Economical Pt-free catalysts for counter electrodes of dye-sensitized solar cells. *Journal of the American Chemical Society*, *134*(7), 3419–3428. Available from https://doi.org/10.1021/ja209657v.

Wu, M.-S., & Lin, J.-C. (2019). Dual doping of mesoporous carbon pillars with oxygen and sulfur as counter electrodes for iodide/triiodide redox mediated dye-sensitized solar cells. *Applied Surface Science*, *471*, 455–461. Available from https://doi.org/10.1016/j.apsusc.2018.12.043.

Wu, K., Zhao, Q., Chen, L., Liu, Z., Ruan, B., & Wu, M. (2019). Effect of transition-metal ion doping on electrocatalytic activities of graphene/polyaniline-M^{2+} (Mn^{2+}, Co^{2+}, Ni^{2+}, and Cu^{2+}) composite materials as Pt-free counter electrode in dye-sensitized solar cells. *Polymer-Plastics Technology and Materials*, *58*(1), 40–46.

Xia, J., Masaki, N., Jiang, K., & Yanagida, S. (2007). Sputtered Nb_2O_5 as a novel blocking layer at conducting glass/TiO_2 interfaces in dye-sensitized ionic liquid solar cells. *The*

Journal of Physical Chemistry C, *111*(22), 8092–8097. Available from https://doi.org/10.1021/jp0707384.

Xu, C., Wang, X., & Zhu, J. (2008). Graphene – metal particle nanocomposites. *The Journal of Physical Chemistry C*, *112*(50), 19841–19845. Available from https://doi.org/10.1021/jp807989b.

Xu, H., Zhang, C., Yao, J., Pang, S., Zhou, X., & Cui, G. (2018). Graphene-wrapped iron carbide nanoparticles as Pt-free counter electrode towards dye-sensitized solar cells via magnetic field induced self-assembly. *Journal of Photochemistry and Photobiology A: Chemistry*, *355*, 48–54. Available from https://doi.org/10.1016/j.jphotochem.2017.11.030.

Xue, Y., Liu, J., Chen, H., Wang, R., Li, D., Qu, J., & Dai, L. (2012). Nitrogen-doped graphene foams as metal-free counter electrodes in high-performance dye-sensitized solar cells. *Angewandte Chemie International Edition*, *51*(48), 12124–12127. Available from https://doi.org/10.1002/anie.201207277.

Yang, Y., Hu, H., Zhou, C.-H., Xu, S., Sebo, B., & Zhao, X.-Z. (2011). Novel agarose polymer electrolyte for quasi-solid state dye-sensitized solar cell. *Journal of Power Sources*, *196*(4), 2410–2415. Available from https://doi.org/10.1016/j.jpowsour.2010.10.067.

Yen, M.-Y., Teng, C.-C., Hsiao, M.-C., Liu, P.-I., Chuang, W.-P., Ma, C.-C. M., … Tsai, C.-H. (2011). Platinum nanoparticles/graphene composite catalyst as a novel composite counter electrode for high performance dye-sensitized solar cells. *Journal of Materials Chemistry*, *21*(34), 12880–12888. Available from https://doi.org/10.1039/C1JM11850A.

Yoon, I.-N., Song, H.-k, Won, J., & Kang, Y. S. (2014). Shape dependence of SiO_2 nanomaterials in a quasi-solid electrolyte for application in dye-sensitized solar cells. *The Journal of Physical Chemistry C*, *118*(8), 3918–3924. Available from https://doi.org/10.1021/jp4104454.

Yu, Z., Vlachopoulos, N., Gorlov, M., & Kloo, L. (2011). Liquid electrolytes for dye-sensitized solar cells. *Dalton Transactions*, *40*(40), 10289–10303. Available from https://doi.org/10.1039/C1DT11023C.

Yusuf, S. N. F., Azzahari, A. D., Selvanathan, V., Yahya, R., Careem, M. A., & Arof, A. K. (2017). Improvement of *N*-phthaloylchitosan based gel polymer electrolyte in dye-sensitized solar cells using a binary salt system. *Carbohydrate Polymers*, *157*, 938–944. Available from https://doi.org/10.1016/j.carbpol.2016.10.032.

Zakeeruddin, S. M., & Grätzel, M. (2009). Solvent-free ionic liquid electrolytes for mesoscopic dye-sensitized solar cells. *Advanced Functional Materials*, *19*(14), 2187–2202. Available from https://doi.org/10.1002/adfm.200900390.

Zhang, D. W., Li, X. D., Li, H. B., Chen, S., Sun, Z., Yin, X. J., & Huang, S. M. (2011). Graphene-based counter electrode for dye-sensitized solar cells. *Carbon*, *49*(15), 5382–5388. Available from https://doi.org/10.1016/j.carbon.2011.08.005.

Zhang, J., Peng, W., Chen, Z., Chen, H., & Han, L. (2012). Effect of cerium doping in the TiO_2 photoanode on the electron transport of dye-sensitized solar cells. *The Journal of Physical Chemistry C*, *116*(36), 19182–19190. Available from https://doi.org/10.1021/jp3060735.

Zhang, J., Zhou, P., Liu, J., & Yu, J. (2014). New understanding of the difference of photocatalytic activity among anatase, rutile and brookite TiO_2. *Physical Chemistry Chemical Physics*, *16*(38), 20382–20386. Available from https://doi.org/10.1039/C4CP02201G.

Zhang, L., Yang, Y., Fan, R., Chen, H., Jia, R., Wang, Y., … Wang, Y. (2012). The charge-transfer property and the performance of dye-sensitized solar cells of nitrogen doped zinc oxide. *Materials Science and Engineering: B*, *177*(12), 956–961. Available from https://doi.org/10.1016/j.mseb.2012.04.026.

Zhang, Q., Dandeneau, C., Zhou, X., & Cao, G. (2009). ZnO nanostructures for dye-sensitized solar cells. *Advanced Materials*, *21*, 4087–4108. Available from https://doi.org/10.1002/adma.200803827.

Zhang, S., Yang, X., Numata, Y., & Han, L. (2013). Highly efficient dye-sensitized solar cells: Progress and future challenges. *Energy & Environmental Science*, *6*(5), 1443–1464. Available from https://doi.org/10.1039/C3EE24453A.

Zhang, X., Liu, F., Huang, Q.-L., Zhou, G., & Wang, Z.-S. (2011). Dye-sensitized W-doped TiO_2 solar cells with a tunable conduction band and suppressed charge recombination. *The Journal of Physical Chemistry C*, *115*(25), 12665–12671. Available from https://doi.org/10.1021/jp201853c.

Zhao, J., Jin, T., Islam, A., Kwon, E., Akhtaruzzaman, M., Asao, N., ... Yamamoto, Y. (2014). Thieno[2,3-a]carbazole-based donor–π–acceptor organic dyes for efficient dye-sensitized solar cells. *Tetrahedron*, *70*(36), 6211–6216. Available from https://doi.org/10.1016/j.tet.2014.01.001.

Zheng, Y.-Z., Ding, H., Tao, X., & Chen, J.-F. (2014). Investigation of iodine dopant amount effects on dye-sensitized hierarchically structured ZnO solar cells. *Materials Research Bulletin*, *55*, 182–189. Available from https://doi.org/10.1016/j.materresbull.2014.04.027.

Zheng, Y.-Z., Tao, X., Hou, Q., Wang, D.-T., Zhou, W.-L., & Chen, J.-F. (2011). Iodine-doped ZnO nanocrystalline aggregates for improved dye-sensitized solar cells. *Chemistry of Materials*, *23*(1), 3–5. Available from https://doi.org/10.1021/cm101525p.

Zhou, H., Wu, L., Gao, Y., & Ma, T. (2011). Dye-sensitized solar cells using 20 natural dyes as sensitizers. *Journal of Photochemistry and Photobiology A: Chemistry*, *219*(2), 188–194. Available from https://doi.org/10.1016/j.jphotochem.2011.02.008.

Zhu, S., Tian, X., Shan, L., Ding, Z., Kan, Z., Xu, X., ... Wang, L. (2013). Effect of Al3 + on the growth of ZnO nanograss film and its application in dye-sensitized solar cells. *Ceramics International*, *39*(8), 9637–9644. Available from https://doi.org/10.1016/j.ceramint.2013.05.085.

Zistler, M., Wachter, P., Wasserscheid, P., Gerhard, D., Hinsch, A., Sastrawan, R., & Gores, H. J. (2006). Comparison of electrochemical methods for triiodide diffusion coefficient measurements and observation of non-Stokesian diffusion behaviour in binary mixtures of two ionic liquids. *Electrochimica Acta*, *52*(1), 161–169. Available from https://doi.org/10.1016/j.electacta.2006.04.050.

SUBCHAPTER

Quantum dot-sensitized solar cells 4.2

Muhammad Rizwan[1], Muhammad Ammar Bin Mingsukang[2] and Md. Akhtaruzzaman[3]

[1]*Department of Chemistry, Faculty of Science, The University of Lahore, Lahore, Pakistan*
[2]*Center for Ionics, Faculty of Science, Department of Physics, University of Malaya, Kuala Lumpur, Malaysia*
[3]*Solar Energy Research Institute, Universiti Kebangsaan Malaysia, Bangi, Malaysia*

4.2.1 Introduction

Renewable energy, particularly green energy, is believed to be a reliable solution to counter the issues arising from energy crises and global warming. The definition of renewable energy is energy harnessed from natural resources that can be naturally renewed and regenerated. Examples of renewable energy sources are solar power, hydropower, tidal power, and wind power. The production of electricity through renewable energy causes less harm to the environment and requires less production/maintenance cost.

Among all available renewable sources, solar energy is the best and most abundant alternate to conventional energy resources, that is, the amount of solar energy reaching the earth is 3×10^{24} J/year while the global energy consumption per year was 4.25×10^{20} J in 2001 (Grätzel, 2005). Thus solar energy is the most promising energy source for the future and the most appropriate to replace conventional energy sources that will be eventually exhausted.

Photovoltaic devices, or solar cells, were invented in the 19th century and are devices that can convert sunlight into electricity. Currently, there is a lot of ongoing research and development for solar cells. Until now, solar cell technology has been categorized into three types that is, 1st generation, 2nd generation, and 3rd generation. The 1st developed solar cells technologies that is, 1st generation solar cells are based on pure bulk crystalline silicon for which the energy conversion efficiency has reached a maximum of 25% (Blakers et al., 2013). However, the high cost of the 1st generation solar cell has led to the development of 2nd generation solar cells which offer lower cost. Second generation solar cells are based on thin film technologies that utilize cheaper material compared to bulk pure silicon. The materials that have been used for 2nd generation solar cells are amorphous crystalline silicon, cadmium telluride, and copper indium diselenide.

Comprehensive Guide on Organic and Inorganic Solar Cells. DOI: https://doi.org/10.1016/B978-0-323-85529-7.00003-7

Third generation solar cells are technologies developed to further reduce fabrication cost with better performance to price ratios and simple preparation procedures (Späth et al., 2003). Examples of third generation solar cells are dye-sensitized solar cells (DSSCs), quantum dot-sensitized solar cells (QDSSCs), and perovskite solar cells. All three electrochemical solar cells directly convert solar power into electricity. The power conversion efficiency (PCE) of all three cells, by year, are shown in Fig. 4.2.1.

QDSSCs are one of the 3rd generation solar cell technologies that have been rigorously being studied by researchers around the globe. Compared to conventional silicon-based solar cells, QDSSCs have great potential and provide outstanding performance with cheap production costs. Many researchers have strived to improve the performance and stability of the QDSSC owing to its good performance to production cost ratio. The idea of QDSSCs emerged as a result of efforts to replace organic dyes with inorganic nanosized semiconductors (quantum dots) having the same mechanism as that of the DSSC. The components of the QDSSC include the photoanode (anode), electrolyte, and counter electrode (cathode). A large number of trials and testing is required for the development of each component to maximize the performance and stability of the QDSSC.

4.2.2 Device mechanism

The function of each component in the QDSSC consists of the photoanode generating electrons, the counter electrode as the cathode, and the electrolyte acting as an electron transport medium and junction between the photoanode and counter electrode. The

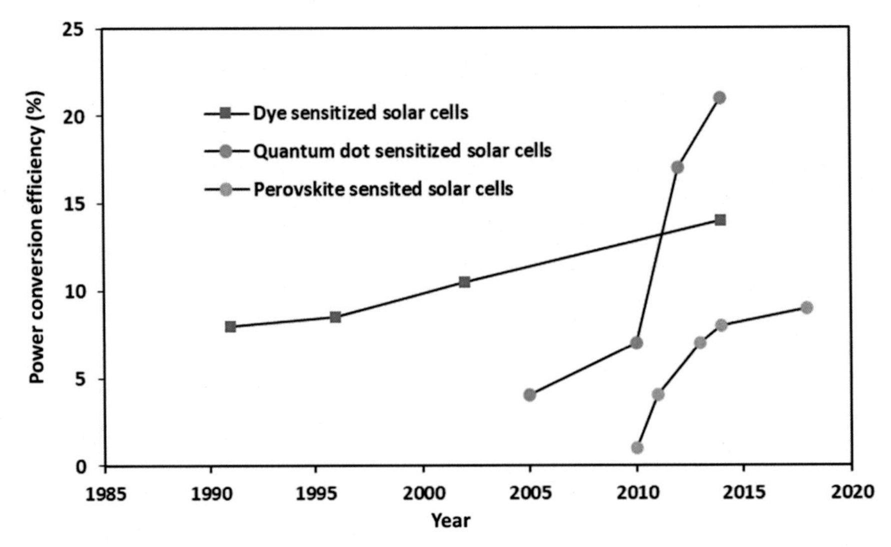

FIGURE 4.2.1

The power conversion efficiency of 3rd generation solar cells by year.

mechanism for electricity generation of the QDSSC is the same as that of the DSSC with the only difference being the electron producer, quantum dot in the former and dye in the latter (Peter, 2011; Yeh et al., 2011). Fig. 4.2.2 shows a schematic illustration for better understanding of the QDSSC working mechanism (Jun, Careem, & Arof, 2013).

The generation of electricity in the QDSSC starts with electron excitation from the valence band of the QD to the conduction band when exposed to light. From the conduction band of QD, electrons are pushed to the conduction band of semiconductor for example, TiO_2 with the aid of a certain amount of energy. This results in the creation of a hole in the valence band of the QD and it is said to be in the oxidized state. The electrons in the conduction band of TiO_2 are then transferred to the conducting layer of glass and then the counter electrode through an external circuit. These electrons are collected by redox mediator ($S_x{}^{2-}$ ions) to form a S^{2-} counter electrode electrolyte interface. Chemical Eq. (4.2.1) describes the process that occurs at the interface of the counter electrode and electrolyte (Yang et al., 2011):

$$S_x^{2-} + 2e^{2-} \rightarrow S_{x-1}^{2-} + S^{2-} \tag{4.2.1}$$

From here, electrons are transferred to the QD to fill the holes previously created by electron excitation.

Chemical Eqs. (4.2.2), (4.2.3), and (4.2.4) describe the oxidation processes that happens at the electrolyte/photoanode interface (Feng et al., 2016). Eq. (4.2.2) shows that S^{2-} ions donate two electrons to the QD and transform into S (intermediate form), Eq. (4.2.3) shows the same process which involves two holes in the process. Eq. (4.2.4) describes that S (intermediate form) then transforms into S_x^{2-} ions in combination with S_{x-1}^{2-} ions.

$$S^{2-} \rightarrow S + 2e^{-} \tag{4.2.2}$$

$$S^{2-} + 2h^{+} \rightarrow S \tag{4.2.3}$$

$$S + S_{x-1}^{2-} \rightarrow S_x^{2-} (x = 2 - 5) \tag{4.2.4}$$

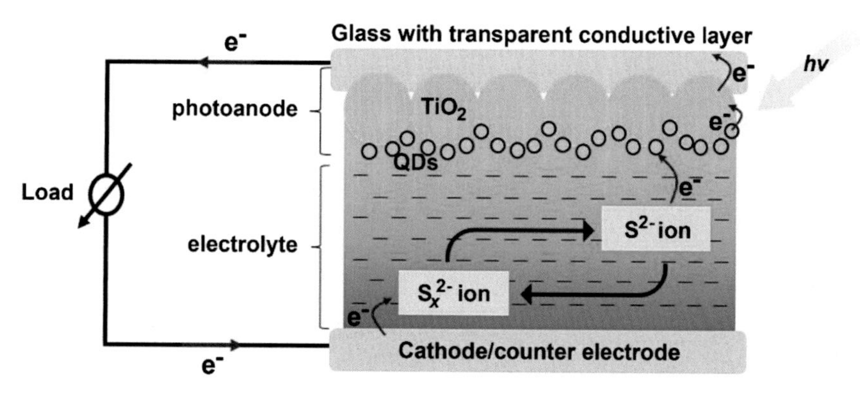

FIGURE 4.2.2

The mechanism of electron generation and electron transportation in the QDSSC.

Through this transport of electrons through the QD-TiO_2-glass-external circuit-counter electrode-redox couple-QD, the circuit is completed and electricity is generated. The production of electricity can be verified using a voltmeter. The number of electrons produced in the photocurrent produced depends upon various factors such as QD morphology, nature of the counter electrode, conductivity of the electrolyte, and recombination at the electrolyte/photoanode interface. Overall, the amount of photocurrent is decided by how effectively electrons are ejected from QDs followed by the completion of the circuit (Feng et al., 2016; Radich, Dwyer, & Kamat, 2011; Shen et al., 2008; Sun et al., 2008). Photovoltage generated in the QDSSC depends on many factors, but the major contribution to the photovoltage in this cell is due to the distinction between the quasi-Fermi level of the electron in the photoanode and the redox potential of the polysulfide electrolyte (Sayama, Sugihara, & Arakawa, 1998). An illustration on the distinction between the Fermi level of the photoanode and electrolyte can be seen in Fig. 4.2.3:

4.2.3 Quantum dot-sensitized solar cells components and materials selection

The essence of electricity production in QDSSCs is the production of electrons by QDs as described previously, where the rest of the components and working

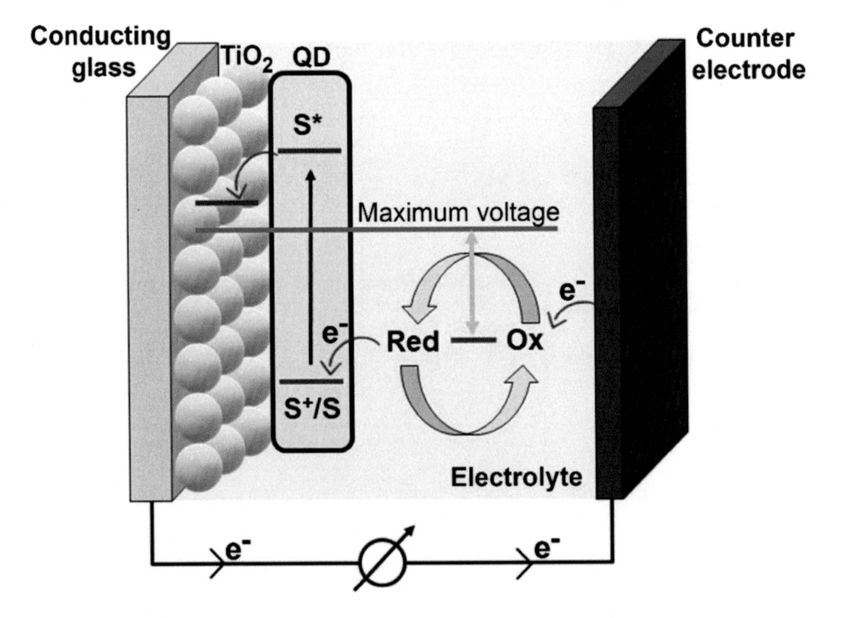

FIGURE 4.2.3

Schematic of voltage generation in the QDSSC.

principle for the QDSSC and DSSC are similar. The QDSSC has three main components: photoanode, electrolyte, and counterelectrode. Although QDSSC and DSSC have the same principle, the materials used in both cells are different based on differences in the compatibility of materials selected in both cells. For instance, an iodide-based electrolyte will give an excellent performance in DSSCs performance but it has a negative effect on QDSSC performance since iodide solution is corrosive to QD materials (Chi-Hsiu & Yuh-Lang, 2007; Shalom et al., 2009). Extensive research is required on all three QDSSC components to obtain the high performance and stability of the device.

4.2.3.1 Photoanode

Generally, the structure of the QDSSC photoanode resembles a similar structure to the DSSC photoanode introduced by Grätzel and O'Regan (O'Regan & Grätzel, 1991), which consists of fluorine doped tin oxide (FTO) glass with a mesoporous layer of metal oxide with properties of a wide bandgap semiconductor (for example TiO_2, ZnO and SnO_2) and a QD sensitizer (for example CdS, CdSe, CdTe, Ag_2S, etc.) attached to the metal oxide surface. Fig. 4.2.4 shows a simple illustration for the QDSSC photoanode for better understanding. There is an additional subcomponent in the photoanode called the passivation layer. It has been widely proven that the passivation layer improves QDSSC performance.

The mesoporous metal oxide with a larger surface area (wide bandgap semiconductor) in the photoanode of the QDSSC, such as TiO_2, ZnO, Nb_2O_5, or SnO_2 is crucial for quantum dot sensitizer adsorption. In addition, the mesoporous nature of the photoanode will provide more surface contact at the photoanode/electrolyte interface. Larger surface contact between the electrolyte and photoanode is beneficial for QDSSC performance as it will facilitate charge transfer, hence increasing electricity production and eventually PCE. The most widely

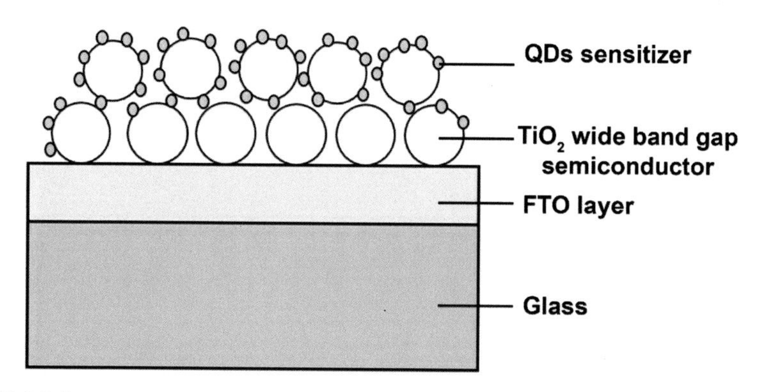

FIGURE 4.2.4

Schematic of the structure of the QDSSC photoanode.

used metal oxide in sensitized solar cells is TiO_2. This is because of its excellent properties, such as a mesoporous structure, perfect energy level band for electron transfer/injection from the quantum dot sensitizer (Fig. 4.2.5), and fast electron mobility for the electron generated by the photon of light (Baraton, 2012).

QDs are a nanosized semiconductor material with a narrow bandgap, which is essential for light absorption. The sizes of quantum dots usually range from 1–10 nm and their properties, such as optical and electronic, are different from their bulk properties. Because of their properties, quantum dots have been widely studied in numerous applications such as light emitting diodes, photodetectors, photocatalysis, and photovoltaics (Caputo et al., 2017; De Iacovo et al., 2016; Gong et al., 2016; Kosyachenko, 2015). In the sensitized solar cell field, after the introduction of highly porous TiO_2 in DSSC, new work has begun to replace organic dye sensitizer with quantum dot sensitizer (Liu et al., 2010; Vogel, Pohl, & Weller, 1990; Vogel, Hoyer, & Weller, 1994). The motivation in replacing organic dye sensitizer with inorganic quantum dots is due to the advantages described below:

1. Quantum dots are easier to fabricate and exhibit good durability (Lee et al., 2009b).
2. Ionization impact: production of two excitons per photon with hot electrons (Nozik, 2005).

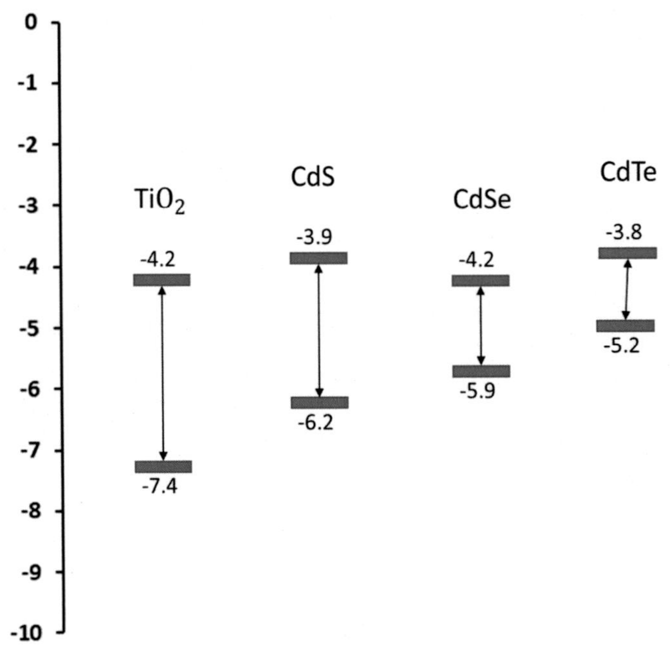

FIGURE 4.2.5

Energy level bandgap of CdS, CdSe, and CdTe QDs versus TiO_2.

3. Reduction of dark current with a high extinction coefficient (Lee et al., 2009b).
4. High theoretical PCE of 44% by considering the multiplication of the carrier by ionization (Hanna & Nozik, 2006).

The most frequently used quantum dot sensitizer in QDSSCs is the cadmium chalcogenide family, consisting of CdS, CdSe, and CdTe (Rhee, Chung, & Diau, 2013). Cadmium chalcogenide preparation/synthesis methods are easy to perform and the bandgaps of cadmium chalcogenides can be adjusted by manipulating their size. Utilization of more than one quantum dots sensitizer in QDSSCs has been reported in many studies, for example CdS/CdSe, CdTe/CdSe, and CdTe/CdS (Lee et al., 2009b; McElroy et al., 2014; Yu et al., 2011). The combination of two quantum dots sensitizers effects electron distribution, which in turn, causes the band gap to shift to either a more positive potential side or a more negative potential. This effect of energy bandgap shifting is based on Fermi level alignment (Nozik, 2005). Another benefit of more than one quantum dot sensitizer on QDSSC is that it can widen the wavelength range of light, thereby increasing the device performance (Pan et al., 2012). Various CdS-based QD sensitizers have been listed in Table 4.2.1. Fig. 4.2.6 below shows the electron injection from a combination of CdTe/CdSe and CdS/CdSe.

Table 4.2.1 List of CdS-based photoanode compositions along with power conversion efficiency value.

Electrolyte	Photoanode arrangement	Cathode	Deposition method	PCE (%)	References
H_2O/methanol +Na_2S+S+KCl	TiO_2/CdS	Platinum (Pt)	Chemical bath deposition (CBD)	1.15	Li et al. (2014)
H_2O+ Na_2S+S	TiO_2/CdS	Pt	CBD	1.13	Hossain, Biswas, and Takahashi, (2009)
MPN+LiI +I_2+DMPII +TBP	TiO_2/CdS	Pt	ML	0.30	Yu-Jen and Yuh-Lang (2008)
Polysulfide	TiO_2/CdS	Pt	CBD	1.28	Sudhagar et al. (2009)
H_2O+ Na_2S +Na_2SO_3	TiO_2/CdS	Pt	CBD	1.91	Chen et al. (2010a)
H_2O+ Na_2S+S	TiO_2/CdS	Carbon (C)	CBD	1.47	Zhang et al., (2010)
H_2O/methanol +Na_2S+S+KCl	TiO_2/CdS	Gold (Au)	SILAR	1.62	Zhu et al. (2010)
H_2O/methanol +Na_2S+S+KCl +GuSCN+SiO_2	TiO_2/CdS	Au	SILAR	2.01	Chou et al. (2011)
H_2O/methanol +Na_2S+S+KCl	TiO_2/CdS	PEDOT	SILAR	1.35	Yeh et al. (2011)

(*Continued*)

Table 4.2.1 List of CdS-based photoanode compositions along with power conversion efficiency value. *Continued*

Electrolyte	Photoanode arrangement	Cathode	Deposition method	PCE (%)	References
H_2O/methanol $+Na_2S+S+KCl$	TiO_2/CdS	Au	CBD	1.03	Zhu et al. (2011c)
H_2O/methanol $+Na_2S+S+KCl$	TiO_2/CdS	Pt	SILAR	1.56	Jang et al. (2012)
Methylcellulose $+H_2O+$ Na_2S $+S$	TiO_2/CdS/ ZnS/SiO_2	Pt	SILAR	1.42	Mingsukang, Buraidah, and Careem (2016)
H_2O/methanol $+Na_2S+S+KCl$	TiO_2/CdS/ ZnS	Pt	SILAR	1.72	Jung et al. (2012)
H_2O/methanol $+Na_2S+S+KCl$	TiO_2/Mn-CdS	CuxS	SILAR	3.55	Ghosh et al. (2016)
H_2O+Na_2S+S $+KCl$	TiO_2/CdS/ ZnS	CuxS	SILAR	2.36	Amr et al. (2016)
Aqueous polysulfide	TiO_2-nanorod/CdS	Pt	SILAR	1.75	Pawar et al. (2016)
H_2O+ Na_2S+S $+NaOH$	TiO_2/CdS/ ZnS/P_3HT	Cu2S	SILAR	3.07	Cerdán-Pasarán et al. (2016)
Aqueous polysulfide	TiO_2-nanorod/ CdS/ZnS	CuS	SILAR	3.22	Lan et al. (2016)
ethylene glycol $+LiI+I_2$	TiO_2/RGO/ CdS	Pt	SILAR	0.37	Badawi (2015)

CBD, *Chemical bath deposition;* ML, *Molecular linker;* SILAR, *Successive ionic layer absorption and reaction.*

One of the major problems in QDSSCs is a very high recombination rate at the electrolyte/photoanode interface (Tubtimtae & Lee, 2012). The recombination process is a parasitic process in the QDSSC that will reduce the performance of the cell. This process happens when the electron produced does not result in electricity generation. A passivation layer is a successful strategy to reduce the recombination rate in QDSSCs (Guijarro et al., 2011). Examples of passivation layers that has been used to overcome the recombination problem in QDSSC are zinc sulfide (ZnS), silicon dioxide (SiO_2), a thin layer of amorphous TiO_2 and many others (Yang et al., 2015). Thus the passivation layer is considered as an important component in the photoanode of QDSSCs.

4.2.3.2 Counterelectrode (cathode)

The third major component of the QDSSC is the counterelectrode or cathode. The electrons produced at the photoanode that successfully travel along the external circuit will reach the counterelectrode. From the counterelectrode, electrons will be transferred back to the photoanode through the mediator ion in the electrolyte as discussed in Chapter 3.2.2, Electron transport properties at the counter

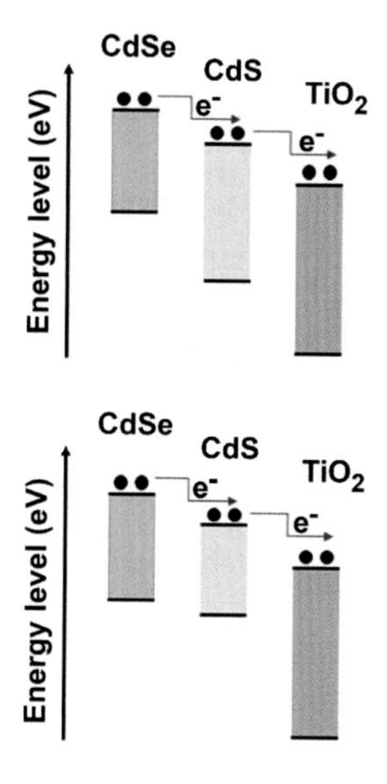

FIGURE 4.2.6

Bandgap edge level of QD sensitizers after electron distribution in: (A) CdTe/CdSe and (B) CdS/CdSe due to Fermi level alignment favoring electron injection into TiO_2.

electrode/electrolyte interface influence the performance of the sensitized solar cells that is, good interfacial properties will benefit the performance of the QDSSC (Giménez et al., 2009). To provide the best performance to the sensitized solar cells, materials used in the counterelectrode must be highly conducting and possess good catalytic properties (Lee et al., 2014). Materials that possess high conductivity will facilitate electron transport and the completion of the whole circuit of the QDSSC. With highly conducting counter electrodes, the series resistance (R_s) of the device will be lowered and cause an improvement in fill factor and power conversion (Meng, Chen, & Thampi, 2015). The catalytic properties of the counterelectrode determine the rate of electron transfer from counterelectrode to the mediator ions (S_x^{2-} ions) that is, the highly catalytic counterelectrode will result in fast electron transfer from the counterelectrode to the S_x^{2-} ions.

In DSSC studies, platinum has been widely used as the counterelectrode material. This is because Pt is highly catalytic toward iodide-based electrolytes that are commonly utilized in DSSCs. Utilization of the platinum counterelectrode will make the process of electron transfer from CE to the mediator ion fast and stable, thereby improving device performance. However, the Pt counter electrode in the QDSSC will degrade the performance of the device. This has been reported in many studies where the QDSSC consisting of a combination of polysulfide electrolyte and platinum counterelectrode will give bad performance that is, low fill factor and low PCE (Jun et al., 2013; Lee & Chang, 2008). The low fill factor of the QDSSC with polysulfide electrolyte and platinum counterelectrode is due to the fact that the platinum counter electrode is not catalytic to the polysulfide electrolyte (Chen et al., 2014; Radich et al., 2011; Tachan et al., 2011; Yeh et al., 2011). This can be explained by the "Pt poisoning effect" where S compounds in electrolytes adsorb onto the platinum surface creating a layer of resistance at the counterelectrode/electrolyte interface (Hodes, Manassen, & Cahen, 1977; Hodes, Manassen, & Cahen, 1980). The Pt poisoning effect will eventually hinder the transportation of electrons to the redox mediator. In addition, there have been reports noting that the sulfur in the polysulfide electrolyte also corrodes the platinum (Seo et al., 2011).

Researchers have tried to find replacements to Pt counterelectrodes for QDSSCs, such as noble metals, conducting polymers, carbon-based materials, and metal chalcogenides (Seo et al., 2011). The replacement of platinum started with the introduction of gold as the counterelectrode owing to its high conductivity and catalytic activity toward the iodide electrolyte. An example of an Au counterelectrode study conducted by Seo and coworker showed that a QDSSC fabricated with a Au counterelectrode showed 42% improvement in PCE as compared to the QDSSC with a platinum counter electrode (Seo et al., 2011).

The most fascinating material used for the QDSSC counterelectrode is the metal sulfide, and it have been proven that metal sulfide counterelectrodes provide positive impacts to the performance of the device. The reason behind this is that metal sulfide is more catalytic toward the reduction process of the mediator ion at the counter electrode/polysulfide electrolyte interface. This allows easier electron transfer at the interface with less resistance, which eventually improves the current density of the device. Examples of metal sulfides used as counterelectrodes for QDSSCs include:

1. Copper monosulfide (CuS) (Lin et al., 2013)
2. Lead sulfide (PbS) (Seo et al., 2011)
3. Copper(I) sulfide (Cu_2S) (Radich et al., 2011)
4. Cobalt sulfide (CoS) (Faber et al., 2013)
5. Nickel sulfide (NiS) (Hodes et al., 1977).

PbS can absorb light in the near infrared region owing to its semiconductor and small bandgap properties. This property of PbS is beneficial for QDSSC as it can absorb the remaining light not absorbed by the photoanode, hence leading to the generation of additional electrons and additional potential (V_{oc}) as shown in the Fig. 4.2.7. This effect is called the "auxiliary tandem effect" whereby the counterelectrode functions as a photocathode, providing additional electron sources and potentials to the device. According to the auxiliary tandem effect, the V_{oc} of the cell is the energy difference between the quasi-Fermi levels ($_qE_f$) of the photoanode (n-type) and cathode (p-type) that is V_{oc} = V_{oc} 1 + V_{oc} 2 (Lee & Chang, 2008). The quasi-Fermi level of an electrode, either the anode or the cathode, depends on the equilibrium between the electrode to the electrolyte. Illuminating the photoactive electrode shifts the quasi-Fermi level toward the vacuum Fermi level.

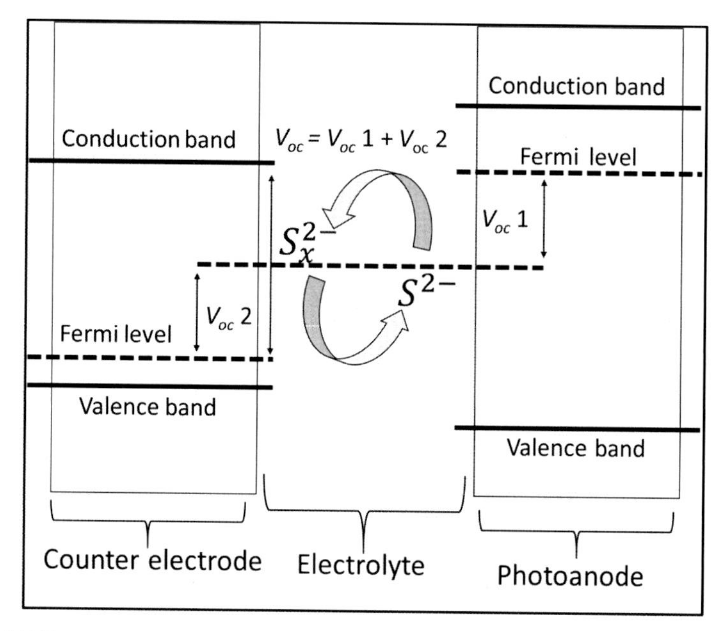

FIGURE 4.2.7

Diagram of the auxiliary tandem effect with the movement of electrons, the energy bandgap, and Fermi level of the QDSSC with PbS CE (Lin et al., 2013).

4.2.3.3 Electrolyte

The redox couple is used as an electrolyte or hole transporting material (HTM) and is significantly important to shuttle the photogenerated holes to the counter-electrode and regenerate quantum dots. The main characteristics of materials used as electrolytes are: (1) a suitable redox potential is required for the regeneration of quantum dots as well as to maintain a high V_{oc} value; (2) the material should be less corrosive so as to maintain long term stability of the device; (3) highly conductivity to ions to ease hole transfer; (4) completely regenerative; (5) high level of transparency and stability in the visible light range (Wu et al., 2015). The solar cell performance and its long-range stability are remarkably influenced by the choice of electrolyte. The electrolyte can be categorized into solid-state, quasi-solid-state, and liquid-state based on their characteristics and physical states at the time of application. The different types of electrolytes in QDSC-based photovoltaic performances are summarized in Table 4.2.2.

A redox couple such as I^-/I_3^- and/ or S^{2-}/S_n^{2-} represents a typical liquid electrolyte. The most typical and ideal redox couple for DSCs is I^-/I_3^- but its corrosive action to QDs makes it unfit for later applications. The most suitable and stable redox couple for QDs was found to be S^{2-}/S_n^{2-} with a maximum PCE (Lee & Chang, 2008) as indicated in the findings of many researchers (Lee & Chang, 2008; Licht,

Table 4.2.2 Effect of various redox couples and solid-state HTMs on the photovoltaic properties of the QDSSC.

Redox couple/ hole-transporting materials (HTMs)	Quantum dots (QDs)	CE	J_{sc} (mA/cm^2)	V_{oc} (V)	FF	PCE (%)	References
Polysulfide	ZCISe-CdSe	MC/Ti	27.39	0.752	0.619	12.75	Wang et al. (2018)
$[(CH_3)_4N]_2S/[(CH_3)_4N]_2S_n$	CdS	Pt	3.0	1.2	0.89	3.2	Li et al. (2011)
Polysulfide/IL	CdSe	Pt	13.85	0.42	0.32	1.86	Jovanovski et al. (2011)
$[Fe(CN)_6]^{3-/4-}$	CdS	Pt	3.8	0.8	0.66	2.0	Evangelista et al. (2016)
$[Co(bpy)_3]^{2+/3+}$	CdS	Pt	2.34	0.704	0.62	1.01	Mora-Seró et al. (2008)
Polymer gel electrolyte/PANI[a]	CdS	Pt	2.20	0.60	0.78	1.25	Kumar et al. (2018)
Polysulfide	CdS	Pt	3.63	0.43	0.17	1.15	Lee and Chang (2008)
$[Co(bpy)_3]^{2+/3+}$	Au_x	Pt	3.96	0.832	0.716	2.36	Chen, Choi, and Kamat (2013)
$[Co(o\text{-}phen)_3]^{2+/3+}$	$CdSe_xTe_{1-x}$	Pt	4.94	0.67	0.54	1.77	Lee et al. (2009c)
Tetramethylthiourea	ZnSe/CdS	Pt	2.25	0.66	0.58	0.86	Ning et al. (2012)
[DHexBIm][SCN][b]	CdS/CdSe	PbSe	12.58	0.60	0.56	4.26	Dang et al. (2017)
Polysulfide	CdS/CdSe	Au	16.1	0.56	0.34	–	Zhu et al. (2011b)
CuSCN	CdSe	Au	4.00	0.50	–	2.3	Lévy-Clément et al. (2005)
Spiro-OMeTAD[c]	PbS	Au	13.56	0.52	0.579	4.10	Zhang et al. (2017)
P3HT[d]	PbS/ CuS	Au	20.7	0.60	0.65	8.07	Park et al. (2016)
PCPDTBT[e]	Sb_2S_3	Au	15.3	0.616	0.657	6.18	Park et al. (2016)

[a]*PANI = Polyaniline.*
[b]*1,3-dihexylbenzimidazolium cation combined with the SCN anion.*
[c]*2,2′,7,7′-Tetrkis-(N,N-di-p-methoxyphenylamine)-9,9′-spirobifluorene.*
[d]*Poly(3-hexylthiophene).*
[e]*Poly(2,6-(4,4)-bis-(2-ethyhexyl)-4H-cyclopenta[2,1-b;3,4-b′]dithiophene)alt-4,7(2,1,3-benzothiadiazol).*

1995; Vogel et al., 1994). However, polysulfide electrolytes also show disadvantages such as (1) low V_{oc} owing to its relatively high redox potential which results in energy loss and (2) decreased QDs stability and increased recombination rate which is due to the slow QDs regeneration and accumulation of holes (Kamat, 2012). However, these shortcomings can be overcome by exploring new redox couples or electrochemical properties of the existing redox couple can be improved by using additives to the polysulfide electrolyte.

The development of novel redox couples for QDSCs has been the main focus of researchers over the years (Evangelista et al., 2016; Jovanovski et al., 2011; Lee et al., 2009c; Li et al., 2011; Ning et al., 2012). Compared to polysulfide electrolyte redox couples, organic electrolyte based cells provide significantly less charge transfer resistance at the electrolyte/ counter electrode interface. Based on this observation, Ning et al. introduced tetramethylthiourea-based organic electrolytes for QDSCs (Ning et al., 2012). Also, Sun et al. developed a redox couple based on organic-polysulfide $[(CH_3)_4N]_2S/[(CH_3)_4N]_2Sn$ as an electrolyte with CdS as the QD in QDSCs (Li et al., 2011). This redox couple exhibited an oxidation potential of 1.045 V versus normal hydrogen electrode, and it was 0.5 V lower as compared to that of a neat polysulfide redox couple with V_{oc} and fill factor 1.2 V and 0.89, respectively. Based on the known efficient cobalt complex-based electrolyte for DSSC, Grätzel et al. utilized a $[Co(o\text{-phen})_3{}^{2+/3+}]$ complex as an electrolyte in QDSCs as well (Lee et al., 2009c). The cell using a cobalt complex and CdSeTe QD sensitizer exhibited a PCE of 4.18% with 0.1 sun illumination. Mora Sero et al. designed an innovative electrolyte based on pyrrolidinium ionic liquid which is stable for more than ten days with CdSe as the QDs in QDSCs having a PCE of 1.86%. A ferricyanide/ferrocyanide $[Fe(CN)_6{}^{3-/4-}]$ redox couple as an electrolyte in QDSCs was prepared by Tachibana et al. This couple showed positive redox potential and is responsible for an enhanced V_{oc} value of 0.8 V (Evangelista et al., 2016).

Despite the improved V_{oc} values by employing the above mentioned novel redox couples, these electrolytes are still far behind polysulfide electrolyte systems in terms of PCE. The low PCE value may be attributed to either slow QDs regeneration rate or the limited stability of QDs in the electrolyte. So, it remains a challenging target to develop a redox couple with a PCE value higher than that reported for polysulfide electrolytes. An alternative way to improve PCE and overall performance of cell and electrochemical properties is the modification of polysulfide-based electrolytes using additives. Researchers already employed this strategy in DSCs to modify the $I^-/I_3{}^-$ redox couple by incorporating additives, such as 4-tert-butylpyridine, guanidine thiocyanate, lithium ions, etc. The incorporation of these additives has significantly improved the photovoltaic performances by either changing conduction properties of TiO_2 and shifting the conduction band edge or hindering the charge recombination at the photoanode/electrolyte interface. This same strategy has been recently employed by researchers to modify the polysulfide electrolyte to improve the photovoltaic performance of QDSCs (Du et al., 2015; Jiang et al., 2016; Wei et al., 2016; Yu et al., 2017).

Various groups attempted to improve the photovoltaic properties of polysulfides by using additives. Zhong et al. observed remarkable increases in the PCE by modifying polysulfide with different water soluble polymers as additives such as poly (vinyl pyrrolidone) and poly(ethylene glycol) (Du et al., 2015; Jiang et al., 2016). Another way to enhance the PCE efficiency is by creating an energy barrier at the electrode/electrolyte interface with the incorporation of fumed silica nanoparticles into the polysulfide electrolyte (Wei et al., 2016). In another attempt, Zhong et al. modified polysulfide with the addition of tetraethyl orthosilicate and observed an improvement in PCE from 11.75% to 12.34% (Yu et al., 2017).

The long term stability of QDSCs can be actualized using a solid-state HTM (so-called all-solid-state devices) to overcome the drawbacks of volatilization and leakages of liquid electrolytes (Wu et al., 2015). Grätzel, the pioneer of all-solid-state QDSCs, for the first time reported spiro-OMeTAD as a hole transporting medium having an efficiency of 0.49% with PbS as QDs (Plass et al., 2002). The pore size in TiO_2 films was optimized along with the PbS deposition cycles which led to an increased efficiency of 1.46% (Lee et al., 2009a). Seok et al. used poly (3-hydroxythiophene) as a hole transporting medium with multilayered deposition of PbS on TiO_2 electrode in solid-state QDSCs and achieved improved performance of the cell with a PCE of 2.9% (Im, 2011).

4.2.4 Fabrication techniques

QDs as sensitizers play a key role in QDSCs with one sensitizer moiety by single-handedly absorbing a wide range of the solar spectrum (Semonin et al., 2010). Promising QD sensitizers must have the following features for effective electron injection and high photo conversion efficiency: (1) an elevated conduction band edge as compared to metal oxide electron-transporting materials (ETMs); (2) a narrow band gap, tunable absorption range, and increased absorption coefficient; (3) good thermal, high light, and moisture stability; (4) facile fabrication, low toxicity, and low cost availability. However, over the last few years, a comeback has been observed in the facile fabrication development and sensitization of QDs (Jun et al., 2013; Kouhnavard et al., 2014; Pan et al., 2012). Instead of the structure, the fabrication routes for the surface modification of ETMs with sensitizing QDs are also responsible for the improved photovoltaic performance of QDSCs. Unlike dyes, QDs are larger-sized inorganic semiconducting light-harvesting materials. Due to the absence of anchoring positions on the ETM layer, it is difficult to bind a QD moiety to make a completely covered monolayer. Thus it is an incredibly challenging task to design a QD-sensitized electrode with a higher QD load for efficient light-harvesting (Li & Zhong, 2015; Watson, 2010). In general, for the fabrication of QDSCs, QD sensitization over the ETMs layer can be introduced via two basic techniques: in situ and ex situ methods (Kamat, 2013).

4.2.4.1 In situ methods

In situ deposition causes direct growth of QDs on the ETMs layer from precursor solutions. Chemical bath deposition (CBD) and successive ionic layer adsorption and reaction (SILAR) are widely used as simple and low cost in situ techniques for QD sensitization on nanostructured wide bandgap ETMs. In this approach, polydisperse distribution of QD growth strongly depends on the nucleation mechanism that takes place inside the nanostructured semiconductor, and, consequently, efficient charge injection could happen from the QD to semiconductor (Rühle, Shalom, & Zaban, 2010). Due to high surface coverage, good anchoring to the electrodes, simple processability, increased QD loading amount, and good reproducibility, in situ deposition methods are widely used. However, in these techniques, the size and distribution of QDs and density of trap state defects are not precisely controlled. As a result, severe charge recombination is observed due to the high density of defects. Therefore only about 7% PCE is reported for QDSCs derived from in situ QD fabrication (Radich et al., 2014).

4.2.4.1.1 Chemical bath deposition

CBD is a relatively easy, cost-effective and reproducible technique to attach QD sensitizer onto the surface of semiconductors. In this method, QD sensitization occurs in a bath (i.e., container) by dipping the electrode for a defined period into the solution containing cationic and anionic precursor for slow nucleation and growth of QD (Gorer & Hodes, 1994). However, the fabrication of the QDs is controlled by immersion time, temperature, and composition of precursor solutions, and nature of electrode. In the literature, the attachment of CdS and CdSe QDs to semiconductors has been reported by the CBD method, and improved photovoltaic performance has been observed (Rawal et al., 2012; Sudhagar et al., 2011). This technique helps in dense deposition, where, subsequently, the recombination of charge carriers is reduced but sometimes, deposition takes place in a nanocrystalline film pattern instead of nanoparticles which can hinder the development of the narrow-sized pore channel. Thus CBD is more appropriate for nanostructured semiconductors which have large porosity of the ETMs layer for example nanotubes, nanorods, or 3D hierarchical structures (Sharma, Jha, & Kumar, 2016). Different research groups have modified the CBD method in a controlled sequential manner to sensitize QDs over the semiconductor electrode for efficient and stable QDSCs, known as sequential CBD (Chen et al., 2010b). Similarly, the microwave assisted CBD method was also observed for use to spur faster nucleation and growth of cosensitized CdS/CdSe QDs on TiO_2-deposited electrodes. The microwave irradiation assisted in suppressing the charge carrier recombination, and easy attachment of QDs due the wettability of TiO_2 surface, therefore high short circuit current density (J_{sc} = 16.1 mA/cm^2) and PCE (3.06%) was measured (Zhu et al., 2011b). In another approach known as electric field assisted CBD, a low electric field was employed to control CBD for PbS QDs fabrication on TiO_2 nanotube arrays, which increased the QD loading amount and resulted in high J_{sc} (8.48 mA/cm^2) and 3.41% PCE (Tao et al., 2014).

4.2.4.1.2 Successive ionic layer adsorption and reaction

SILAR is an inexpensive, easy, and suitable method for large-area thin film fabrication. In this approach, cationic and anionic precursor solutions are placed in separate containers. For one complete deposition or SILAR cycle, a nanostructured semiconductor-coated electrode is immersed into the two solutions alternately, and is then followed by washing and drying (Lee & Lo, 2009). The development of QDs can be monitored by the number of deposition cycles, composition, and pH of the precursor solutions, rate of dipping, immersion time, rinsing time duration, and reaction temperature (Hu et al., 2013). The growth mechanism of QDs in SILAR is mostly based on the successive adsorption on the surface of the semiconductor, reaction of the cations/anions from the precursor solutions, and washing with deionized water between every dipping to enable a heterogeneous reaction (between solid phase and the solvated ions) rather to avoid homogeneous precipitation. This method is designed such that in a deposition cycle, every dipping period can be optimized to obtain a desired particle size, where a monolayer will be formed after every cycle. Therefore a thin film can be developed layer by layer. Due to shorter time and close stoichiometry, the SILAR method has been reported as an improved approach to CBD (Senthamilselvi et al., 2012).

In particular, this scheme has been reported to fabricate metal sulfide QDs (Toyoda, Sato, & Shen, 2003), while also successfully utilized to synthesize metal selenides and tellurides (Lee et al., 2009c) onto TiO_2 and ZnO electrodes. In one work, a new form of SILAR, known as potential induced ionic layer adsorption and reaction (PILAR), has been emerged as an in situ fabrication method. Enhanced ion adsorption and high QD load were observed while using this technique for CdSe QDs sensitization into a mesoporous TiO_2 electrode. In the first step, cation adsorption was carried out under an applied bias accompanied by washing and drying. Afterwards, the electrode was immersed into anionic precursor solution to assemble QDSCs. The CdSe/TiO_2 assembly designed via PILAR method has revealed PCE values up to 4.30% (Liu et al., 2014).

4.2.4.2 Ex situ methods

High-quality monodisperse QDs can be synthesized ex situ via organometallic high-temperature well-developed QD synthetic routes (Resch-Genger et al., 2008; Zrazhevskiy, Sena, & Gao, 2010). So, the ex situ synthesis approach can take the benefits of the huge developments in controlling the growth of highly crystalline and monodisperse QDs of various semiconductor materials. It is evident that high-quality presynthesized QDs play a significant role in assembling highly efficient QDSCs. However, it is very challenging to incorporate QDs at high concentrations to accomplish efficient QD-electrode junctions with enhanced charge separation and minimum surface trapping. For the ex situ approach, methods used to immobilize presynthesized colloidal QDs include direct adsorption (DA), electrophoretic deposition (EPD), and a linker-molecule assisted self-assembly. Colloidal QDSCs have two limitations: first, charge separation, electron transfer, and PCE of the photovoltaic cell strongly depend on the chemical nature of the bifunctional molecular linker (Chen et al., 2009; Shen & Lee, 2008). Second, the surface coverage of QDs onto the semiconductor substrate is generally not good enough when ex situ methods are used (Guijarro et al., 2009). For proficient and outstanding development of solar devices, strong light absorption and enhanced charge separation are needed, which has been challenging for QDSCs based on ex situ routes.

4.2.4.2.1 Direct adsorption

In the DA method, monodisperse QDs are deposited onto the surface of the semiconductor without the aid of any molecular linker or external DC voltage. It is usually performed by immersing a semiconductor-coated electrode into a colloidal QD solution for a specified period of time. This working phenomenon involves solvent/nonsolvent precipitation of QDs from the solution onto the mesoporous semiconductor electrode. However, precipitation leads to a high amount of agglomeration in addition to a low, uneven, and polydisperse coverage by aggregates at approximately 14% (Giménez et al., 2009; Guijarro et al., 2009). Guijarro et al. reported that in spite of the low surface coverage in DA, an IPCE of 36% was obtained at an excitonic peak wavelength higher than the linker-assisted adsorption (Guijarro et al., 2009). A less investigated technique known as physisorption has been studied by Rühle et al. (2010). In this deposition process, the semiconductor electrode was immersed into a solution of QDs for extended periods of time, for example, 100 h. In literature, research on this method is rather limited. Nevertheless, QDSCs assembled via physisorption show better performance than solar cells prepared through linker-assisted adsorption. In another study by Wijayantha, Peter, and Otley (2004), a pressing route was utilized to fabricate light-harvesting electrodes where CdS QDs were self-assembled onto the surface of dispersed nanocrystalline TiO_2 particles. The findings about the work signify that electrode fabrication led to partial loss or scratch of CdS layer, and the formation of inaccessible parts to the redox electrolyte. However, the pressing route may provide a low cost method to fabricate photoelectrodes in QDSCs.

4.2.4.2.2 Electrophoretic deposition

An electrophoretic bath is used to sensitize the FTO/semiconductor-coated electrode with presynthesized QDs prepared using polar/nonpolar solvents and colloidal solution of QDs. An FTO/semiconductor as a positive terminal and a blank FTO-coated glass substrate as a negative terminal are dipped in EPD bath and a specified distance is maintained. A bias voltage is provided between the semiconductor electrode and FTO for a fixed period of time to deposit QDs onto the semiconductor. This method has advantages to maintain the size and shape of QD particles without any significant aggregation. Several factors such as applied voltage, deposition time, and concentration of bath solution can determine the yield of the EPD process (Jara et al., 2014).

Previously, EPD was used to deposit metallic, semiconductor, and insulating particles on polymers and conductive substrates. Smith, Emmett, and Rosenthal (2008) and Kamat et al. (Brown & Kamat, 2008; Farrow & Kamat, 2009) have used EPD for solar cell fabrication, yielding low conversion efficiencies. Salant et al. reported high PCE (1.7%) under one sun illumination condition in CdSe-sensitized QDSCs via EPD deposition compared with solar cells prepared with a molecular linker approach. It was also investigated that PCE was not dependent on size signifying enhanced electron injection even for large size QDs (Salant et al., 2010). In another study, Yufeng et al. used a combined method of functional linker with EPD to fabricate electrodes based on CdSe QDs or CdSe(S) nanoplates (NPLs)/TiO_2. Upon one sun light, the photocurrent densities via EPD were found to be higher than those fabricated with a linker technique. Therefore the results obtained from this study open the doors to engineer the surface/dimension/composition of semiconductors, adsorption of QDs on porous electrodes, and scheming highly efficient photoelectrochemical devices (Zhou et al., 2020).

4.2.4.2.3 Molecule linker attachment

In MLA, the capping-ligand-induced self-assembly allows the surface attachment of QDs onto metal oxide semiconductor electrode with a higher QD load (Jiao et al., 2015; Wang et al., 2013). The linking of QDs is carried out using a linker moiety, which is a bifunctional material that anchors the QD to semiconductor electrode, acting as a molecular cable. The chemical nature and length of the linker plays a decisive role in the efficient electron injection from QDs to the matrix of semiconductor. Various organic capping agents [mercaptoproponic acid (MPA), trioctylphosphine oxide, trioctylphosphine, oleic acid] have been used to cover the nascent nucleus to control the size, shape, and optical properties of presynthesized monodisperse QDs. High-quality QDs synthesis is performed via well-developed organometallic high-temperature injection method, followed by a ligand exchange step to acquire linker-molecule-capped-water-dispersible QDs, which can be attached onto the electrode. The carboxyl group in the bifunctional linker (COOH)-R-SH is attached to the semiconductor surface while the thiol group is available to attach to QDs. It should be noted that functional linkers can effectively disperse and stabilize QDs (Qian et al., 2011). Subsequent immersion into linker-capped QD solution is needed to perform the adsorption of QDs onto the surface of the semiconductor which may last from a few hours to few days. However, the electron transport rate can be reduced due to a long distance between the QDs and the semiconductor surface which can limit the resulting PCE of QDSCs.

In 1990s, Alivisatos et al. immobilized CdS and CdSe QDs on metal (Au, Al) surfaces using a bifunctional molecular linker containing carboxylate and thiol groups (Colvin, Goldstein, & Alivisatos, 1992). Later, this technique was borrowed by other researchers to fabricate QDs on TiO_2 electrodes for QDSCs assembly (Lee et al., 2008; Leschkies et al., 2007; Zhang et al., 2012). Despite the benefits of the MLA technique, researchers have been unsuccessful in obtaining a high load of QD sensitizers on the electrode and highly efficient QDSCs. It is worthwhile to note that some parameters, including the choice of capping agent, excess molecular linker, and the pH of solution, were found to be effective in developing a fast, homogeneous, and high loading of QD deposition. It has been observed that due to improvements in ligand exchange methods and optimization of QD deposition conditions, the PCE has rapidly improved to nearly 13% (Wang et al., 2018). The design of aqueous QDs (capped with a short-chain mercaptoalkylcarboxyl ligand in aqueous media) provides an easy way to immobilize QD sensitizers onto the TiO_2 electrode due to obviation of the additional step for ligand exchange. Meng and coworker reported aqueous $CuInS_2$ QDs sensitization with a PCE of 1.47% (Hu et al., 2011). Through optimizing the fabrication conditions and engineering the electronic structure of aqueous QD sensitizers, 8.0% PCE has been reported under one full sun irradiation, while a PCE of 8.15% has been achieved under 30 mW/cm^2 light intensity (Chiang et al., 2018; Raevskaya et al., 2016). MPA is reported as a superior linker moiety than thiolacetic or mercaptohexadecanoic acid (Raevskaya et al., 2016), and in a similar way, cysteine as a molecular linker gives rise to more efficient electrodes than using thioglycolicacid or MPAs (Kongkanand et al., 2008).

4.2.4.3 Other methods

Besides in situ and ex situ techniques, QD fabrication has been reported by other methods, such as electrospray-assisted techniques (Zhu et al., 2012), ultrasonic spray pyrolysis deposition (Zhu et al., 2011a), electrochemical deposition (Rao et al., 2014), and chemical vapor deposition (Seo et al., 2014). Each technique discussed has its own drawbacks and advantages. Strong electronic coupling between the semiconductor and QD sensitizer is observed in in situ fabrication whereas ex situ methods provide an option to control the size and shape of QDs. However, effective charge transportation could be limited due to the presence of linker molecules that may lead to less efficient photovoltaic devices. Therefore the selection of fabrication methods should be strongly focused on the objectives of the specified research investigation.

4.2.5 Challenges

The main challenge in QDSSC devices is its low PCE. There are several reasons causing QDSSC to exhibit low PCE such as electron recombination problems at the photoanode and electrolyte issues: low charge transportation, volatility, and stability. One of the major challenges in QDSSCs is electron recombination. Electron recombination happens when electrons produced in the photoanode combine with S_x^{2-} ions in the electrolyte. S_x^{2-} ions are formed when the sulfide ions (S^{2-}) react with the sulfurs (S) contained in the electrolyte. The recombination process is a parasitic process that will reduce the cell's performance. Recombination will cause produced electrons to not take part in electricity generation.

Other major challenges in QDSSCs are caused by the electrolyte. One of the issues is that water is used as the solvent for sulfides. This is because all sulfide-based salts are insoluble except for group one-based sulfide salts, which are very soluble in water. Utilization of water in the electrolyte of the QDSSC will cause volatilization problems due to the heat from the light. Electrolyte volatilization will lead to low charge transportation and instability in the QDSSC.

4.2.6 Future prospect

The majority of QDSSC studies use cadmium- and lead-based QDs. Cadmium and lead are toxic to the human health and environment. Due to the toxicity, several studies have focused on green-QDSSC, where the QD used is not based on Cd. However, the performance of green-QDSSC is still low compared to cadmium- and lead-based QDSSCs. There have also been efforts to synthesize green QDs. Due to the hazardous and toxic preparation of QDs, there are current trends and developments that avoid the use of some of these dangerous chemicals during the synthesis of QDs.

References

Amr, H., et al. (2016). One-step fabrication of copper sulfide nanoparticles decorated on graphene sheets as highly stable and efficient counter electrode for CdS-sensitized solar cells. *Japanese Journal of Applied Physics*, *55*(11), 112301.

Badawi, A. (2015). Decrease of back recombination rate in CdS quantum dots sensitized solar cells using reduced graphene oxide. *Chinese Physics B*, *24*(4), 047205.

Baraton, M.-I. (2012). Nano-TiO_2 for dye-sensitized solar cells. Recent Patents on Nanotechnology *6*(1), 10–15.

Blakers, A., et al. (2013). High efficiency silicon solar cells. *Energy Procedia*, *33*, 1–10.

Brown, P., & Kamat, P. V. (2008). Quantum dot solar cells. Electrophoretic deposition of CdSe – C60 composite films and capture of photogenerated electrons with nC60 cluster shell. *Journal of the American Chemical Society*, *130*(28), 8890–8891.

Caputo, J. A., et al. (2017). General and efficient C–C bond forming photoredox catalysis with semiconductor quantum dots. *Journal of the American Chemical Society*, *139*(12), 4250–4253.

Cerdán-Pasarán, A., et al. (2016). Photovoltaic study of quantum dot-sensitized TiO_2/CdS/ZnS solar cell with P3HT or P3OT added. *Journal of Applied Electrochemistry*, *46*(9), 975–985.

Chen, H., et al. (2010a). Photosensitization of TiO_2 nanorods with CdS quantum dots for photovoltaic devices. *Electrochimica Acta*, *56*(2), 919–924.

Chen, H., et al. (2010b). A suitable deposition method of CdS for high performance CdS-sensitized ZnO electrodes: Sequential chemical bath deposition. *Solar Energy*, *84*(7), 1201–1207.

Chen, H., et al. (2014). Efficient iron sulfide counter electrode for quantum dots-sensitized solar cells. *Journal of Power Sources*, *245*, 406–410.

Chen, J., et al. (2009). An oleic acid-capped CdSe quantum-dot sensitized solar cell. *94* (15), 153115.

Chen, Y.-S., Choi, H., & Kamat, P. V. (2013). Metal-cluster-sensitized solar cells. A new class of thiolated gold sensitizers delivering efficiency greater than 2%. *Journal of the American Chemical Society*, *135*(24), 8822–8825.

Chiang, Y.-H., et al. (2018). Aqueous solution-processed off-stoichiometric Cu–In–S QDs and their application in quantum dot-sensitized solar cells. *Journal of Materials Chemistry A*, *6*(20), 9629–9641.

Chi-Hsiu, C., & Yuh-Lang, L. (2007). Chemical bath deposition of CdS quantum dots onto mesoscopic TiO_2 films for application in quantum-dot-sensitized solar cell. *Applied Physics Letters*, *91*(5), 053503.

Chou, C.-Y., et al. (2011). Efficient quantum dot-sensitized solar cell with polystyrene-modified TiO_2 photoanode and with guanidine thiocyanate in its polysulfide electrolyte. *Journal of Power Sources*, *196*(15), 6595–6602.

Colvin, V. L., Goldstein, A. N., & Alivisatos, A. P. (1992). Semiconductor nanocrystals covalently bound to metal surfaces with self-assembled monolayers. *Journal of the American Chemical Society*, *114*(13), 5221–5230.

Dang, R., et al. (2017). Benzimidazolium salt-based solid-state electrolytes afford efficient quantum-dot sensitized solar cells. *Journal of Materials Chemistry A*, *5*(26), 13526–13534.

De Iacovo, A., et al. (2016). PbS colloidal quantum dot photodetectors operating in the near infrared. *Scientific Reports*, *6*(1), 37913.

Du, J., et al. (2015). Performance enhancement of quantum dot sensitized solar cells by adding electrolyte additives. *Journal of Materials Chemistry A*, *3*(33), 17091–17097.

Evangelista, R. M., et al. (2016). Semiconductor quantum dot sensitized solar cells based on ferricyanide/ferrocyanide redox electrolyte reaching an open circuit photovoltage of 0.8 V. *ACS Applied Materials & Interfaces*, *8*(22), 13957–13965.

Faber, M. S., et al. (2013). Earth-abundant cobalt pyrite (CoS_2) thin film on glass as a robust, high-performance counter electrode for quantum dot-sensitized solar cells. *The Journal of Physical Chemistry Letters*, *4*(11), 1843–1849.

Farrow, B., & Kamat, P. V. (2009). CdSe quantum dot sensitized solar cells. shuttling electrons through stacked carbon nanocups. *Journal of the American Chemical Society*, *131*(31), 11124–11131.

Feng, W., et al. (2016). Highly efficient and stable quasi-solid-state quantum dot-sensitized solar cells based on a superabsorbent polyelectrolyte. *Journal of Materials Chemistry A*, *4*(4), 1461–1468.

Ghosh, D., et al. (2016). A microwave synthesized CuxS and graphene oxide nanoribbon composite as a highly efficient counter electrode for quantum dot sensitized solar cells. *Nanoscale*, *8*(20), 10632–10641.

Giménez, S., et al. (2009). Improving the performance of colloidal quantum-dot-sensitized solar cells. *Nanotechnology*, *20*(29), 295204.

Gong, X., et al. (2016). Highly efficient quantum dot near-infrared light-emitting diodes. *Nature Photonics*, *10*(4), 253–257.

Gorer, S., & Hodes, G. (1994). Quantum size effects in the study of chemical solution deposition mechanisms of semiconductor films. *The Journal of Physical Chemistry*, *98*(20), 5338–5346.

Grätzel, M. (2005). Solar energy conversion by dye-sensitized photovoltaic cells. *Inorganic Chemistry*, *44*(20), 6841–6851.

Guijarro, N., et al. (2009). CdSe quantum dot-sensitized TiO_2 electrodes: Effect of quantum dot coverage and mode of attachment. *The Journal of Physical Chemistry C*, *113*(10), 4208–4214.

Guijarro, N., et al. (2011). Uncovering the role of the ZnS treatment in the performance of quantum dot sensitized solar cells. *Physical Chemistry Chemical Physics*, *13*(25), 12024–12032.

Hanna, M.C., & Nozik, A.J. (2006). Solar conversion efficiency of photovoltaic and photoelectrolysis cells with carrier multiplication absorbers. *Journal of Applied Physics* 100 (7), 074510.

Hodes, G., Manassen, J., & Cahen, D. (1977). Photo-electrochemical energy conversion: Electrocatalytic sulfur electrodes. *Journal of Applied Electrochemistry*, *7*(2), 181–182.

Hodes, G., Manassen, J., & Cahen, D. (1980). Electrocatalytic electrodes for the polysulfide Redox system. *Journal of the Electrochemical Society*, *127*(3), 544–549.

Hossain, M. F., Biswas, S., & Takahashi, T. (2009). Study of CdS-sensitized solar cells, prepared by ammonia-free chemical bath technique. *Thin Solid Films*, *518*(5), 1599–1602.

Hu, X., et al. (2011). Aqueous colloidal $CuInS_2$ for quantum dot sensitized solar cells. *Journal of Materials Chemistry*, *21*(40), 15903–15905.

Hu, Y., et al. (2013). Synthesis and photoelectrochemical response of CdS quantum dot-sensitized TiO_2 nanorod array photoelectrodes. *Nanoscale Research Letters*, *8*(1), 222.

Im, S. H., Kim, H.-j., Kim, S. W., Kim, S.-W., & Seok, S. I. (2011). All solid state multiply layered PbS colloidal quantum-dot-sensitized photovoltaic cells. *Energy & Environmental Science*, *4*(10), 4181–4186.

Jang, W., et al. (2012). Liquid carbon dioxide coating of CdS quantum-dots on mesoporous TiO2 film for sensitized solar cell applications. *The Journal of Supercritical Fluids*, *70*, 40–47.

Jara, D. H., et al. (2014). Size-dependent photovoltaic performance of $CuInS_2$ quantum dot-sensitized solar cells. *Chemistry of Materials*, *26*(24), 7221–7228.

Jiang, G., et al. (2016). Poly(vinyl pyrrolidone): A superior and general additive in polysulfide electrolytes for high efficiency quantum dot sensitized solar cells. *Journal of Materials Chemistry A*, *4*(29), 11416–11421.

Jiao, S., et al. (2015). Band engineering in Core/Shell ZnTe/CdSe for photovoltage and efficiency enhancement in exciplex quantum dot sensitized solar cells. *ACS Nano*, *9*(1), 908–915.

Jovanovski, V., et al. (2011). A sulfide/polysulfide-based ionic liquid electrolyte for quantum dot-sensitized solar cells. *Journal of the American Chemical Society*, *133*(50), 20156–20159.

Jun, H. K., Careem, M. A., & Arof, A. K. (2013). Quantum dot-sensitized solar cells—Perspective and recent developments: A review of Cd chalcogenide quantum dots as sensitizers. *Renewable and Sustainable Energy Reviews*, *22*, 148–167.

Jung, S. W., et al. (2012). ZnS overlayer on in situ chemical bath deposited CdS quantum dot-assembled TiO_2 films for quantum dot-sensitized solar cells. *Current Applied Physics*, *12*(6), 1459–1464.

Kamat, P. V. (2012). Boosting the efficiency of quantum dot sensitized solar cells through modulation of interfacial charge transfer. *Accounts of Chemical Research*, *45*(11), 1906–1915.

Kamat, P. V. (2013). Quantum dot solar cells. The next big thing in photovoltaics. *The Journal of Physical Chemistry Letters*, *4*(6), 908–918.

Kongkanand, A., et al. (2008). Quantum dot solar cells. Tuning photoresponse through size and shape control of $CdSe - TiO_2$ architecture. *Journal of the American Chemical Society*, *130*(12), 4007–4015.

Kosyachenko, L.A. (2015). *Solar cells: New approaches and reviews*. IntechOpen.

Kouhnavard, M., et al. (2014). A review of semiconductor materials as sensitizers for quantum dot-sensitized solar cells. *Renewable and Sustainable Energy Reviews*, *37*, 397–407.

Kumar, S., et al. (2018). Functionalized thermoplastic polyurethane as hole conductor for quantum dot-sensitized solar cell. *ACS Applied Energy Materials*, *1*(9), 4641–4650.

Lan, Z., et al. (2016). Preparation of high-efficiency CdS quantum-dot-sensitized solar cells based on ordered TiO_2 nanotube arrays. *Ceramics International*, *42*(7), 8058–8065.

Lee, H., Leventis, H. C., Moon, S.-J., Chen, P., Ito, S., Haque, S. A., Torres, T., Nüesch, F., Geiger, T., Zakeeruddin, S.M., Grätzel M., & Nazeeruddin M. K., (2009a). PbS and CdS Quantum Dot-Sensitized Solid-State Solar Cells: Old Concepts, *New Results*, *19*(17): 2735–2742.

Lee, H., et al. (2009b). PbS and CdS quantum dot-sensitized solid-state solar cells: "Old Concepts, New Results." *19*(17), 2735–2742.

Lee, H., et al. (2009c). Efficient CdSe quantum dot-sensitized solar cells prepared by an improved successive ionic layer adsorption and reaction process. *Nano Letters*, *9*(12), 4221–4227.

Lee, H. J., et al. (2008). CdSe quantum dot-sensitized solar cells exceeding efficiency 1% at full-sun intensity. *The Journal of Physical Chemistry C*, *112*(30), 11600–11608.

Lee, L. T. L., et al. (2014). Few-layer MoSe2 possessing high catalytic activity towards iodide/tri-iodide redox shuttles. *Scientific Reports*, *4*(1), 4063.

Lee, Y.-L., & Chang, C.-H. (2008). Efficient polysulfide electrolyte for CdS quantum dot-sensitized solar cells. *Journal of Power Sources*, *185*(1), 584–588.

Lee, Y.-L., & Lo, Y.-S. (2009). Highly efficient quantum-dot-sensitized solar cell based on co-sensitization of CdS/CdSe. *19*(4), 604–609.

Leschkies, K. S., et al. (2007). Photosensitization of ZnO nanowires with CdSe quantum dots for photovoltaic devices. *Nano Letters*, *7*(6), 1793–1798.

Lévy-Clément, C., et al. (2005). CdSe-sensitized p-CuSCN/nanowire n-ZnO heterojunctions. *17*(12), 1512–1515.

Li, L., et al. (2011). Highly efficient CdS quantum dot-sensitized solar cells based on a modified polysulfide electrolyte. *Journal of the American Chemical Society*, *133*(22), 8458–8460.

Li, L., et al. (2014). Cu-Doped-CdS/In-Doped-CdS cosensitized quantum dot solar cells. *Journal of Nanomaterials*, *2014*, 314386.

Li, W., & Zhong, X. (2015). Capping ligand-induced self-assembly for quantum dot sensitized solar cells. *The Journal of Physical Chemistry Letters*, *6*(5), 796–806.

Licht, S. (1995). Electrolyte modified photoelectrochemical solar cells. *Solar Energy Materials and Solar Cells*, *38*(1), 305–319.

Lin, C.-Y., et al. (2013). Photoactive p-type PbS as a counter electrode for quantum dot-sensitized solar cells. *Journal of Materials Chemistry A*, *1*(4), 1155–1162.

Liu, I. P., et al. (2014). Performance enhancement of quantum-dot-sensitized solar cells by potential-induced ionic layer adsorption and reaction. *ACS Applied Materials & Interfaces*, *6*(21), 19378–19384.

Liu, Z., et al. (2010). Enhancing the performance of quantum dots sensitized solar cell by SiO_2 surface coating. Applied Physics Letters, 96(23), 233107.

McElroy, N., et al. (2014). Comparison of solar cells sensitised by CdTe/CdSe and CdSe/CdTe core/shell colloidal quantum dots with and without a CdS outer layer. *Thin Solid Films*, *560*, 65–70.

Meng, K., Chen, G., & Thampi, K. R. (2015). Metal chalcogenides as counter electrode materials in quantum dot sensitized solar cells: A perspective. *Journal of Materials Chemistry A*, *3*(46), 23074–23089.

Mingsukang, M. A., Buraidah, M. H., & Careem, M. A. (2016). Development of gel polymer electrolytes for application in quantum dot-sensitized solar cells. *Ionics*, 1–9.

Mora-Seró, I., et al. (2008). Factors determining the photovoltaic performance of a CdSe quantum dot sensitized solar cell: The role of the linker molecule and of the counter electrode. *Nanotechnology*, *19*(42), 424007.

Ning, Z., et al. (2012). Type-II colloidal quantum dot sensitized solar cells with a thiourea based organic redox couple. *Journal of Materials Chemistry*, *22*(13), 6032–6037.

Nozik, A. J. (2005). Exciton multiplication and relaxation dynamics in quantum dots: Applications to ultrahigh-efficiency solar photon conversion. *Inorganic Chemistry*, *44*(20), 6893–6899.

O'Regan, B., & Grätzel, M. (1991). A low-cost, high-efficiency solar cell based on dye-sensitized colloidal TiO_2 films. *Nature*, *353*(6346), 737–740.

Pan, Z., et al. (2012). Highly efficient inverted type-I CdS/CdSe core/shell structure QD-sensitized solar cells. *ACS Nano*, *6*(5), 3982–3991.

Park, J. P., et al. (2016). Highly efficient solid-state mesoscopic PbS with embedded CuS quantum dot-sensitized solar cells. *Journal of Materials Chemistry A*, *4*(3), 785–790.

Pawar, S. A., et al. (2016). Chemical synthesis of CdS onto TiO_2 nanorods for quantum dot sensitized solar cells. *Optical Materials*, *58*, 46–50.

Peter, L. M. (2011). The Grätzel cell: Where next? *The Journal of Physical Chemistry Letters*, *2*(15), 1861–1867.

Plass, R., et al. (2002). Quantum dot sensitization of organic – inorganic hybrid solar cells. *The Journal of Physical Chemistry. B*, *106*(31), 7578–7580.

Qian, S., et al. (2011). An enhanced CdS/TiO_2 photocatalyst with high stability and activity: Effect of mesoporous substrate and bifunctional linking molecule. *Journal of Materials Chemistry*, *21*(13), 4945–4952.

Radich, J. G., et al. (2014). Charge transfer mediation through CuxS. The hole story of CdSe in polysulfide. *The Journal of Physical Chemistry C*, *118*(30), 16463–16471.

Radich, J. G., Dwyer, R., & Kamat, P. V. (2011). Cu2S reduced graphene oxide composite for high-efficiency quantum dot solar cells. Overcoming the Redox limitations of S2–/Sn2– at the counter electrode. *The Journal of Physical Chemistry Letters*, *2*(19), 2453–2460.

Raevskaya, A., et al. (2016). Non-stoichiometric Cu–In–S@ ZnS nanoparticles produced in aqueous solutions as light harvesters for liquid-junction photoelectrochemical solar cells. RSC Advances *6*(102), 100145–100157.

Rao, H.-S., et al. (2014). CdS/CdSe co-sensitized vertically aligned anatase TiO_2 nanowire arrays for efficient solar cells. *Nano Energy*, *8*, 1–8.

Rawal, S. B., et al. (2012). Optimization of CdS layer on ZnO nanorod arrays for efficient CdS/CdSe co-sensitized solar cell. *Materials Letters*, *82*, 240–243.

Resch-Genger, U., et al. (2008). Quantum dots vs organic dyes as fluorescent labels. *Nature Methods*, *5*(9), 763–775.

Rhee, J. H., Chung, C.-C., & Diau, E. W.-G. (2013). A perspective of mesoscopic solar cells based on metal chalcogenide quantum dots and organometal-halide perovskites. *NPG Asia Materials*, *5*(10), p. e68-e68.

Rühle, S., Shalom, M., & Zaban, A. (2010). Quantum-dot-sensitized solar cells. *Chemphyschem*. 11, 2290–2304.

Salant, A., et al. (2010). Quantum dot sensitized solar cells with improved efficiency prepared using electrophoretic deposition. *ACS Nano*, *4*(10), 5962–5968.

Sayama, K., Sugihara, H., & Arakawa, H. (1998). Photoelectrochemical properties of a porous Nb2O5 electrode sensitized by a Ruthenium dye. *Chemistry of Materials*, *10*(12), 3825–3832.

Semonin, O. E., et al. (2010). Absolute photoluminescence quantum yields of IR-26 Dye, PbS, and PbSe quantum dots. *The Journal of Physical Chemistry Letters*, *1*(16), 2445–2450.

Senthamilselvi, V., et al. (2012). Photovoltaic properties of nanocrystalline CdS films deposited by SILAR and CBD techniques—A comparative study. *Journal of Materials Science: Materials in Electronics*, *23*(1), 302–308.

Seo, H., et al. (2014). Performance dependence of Si quantum dot-sensitized solar cells on counter electrode. *Japanese Journal of Applied Physics*, *53*(5S1), 05FZ01.

Seo, M., et al. (2011). Improvement of quantum dot-sensitized solar cells based on Cds and CdSe quantum dots. In *Proceedings of the thirty-seventh IEEE photovoltaic specialists conference*.

Shalom, M., et al. (2009). Core/CdS quantum dot/shell mesoporous solar cells with improved stability and efficiency using an amorphous TiO_2 coating. *The Journal of Physical Chemistry C*, *113*(9), 3895–3898.

Sharma, D., Jha, R., & Kumar, S. (2016). Quantum dot sensitized solar cell: Recent advances and future perspectives in photoanode. *Solar Energy Materials and Solar Cells*, *155*, 294–322.

Shen, Q., et al. (2008). Effect of ZnS coating on the photovoltaic properties of CdSe quantum dot-sensitized solar cells. *Journal of Applied Physics 103*(8), 084304.

Shen, Y.-J., & Lee, Y.-L. (2008). Assembly of CdS quantum dots onto mesoscopic TiO_2 films for quantum dot-sensitized solar cell applications. *Nanotechnology*, *19*(4), 045602.

Smith, N. J., Emmett, K. J., & Rosenthal, S. J. (2008). Photovoltaic cells fabricated by electrophoretic deposition of CdSe nanocrystals. *Applied Physics Letters 93*(4), 043504.

Späth, M., et al. (2003).Reproducible manufacturing of dye-sensitized solar cells on a semi-automated baseline. *Progress in Photovoltaics*, *11*(3), 207–220.

Sudhagar, P., et al. (2009). Self-assembled CdS quantum dots-sensitized TiO_2 nanospheroidal solar cells: Structural and charge transport analysis. *Electrochimica Acta*, *55*(1), 113–117.

Sudhagar, P., et al. (2011). Robust mesocellular carbon foam counter electrode for quantum-dot sensitized solar cells. *Electrochemistry Communications*, *13*(1), 34–37.

Sun, W.-T., et al. (2008). CdS Quantum dots sensitized TiO_2 nanotube-array photoelectrodes. *Journal of the American Chemical Society*, *130*(4), 1124–1125.

Tachan, Z., et al. (2011). PbS as a highly catalytic counter electrode for polysulfide-based quantum dot solar cells. *The Journal of Physical Chemistry C*, *115*(13), 6162–6166.

Tao, L., et al. (2014). High performance PbS quantum dot sensitized solar cells via electric field assisted in situ chemical deposition on modulated TiO_2 nanotube arrays. *Nanoscale*, *6*(2), 931–938.

Toyoda, T., Sato, J., & Shen, Q. (2003). Effect of sensitization by quantum-sized CdS on photoacoustic and photoelectrochemical current spectra of porous TiO_2 electrodes. *Review of Scientific Instruments 74*(1), 297–299.

Tubtimtae, A., & Lee, M.-W. (2012). Effects of passivation treatment on performance of CdS/CdSe quantum-dot co-sensitized solar cells. *Thin Solid Films*, *526*, 225–230.

Vogel, R., Pohl, K., & Weller, H. (1990). Sensitization of highly porous, polycrystalline TiO_2 electrodes by quantum sized CdS. *Chemical Physics Letters*, *174*(3), 241–246.

Vogel, R., Hoyer, P., & Weller, H. (1994). Quantum-sized PbS, CdS, Ag_2S, Sb_2S_3, and Bi_2S_3 particles as sensitizers for various nanoporous wide-bandgap semiconductors. *The Journal of Physical Chemistry*, *98*(12), 3183–3188.

Wang, J., et al. (2013). Core/shell colloidal quantum dot exciplex states for the development of highly efficient quantum-dot-sensitized solar cells. *Journal of the American Chemical Society*, *135*(42), 15913–15922.

Wang, W., et al., (2018). Cosensitized quantum dot solar cells with conversion efficiency over 12%. *Advanced Materials 30*(11), 1705746.

Watson, D. F. (2010). Linker-assisted assembly and interfacial electron-transfer reactivity of quantum dot – substrate architectures. *The Journal of Physical Chemistry Letters*, *1* (15), 2299–2309.

Wei, H., et al. (2016). Fumed SiO2 modified electrolytes for quantum dot sensitized solar cells with efficiency exceeding 11% and better stability. *Journal of Materials Chemistry A*, *4*(37), 14194–14203.

Wijayantha, K. G. U., Peter, L. M., & Otley, L. C. (2004). Fabrication of CdS quantum dot sensitized solar cells via a pressing route. *Solar Energy Materials and Solar Cells*, *83*(4), 363–369.

Wu, J., et al. (2015). Electrolytes in dye-sensitized solar cells. *Chemical Reviews*, *115*(5), 2136–2173.

Yang, J., et al. (2015). CdSeTe/CdS Type-I core/shell quantum dot sensitized solar cells with efficiency over 9%. *The Journal of Physical Chemistry C*, *119*(52), 28800–28808.

Yang, Z., et al. (2011). Quantum dot-sensitized solar cells incorporating nanomaterials. *Chemical Communications*, *47*(34), 9561–9571.

Yeh, M.-H., et al. (2011). Conducting polymer-based counter electrode for a quantum-dot-sensitized solar cell (QDSSC) with a polysulfide electrolyte. *Electrochimica Acta*, *57*, 277–284.

Yu, J., et al. (2017). Quantum dot sensitized solar cells with efficiency over 12% based on tetraethyl orthosilicate additive in polysulfide electrolyte. *Journal of Materials Chemistry A*, *5*(27), 14124–14133.

Yu, X.-Y., et al., (2011). Highly efficient CdTe/CdS quantum dot sensitized solar cells fabricated by a one-step linker assisted chemical bath deposition. *Chemical Science 2*(7), 1396–1400.

Yu-Jen, S., & Yuh-Lang, L. (2008). Assembly of CdS quantum dots onto mesoscopic TiO 2 films for quantum dot-sensitized solar cell applications. *Nanotechnology*, *19*(4), 045602.

Zhang, H., et al. (2012). Efficient CdSe quantum dot-sensitized solar cells prepared by a postsynthesis assembly approach. *Chemical Communications*, *48*(91), 11235–11237.

Zhang, Q., et al. (2010). Application of carbon counterelectrode on CdS quantum dot-sensitized solar cells (QDSSCs). *Electrochemistry Communications*, *12*(2), 327–330.

Zhang, Z., et al. (2017). Combination of short-length TiO_2 nanorod arrays and compact PbS quantum-dot thin films for efficient solid-state quantum-dot-sensitized solar cells. *Applied Surface Science*, *410*, 8–13.

Zhou, Y., et al. (2020). Electron transfer in a semiconductor heterostructure interface through electrophoretic deposition and a linker-assisted method. *CrystEngComm*, *22*(9), 1664–1673.

Zhu, G., et al. (2010). Au nanoparticles as interfacial layer for CdS quantum dot-sensitized solar cells. *Nanoscale Research Letters*, *5*(11), 1749.

Zhu, G., et al. (2011a). All spray pyrolysis deposited CdS sensitized ZnO films for quantum dot-sensitized solar cells. *Journal of Alloys and Compounds*, *509*(2), 362–365.

Zhu, G., et al. (2011b). CdS/CdSe-cosensitized TiO_2 photoanode for quantum-dot-sensitized solar cells by a microwave-assisted chemical bath deposition method. *ACS Applied Materials & Interfaces*, *3*(8), 3146–3151.

Zhu, G., et al. (2011c). Microwave assisted chemical bath deposition of CdS on TiO_2 film for quantum dot-sensitized solar cells. *Journal of Electroanalytical Chemistry*, *659*(2), 205–208.

Zhu, L., et al. (2012). Linker-free quantum dot sensitized TiO_2 photoelectrochemical cells. *International Journal of Hydrogen Energy*, *37*(8), 6422–6430.

Zrazhevskiy, P., Sena, M., & Gao, X. (2010). Designing multifunctional quantum dots for bioimaging, detection, and drug delivery. *Chemical Society Reviews*, *39*(11), 4326–4354.

SUBCHAPTER

Organometal halide perovskite photovoltaics 4.3

Md. Shahiduzzaman[1], Mohammad Ismail Hossain[2,3], Md. Akhtaruzzaman[4], Masahiro Nakano[1], Makoto Karakawa[1], Jean-Michel Nunzi[1,5] and Tetsuya Taima[1]

[1]*Nanomaterials Research Institute, Kanazawa University, Kanazawa, Japan*
[2]*Department of Materials Sciences and Engineering, City University of Hong Kong, Kowloon, Hong Kong, P.R. China*
[3]*Department of Electrical and Computer Engineering, University of California, Davis, CA, United States*
[4]*Solar Energy Research Institute, Universiti Kebangsaan Malaysia, Bangi, Malaysia*
[5]*Department of Physics, Engineering Physics and Astronomy, Department of Chemistry, Queens University, Kingston, ON, Canada*

4.3.1 Introduction

Organic–inorganic hybrid halide perovskite solar cells (PSCs) are considered the most promising photovoltaic (PV) technology due to their lower manufacturing costs, simple fabrication process, ecofriendliness, and high power conversion efficiency (PCE). The advantageous properties of the perovskite absorber layer include a tunable bandgap (1.2–2.3 eV) (Yang, Fiala, Jeangros, & Ballif, 2018), high charge carrier lifetime ($>15\,\mu s$) (Bi, Y., et al., 2016), long diffusion length ($\sim 175\,\mu m$) (Dong, Yuan, et al., 2015), large absorption coefficient ($>0.5 \times 104$/cm) (Park, 2015), and high carrier mobility ($\sim 20\,cm^2/V/S$). PSCs stem from dye-sensitized solar cells (DSSC). In 2009, T. Miyakasa and his colleagues, for the first time, introduced perovskite material as a photoactive absorber layer in the DSSC and termed it a perovskite-sensitized solar cell. The team used methylammonium lead iodide ($CH_3NH_3PbI_3$) and methylammonium lead bromide ($CH_3NH_3PbBr_3$) as the photoactive layer, and the DSSC exhibited PCE values of 3.1% and 3.8%, respectively (Kojima, Teshima, Shirai, & Miyasaka, 2009). At a given nanocrystalline titanium oxide (TiO_2) film thickness ($3.6\,\mu m$), $CH_3NH_3PbI_3$ perovskite exhibited 10 times greater absorption coefficient than that of the conventional ruthenium-based molecular dye (Im, Lee, Lee, Park, & Park, 2011). Since organic-inorganic hybrid halide perovskite is an ionic crystal, it easily dissolves in a polar solvent, particularly liquid electrolyte-based sensitized solar cells. Therefore perovskite material is unsuitable for liquid electrolyte-based sensitized solar cells as it possesses chemical instability due to dissolution in the liquid electrolyte. In 2012, Park

Comprehensive Guide on Organic and Inorganic Solar Cells. DOI: https://doi.org/10.1016/B978-0-323-85529-7.00007-4

and his colleagues (Kim et al., 2012) solved this instability problem by replacing the liquid electrolyte with solid hole transport material (HTM) and reported long-term stable PSCs with a PCE of 9.7%. A significant scientific breakthrough of PV research was made over the past decade, where the PSC achieved solid-state formation in 2012 by M. Grätzel and N.-G. Park (Kim et al., 2012). The PCE of PSC has been boosted tremendously from 3.8% to 25.5% at laboratory scale (Shahiduzzaman et al., 2020; Shahiduzzaman et al., 2021). This value approaches or exceeds the PCEs of other PV devices, including silicon solar cells (26.7%) and perovskite–silicon tandem solar cells (29.1%) (NREL, 2019).

PSCs are promising in terms of both high efficiency and low manufacturing costs, with the manufacturing costs estimated to be half those of cells made from crystalline silicon (Meng, You, & Yang, 2018). Thus significant efforts have been devoted to developing less costly, lighter weight, and more ecofriendly alternatives to silicon solar cells. However, the cost of a perovskite absorber layer is low (~United States $ 2/m^2 of a module), but the cost of the low-resistance (<10/Ω^2) transparent conductive substrate is relatively still high (~United States $2/m^2). The estimated value of PSCs, including all material costs, is ~United States $20m^2 (Cai et al., 2017). This amount can be further reduced by changing the charge carrier transports and conductive layers. The most significant part of PSC cost reduction is the high-throughput solution process that minimizes total material cost for mass production. To date, lead (Pb)-based PSCs have appeared as one of the most promising designs for next-generation thin-film (TF) PVs. However, Pb-based PSCs suffer from poor stability and high toxicity of Pb metal. Additionally, the environmental impact of heavy metal like Pb should be assessed prior to large-scale production. The European restriction of hazardous substances restricts the commercial use of Pb for the manufacturing of electronic and electrical equipment. In addition, other important challenges of PSCs are long-term stability and reproducibility due to the unstable behavior of perovskite materials in ambient conditions and the synergistic factors of the functional components of PSCs (Xu, Chen, Guo, & Ma, 2016). Salhi et al. reported that shortcomings in the stability of PSCs is due to the effects of moisture, UV light, temperature, and exposure to ambient conditions (Salhi, Wudil, Hossain, Al-Ahmed, & Al-Sulaiman, 2018). The structure and properties of perovskite materials and deposition techniques also pose challenges to PSCs (Ansari, Qurashi, & Nazeeruddin, 2018). Understanding the role of the fabrication procedure of perovskite absorber layers enhances the performance of PSCs and reduces the overall production cost for roll-to-roll production. Moreover, bandgap, optical constant, and optical absorption are affected by the fabrication technique. Low cost, ease of fabrication process, and facile deposition techniques are preferable for depositing perovskite films to obtain highly efficient PSCs. High-quality perovskite films with suitable morphology, high phase purity, high crystallinity, and minimal structural defects are necessary to increase the PCE of PSCs. In terms of improving the operational stability of PSCs, simultaneous device engineering improvements can be considered to realize their full potentials.

4.3.1.1 Structure and materials properties of perovskite materials

The general chemical formula for a perovskite material is ABX_3. It was named after the Russian scientist Lev Perovski after it was discovered by Gustav Rose in the Ural Mountains in 1839.

It comprises two cations of different sizes ("A" and "B") and one anion ("X"), where A stands for a monovalent organic cation typically methylammonium $CH_3NH_3^+$ (MA^+), formamidinium $HC(NH_2)_2^+$ (FA^+), or cesium (Cs^+); B is a divalent metal cation, such as Pb^{2+}, Sn^{2+}, Ge^{2+}or Bi^{3+}; and X is a halogen anion (i.e., I^-, Br^-, Cl^-, or a mixture of these) as shown in Fig. 4.3.1A. A perfect perovskite crystal structure is organized in such a way that the B-cation has six-fold coordination while covered by an octahedral arrangement of anions expressed in Fig. 4.3.1A (Tablero Crespo, 2019). The A cation remains in 12-fold cuboctahedra coordination. The stability (cubic structure) of this type of material is dependent and sensitive on the comparative A and B ion sizes. Because a slight distortion can produce a huge change in the A and B cations coordination

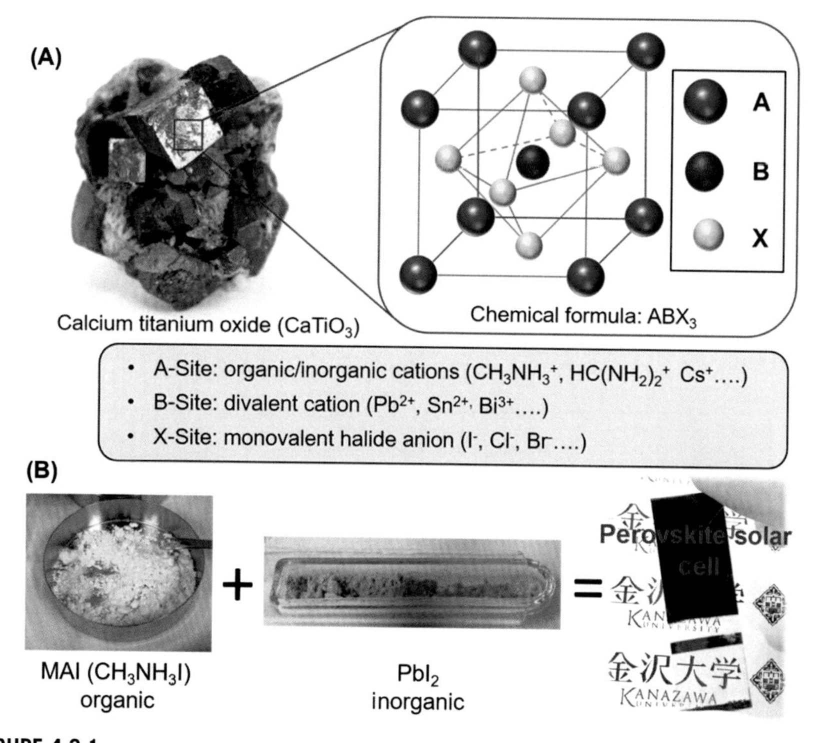

FIGURE 4.3.1

(A) General ABX_3 perovskite structure; (B) Photograph of MAI and PbI_2 with a perovskite solar cell.

number, as shown in Fig. 4.3.1A. Photographs of organic MAI and inorganic PbI_2 materials used to fabricate PSCs (Fig. 4.3.1B).

The coexistence of an amine-based organic cation molecule (e.g., $CH_3NH_3^+$), metal cation (e.g., Pb^{2+}), and anion halide (e.g., I^-) together improvise the exceptional materialistic properties which lead to achieving greater consideration of the scientific community of the world (Ozin, 1970). Weber et al. first synthesized the most commonly used organo-metallic halide perovskite framework, namely, $CH_3NH_3PbI_3$ ($MAPbX_3$) (Weber, 1978). The change in compositional stoichiometry could offer different color variations. The role of ion A is to provide a framework's structural stability by ensuring charge neutrality without jeopardizing the bandgap. But the size of A site can influence the structural deformation, which may eventually hamper the bandgap, as seen in Table 4.3.1 (Kulkarni et al., 2014). The insertion of Sn to replace Pb in the framework gradually decreases the bandgap. Since there are at least three different component materials, the simple and commonly used $MAPbX_3$ has a tunable optical bandgap from 1.2 to 2.3 eV, which could be attained by changing halide content and the halide itself, as illustrated in Fig. 4.3.2. Moreover, the A site can manipulate the bandgap (Cs, FA) between ~1.2 to ~4.0 eV.

Table 4.3.1 Effect of B cations on perovskite bandgap from (Ogomi et al., 2014).

Material	LUMO/eV	HOMO/eV	E_g/eV
$CH_3NH_3PbI_3$	−3.88	−5.39	1.51
$CH_3NH_3Sn_{0.3}Pb_{0.7}I_3$	−3.81	−5.12	1.31
$CH_3NH_3Sn_{0.5}Pb_{0.5}I_3$	−3.67	−4.95	1.28
$CH_3NH_3Sn_{0.7}Pb_{0.3}I_3$	−3.69	−4.92	1.23
$CH_3NH_3Sn_{0.9}Pb_{0.1}I_3$	−3.57	−4.75	1.18
$CH_3NH_3SnI_3$	−3.63	−4.73	1.10

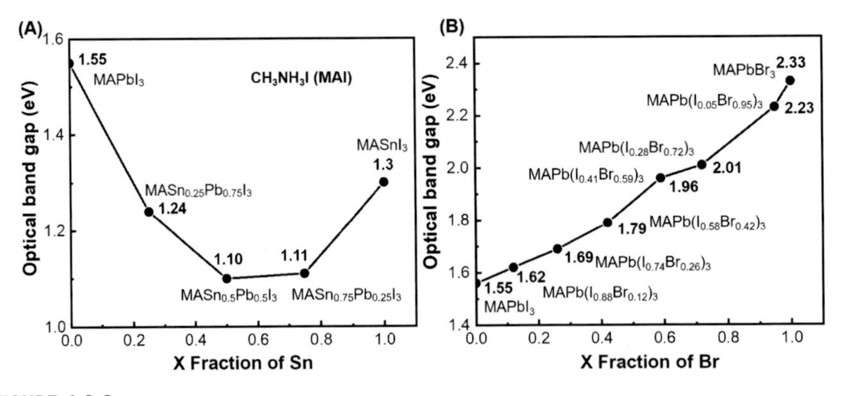

FIGURE 4.3.2

Effects of (A) B cations and (B) X anions on the perovskite bandgap (Zhu & Choy, 2018).

Other mixed lead halide perovskite compounds include $CH_3NH_3PbI_{3-x}Cl_x$, which has direct bandgaps within the range of 1.61–1.96 eV (Colella et al., 2013), and $CH_3NH_3PbI_{3-x}Br_x$, which has direct bandgaps of 1.58–2.28 eV (Noh, Im, Heo, Mandal, & Seok, 2013). Band gaps depend on the value of X. Hence, these materials are outstanding light absorbers, and they can efficiently cover a wide range of solar spectrum from the ultraviolet (UV) to the inferred reason (IR) regions. $CH_3NH_3PbI_3$ exhibits a weak bond exciton binding energy ranging from 37 to 50 meV and offers an easy way for the movement of free carriers (Hirasawa, Ishihara, Goto, Uchida, & Miura, 1994). It is a popular choice for fabricating optoelectronic devices because of its exceptional charge carrier transport properties. As a single crystal (SC), $CH_3NH_3PbI_3$ has a large diffusion length that exceeds 175 μm. By contrast, $CH_3NH_3PbI_{3-x}Cl_x$ in the form of TF has a diffusion length of up to 1 μm, as indicated in Table 4.3.2.

Perovskite materials can be prepared using a small cation, such as PbX_2, SnX_2, or CH_3NH_2, or a big cation, such as $C_{10}H_{21}NH_2$, $CH_3C_6H_4CH_2NH_2$, or $C_{10}H_7CH_2NH_2$ (Goto, Ohshima, Mousdis, & Papavassiliou, 2000), as starting materials. The reactions involving small and large cations are shown in Eqs. (4.3.1) and (4.3.2), respectively, and Eq. (4.3.3) expresses the overall reaction.

$$(SC)X + MX_2 = (SC)MX_3, \tag{4.3.1}$$

$$2(BC)X + MX_2 = (BC)_2MX_4, \tag{4.3.2}$$

and

$$(SC)_{(n-1)}MX_3 + (BC)_2MX_4 = (SC)_{n-1}(BC)_2MnX_{3n+1}, \tag{4.3.3}$$

where $n = 1, 2, \ldots, \infty$ inorganic layers.

The perovskite structure stability and distortion are figured by the Goldschmidt tolerance factor (t) that can be calculated using the relationship $t = rA + rX / \sqrt{2}(rB + rX)$,

Table 4.3.2 Charge carrier properties of organic–inorganic lead halide perovskite compounds.

Perovskites	Carrier mobility (cm^2 /V/S)	Carrier recombination (μs)	Diffusion length, L_D (μm)	Reference
$CH_3NH_3PbI_3$ SC	24.8 ± 4.1	95	~175	Dong, Fang, et al. (2015)
$CH_3NH_3PbBr_3$ SC	2.5	0.357	~0.3–1.7	Shi, Li, Li, and Wang (2015)
$CH_3NH_3PbI_3$ TF	–	0.0096	0.129 ± ˙41	Stranks et al. (2013)
$CH_3NH_3PbI_{3-x}Cl_x$ TF	–	0.2727	1.069 ± 0.204	Stranks et al. (2013)

SC *Single crystal,* TF *Thin film.*

where r denotes the ionic radius of the corresponding ions (Zhang et al., 2017). The t is a dimensionless number, which is calculated as the ratio of the ionic radii. Stabilized perovskite t values should be greater than 1.0 or less than 0.8. With a t value of 0.91, MA lead triiodide ($CH_3NH_3PbI_3$ or $MAPbI_3$) perovskite is the most-studied perovskite material; it has light absorption up to 800 nm for PV devices whereas its easily degrades under direct exposure to sunlight, moisture, oxygen, and heat, which tends to limit its use as a light absorber in PSCs (Lee, Teuscher, Miyasaka, Murakami, & Snaith, 2012). The absorption range (800 nm) can be increased in the infrared region to 840 nm with the replacement of the MA cation by the FA cation (Zhang, Grancini, Feng, Asiri, & Nazeeruddin, 2017). Conversely, $FAPbI_3$ perovskite shows a lower band gap, which can be extended to light absorption range, better stability, and reduced $I-V$ hysteresis. In terms of reducing t, several studies reported that the use of FA and MA mixed cation-based perovskites stabilized the perovskite structures and thus improved the efficiency and stability of PSCs. Another approach adopted to increase absorption is introducing bivalent cations Sn^{2+} at the B position; this extends the harvesting of solar cell photons beyond 900 nm (Samiul Islam et al., 2021; Zhang et al., 2017). Interestingly, by sharing the B position between Pb^{2+} and Sn^{2+}, the absorption can be extended to 1060 nm toward the infrared region (Zhao et al., 2017). As a result, the broad absorption ranges of $Sn-Pb$ mixed PSCs have high-photocurrent characteristics when compared with those of the pure Pb counterparts. Thus $Sn-Pb$ mixed cation strategy is a promising way of tuning the optical and electronic properties of perovskite, particularly the bandgap, charge diffusion length, and absorption efficiency.

4.3.1.2 Device structures and working principle

When PSCs were first realized through their deployment in DSSCs, their working mechanism was assumed to be similar to that of DSSCs (Jena, Kulkarni, & Miyasaka, 2019; Shimada, Shahiduzzaman, & Taima, 2020). It was later found that a perovskite absorber still transfers charge when encapsulated with an insulating layer, such as Al_2O_3, without a similar semiconducting layer (Lee et al., 2012). This finding indicates that the PSC working principle is more like that of a p-n junction semiconducting solar cell for which the device structure determines the charge transfer pathway. The perovskite photosensitive layer is sandwiched between metal oxide (MOx) layers, namely an electron-transporting layer (ETL) and a hole-transporting layer (HTL), which can be n–i–p (Fig. 4.3.3A) or p–i–n (Fig. 4.3.3B) device architectures.

The corresponding layers are sandwiched between transparent electrodes and metal electrodes in fabricated PSCs. The n–i–p architecture is realized when the ETL is positioned on the side where light is incident, while the p–i–n architecture is realized when the HTL is on the side where light is incident. The charge carrier electron-hole (e-h) pairs are generated in perovskite absorber upon successful photon absorption from solar radiation carrying sufficient energy. In the case of p–i–n and n–i–p structures, the electrons are carried to n-type ETL and pushed to the anode for completing the cycle by reaching the cathode through load meanwhile, and the holes are extracted and migrated to the cathode by HTM. The evolution of PSCs device structure is schematically shown in Fig. 4.3.4. Since 2013, there have been two leading types of device architecture, such as mesoporous (mp) structures (Fig. 4.3.4D) and planar heterojunctions

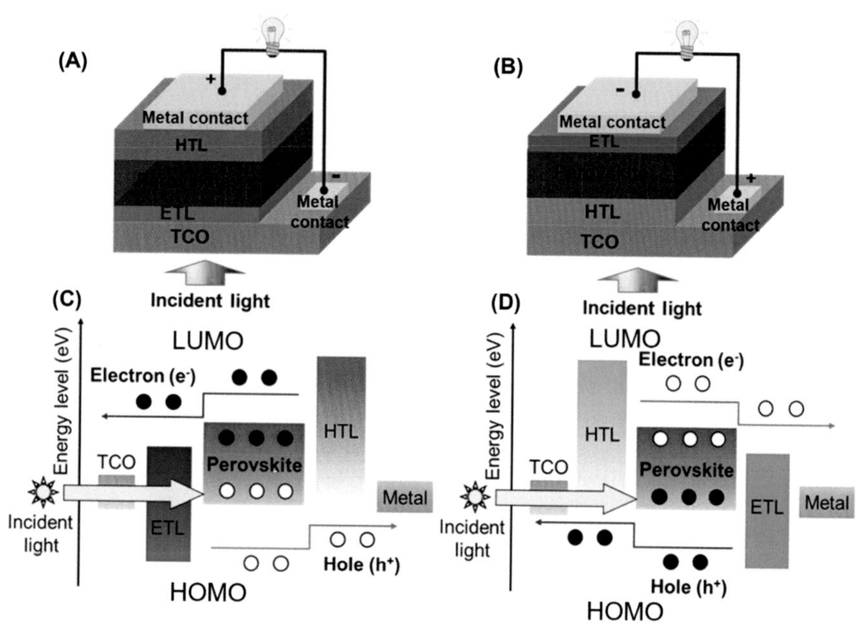

FIGURE 4.3.3

Schematic illustrations of (A) conventional n–i–p, and (B) inverted p–i–n. Energy band structures of (C) conventional n–i–p and (D) inverted p–i–n devices.

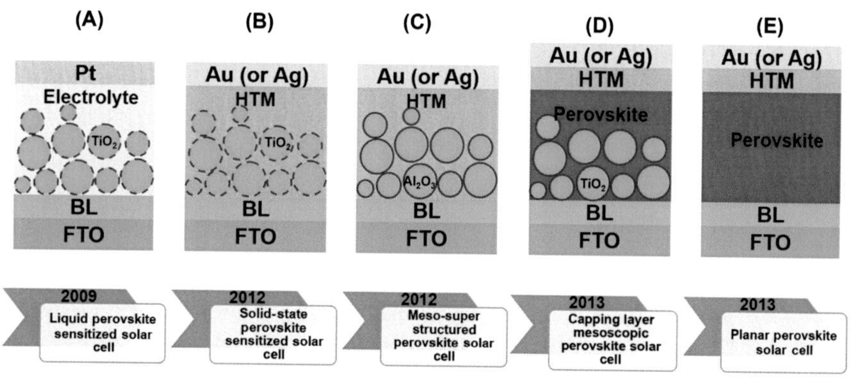

FIGURE 4.3.4

Structure evolution of perovskite solar cells.

(Fig. 4.3.4E), which have been extensively investigated. A mesoporous material retain pores with diameters between 2 and 50 nm.

Many points of view have been considered to explain $MAPbI_3$ perovskite operation through the microscopic mechanism, but it has not been fully defined yet. The solid-state PSCs working mechanism is similar to that of inorganic thin film PV cells with $p-i-n$ junction structure. Conceptually, the $MAPbI_3$ perovskite active layer in PSC generates excitons (e-h pair) upon photon absorption. Any energy mismatch between interface layers will insist on delaying the carrier or introduce recombination. As a result, a continuous research effort is demanded for the successful integration among ETL and HTL with perovskite. Additional transporters (ETL for electron and HTM for holes) execute the migration of charge carriers to the corresponding circuit/electrodes. This mechanism shows that once generated, the exciton diffuses through the perovskite toward the carrier transporter. The electrons and holes move toward the ETL and HTL as shown in the reaction below:

Electron injection:

$$\left(e^{-}\ldots\ldots h^{+}\right)\text{Perovskite} \rightarrow e^{-}_{(\text{ETL})} + h^{+}_{(\text{Perovskite})} \tag{4.3.4}$$

Hole injection:

$$\left(e^{-}\ldots\ldots h^{+}\right)\text{Perovskite} \rightarrow h^{+}_{(\text{HTL})} + e^{-}\text{Perovskite} \tag{4.3.5}$$

To exhibit efficient PSC performance, the collection of charge carriers must remain interrupted at the respective destination. When an exciton comes near to ETL, the electron is captured by ETL, and accompanying holes return to HTL and vice versa for a hole. Besides these expected forward reactions, there are some unwanted charge neutralization reactions that take place, as shown below:

Charge extinction:

$$\left(e^{-}\ldots\ldots h^{+}\right)_{\text{Perovskite}} \rightarrow hv \tag{4.3.6}$$

$$\left(e^{-}\ldots\ldots h^{+}\right)_{\text{Perovskite}} \rightarrow \nabla \tag{4.3.7}$$

Charge recombination:

$$h^{+}(\text{HTL}) + e^{-}_{(\text{Perovskite})} \rightarrow \nabla \tag{4.3.8}$$

$$e^{-}(\text{ETL}) + h^{+}_{(\text{Perovskite})} \rightarrow \nabla \tag{4.3.9}$$

$$e^{-}(\text{ETL}) + h^{+}_{(\text{HTL})} \rightarrow \nabla \tag{4.3.10}$$

It is very crucial to achieve proper band positioning for both ETL and HTL as illustrated in figure 4.25C, D. The conduction band edge (CBE) and valance band edge (VBE) of ETL must be lower than CBE or the lowest unoccupied molecular orbital (LUMO) and the highest occupied molecular orbital (HOMO) of perovskite, respectively, for successful electron migration process. Meanwhile, the VBE and CBE of HTL must be greater than the VBE (HOMO) and CBE (LUMO) of the perovskite absorber, respectively, for successful hole migration.

This is how the ETL and HTL accept and reject the charge carrier accordingly (Mahmood, Sarwar, & Mehran, 2017). Holes should be able to reach the VB of the HTL from the VB of perovskite easily, but electrons should be prevented from reaching the CB of the HTL to eliminate chances of recombination. This also activates the PSC and protects the perovskite absorber layer from being exposed to the environment, which is accredited to the exceptional stability of PSCs (Shahiduzzaman, Kulkarni et al., 2020; You et al., 2016).

The simplest and most used perovskite is the $MAPbI_3$ compound, which possesses a nearly ideal direct energy bandgap of 1.55 eV. The bandgap $MAPbI_3$ is easily alterable between 1.55 and 2.3 eV by changing the halide anion with mixed halide anions, such as $CH_3NH_3PbI_{3-x}Br_x$ perovskite, or insertion single valence cations, such as formamidinium (FA-$HC_3(NH_2)_2^+$) (Eperon et al., 2014). Metallic cation Pb and halide anion I_2 determine the bandgap of $MAPbI_3$, whereas MA does not influence the bandgap. However, it can form a 3D crystal structure, which eventually affects the optical properties of such compounds (Marinova, Valero, & Delgado, 2017).

4.3.1.3 Characterization of solar cells

4.3.1.3.1 Current–voltage characteristics

The current density-*vs*-voltage ($J-V$) relationship of a solar cell is measured to evaluate the performance of the device (Fig. 4.3.5).

The short-circuit current density J_{SC} is the current density at zero voltage (mA/cm^2), the open-circuit voltage V_{oc} is the voltage at zero current (V), J_{max} is the current density at the time of the maximum output of the element (mA/cm^2), and V_{max} is the maximum output. The maximum output of the element P_{max} is given by $P_{max} = J_{max} \times V_{max}$ (W/cm^2). The energy of the incident light

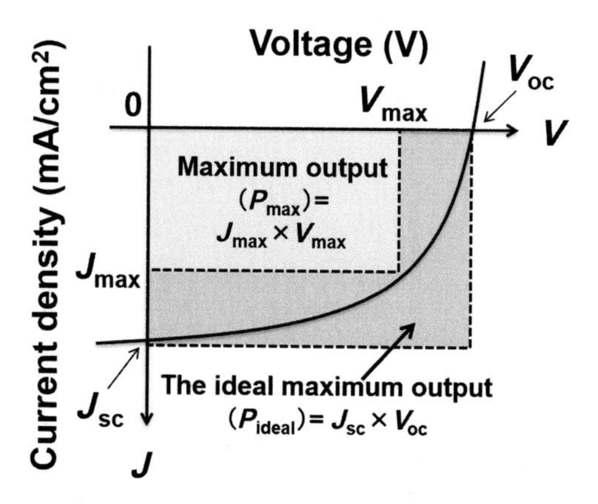

FIGURE 4.3.5

Schematic of the current-vs-voltage relationship.

(AM1.5 G) is $E = 100$ (mW/cm^2). The fill factor (*FF*) and PCE (*PCE*) are obtained as follows.

$$FF = J_{max} \times V_{max}/(J_{sc} \times V_{oc})$$

$$\begin{aligned} PCE &= P_{max}/E \\ &= J_{max} \times V_{max}/E \\ &= (J_{sc} \times V_{oc}) \times FF/E \end{aligned}$$

4.3.1.4 Chronological development of perovskite solar cells

$MAPbI_3$ perovskite was introduced for the first time by T. Miyasaka for the application of perovskites in perovskite-sensitized solar cells in 2009 (Kojima et al., 2009). The currently studied solid-state PSCs are based on the invention of the perovskite-sensitized solar cell in 2012 by N.-G. Park and M. Grätzel (Kim et al., 2012). They replaced dye 719 with a perovskite and liquid electrolyte with HTM (spiro-OMeTAD) to fabricate an all-solid-state PSC, which showed 9.7% efficiency (Kim et al., 2012). The PSC device in a configuration with compact TiO_2/mesoporous TiO_2 as the ETL, perovskite as the active layer, and spiro-OMeTAD as HTM was first named as an mp-PSC. This gave birth to a new technology that has progressed remarkably rapidly. Meanwhile, Snaith et al. proposed the replacement of mp-TiO_2 with mp-Al_2O_3. The mp-Al_2O_3 was considered to be a blocking layer for hole generated into perovskite ($CH_3NH_3PbI_{3-X}Cl_X$) exhibited an outstanding PCE of 10.9% (Hawash, Ono, Raga, Lee, & Qi, 2015). The mp-Al_2O_3 diminished the existence of the density of states near bandgap better than mp-TiO_2. Though this insertion of Al_2O_3 aided in the enhancement of open-circuit voltage (V_{oc}) to 1.12 V but the mesoporous layer possessed higher CB (~ -2.1 V) (Robertson & Falabretti, 2006) than perovskite CB (~ -3.9 V) resulting a band mismatch which lead the electron to stay back at perovskite layer itself (Park, Miyasaka, & Grätzel, 2016). Within 10 years, a significant improvement in the PCE of PSCs from 3.8% to 25.5% was obtained with advances in basic scientific understanding, as illustrated in Fig. 4.3.6.

Perovskite compound has such an attribute that it could facilitate hole transportation and needs no additional HTL (Etgar et al., 2012). Several perovskite layer deposition techniques have been proposed to fabricate mp-PSC and planar PSC (Burschka et al., 2013; Chen et al., 2014). The sequential deposition method (SDM) was first proposed by Grätzel et al. In this method, PbI_2 was spin-coated followed by annealing on top of mp-TiO_2, and then the film was dipped into methylammonium iodide (MAI) to form $MAPbI_3$, which showed 15% efficiency for mesoporous PSC (Burschka et al., 2013). Meanwhile, with the same technique, an inverted device structure like FTO/NiO/mp-Al_2O_3/$CH_3NH_3PbI_3$/PCBM/BCP/Ag displayed an efficiency of 13% (Sha, Ren, Chen, & Choy, 2015). Planar PSC reached up to 15.4%, where perovskite was deposited by vapor-assisted deposition technique directly on

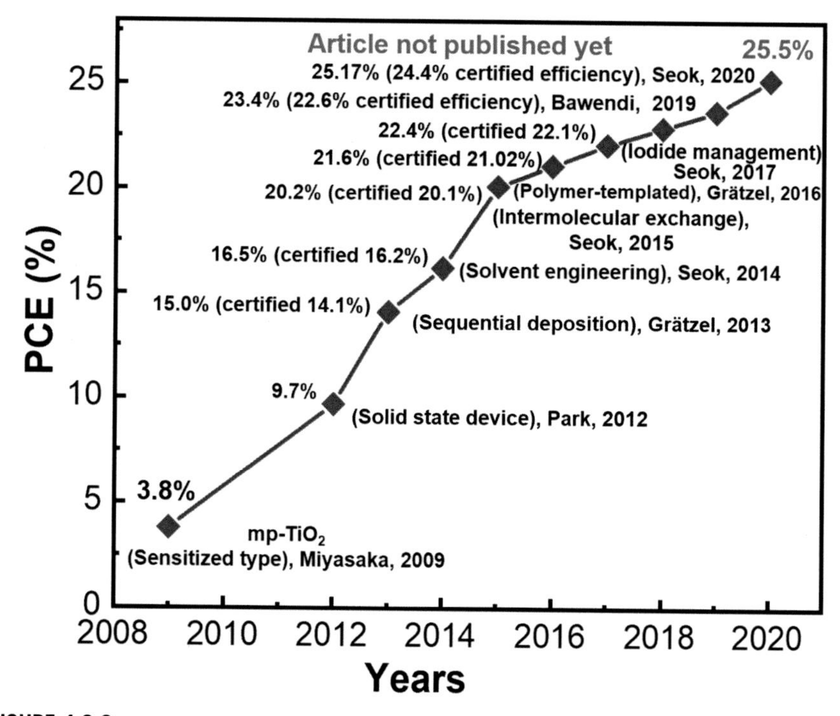

FIGURE 4.3.6

History of improvements to the conversion efficiency of perovskite solar cells certified by the NREL (Bi, D., et al., 2016; Burschka et al., 2013; Jeon et al., 2014; Kim et al., 2020; Yang et al., 2015; Yang et al., 2017; Yoo et al., 2019). The power conversion efficiency of 25.5% shown in blue has not yet been reported.

Reproduced with permission Shahiduzzaman, M., Muslih, E. Y., Hasan, A. K. M., Wang, L., Fukaya, S., Nakano, M., ... Taima, T. (2021). The benefits of ionic liquids for the fabrication of efficient and stable perovskite photovoltaics. Chemical Engineering Journal, *128461. Copy right 2021, Elsevier, Chemical Engineer Journal.*

top of the compact TiO_2 layer avoiding the mp-TiO_2 layer (Liu, Johnston, & Snaith, 2013). Seok et al. testified 16.4% efficiency while prepared the perovskite precursor solution by dissolving into dimethylsulfoxide (DMSO) and γ-butyrolactone mixed solvent (Jeon et al., 2014). Toluene was drop cast on top of the perovskite solution during spin coating. The efficiency continued to increase further by altering ETL properties helped by doping. As for an example, TiO_2 was doped with yttrium, which showed an immense efficiency improvement (19.3%) (Zhou et al., 2014). Even greater efficiency was achieved by incorporating formamidinium iodide (NH_2CHNH_2I, FAI) as an organic part of the ABX_3 structure. Yang et al. dissolved FAI and PbI_2 into DMSO to fabricate mp-PSC, which crossed 20% efficiency for the first time (Yang et al., 2015). In 2016, Grätzel and the team proposed a concept of engineering on the cation and anion parts of ABX_3 perovskite structure by the inclusion of cesium (Cs) and made a triple cation perovskite. Not only that, they also included Br_2 beside I_2, and the resulting perovskite compound took the formula like $Cs_x(MA_{0.17}FA_{0.83})_{(100-x)}Pb(I_{0.83}Br_{0.17})_3$. The mp-PSC with such a perovskite compound produced 21.1% efficiency and operated for greater than ten days. The achieved stability might be resulting from the reduction of free energy about 200K−300K into triple cation perovskite compound (Saliba et al., 2016). The higher stability of 500 hours was exhibited by mp-PSC, while Ru was inserted into the CsMAFA framework to form RbCsMAFA, which possessed ionic defects lower than the former, restricting its movement through the crystal lattice (Saliba et al., 2016).

Most high performance PSC are mesoporous structured, but the inverted (p−i−n) structured device exhibited better performance in terms of stability (Kim et al., 2015). In a p−i−n structure, NiO was used as an HTM deposited on top of FTO, then perovskite layer, and an ETL followed by related electrodes like Ag or Au. The p−i−n structure allows device fabrication at low temperatures, which was the first time reported with 12.04% and 7.4% efficiency in 2014 (Malinkiewicz et al., 2014; Sun et al., 2014). It was determined that defects had been framed in a fewer number within the perovskite active layer, which resulted in almost hysteresis-free attributes during J_{SC} and V_{oc} measurement (Wu et al., 2015). The inverted structure seems to be shown as one of the superlative configurations for successful suppression of hysteresis behavior, which a great obstacle for further commercialization of PSC. An efficiency of 18.1% was achieved by Heo et al. where PEDOT: PSS and PCBM were used as HTL and ETL, respectively (Kim et al., 2015). Chen et al. (2015). More than 15% cell efficiency in 1.02 cm^2 active area was achieved by using an all-inorganic charge carrier layer. Table 4.3.3

Table 4.3.3 Chronological development of perovskite solar cells.

Device structure	Efficiency (%)	Reference
As certified by NREL	25.5	Green et al. (2019)
FTO/c-TiO_2/mp-TiO_2/$(FAPbI_3)_{1-x}(MDA,Cs)_x$/Spiro-OMeTAD/Au	24.4	Kim et al. (2020)
Glass/FTO/bl-TiO_2/mp-TiO_2/Li-TFSI/ $(FAPbI_3)_{0.92}(MAPbBr_3)_{0.08}$/Spiro-OMeTAD/Au	22.6	Yoo et al. (2019)
FTO/bl-TiO_2/mp-TiO_2: perovskite composite/perovskite /TBP/PTAA/Au	22.1	Yang et al. (2017)
Glass/FTO/bl-TiO_2/mp-TiO_2/ $(FAI)_{0.81}(PbI_2)_{0.85}(MAPbBr_3)_{0.15}$/PTAA/Au	21.02	Bi, D., et al. (2016)
FTO/bl-TiO_2/mp-TiO_2/$(FAPbI_3)1-x(MAPbBr_3)_x$/PTAA/Au	20.1	Yang et al. (2015)
Glass/FTO/bl-TiO_2/mp-TiO_2/ $MAPb(I_{1-x}Br_x)_3$/PTAA/Ag	16.2	Jeon et al. (2014)
FTO/bl-TiO_2/mp-TiO_2/ $CH_3NH_3PbI_3$/spiro-OMeTAD/Au	14.1	Burschka et al. (2013)
FTO/bl-TiO_2/mp-TiO_2/$CH_3NH_3PbI_3$/spiro-OMeTAD/Au	9.7	Kim et al. (2012)
FTO/bl-TiO_2/mp-TiO_2/$CH_3NH_3PbI_3$/redox liquid electrolyte/Pt	6.5	Im et al. (2011)
FTO/bl-TiO_2/mp-TiO_2/$CH_3NH_3PbI_3$/redox liquid electrolyte/Pt	3.8	Kojima et al. (2009)

represents the sequential development of PSCs since 2009. Retaining 97% of the initial PV performance without hysteresis after 1000 hours of illumination was achieved by p–i–n PSCs. You et al. reported p–i–n PSCs with p-type NiOx HTM and n-type ZnO nanoparticles as ETM and provided with 60 days stability against water and oxygen degradation while maintaining 90% of its initial PCE of 14.6% (You et al., 2016). Jung et al. achieved efficient, stable, and scalable PSCs showing PCE of 24.1% in a FTO/d-TiO_2/mp-TiO_2/$(FAPbI_3)_{0.95}$ $(MAPbBr_3)_{0.05}$/P3HT/Au device structure (Jung et al., 2019).

4.3.1.5 Charge transporting materials for perovskite solar cells

To realize highly efficient PSCs, the charge transporting materials should satisfy the following criteria as the transport layers.

1. It should have a well-matched band alignment with the photoactive perovskite layer, which helps to transfer and efficiently block electrons or holes, then improve the PV properties.
2. High charge mobility is required to transport electrons and holes, and thus reducing charge recombination in a device.
3. A wide bandgap is beneficial in maximizing the light absorption of the perovskite layer by permitting more light to pass through them.
4. The crystallinity and morphology of the transport layers affect the PSCs performance.

4.3.1.5.1 The role of the electron transport layer

Even though few research groups have suggested ETL-free PSC, this layer is a must-have component to achieve higher device efficiency and stability. Since ETL plays a very crucial part in charge carrier collection and transportation to the respective photoanode. A standard ETL should possess key attributes that perform the designated responsibilities such as low optical absorption, matching band alignment between the adjacent active layer and TCO, higher electron mobility, acting as hole blocker, last but not least optimum physical robustness in the presence of high energy or UV light spectrum. Higher transparency of ETL is much expected to allow maximum photon to reach the perovskite absorber layer. Moreover, ETL should be chemically inert with the corresponding attached layer. TiO_2 is used as ETLs for efficient PSCs owing to its tunable electronic properties, cost effective, and conduction band, which is well-matched with that of perovskites, therefore leading to efficient electron injection and collection. Besides, the thickness of the TiO_2 compact layer (CL) desires to be optimized to exploit electron transport from the perovskite to the fluoride-doped tin oxide (FTO) electrode. When the TiO_2 CL is too thick, then increasing the distance of electron transport from the perovskite to FTO drops the efficiency of charge transport (Mohammadian-Sarcheshmeh & Mazloum-Ardakani, 2018). On the other hand, a TiO_2 CL, which is too thin cannot effectively coat the FTO substrate. But it has already been determined that the rate of recombination is high due to lower electron mobility and charge transport characteristics of TiO_2. However, to date, high performing mp-PSC has been still fabricated with mesoscopic-TiO_2 requiring high process temperature (Jeon et al., 2015; Yang et al., 2015).

4.3.1.5.2 TiO_2 preparation methods

Many deposition techniques have been used to pattern TiO_2 CLs in PSCs such as, the sol–gel method (Liu et al., 2013), oblique electrostatic inkjet (OEI) deposition (Shahiduzzaman, Sakuma et al., 2019), chemical bath deposition (CBD) (Shahiduzzaman et al., 2015a; Shahiduzzaman, Karakawa et al., 2018), spray pyrolysis (Noh et al., 2013; Shahiduzzaman, Hossain et al., 2021), atomic layer deposition (ALD) (Yongzhen et al., 2014), thermal oxidation (Ke et al., 2014), sputtering (Chen, Cheng, Dai, & Song, 2015), chemical vapor deposition (CVD) (Maruyama & Arai, 1992), and electrodeposition (Su, Hsieh, Hong, & Wei, 2015). Spin coating is a simple method frequently used to pattern TiO_2 CLs. However, this technique cannot fabricate uniform TiO_2 CLs, which also has limited suitability for large-scale production. CBD is a simple and cost effective technique used to form TiO_x CLs from aqueous solutions, forming it beneficial for the large-scale PSCs and other PV cells (Shahiduzzaman, Kuwahara et al., 2020). Previously, we have been conducted with CBD TiO_2-based PSCs (Adhikari, Shahiduzzaman, Yamamoto, Lebel, & Nunzi, 2017; Shahiduzzaman et al., 2015a; Shahiduzzaman et al., 2015b; Shahiduzzaman, Karakawa et al., 2018). Recently, we fabricated PSCs with low temperatures ($< 150°C$) and high temperatures ($> 450°C$) treated CBD TiO_2 CL and subsequently achieved PCEs of 14.51% and 15.50%, respectively (Shahiduzzaman, Kuwahara et al., 2020). Photos of the hydrolysis stages of CBD are shown in Fig. 4.3.7 (Taima et al., 2016). TiOx film with 30-nm thickness is formed in a single operation. By repeating the operation, the thickness of TiOx can be increased. Controlling morphology and thickness of TiO_2 CLs is difficult and can lead to poor reproducibility.

Annealing at 500°C transforms amorphous to crystalline (anatase or rutile) of TiO_2, ascertaining higher electrical properties such as conductivity and better charge transportation. So, it is very important to achieve those attributes in TiO_2 in the low-temperature process for efficient PSC. Crystalline, nanosized TiO_2 has been deposited by CBD at 70°C to achieve outstanding PSC efficiency (Yella, Heiniger, Gao, Nazeeruddin, & Grätzel, 2014).

4.3.1.5.3 Commonly used TiO_2, SnO_2, ZnO, and fullerene derivatives as ETLs

4.3.1.5.3.1 TiO_2 thin film

The three most broadly used different crystalline polymorphs of TiO_2 [anatase (tetragonal), brookite (orthorhombic), and rutile (tetragonal)] possess distinct structures and sole properties. Previously, we introduced anatase TiO_2 NPs between perovskite and TiO_2 CL and subsequently improved a PCE of 17.05%

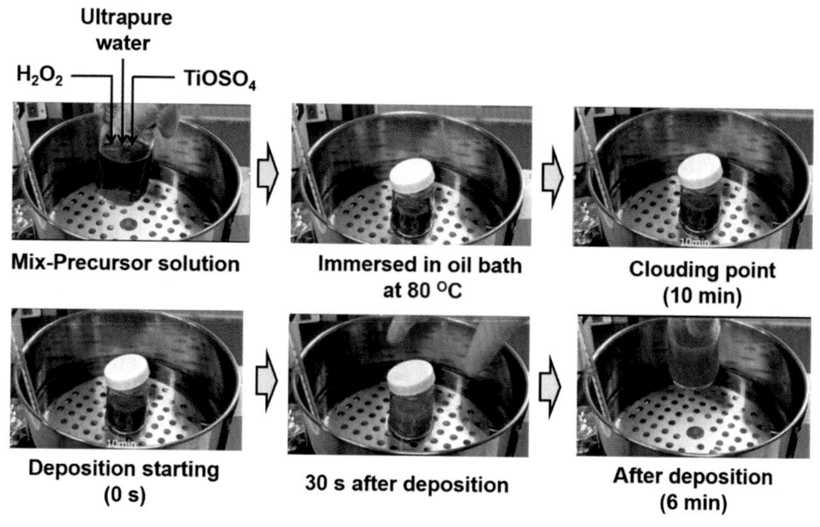

FIGURE 4.3.7

Photographs illustrating the hydrolysis steps in the chemical bath deposition (CBD) of TiO_x films.

Reproduced with permission Taima, T., Shahiduzzaman, M., Yamamoto, K., Furumoto, Y., Kuwabara, T., & Takahashi, K. (2016). Planar heterojunction type perovskite solar cells based on TiOx compact layer fabricated by chemical bath deposition. SPIE. Copy right 2021, SPIE.

(Shahiduzzaman, Ashikawa et al., 2018). Compared with anatase TiO_2, brookite TiO_2 shows well-matched band gap alignment with the perovskite layer, which leads efficient electron transfer from brookite to the anatase phase. Kogo et al. (Kogo, Sanehira, Numata, Ikegami, & Miyasaka, 2018). reported a TiO_2 CL/ mesoporous brookite TiO_2 as the ETLs-based PSCs with a PCE of 21.6% and improved V_{oc} owing to the closer CBE of the brookite to that of the perovskite. Furthermore, we demonstrated TiO_2 heterophase junctions, namely anatase-brookite or brookite-anatase, and found PSCs with anatase-brookite phase junction as an ETL had a PCE of 16.82% (Shahiduzzaman, Visal et al., 2019). This study also implied that single-phase brookite TiO_2 NPs (size ranged from 30 to 50 nm in diameter) without having a TiO_2 CL can perform as an advantageous ETL in PSCs. These PSCs exhibited a PCE of 14.92% that was as good as to the results of the first work reported. Single-phase brookite TiO_2 NPs can block holes effectively owing to their suitable internecking particle-to-particle structure. Brookite TiO_2 NPs showed large specific surface area, which allows efficient electron collection and following transport that helps to balance the electron and hole current densities, then intensely influences the performance of PSCs. The efficiency and stability of PSCs are strongly influenced by the bottom substrate of perovskite film. We reported a PSC by introducing brookite TiO_2 NPs layer between the TiO_2 CL and perovskite layer and achieved a PCE of 18.2%

(Shahiduzzaman, Kulkarni et al., 2020). Brookite TiO_2 NPs phase can be considered an alternative to anatase TiO_2 phase based materials for improving performance of PSCs. Besides, Park et al. demonstrated a PSC having the rutile TiO_2 phase (Lee et al., 2014). Rutile TiO_2 shows more positive CBE potential than that of anatase and brookite TiO_2 phase, ensuing in lower V_{oc}. A mesoporous material holds pores with diameters between 2 and 50 nm, corresponding to International Union of Pure and Applied Chemistry (IUPAC) nomenclature. The IUPAC describes the microporous material possess pores smaller than 2 nm in diameter and macroporous material containing pores greater than 50 nm in diameter. Commonly, the fabrication of mesoporous TiO_2 films involves a complex and time-consuming procedure, which contains the patterning a TiO_2 CL followed by the synthesis of mesoporous TiO_2. Owing to changing the crystalline phase (anatase) of the amorphous oxide film, mesoporous TiO_2 needs a sintering procedure at high temperatures ($> 450°C$) to enhance electron transport properties and eliminate polymer template molecules, which limits the application of mesoporous TiO_2 in flexible PSCs. Trilok et al. (Singh et al.). demonstrated the efficient PSCs with a PCE of 21.1% by inserting alkali-metal dopants into mesoporous TiO_2. Several other mesoporous TiO_2-based PSCs have also been described to date (Shin, Lee, & Seok, 2019; Singh, Singh, & Miyasaka, 2016). Furthermore, achieved 15.6% efficiency of PSC usingTiO_2 nanoparticle and graphene nanocomposite ETL (Wang et al., 2013). A low-temperature ETL process was reported by Snaith et al. (Burschka et al., 2013). The team fabricated multifunction PSC device on flexible organic substrate inserting Al_2O_3 as scaffolding layer and TiO_2 CL. PSC devices with TiO_2 of different nanostructure (e.g., nanorods, nanotubes, and core-shell nanofibers) as ETL have also been fabricated (Dharani et al., 2014; Gao et al., 2014; Kim et al., 2013; Salazar et al., 2015) due to possessing better charge transportation attributes to the expected direction. Additionally, these hollow type nanostructures exhibited pore filling properties opening more active surface area for perovskite crystal attachment. The anatase phase of TiO_2 nanowires with high forking showed enhanced light harvesting efficiency (LHE), efficient hole blocking, and superior electron-transporting properties in mp-PSC (Wu, Huang, Chen, Cheng, & Caruso, 2015). Hydrothermal synthesis includes the various techniques of crystallizing substances from high temperature aqueous solutions at high vapor pressures, also termed "hydrothermal method." The hydrothermal route is a convenient way of achieving TiO_2 nanorods with improved optoelectronic properties and morphology, producing the observable PV performance of PSC devices (Li, Zhang, Gao, Zhang, & Mao, 2015). It is very important in the case of an ETL to consider attributes such as absorbance, crystallographic orientation, higher carrier mobility and morphology of films, diameters of nanoparticles, and trap states on the surface. So, the PSC performance based on crystal phase and films morphology of ETL also were studied (Lee et al., 2014). Putting SnO_2 SC nanowires into TiO_2 epitaxial shell to form an effective ETL have been incorporated in PSC by Han et al. (2015). which exhibited outstanding performance in electron-transporting along with pore-filling attributes.

4.3.1.5.3.2 SnO_2 thin film

SnO_2 have also been studied as ETLs because of their superior optoelectronic properties. Song et al. (2015) deposited SnO_2 by spin coating in a two-step perovskite deposition method while controlling $CH_3NH_3PbI_3$ crystal growth by solvent evaporation followed by annealing but the full conversion of PbI_2 into $CH_3NH_3PbI_3$was not achieved. On the other hand, the position of conduction band minimum (CBM) was in between CBM of SnO_2 and $CH_3NH_3PbI_3$, accelerating favorable conditions for effective and successful electron transfer between the absorber layer and ETL. The fabricated device exhibited 13.0% efficiency. A further study was performed following the same device fabrication process to assess the stability between SnO_2- and TiO_2-based PSC devices with a time frame of 700 hours. It was noticed that SnO_2 devices degraded slower than the device with c-TiO_2 devices. In a consultation, the authors stated that the presence of an unconverted PbI_2 layer elevated the SnO_2 device stability by inhibiting contact between SnO_2 and $CH_3NH_3PbI_3$. Kogo et al. conducted further studies introducing brookite (TiO_2) in the SnO_2-TiO_2 bilayer ETL framework and achieved 13.4% efficiency (Kogo, Ikegami, & Miyasaka, 2016). Sputtered SnO_2 at room temperature as ETL for air-stable and efficient PSC on rigid and flexible substrates were made out by Kam, Zhang, Zhang, and Fan (2019). PSC on rigid glass and flexible PEN substrates produced 12.82% and 5.88% PCE, respectively. The PSC device on glass substrate retained 93% of its initial PCE after 192-hour exposure to dry air while the PEN substrate device maintained over 90% of its initial PCE after 100 consecutive bending cycles.

4.3.1.5.3.3 ZnO thin film

Due to having better charge carrier mobility of ZnO (205–300 $cm^2/V/s$) compared to TiO_2, the former one has received higher attention to be used as electron–selective contact (Zhang, Dandeneau, Zhou, & Cao, 2009) in mesoporous architecture for the first time resulting 5.0% efficiency (Bi et al., 2013). The metal nanorod arrays (NRA) that were grown hydrothermally showed rapid electron injection attributes along with stability than TiO_2. The noncapsulated device with ZnO NRA lasted for 500 hours. But the major loss seemed to have arisen from recombination that took place into oxide NRAs. Park et al. achieved 11.13% efficient PSC by deploying ZnO nanorods, which were 10^{-6} m long (Son, Im, Kim, & Park, 2014). Films of upright allied nanosheets array of mesoporous ZnO synthesized by the hydrothermal procedure were used in a double layer as ETL to fabricate PSC by Mahmood, Swain, and Amassian (2014). The PSC with double-layered ZnO acted as compact, and the mesoporous layer produced 10.35%. This efficiency was able to sustain 80% of its J_{SC} for up to 10 days. Enlargement of work function tuned at the ITO/ZnO interface layer by positioning organic dipole layers such as polyethyleneimine (PEI) and N_2-doped ZnO NRs was performed (Mahmood et al., 2014). The authors claimed a hysteresis-free PCE of 16.1% by virtue of better electron-transporting properties due to enhanced work function.

This technique of infestation dipole materials in oxide and active layers interface also could reduce work function at the junction layer eventually enhances the device performance. A range of 10%–13% PCEs of PSCs were achieve by electrospray deposited ZnO nanostructure (Mahmood, Swain, & Jung, 2014), and magnetron sputtered ZnO nanorods (Liang et al., 2014). Mesoporous ZnO nanofiber doped with indium (In) was used as ETL layer in PSCs offered 17.18% (Mahmood, Khalid, Ahmad, & Mehran, 2018) since doping of In into MOx framework could also increase the carrier concentration of ZnO providing the basis for superior conductivity (Ahmad, Sun, & Zhu, 2011). Other than commonly used TiO_2, SnO_2, and ZnO, there are a few MOxs like zinc stannate (Zn_2SnO_4) [28], barium tin oxide ($BaSnO_3$)[32], barium titanate ($BaTiO_3$)[27], zirconium dioxide (ZrO_2) (Mejía Escobar, Pathak, Liu, Snaith, & Jaramillo, 2017), tungsten oxide (WO_3)[25], niobium pentoxide (Nb_2O_5) (Ling et al., 2017), indium (III) oxide (In_2O_3)[27], and strontium titanate ($SrTiO_3$)[29] that can be used as ETLs for PSCs.

4.3.1.5.3.4 Organic fullerene derivatives

Device architecture is a very important factor in determining effective ETL. Fullerenes containing C60, C70, and phenyl-C61-butyric acid methyl ester (PCBM), which is a soluble form, are widely recognized and used as ETL in inverted planar PSC. Though, PCBM possesses a different passivation mode, which was reported earlier (Xu et al., 2015). PCBM is a Lewis acid (electron acceptor) that can make a network with M-X antisite defect MI^{-3}; thus passivation of the perovskite layer occurs. The highest PCE of 3.9% was achieved using many fullerene derivatives as ETMs by Jeng et al. (2013) in a p–i–n device architecture. Once these issues were addressed, 20.3% PCE was achieved within a few years deploying PCBM as a successful ETL for PSC (Luo et al., 2018). Snaith et al. (Wehrenfennig, Eperon, Johnston, Snaith, & Herz, 2014), for the first time, determined charge diffusion length within a simple $CH_3NH_3PbI_3$ layer, which was assisted by PCBM consistency. Lam and co-workers (Sun et al., 2014) were trying to understand the high performance exhibited by PCBM as ETL and discovered that PCBM could restrict the relocation of halide-ion by the formation of halide-π noncovalent bonding from PbI_2 framework during perovskite self-assembly. This interaction eventually promoted better electron transportation in PCBM (Sun et al., 2017). Furthermore, the deviation in PCE due to forward and reverse scan during PCE measurements, which is known as "hysteresis," has been considered as one of the major constraints for further commercialization. It is though still to determine the source of such deviation it's related to the movement of ion and ferroelectric characteristics of $CH_3NH_3PbI_3$ absorber layer. The existence of defect states, charge traps, and the surface of $CH_3NH_3PbI_3$ film are also responsible for such behavior (Chen, Yang, Priya, & Zhu, 2016). Fullerene-based PCBM can lower the hysteresis, as mentioned earlier attributes by forming an outer shield on the charge trapped locations resulting in dramatic reduction of interface charge recombination and increase photocurrent response speed (Xu et al., 2015). The effectiveness of PCBM in suppressing hysteresis by trap states passivation was first called by Huang and coworkers, which was validated through further photoluminescence (PL) spectroscopy of the surface and inside of the film. Heat treatment could alter some properties of PCBM, which eventually were beneficial for enhanced PSC performance in the p–i–n device structure resulting from better J_{SC} and FF. During the heat treatment, the PCBM decomposed into the grain of the perovskite absorber layer by increasing diffusion attributes (Li et al., 2013). Likewise, it was observed by the same group that annealing at 100°C just after PCBM spin coating on the active layer with additional dichlorobenzene (DCB) could significantly improve the V_{oc}, since this heat treatment allows postdeposition thermal treatment and eliminates trap densities and band alignment mismatch (Shao, Yuan, & Huang, 2016). Chiang, Tseng, and Wu (2014) introduced bulk heterojunction PSC in p–i–n device assembly fabricated by the two-step deposition technique. The group predoped PbI_2 with 1wt.% PCBM and spin-coated on hole selective layer in the first step. Then MAI was deposited on top of the PCBM-PbI_2 hybrid layer in the second step. The achieved $MAPBI_3$ was found to be very smooth in morphology. The PSC device showed high FF and hysteresis-free attributes might be due to the pore and grain boundary filling aspects establishing better mobility of electron and hole toward respective electrode (Chiang & Wu, 2016). A mixture of C60 and C70 in a 1:1 ratio was able to prohibit quick crystallization of individual PCBM fullerene. The mixed fullerene device showed higher PCE of 14.04% than identical PCBM device (13.74%) (Dai et al., 2017), resulting from enhanced charge carrier mobility. Doping of PCBM with Ag is a new widow to enhance LHE and reduce scrics resistance due to localized surface plasmon resonance. Inverted PSC with and without 5% Ag in PCBM produced 11.9% and 7.43% PCE, respectively (Higgins et al., 2018). It is evident that fullerene and its different imitative materials can produce highly efficient PSC in p–i–n device configuration but there are bottlenecks that are obstacles to mass production. The major hurdles are complex synthesis and purification route which elevate production cost, instability due to hygroscopic attributes, uncontrollable surface morphology, and lack of conjugation for proper charge transportation (Ala'a, Sun, Hill, & Welch, 2014; Anctil, Babbitt, Raffaelle, & Landi, 2011; Chen & Zhang, 2017; Xie, Zhao, Lin, Gu, & Zhang, 2016). Many steps have been taken to fix all those issues by doping of PCBM with other materials, introduction of third material in-between PCBM and adjacent electrode component interface and adaption of altered fullerene. Adding other martial into PCBM is also a phenomenon that can offer better electrical properties like conductivity and carrier mobility and durability than individual PCBM. With innovations in the perovskite layer, other functional materials, and device structure, fullerene-based PSCs will advance in efficiency and stability further.

4.3.1.5.4 The role of hole transport layer

A PSC device's PV performance and stability are those aspects that shapely depends on the deployed charge carrier layer based on device structure. So, these materials should possess attributes such as (Salim et al., 2015) (1) blocking the electron transfer from ETL and perovskite absorber layer due to hole majority; (2) increasing majority charge carrier (hole) transportation due to proper energy level matching between corresponding HTM and electrode; (3) helping to avoid decomposition and corrosion that occurs at perovskite and metal electrode junction layer; (4) suffocating charge recombination by isolating the top contact form the base layer or contact layers prompting improved performance. The PCE can be additionally improved by streamlining judicious HTMs with settled device design of PSC. HTM plays a very crucial part while it comes to determining the device performance and stability, requiring special considerations such as (Bakr et al., 2017):

1. An ideal HTM ought to possess such an energy barrier that can generate sufficient driving force to allow the hole to move to the electrode. The perovskite material's VB must be at a lower energy level than the HOMO level of HTM. Meanwhile, the HTM should have a higher LUMO level than CB of perovskite material so that it can act as an electron blocker.
2. The expected hole mobility should be greater than 10^{-3} $cm^2/V/s$ for smooth and effective hole transport toward the TCO anode and metallic cathode in n−i−p and p−i−n PSC device, respectively.
3. It should allow the perovskite quasi fermi-energy level degeneration to support the V_{oc}.
4. An ideal HTM should not diffuse into back metallic contact and perovskite active layer.
5. It should have a lower absorption profile to allow maximum photon to reach toward perovskite absorber layer for maximum LHE especially for a p−i−n PSC device. To avoid the waste of photon energy in the HTM, the transmission of light through the HTMs should be as strong as possible.
6. In addition to physical robustness, higher thermal, chemical stability, and degradation resistance toward peripheral influences like humidity and oxygen, ecofriendliness is also a prerequisite for an ideal HTM that can offer longer device stability.
7. Therefore choosing an appropriate HTL that improves the stability of PSCs and maintains its efficiency is important.

4.3.1.5.4.1 Organic spiro-OMeTAD as the hole-transporting layer

Generally, the organic polymer contains a repetitive unit of monomers. The location of carbon molecules determines the difference between small and polymer molecules. Therefore it is quite possible to tune the structure of such molecules to achieve expected optoelectronic properties while synthesizing for PV applications (Lattante, 2014; Wang, Shahiduzzaman, Fukaya et al., 2021). The most commonly used high performer organic HTMs are spiro-OMeTAD, PEDOT: PSS, P3HT, PTAA & cross-linked polymers spiro-OMeTAD (2,2',7,7'-Tetrakis-(N,N-di-4-methoxyphenylamino)-9,9-spirobifluorene) have been used as a higher efficiency producer HTM in PSC. It possesses outstanding advantages due to its molecular structure. The spiro center is responsible for higher glass transition temperature (T_g), and two arylamine radicals allow proper hole transportation. Moreover, better solubility, the potential energy of ionization, and, most importantly, solid-state (ss) characteristics are added features of spiro-OMeTAD (Huang et al., 2016). Batch et al. inserted spiro-OMeTAD as a hole conductor in DSSC for achieving a decent PCE while forming an effective heterojunction with an active dye layer (Bach et al., 1998). After them, spiro-OMeTAD was used in many ss-DSSC by different research groups as HTM, but Hawash et al. brought the most remarkable PCE, over 22% in 2018, by deploying this HTM in PSC (Hawash, Ono, & Qi, 2018). These results were inspiring, but the unsteadiness of spiro-OMeTAD due to its amorphous and hygroscopic nature also had a significant impact on whole device stability. Since these properties of spiro-OMeTAD exposed the organohalide perovskite to humid, UV light, and temperature, which lead to ultimate device degradation. Furthermore, low conductivity (4.4×10^{-5} S/cm) (Leijtens et al., 2012) and low hole mobility (4×10^{-5} cm^2/V/s) (Leijtens, Lim, Teuscher, Park, & Snaith, 2013) decrease the J_{SC} and FF of the overall device performance (Fabregat-Santiago, Bisquert, Palomares, Haque, & Durrant, 2006; Fabregat-Santiago et al., 2009). Burschka et al. (2013). introduced $Li[(CF_3SO_2)_2N]$ (Li-TFSI) salt in spiro-OMeTAD as an additive in ss-DSSC for the first time assuming that antimony would provide sufficient free charge transporter by oxidizing spiro-OMeTAD$^+$ (Huang et al., 2016). Meanwhile, Li-TFSI dopant would supply Li^+, which might reduce charge transport resistance with a corresponding adjacent layer by offering higher conductivity (Im et al., 2011) exhibiting 7.2% PCE. Li-TFSI salt responds to available moisture due to possessing hygroscopic phenomena, which is responsible for perovskite layer degradation that leads to device poor PCE. Meanwhile, the migration of Li^+ through the spiro-OMeTAD layer introduces pinholes endorsing possible degradation. An additional material is required to realize the expected higher performance from spiro-OMeTAD, which is TBP. TPB significantly improvises to maintain homogeneity of the precursor solution restricting agglomeration or dispersion of Li-salt (Zhang et al., 2016). But it initiates corrosion again, affecting device stability adversely. Li-TFSI in acetonitrile helps to enhance PSC's electronic conductivity (Krishnamoorthy et al., 2014). However, various effort has been considered to enhance PSC performance in terms of conductivity, stability, and lowering the cost of total device fabrication (Hua et al., 2016). Many different additives, as well as doping materials, have been prescribed by the different research groups. These materials ought to exhibit such PV performance along with robustness that is competitive or outperforms that of pristine spiro-OMeTAD (Ou, Li, Wang, Li, & Tang, 2017). PEDOT: PSS is comprised of poly (3,4-ethylene dioxythiophene) and polystyrene sulfonate polymer system. PEDOT: PSS is a suitable material for optoelectronic devices since it allows to perform tuning of its electrical charge carrier properties and exhibit low absorption attributes in the strong energy region of the solar spectrum (Li et al., 2013). It is a high performing HTM in PSC at an inverted device architecture. But its acidic nature refraining from its' use in a commercial production line. This acidic behavior may be responsible for corrosion of transparent conduction oxide anode (ITO/FTO) and opens the pinholes that allow transportation of In to into PEDOT: PSS layer (Zhao et al., 2015). Not only that, but it also goes for chemical reaction with organic part (base) of perovskite components.

4.3.1.5.4.2 Organic PEDOT: PSS as the hole-transporting layer

Usually PEDOT: PSS is deposited by spin-casting from its liquid solution. Once PEDOT: PSS is deposited on top of perovskite in p–i–n architecture, the perovskite is unstable due to the hygroscopic nature of PEDOT: PSS. Furthermore, the band PEDOT: PSS a work function of -4.9 eV but the HOMO level of simple perovskite film is -5.3 eV, so there is a huge loss of driving force or potential energy loss for hole migration due to the significant energy band incongruity (Lim et al., 2014). This low WF may exhibit lower V_{oc} (<1.00 eV) (Jeng et al., 2014) in the PSC of inverted structure though it has been also reported that the V_{oc} is mostly dependent of the intrinsic property of perovskite film. However, it is has been reported that some high V_{oc} was achieved through modification of PEDOT: PSS or other materials of the device of inverted PSC. As an example, MoO_3 and PEDOT: PSS double sheet was used as an HTM in a p–i–n architecture PSC device showed PCE of 12.78% resulting from better hole migration at PEDOT: PSS/ITO interface (Hou et al., 2015). The device was stable for about 11 days without major loss of PV performance while competing for the device with pristine PEDOT: PSS HTM. Though p-type HTM Poly (3-hexylthiophene) (P3HT) polymer possesses expected charge carrier properties, the corresponding mismatch in band alignment with perovskite is a major concern in terms of deploying into mass production since the V_{oc} is severely affected. Currently, organic semiconductors are in great scrutiny for deployment as HTM in inverted architecture PSC devices. The main advantage of such materials is their eco-friendliness, the ability to tune their optical and electrical properties by changing their functional groups, large-scale production and the ease of selection for a broad spectrum of HTM elements.

4.3.1.5.4.3 Poly[bis(4-phenyl) (2,4,6-trimethylphenyl)] amine (PTAA) as the hole-transporting layer

Solution-processed perovskite films can be even though deposited on top of hydrophobic PTAA layer since it possesses a great attribute of surface moistening property. An immense height and width ratio in perovskite grains on PTAA was reported as improvised by PTAA's ideal wettability (Bi et al., 2015). The inverted PSC device displayed staggering 18.9% efficiency using PTAA as HTM. Even higher efficiencies (19.4%) have been reported by Dong et al. (Park, Seo, Park, Shin, Kim, Jeon, ... Seok, 2015) for p–i–n structured PSC devices. This outcome brings a great possibility for organic semiconductor to be a strong contender in the HTM of the PSCs device component.

4.3.1.5.4.4 Inorganic and other types of the hole-transporting layers

Inorganic HTMs have become the center of attention for inverted PSCs, especially MOxs such as NiOx (Hasan et al., 2020; Yin et al., 2017; Mahmud Hasan et al., 2020), CuI (Christians, Fung, & Kamat, 2013), CuSCN (Ye et al., 2015), CuxO (Sun et al., 2016), and MO_3 (Murase & Yang, 2012). Using MOxs, including MoO_3, WO_3, NiOx, and Cr_2O_3, as HTMs have promising outcomes such as improving the carrier transporting properties within solar cell devices (Mahjabin et al., 2021). MOxs can also serve as protection for the photo absorber layers against unfavorable environmental conditions (Kaltenbrunner et al., 2015; Liu et al., 2015) though these inorganic oxides still need to go further to catch up organic materials in terms of PCE. Among these MOxs, NiOx is an ideal choice for holding an adequately large bandgap (>3.7 eV) (Boschloo & Hagfeldt, 2001), allowing high transparency and suitable and favorable energy levels for effective hole transport. With upper valence band maximum, this HTM can facilitate hole movement to the photoanodes, unlike perovskite absorber layers. NiOx also has a higher CBM compared with perovskite absorber layers, which restrict electron movement to the photoanodes of perovskite, thus providing good chemical stability (Yin et al., 2016). Different approaches have been considered for the successful preparation of NiOx as an HTM in PSC. Seo et al. (2016) deposited NiOx as HTM by ALD, which achieved 16.6% efficiency. Park et al. (2015) used pulsed laser deposition, which produced 17.3% device efficiency. Both of these costly processes are known for small-scale deployment, and they may be unsuitable for complete device fabrication in commercial production. The deposition of Li-doped NiOx (LiNiOx) through the hot-casting method was introduced by Nie et al. (2018) and they achieved a highly efficient, hysteresis-free, and stable PSC with a PCE of approximately 18%. Guo et al. (2018) utilized thermolysis, which requires a high temperature of nickel hydroxide to form NiOx HTM layers for inverted PSCs. This method achieved a PCE of 14.55%. Solution-processed NiOx thin films were prepared by spin coating nickel (II) acetylacetonate in ethanol and HCl. The PSC obtained a PCE of 18.8%, retaining 16.92% of the initial PCE upon exposure to environmental conditions ($>70\%$ relative humidity) for 720 hours (Zhu et al., 2016). Sol–gel processed NiOx nanocrystals showed a PCE of 9.11% (Zhu et al., 2014). All these studies indicated the great potential of NiOx as a successful HTM for PSC. The characteristic properties of NiOx are largely influenced by NiOx preparation methods, which allow for versatile fabrication of full device structure. Although the solution process generated expected results, the conversion of crystallize source to polycrystalline NiOx demanded high. Alternative small molecule-based HTMs are currently in the process of replacing spiro-OMeTAD in PSCs. Multistep synthetic steps, high purity, UV sensitivity, low hole mobility, and the risk of crystallization during the annealing process makes them suitable for wide-scale applications. Alternative HTMs for p–i–n PSCs are under investigation, although envisioning that a new molecule can possess all the desired requirements to be an efficient HTM is difficult. Nevertheless, a balance of properties in inorganic p-type compounds for p–i–n PSCs can be attained for efficient hole extraction, and efficient HTMs can be produced. Considering the prospect for PSC mass production, the deposition of HTMs should be simple.

4.3.1.6 Fabrication methods for a perovskite absorber layer

The complete fabrication processing has to be carried out in a glove box under a controlled atmosphere without exposing the device to moisture. Fabrication methods of perovskite films are generally classified into (I) wet/or solution processing (spin coating) and (II) dry processing (vacuum deposition). Meanwhile, vacuum deposition processes comprise the vapor-assisted solution process and thermal vapor deposition process.

4.3.1.6.1 *Wet/or solution processing (spin coating)*

A simple, low-cost spin coating is a technique to pattern a uniform thin-film onto a solid surface by applying centrifugal force and involves a liquid–vapor interface. In general process, a liquid is placed at the center of a circular surface and is fast rotated to fabricate homogeneous films with a thickness of several micrometers. A spin coating technique was applied to pattern perovskite films. Fabrication of perovskite film using spin coating can be done in two ways one-step spin coating and two-step spin coating (Fig. 4.3.8). For one-step coating, the

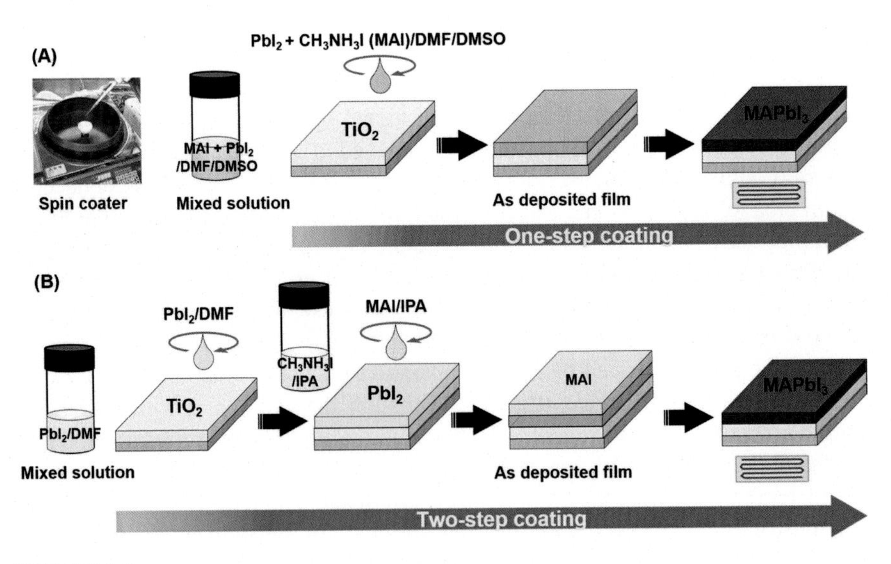

FIGURE 4.3.8

(A) One-step and (B) two-step spin coating procedures for $CH_3NH_3PbI_3$ perovskite formation.

perovskite precursors such as inorganic lead iodide (PbI_2) and organic MAI were mixed in anhydrous commonly used mix solvents of dimethylformamide (DMF) and DMSO. The resultant precursor mixtures were stirred for one hours at 60°C in a glove box and filtered through 0.45 μm PTFE filters prior to spin coating. A 100 μL precursor solution was dropped onto a TiO_2 coated FTO substrate and spin-coated at a low spinning rate (1000 rpm) for 10 seconds. Then, the spinning rate was slowly increased up to 6000 rpm for 40 seconds with dripping anti-solvent chlorobenzene (500 μL) 8 seconds after the spin coating started to produce a thin layer of high-quality perovskite (Fig. 4.3.8A) (Shahiduzzaman, Ashikawa et al., 2018). The anti-solvent (toluene) dripping method was introduced for the first time to regulate the kinetics of crystal growth and film quality (Jeon et al., 2014). As-deposited substrates were then allowed to anneal at 100°C on a hot plate for 1 hour to crystallize the perovskite. Researchers commonly used solvent such as DMF, DMSO, γ-butyrolactone (GBL) and so on to dissolve perovskite.

For two-step coating, a PbI_2 powder precursor was dissolved in DMF solvent and resultant solution was first spin-coated on the TiO_2 coated FTO substrate, dried and then a CH_3NH_3I-dissolved isopropyl alcohol solution was spin-coated on the PbI_2 coated substrate, followed by thermal treatment for 1 hour to produce a high-quality perovskite film (Fig. 4.3.8B) (Md et al., 2018; Shahiduzzaman, Hamada et al., 2019). Another approach to produce high-quality perovskite film was achieved by pre-wetting the PbI_2 film in isopropanol for 1 second before dipping it into the MAI solution. Through this approach, morphology can be easily controlled with reduced defects and surface roughness of the perovskite film by controlling the concentration of the MAI solution. Several researchers (Liang et al., 2014; Wang et al., 2014) reported that the formation of pinholes cannot be prevented in a solution-processed film due to its natural mechanism (Emslie, Bonner, & Peck, 1958) and that a special separate treatment is necessary to reduce pinhole formation. Another drawback is the incomplete conversion of the perovskite compound. When PbI_2 film is dipped into the MAI solution, PbI_2 readily reacts with small surface molecules and forms a dense layer that impedes the penetration of MAI into the underlying layer; however, this issue is addressed by advanced device engineering (Song, Watthage, Phillips, & Heben, 2016).

4.3.1.6.2 Dry processing (vacuum deposition)

Vacuum deposition is a simple and easy-to-control thicknesses as well as large area multistack thin-films approach that aims to fabricate high-quality perovskite films compared with those formed using solution-based deposition processes. In general, patterning perovskite is carried out under high-vacuum conditions. Commonly, vacuum deposition techniques involve the sublimation of two kinds of precursors such as a PbI_2 and MAI in a vacuum chamber. High temperature is required to evaporate the solid PbI_2 and MAI into vapor. Due to the gasification behavior of MAI that has a small molecular weight which causes the molecules to randomly diffuse inside the vacuum chamber, followed by makes it difficult to monitor and control the deposition rate using quartz microbalance sensors

(Liu et al., 2013; Malinkiewicz et al., 2014). Therefore to overcome this problem, researchers have fabricated highly efficient vacuum-deposited perovskite films via sequential vacuum deposition and CVD methods (Chen et al., 2014; Leyden et al., 2014; Leyden, Jiang, & Qi, 2016; Tavakoli et al., 2015). There are now some other ways to fabricate perovskite films via vacuum evaporation whereas precisely controlling the thickness and deposition rate of the material without using a precursor with gas-like properties (Chen et al., 2014; Chen et al., 2017). Previously, we reported sequentially-vacuum-deposited $CsPbI_3$ based PSCs with a PCE of 5.71% fabricated with various annealing times (Kyosuke et al., 2017). Furthermore, we replaced MAI with CsI and used an alternating vacuum deposition method (layer-by-layer) to alternately fabricate PbI_2 and CsI layers with a precise intercalation control to yield efficient, inorganic $CsPbI_3$ thin-films and PSCs with a PCE of 6.79% (Fig. 4.3.6B–E) (Shahiduzzaman, Yonezawa et al., 2017) (Fig. 4.3.9).

Furthermore, vapor-assisted solution process was developed to avoid the drawbacks of the solution technique and vapor deposition technique. In this procedure, the pre-deposited inorganic film of PbI_2 (by spin or dip coating) is used as a substrate, followed by the exposure of organic vapors of MAI. The growth of perovskite film took place via in situ induced reaction of the as-deposited film of PbI_2 with MAI, subsequently achieved high-quality perovskite film (Xiao et al., 2017). This study may guide further development of vapor-assisted solution processes based perovskite films with optimal crystallinities, and morphologies that contribute to high PSC performance. Recently, we reported highly stable PSCs with a PCE of 18.43% by using double-layer CsI intercalation into the host commonly

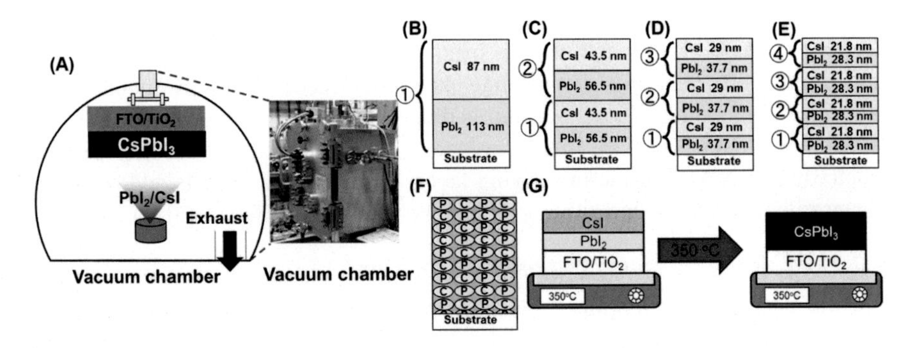

FIGURE 4.3.9

(A) Schematic illustration of vacuum chamber to pattern perovskite film. Inset shows vacuum chamber. Alternate layer deposition method with (B) one, (C) two, (D) three, and (E) four double layers. Panel (F) shows the schematic structure of a coevaporated $CsPbI_3$ thin-film, while (G) shows the final annealing step at 350°C for 1 minute.

Reproduced with permission Shahiduzzaman, M., Yonezawa, K., . . . Taima, T. (2017). Improved Reproducibility and Intercalation Control of Efficient Planar Inorganic Perovskite Solar cells by Simple Alternative Vacuum Deposition of PbI2 and CsI. ACS Omega, 2, 4464–4469. Copyright 2017, from ACS Omega.

used $MAPbI_3$ film (Wang, Shahiduzzaman, Muslih et al., 2021). The double-layer CsI intercalated device retained 83% of its initial efficiency after storing in the dark at relative humidity of 40%−50% more than 4000 hour.

4.3.1.7 The role of ionic liquids to produce high-quality perovskite films

Long-term operational stability is the significantly important key point for the commercialization and market acceptance of PSCs. Up to now, the maximum stable PSC reported is almost 1 year (Grancini et al., 2017) that is far less than the 25 years expected for commercially existing PV technologies. The insufficient lifetimes of PSC PVs devour undoubtedly hindered their commercialization (Rong et al., 2018). A device degradation is occurred by multiple factors associated with the structure and components. Perovskite degrades while it exposed to moisture, oxygen, light and heat (Jena & T., 2016; Yamamoto et al., 2016). Several studies demonstrated that perovskite film degradation is originated mainly at defect sites in grain boundaries and interfaces (Ahn et al., 2016; Selvanathan et al., 2021; Wang et al., 2017; Wang, Su et al., 2018). These regions are sensitive to moisture and oxygen, and lead diffusion can permanently damage the perovskite (Lee et al., 2018; Yamamoto et al., 2016). Thus operationally longer lifetime PV cells are significantly required for the commercialization and market adoption of PSCs. Perovskite constituents are sensitive to moisture and promptly absorb water. UV light produces oxygen vacancies and defects, and the thermal activation of perovskites by UV light origins the diffusion of organic cations, such as $CH_3NH_3^+$ (MA^+), followed by structural degradation (Heo et al., 2019; Nickel, Lang, Brus, Shargaieva, & Rappich, 2017; Shahiduzzaman, Hamada et al., 2019). The intrinsic lifetime of PSCs is influenced by defects in the perovskite layers and at the interfaces between perovskite and the charge transport layers (Hossain et al., 2020; Meng et al., 2018). Hygroscopic nature of the organic cations can be cover from moisture in the environment by capturing the perovskite layer (Wang, Zhang et al., 2018). In general, MA^+ derivatives are initiated near the grain boundaries in perovskite films. Owing to improve the operational stability of PSCs, the corresponding device engineering enhancements should be considered. The standard indoor test criteria should be maintained to ensure the use of PSCs PVs when PSCs are heated at 85 °C and 85% relative humidity for 1000 hours. Over the past few years, significant effort has been pointed to distinct device engineering techniques toward improving the lifetime of PSCs with ionic liquids (ILs) (Shahiduzzaman, Muslih et al., 2021). ILs as additives hold one of the most effective strategies for enhancing the long-term operational stability of PSCs. Snaith et al. (Bai et al., 2019). demonstrated a PSC using 1-butyl-3-methylimidazolium tetrafluoroborate ($BMIMBF_4$) IL, subsequently the primary PCE of 20% reduced by just ~5% under continuous illumination for more than 1800 hours at 70°C−75°C. This result implies that to date it is the most stable PSC performance, which allowing a significant step toward realizing reliable PV technology. Previously, we fabricated a perovskite NPs film by adding a small amount of a 1-hexyl-3-methylimidazolium chloride (HMImCl) (referred to as the IL) into the perovskite precursor solution (Shahiduzzaman, Yamamoto et al., 2015). Recently, we reported a review article on the benefits of ILs for the fabrication of efficient and stable PSCs (Shahiduzzaman, Muslih et al., 2021). We further discussed challenges and the impact of ILs in the fabricating of high-quality perovskite film with large grain and high crystallinity. Fig. 4.3.10A exhibits the chemical structure of HMImCl IL. Fig. 4.3.10B is a schematic illustration of a uniform $CH_3NH_3PbI_3$ suspension in *N,N*-DMF containing 1−3wt.% HMImCl (Md et al., 2016). Fig. 4.3.10C is a schematic of the formation steps of $CH_3NH_3PbI_3$ NPs. According to this study, we assumed that high boiling points and exceedingly low vapor pressure of ILs might slow the formation of $CH_3NH_3PbI_3$ small clusters and aid smooth nucleation during the spin coating

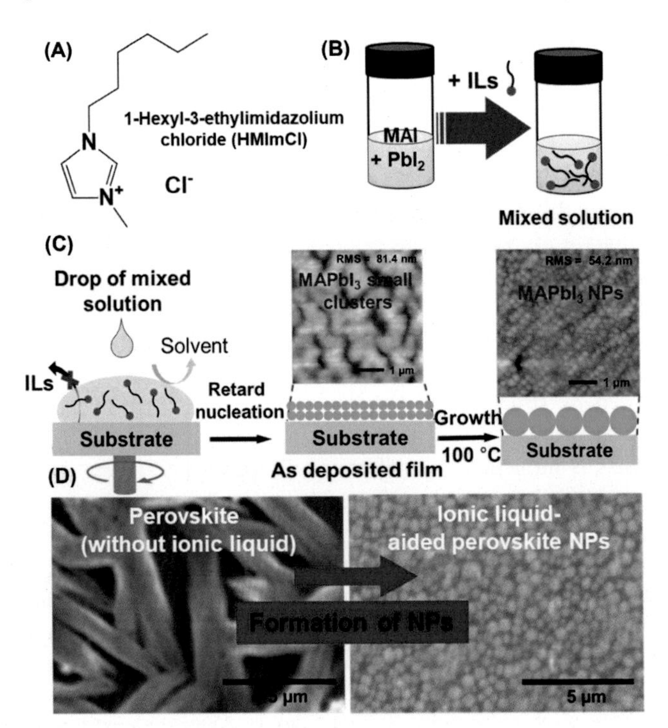

FIGURE 4.3.10

(A) Chemical structure of 1-hexyl-3-ethylimidazolium chloride (HMImCl). (B) Schematic of a homogeneous $MAPbI_3$ suspension in DMF containing 1−3 wt.% HMImCl. (C) Schematic illustration of perovskite NP film fabrication. (D) Scanning electron microscopy images of $MAPbI_3$ films prepared with and without the IL.

Reproduced with permission Shahiduzzaman, M., Muslih, E. Y., Hasan, A. K. M., Wang, L., Fukaya, S., Nakano, M., ... Taima, T. (2021). The benefits of ionic liquids for the fabrication of efficient and stable perovskite photovoltaics. Chemical Engineering Journal, *128461. Copyright 2021, from Chemical Engineering Journal.*

procedure that would result in the organized growth of perovskite NPs with the desired shape and size. As-deposited $CH_3NH_3PbI_3$ small clusters were evident by atomic force microscopy (Fig. 4.3.10C; Inset). The use of additives as a morphological controller should be realized with extremely low vapor pressure and a higher boiling point than that of main solvent of DMF. The additives should be sufficiently miscible with main solvent of DMF. Therefore we choose ILs with boiling points above 300°C, an exceedingly low vapor pressure of 0.1 mm/Hg at room temperature, and better miscibility with DMF, which would make the ILs effective additives. We succeed to fabricate homogeneous spherical shape of $MAPbI_3$ NPs, their corresponding PV performance was quite low because of poor surface coverage, their spherical shape, and a large number of grain boundaries in the film (Fig. 4.3.10D). In addition, we altered the morphologies of perovskite NPs film using ILs with a varying range of viscosities (Shahiduzzaman, Yamamoto et al., 2017). We observed that ILs with low viscosity, and the homogeneous dispersion of perovskite boosted NP formation. A homogeneous perovskite film with the desired morphology was achieved using an IL with low viscosity, whereas the IL was precisely dissolved in the perovskite suspension. Recently, we designed a novel technique in which the pristine $Cs_{0.05}(FA_{0.83}MA_{0.17})_{0.95}Pb(I_{0.83}Br_{0.17})_3$ (termed as CsFAMA) or pristine $MAPbI_3$ perovskite is grown by IL-assisted $MAPbI_3$ NP seeds for efficient light absorption enhancement in PSCs (Shahiduzzaman, Wang et al., 2021). The PSCs with the PCE of 19.5% was achieved by using $MAPbI_3$ NP-seeding growth of $MAPbI_3$ NPs/CsFAMA. In this regard, our result implies that a J_{SC} of 25.3 mA/cm^2 is approaching close to the Shockley–Queisser (S–Q) limit (25.47 mA/cm^2) by introducing a new technique of $MAPbI_3$ NP seeding, with the guest absorber layer (CsFAMA) having a bandgap of 1.6 eV. Additionally, the $MAPbI_3$ NPs/CsFAMA based device (nonencapsulated) remained above 80% of its initial output after 6000 hours storage in the dark at relative humidity of 30%–40%. The use of the IL-aided $MPbI_3$ NPs-seeded growth for PSCs is an important tread toward developing stable perovskite device. Perovskites engineering with ILs boosts the fabrication of high-quality perovskite films. Treating perovskite with ILs also improves the light harvesting capability and operational lifetime of PSCs owing to their hydrophobic nature, which make barriers to protect perovskite films from ambient moisture. It can be concluded that more research is required to perceive the most appropriate IL addition techniques and to define the instability issues of various perovskites, ETLs, and HTLs under operating conditions that take account of elevated temperatures.

According to the S-Q model, the theoretical PCE limit of a single-junction PSC with an absorber band gap (E_g) of 1.6 eV is 30.14%. The device would have J_{SC} of 25.47 mA/cm^2, V_{oc} of 1.309 V, and a fill factor of 0.905 (Hossain et al., 2021; Park & Segawa, 2018). Approaching the S-Q limit of 25.47 mA/cm^2 would enhance light absorption, and J_{SC} can be increased by introducing an efficient light-scattering material. Developing PSCs that both efficiently scatter light and have reasonable charge carrier transport capabilities remains a challenge. Light transmission loss in sandwiched PSCs with J_{SC} values below the S-Q limit is the result of parasitic light absorption or reflection at the interfaces of the transparent conducting electrodes or the incomplete collection of generated carriers. Meanwhile, a reduced V_{oc} or FF reflects unwanted bulk or interfacial carrier recombination, parasitic resistance, or other electrical nonidealities. Furthermore, owing to the theoretical values of V_{oc} and FF during impact, the development of an ideal diode that minimizes nonradiative recombination and interface recombination and trapping passivation is expected to improve the PCE of PSCs (Guo et al., 2020; Wu et al., 2020). Indeed, a recent study reported a high FF value up to 0.86 that can be attributed to the suppression of recombination and defect passivation, suggesting efficient charge tunneling, and the PCE was subsequently improved to 21.5% (Wu et al., 2020). Generally, PSCs hold many heterojunctions, and optimal heterojunctions need to be made by selecting appropriate ETLs or a stable inorganic HTL that will also reduce interface recombination and bulk defects. These outcomes might enhance V_{oc} and FF and afford improved efficiency close to the S-Q limit. The right ILs can be broadened to other perovskites to reach the S-Q single-junction efficiency limits and further enhance lifetime of PSC. However, in realizing a PCE beyond 27%, above the level of 28%, a perovskite photoactive layer with a narrow E_g is required to attain increasing LHE (i.e., photocurrent). This is essentially targeting a performance similar to that of gallium arsenide (GaAs) with E_g of 1.41 eV, which is capable of producing J_{SC} of 28 mA/cm^2. As an example, if the perovskite absorber with a similar Eg (1.4 eV) yields J_{SC} of 28 mA/cm^2, while retaining V_{oc} and FF at 1.2 V and 0.85 respectively, then a PCE of 28.6% can be obtained. Furthermore, tin (Sn)-based perovskite has emerged as a candidate owing to smaller optical bang gap (showing higher J_{SC}) and charge carrier mobilities higher than those of lead-based analogs (Fang, Adjokatse, Shao, Even, & Loi, 2018). These characteristics have been considered in terms of recent developments of Sn-based PSCs and their future potential to further enhance the device performance and stability. Their outstanding device performance means that Sn perovskites are now options for low-toxicity next-generation PV devices (Jiang et al., 2020; Kamarudin et al., 2019). It is IR that Sn-based 2D/3D perovskite structures combined with additives and cocations can be considered for advancing technology toward the S-Q single-junction efficiency limit with improved operational stability. Additionally, mixed Pb/Sn perovskites having lower band gaps (1.2–1.3 eV) where Pb as efficient with unstable Sn can be effective in tuning the band gap to the SQ maximum ($\sim$1.4 eV) (Tang et al., 2017). Suitable approaches of stabilizing Sn^{2+} can be found, and mixed Pb/Sn perovskites will subsequently become material options. Finally, we expect the existing challenges to be addressed in the coming years through the use of ILs, which will improve reliability and long-term operational lifetime for commercialization.

4.3.2 Summary and future outlook

The hybrid organic and inorganic halide perovskite has become the center of solar power-related research due to extreme efforts involved in preparing high-quality perovskite thin-film device structures, elemental optimization, and a basic understanding of charge transportation dynamics. Although PSC efficiency grows every year, the main challenges are reproducibility, stability, and large-scale production. The selection of ETL and HTL with the low-cost fabrication process, appropriate band alignments with perovskite absorber layer, high charge carrier ability, interfacial robustness, higher wettability, etc. and their deployment in device architecture can offer higher PCE and stability of PSC. The idea of alternative HTM can be more effective if an extensive investigation can be carried out to understand the material's intrinsic properties. It is of utmost importance through widespread research to optimize the HTM attributes such as thickness, preparation techniques and temperature, material availability which allows exploring its PV parameter. Moreover, p−i−n PSC has other component layers that take place in the device structure, which also introduces interfacial layers within themselves while stacked into the device. Therefore those components and their interfacial layers should be studied thoroughly that may offer innovative device architecture in future. The active perovskite layer itself commands the device output since the charge carrier are generated into this layer so the formation of high-quality crystalline perovskite is a very important aspect which should be studied and understood clearly. Furthermore, the inclusion of the study of charge carrier transportation from perovskite, the role of defect level and states should be considered in the future study as well. Since it has been observed that the presence of moisture in the atmospheric environment is attacked by organic HTM compounds leading to device instability. Subsequently, consideration for searching alternative inorganic HTM seeks greater attention. It is expected that the current study can offer a superior output expected from PSC once the full potential is exposed.

The recent surge in interest may lead to multiple attempts to commercialize perovskite PV products in the coming years. Perovskites holds varying advantages, which may provide multiple paths to commercialization. For perovskites, such an advantage might be low process prices and high conversion efficiencies relating to the above-mentioned versatile or partially transparent modules. A key necessity of PSCs for commercialization is a compelling advantage over other technically viable solutions, such as efficiency, lifetime (or stability), cost, or sustainability (Fig. 4.3.11). The inset displays a schematic of a perovskite film assembled using IL engineering that intends to address stability issues.

Until now, the record stable PSC was demonstrated using ILs, and it signifies a step headed toward opening innovative business opportunities. Fabricating toxic-free PSCs, whereas retaining their efficiency and stability, will enable the rise of the solar-energy uprising by opening first-hand business opportunities upon addressing efficiency, stability, flexibility and sustainability altogether. Furthermore, with the prompt and impressive advance in performance over recent years, PSCs have come out attention for lightweight power source applications. Implementing mechanically robust PV cells on lightweight, flexible plastic substrates that reduce of production cost of PSCs by using a roll-to-roll (R2R) high-throughput process influences great importance in industrialization. However, by

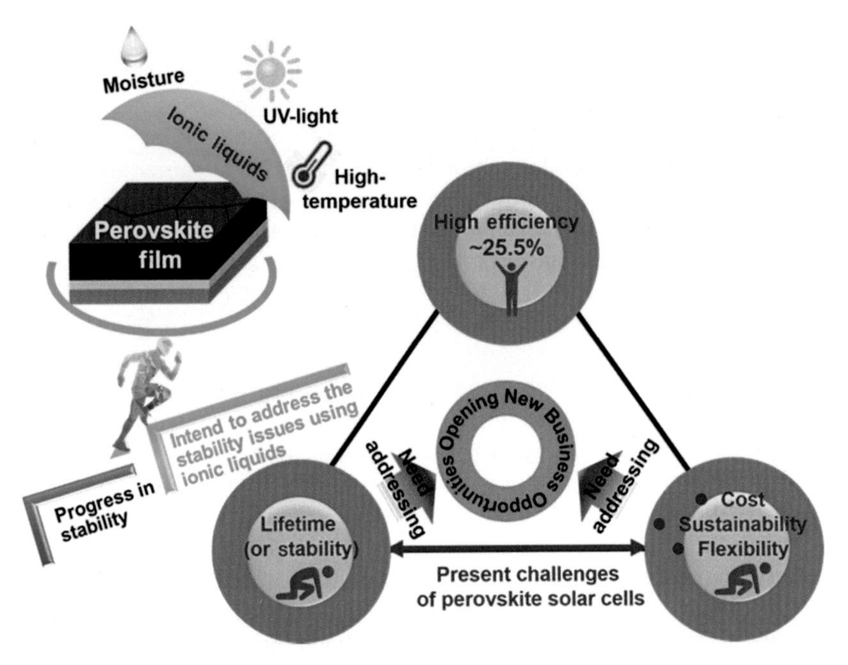

FIGURE 4.3.11

An ideal relationship between perovskite solar cell's efficiency, lifetime, cost, sustainability and flexibility is considered for opening new business opportunities. Inset shows a schematic of a perovskite film assembled using ionic-liquid engineering, which intends to address stability issues.

Reproduced with permission Shahiduzzaman, M., Muslih, E. Y., Hasan, A. K. M., Wang, L., Fukaya, S., Nakano, M., ... Taima, T. (2021). The benefits of ionic liquids for the fabrication of efficient and stable perovskite photovoltaics. Chemical Engineering Journal, *128461. Copyright 2021, from Chemical Engineering Journal.*

benchmarking our findings, experiments should also consider the need to ensure stability, flexibility, sustainability of PSCs, while maintaining its efficiency for high-impact research that may assist in solar-energy uprising by opening innovative business opportunities. We hope that performing PSCs research will be important for the development of renewable energy to address the world's energy issues.

Declaration of competing interests

The authors declare that they have no known competing financial interests.

Acknowledgments

This study was financially supported by a Grant-in-Aid for Scientific Research (Grant Number 20H02838). The authors would also like to acknowledge the support from The National University of Malaysia under Grant LRGS/1/2019/UKM-UKM/6/1.

References

Adhikari, T., Shahiduzzaman, M., Yamamoto, K., Lebel, O., & Nunzi, J.-M. (2017). Interfacial modification of the electron collecting layer of low-temperature solution-processed organometallic halide photovoltaic cells using an amorphous perylenediimide. *Solar Energy Materials and Solar Cells*, *160*, 294–300.

Ahmad, M., Sun, H., & Zhu, J. (2011). Enhanced photoluminescence and field-emission behavior of vertically well aligned arrays of In-doped ZnO nanowires. *ACS Applied Materials & Interfaces*, *3*, 1299–1305.

Ahn, N., Kwak, K., Jang, M. S., Yoon, H., Lee, B. Y., Lee, J.-K., ... Choi, M. (2016). Trapped charge-driven degradation of perovskite solar cells. *Nature Communications*, *7*, 13422.

Ala'a, F. E., Sun, J.-P., Hill, I. G., & Welch, G. C. (2014). Recent advances of non-fullerene, small molecular acceptors for solution processed bulk heterojunction solar cells. *Journal of Materials Chemistry A*, *2*, 1201–1213.

Anctil, A., Babbitt, C. W., Raffaelle, R. P., & Landi, B. J. (2011). Material and energy intensity of fullerene production. *Environmental Science & Technology*, *45*, 2353–2359.

Ansari, M. I. H., Qurashi, A., & Nazeeruddin, M. K. (2018). Frontiers, opportunities, and challenges in perovskite solar cells: A critical review. *Journal of Photochemistry and Photobiology C: Photochemistry Reviews*, *35*, 1–24.

Bach, U., Lupo, D., Comte, P., Moser, J. E., Weissörtel, F., Salbeck, J., ... Grätzel, M. (1998). Solid-state dye-sensitized mesoporous TiO_2 solar cells with high photon-to-electron conversion efficiencies. *Nature*, *395*, 583–585.

Bai, S., Da, P., Li, C., Wang, Z., Yuan, Z., Fu, F., ... Snaith, H. J. (2019). Planar perovskite solar cells with long-term stability using ionic liquid additives. *Nature*, *571*, 245–250.

Bakr, Z. H., Wali, Q., Fakharuddin, A., Schmidt-Mende, L., Brown, T. M., & Jose, R. (2017). Advances in hole transport materials engineering for stable and efficient perovskite solar cells. *Nano Energy*, *34*, 271–305.

Bi, C., Wang, Q., Shao, Y., Yuan, Y., Xiao, Z., & Huang, J. (2015). Non-wetting surface-driven high-aspect-ratio crystalline grain growth for efficient hybrid perovskite solar cells. *Nature Communications*, *6*, 7747.

Bi, D., Boschloo, G., Schwarzmüller, S., Yang, L., Johansson, E. M., & Hagfeldt, A. (2013). Efficient and stable CH_3 NH_3 PbI_3-sensitized ZnO nanorod array solid-state solar cells. *Nanoscale*, *5*, 11686–11691.

Bi, D., Yi, C., Luo, J., Décoppet, J.-D., Zhang, F., Zakeeruddin, S. M., ... Grätzel, M. (2016). Polymer-templated nucleation and crystal growth of perovskite films for solar cells with efficiency greater than 21%. *Nature Energy*, *1*, 16142.

Bi, Y., Hutter, E. M., Fang, Y., Dong, Q., Huang, J., & Savenije, T. J. (2016). Charge carrier lifetimes exceeding 15 μs in methylammonium lead iodide single crystals. *The Journal of Physical Chemistry Letters*, *7*, 923–928.

Boschloo, G., & Hagfeldt, A. (2001). Spectroelectrochemistry of nanostructured NiO. *The Journal of Physical Chemistry. B*, *105*, 3039–3044.

Burschka, J., Pellet, N., Moon, S.-J., Humphry-Baker, R., Gao, P., Nazeeruddin, M. K., & Grätzel, M. (2013). Sequential deposition as a route to high-performance perovskite-sensitized solar cells. *Nature*, *499*, 316–319.

Cai, M., Wu, Y., Chen, H., Yang, X., Qiang, Y., & Han, L. (2017). Cost-performance analysis of perovskite solar modules. *4*, 1600269.

Chen, B., Yang, M., Priya, S., & Zhu, K. (2016). Origin of J–V hysteresis in perovskite solar cells. *The Journal of Physical Chemistry Letters*, *7*, 905–917.

Chen, C., Cheng, Y., Dai, Q., & Song, H. (2015). Radio frequency magnetron sputtering deposition of TiO_2 thin films and their perovskite solar cell applications. *Scientific Reports*, *5*, 17684.

Chen, C.-W., Kang, H.-W., Hsiao, S.-Y., Yang, P.-F., Chiang, K.-M., & Lin, H.-W. (2014). Efficient and uniform planar-type perovskite solar cells by simple sequential vacuum deposition. *Advanced Materials*, *26*, 6647–6652.

Chen, C.-Y., Lin, H.-Y., Chiang, K.-M., Tsai, W.-L., Huang, Y.-C., Tsao, C.-S., & Lin, H.-W. (2017). All-vacuum-deposited stoichiometrically balanced inorganic cesium lead halide perovskite solar cells with stabilized efficiency exceeding 11%. *Advanced Materials*, *29*, 1605290-n/a.

Chen, Q., Zhou, H., Hong, Z., Luo, S., Duan, H.-S., Wang, H.-H., … Yang, Y. (2014). Planar heterojunction perovskite solar cells via vapor-assisted solution process. *Journal of the American Chemical Society*, *136*, 622–625.

Chen, W., & Zhang, Q. (2017). Recent progress in non-fullerene small molecule acceptors in organic solar cells (OSCs). *Journal of Materials Chemistry C*, *5*, 1275–1302.

Chen, W.-Y., Deng, L.-L., Dai, S.-M., Wang, X., Tian, C.-B., Zhan, X.-X., … Zheng, L.-S. (2015). Low-cost solution-processed copper iodide as an alternative to PEDOT:PSS hole transport layer for efficient and stable inverted planar heterojunction perovskite solar cells. *Journal of Materials Chemistry A*, *3*, 19353–19359.

Chiang, C.-H., Tseng, Z.-L., & Wu, C.-G. (2014). Planar heterojunction perovskite/PC71BM solar cells with enhanced open-circuit voltage via a (2/1)-step spin-coating process. *Journal of Materials Chemistry A*, *2*, 15897–15903.

Chiang, C.-H., & Wu, C.-G. (2016). Bulk heterojunction perovskite–PCBM solar cells with high fill factor. *Nature Photonics*, *10*, 196–200.

Christians, J. A., Fung, R. C., & Kamat, P. V. (2013). An inorganic hole conductor for organo-lead halide perovskite solar cells. Improved hole conductivity with copper iodide. *Journal of the American Chemical Society*, *136*, 758–764.

Colella, S., Mosconi, E., Fedeli, P., Listorti, A., Gazza, F., Orlandi, F., … Mosca, R. (2013). MAPbI3-xClx mixed halide perovskite for hybrid solar cells: The role of chloride as dopant on the transport and structural properties. *Chemistry of Materials*, *25*, 4613–4618.

Dai, S.-M., Tian, H.-R., Zhang, M.-L., Xing, Z., Wang, L.-Y., Wang, X., … Huang, R.-B. (2017). Pristine fullerenes mixed by vacuum-free solution process: Efficient electron transport layer for planar perovskite solar cells. *Journal of Power Sources*, *339*, 27–32.

Dharani, S., Mulmudi, H. K., Yantara, N., Trang, P. T. T., Park, N. G., Graetzel, M., … Boix, P. P. (2014). High efficiency electrospun TiO_2 nanofiber based hybrid organic–inorganic perovskite solar cell. *Nanoscale*, *6*, 1675–1679.

Dong, Q., Fang, Y., Shao, Y, Mulligan, P., Qiu, J., Cao, L., & Huang, J. (2015). Electron-hole diffusion lengths >175 μm in solution-grown CH3NH3PbI3 single crystals. *347*, 967–970.

Dong, Q., Yuan, Y., Shao, Y., Fang, Y., Wang, Q., & Huang, J. (2015). Abnormal crystal growth in CH3NH3PbI3 – xClx using a multi-cycle solution coating process. *Energy & Environmental Science*, *8*, 2464–2470.

Emslie, A.G., Bonner, F.T., & Peck, L.G. (1958). Flow of a viscous liquid on a rotating disk. *29*, 858–862.

Eperon, G. E., Stranks, S. D., Menelaou, C., Johnston, M. B., Herz, L. M., & Snaith, H. J. (2014). Formamidinium lead trihalide: A broadly tunable perovskite for efficient planar heterojunction solar cells. *Energy & Environmental Science*, *7*, 982–988.

Etgar, L., Gao, P., Xue, Z., Peng, Q., Chandiran, A. K., Liu, B., ... Grätzel, M. (2012). Mesoscopic $CH_3NH_3PbI_3/TiO_2$ heterojunction solar cells. *Journal of the American Chemical Society*, *134*, 17396–17399.

Fabregat-Santiago, F., Bisquert, J., Cevey, L., Chen, P., Wang, M., Zakeeruddin, S. M., & Grätzel, M. (2009). Electron transport and recombination in solid-state dye solar cell with spiro-OMeTAD as hole conductor. *Journal of the American Chemical Society*, *131*, 558–562.

Fabregat-Santiago, F., Bisquert, J., Palomares, E., Haque, S.A., Durrant, J.R. (2006). Impedance spectroscopy study of dye-sensitized solar cells with undoped spiro-OMeTAD as hole conductor. *100*, 034510.

Fang, H.-H., Adjokatse, S., Shao, S., Even, J., & Loi, M. A. (2018). Long-lived hot-carrier light emission and large blue shift in formamidinium tin triiodide perovskites. *Nature Communications*, *9*, 243.

Gao, X., Li, J., Baker, J., Hou, Y., Guan, D., Chen, J., & Yuan, C. (2014). Enhanced photovoltaic performance of perovskite CH_3 NH_3 PbI_3 solar cells with freestanding TiO_2 nanotube array films. *Chemical Communications*, *50*, 6368–6371.

Goto, T., Ohshima, N., Mousdis, G. A., & Papavassiliou, G. C. (2000). Excitons in a single two-dimensional semiconductor crystal of $H_3N(CH_2)_6NH_3PbI_4$. *Solid State Communications*, *117*, 13–16.

Grancini, G., Roldán-Carmona, C., Zimmermann, I., Mosconi, E., Lee, X., Martineau, D., ... Nazeeruddin, M. K. (2017). One-year stable perovskite solar cells by 2D/3D interface engineering,. *Nature Communications*, *8*, 15684.

Green, M. A., Dunlop, E. D., Levi, D. H., Hohl-Ebinger, J., Yoshita, M., & Ho-Baillie, A. W. (2019). Solar cell efficiency tables (version 54). *Progress in Photovoltaics: Research and Applications*, *27*, 565–575.

Guo, Y., Yin, X., Liu, J., Yang, Y., Chen, W., Que, M., ... Gao, B. (2018). Annealing atmosphere effect on Ni states in the thermal-decomposed NiOx films for perovskite solar cell application. *Electrochimica Acta*, *282*, 81–88.

Guo, Z., Jena, A. K., Takei, I., Kim, G. M., Kamarudin, M. A., Sanehira, Y., ... Miyasaka, T. (2020). VOC Over 1.4V for amorphous tin-oxide-based dopant-free CsPbI2Br perovskite solar cells. *Journal of the American Chemical Society*, *142*, 9725–9734.

Han, G. S., Chung, H. S., Kim, D. H., Kim, B. J., Lee, J.-W., Park, N.-G., ... Jung, H. S. (2015). Epitaxial 1D electron transport layers for high-performance perovskite solar cells. *Nanoscale*, *7*, 15284–15290.

Hasan, A. K. M., Sobayel, K., Raifuku, I., Ishikawa, Y., Shahiduzzaman, M., Nour, M., ... Akhtaruzzaman, M. (2020). Optoelectronic properties of electron beam-deposited NiOx thin films for solar cell application. *Results in Physics*, *17*, 103122.

Hawash, Z., Ono, L. K., & Qi, Y. (2018). Recent advances in spiro-MeOTAD hole transport material and its applications in organic–inorganic halide perovskite solar cells. *5*, 1700623.

Hawash, Z., Ono, L. K., Raga, S. R., Lee, M. V., & Qi, Y. (2015). Air-exposure induced dopant redistribution and energy level shifts in spin-coated spiro-MeOTAD films. *Chemistry of Materials*, *27*, 562–569.

Heo, S., Seo, G., Lee, Y., Seol, M., Kim, S.H., Yun, D.-J., ... Nazeeruddin, M.K. (2019). Origins of high performance and degradation in the mixed perovskite solar cells. *31*, 1805438.

Higgins, M., Ely, F., Nome, R. C., Nome, R. A., Dos Santos, D. P., Choi, H., ... Quevedo-Lopez, M. (2018). Enhanced reproducibility of planar perovskite solar cells by fullerene doping with silver nanoparticles. *Journal of Applied Physics*, *124*, 065306.

Hirasawa, M., Ishihara, T., Goto, T., Uchida, K., & Miura, N. (1994). Magnetoabsorption of the lowest exciton in perovskite-type compound (CH3NH3)PbI3. *Physica B: Condensed Matter*, *201*, 427–430.

Hossain, M.I., Hasan, A.K.M., Qarony, W., Shahiduzzaman, M., Islam, M.A., Ishikawa, Y., ... Tsang, Y.H. (2020). Electrical and optical properties of nickel-oxide films for efficient perovskite solar cells. *4*, 2000454.

Hossain, M. I., Saleque, A. M., Ahmed, S., Saidjafarzoda, I., Shahiduzzaman, M., Qarony, W., ... Tsang, Y. H. (2021). Perovskite/perovskite planar tandem solar cells: A comprehensive guideline for reaching energy conversion efficiency beyond 30%. *Nano Energy*, *79*, 105400.

Hou, F., Su, Z., Jin, F., Yan, X., Wang, L., Zhao, H., ... Li, W. (2015). Efficient and stable planar heterojunction perovskite solar cells with an MoO_3/PEDOT:PSS hole transporting layer. *Nanoscale*, *7*, 9427–9432.

Hua, Y., Zhang, J., Xu, B., Liu, P., Cheng, M., Kloo, L., ... Sun, L. (2016). Facile synthesis of fluorene-based hole transport materials for highly efficient perovskite solar cells and solid-state dye-sensitized solar cells. *Nano Energy*, *26*, 108–113.

Huang, L., Hu, Z., Xu, J., Zhang, K., Zhang, J., Zhang, J., & Zhu, Y. (2016). Efficient and stable planar perovskite solar cells with a non-hygroscopic small molecule oxidant doped hole transport layer. *Electrochimica Acta*, *196*, 328–336.

Im, J.-H., Lee, C.-R., Lee, J.-W., Park, S.-W., & Park, N.-G. (2011). 6.5% efficient perovskite quantum-dot-sensitized solar cell. *Nanoscale*, *3*, 4088–4093.

Jena, A. K., Kulkarni, A., & Miyasaka, T. (2019). Halide perovskite photovoltaics: Background, status, and future prospects. *Chemical Reviews*, *119*, 3036–3103.

Jena, A. K., & T., M. (2016). *Hysteresis characteristics and device stability*. Cham: Springer.

Jeng, J.-Y., Chen, K.-C., Chiang, T.-Y., Lin, P.-Y., Tsai, T.-D., Chang, Y.-C., ... Hsu, Y.-J., (2014). Nickel oxide electrode interlayer in $CH_3NH_3PbI_3$ perovskite/PCBM planar-heterojunction hybrid solar cells. *26*, 4107–4113.

Jeng, J. Y., Chiang, Y. F., Lee, M. H., Peng, S. R., Guo, T. F., Chen, P., & Wen, T. C. (2013). $CH_3NH_3PbI_3$ perovskite/fullerene planar-heterojunction hybrid solar cells. *Advanced Materials*, *25*, 3727–3732.

Jeon, N. J., Noh, J. H., Kim, Y. C., Yang, W. S., Ryu, S., & Seok, S. I. (2014). Solvent engineering for high-performance inorganic–organic hybrid perovskite solar cells. *Nature Materials*, *13*, 897–903.

Jeon, N. J., Noh, J. H., Yang, W. S., Kim, Y. C., Ryu, S., Seo, J., & Seok, S. I. (2015). Compositional engineering of perovskite materials for high-performance solar cells. *Nature*, *517*, 476–480.

Jiang, X., Wang, F., Wei, Q., Li, H., Shang, Y., Zhou, W., ... Ning, Z. (2020). Ultra-high open-circuit voltage of tin perovskite solar cells via an electron transporting layer design. *Nature Communications*, *11*, 1245.

Jung, E. H., Jeon, N. J., Park, E. Y., Moon, C. S., Shin, T. J., Yang, T.-Y., ... Seo, J. (2019). Efficient, stable and scalable perovskite solar cells using poly(3-hexylthiophene). *Nature*, *567*, 511–515.

Kaltenbrunner, M., Adam, G., Głowacki, E. D., Drack, M., Schwödiauer, R., Leonat, L., ... White, M. S. (2015). Flexible high power-per-weight perovskite solar cells with chromium oxide–metal contacts for improved stability in air. *Nature Materials*, *14*, 1032.

Kam, M., Zhang, Q., Zhang, D., & Fan, Z. (2019). Room-temperature sputtered SnO_2 as robust electron transport layer for air-stable and efficient perovskite solar cells on rigid and flexible substrates. *Scientific Reports*, *9*, 6963.

Kamarudin, M. A., Hirotani, D., Wang, Z., Hamada, K., Nishimura, K., Shen, Q., ... Hayase, S. (2019). Suppression of charge carrier recombination in lead-free tin halide perovskite via lewis base post-treatment. *The Journal of Physical Chemistry Letters*, *10*, 5277–5283.

Ke, W., Fang, G., Wang, J., Qin, P., Tao, H., Lei, H., ... Zhao, X. (2014). Perovskite solar cell with an efficient TiO_2 compact film. *ACS Applied Materials & Interfaces*, *6*, 15959–15965.

Kim, G., Min, H., Lee, K.S., Lee, D.Y., Yoon, S.M., & Seok, S.I. (2020). Impact of strain relaxation on performance of α formamidinium lead iodide perovskite solar cells. *370*, 108–112.

Kim, H.-S., Lee, C.-R., Im, J.-H., Lee, K.-B., Moehl, T., Marchioro, A., ... Park, N.-G. (2012). Lead iodide perovskite sensitized all-solid-state submicron thin film mesoscopic solar cell with efficiency exceeding 9%. *Scientific Reports*, *2*, 591.

Kim, H.-S., Lee, J.-W., Yantara, N., Boix, P. P., Kulkarni, S. A., Mhaisalkar, S., ... Park, N.-G. (2013). High efficiency solid-state sensitized solar cell-based on submicrometer rutile TiO2 nanorod and $CH_3NH_3PbI_3$ perovskite sensitizer. *Nano Letters*, *13*, 2412–2417.

Kim, J.H., Liang, P.-W., Williams, S.T., Cho, N., Chueh, C.-C., Glaz, M.S., ... Jen, A.K.-Y. (2015). High-performance and environmentally stable planar heterojunction perovskite solar cells based on a solution-processed copper-doped nickel oxide hole-transporting layer. *27*, 695–701.

Kogo, A., Ikegami, M., & Miyasaka, T. (2016). A sno x–brookite TiO_2 bilayer electron collector for hysteresis-less high efficiency plastic perovskite solar cells fabricated at low process temperature. *Chemical Communications*, *52*, 8119–8122.

Kogo, A., Sanehira, Y., Numata, Y., Ikegami, M., & Miyasaka, T. (2018). Amorphous metal oxide blocking layers for highly efficient low-temperature brookite TiO_2-based perovskite solar cells. *ACS Applied Materials & Interfaces*, *10*(3), 2224–2229.

Kojima, A., Teshima, K., Shirai, Y., & Miyasaka, T. (2009). Organometal halide perovskites as visible-light sensitizers for photovoltaic cells. *Journal of the American Chemical Society*, *131*, 6050–6051.

Krishnamoorthy, T., Kunwu, F., Boix, P. P., Li, H., Koh, T. M., Leong, W. L., ... Mhaisalkar, S. G. (2014). A swivel-cruciform thiophene based hole-transporting material for efficient perovskite solar cells. *Journal of Materials Chemistry A*, *2*, 6305–6309.

Kulkarni, S. A., Baikie, T., Boix, P. P., Yantara, N., Mathews, N., & Mhaisalkar, S. (2014). Band-gap tuning of lead halide perovskites using a sequential deposition process. *Journal of Materials Chemistry A*, *2*, 9221–9225.

Kyosuke, Y., Kohei, Y., Md, S., Yoshikazu, F., Keitaro, H., Teresa, S. R., ... Tetsuya, T. (2017). Annealing effects on $CsPbI_3$ -based planar heterojunction perovskite solar cells formed by vacuum deposition method. *Japanese Journal of Applied Physics*, *56*, 04CS11.

Lattante, S. (2014). Electron and hole transport layers: Their use in inverted bulk heterojunction polymer solar cells. *3*, 132–164.

Lee, J.-W., Bae, S.-H., De Marco, N., Hsieh, Y.-T., Dai, Z., & Yang, Y. (2018). The role of grain boundaries in perovskite solar cells. *Materials Today Energy*, *7*, 149–160.

Lee, J.-W., Lee, T.-Y., Yoo, P. J., Grätzel, M., Mhaisalkar, S., & Park, N.-G. (2014). Rutile TiO_2-based perovskite solar cells. *Journal of Materials Chemistry A*, *2*, 9251–9259.

Lee, M.M., Teuscher, J., Miyasaka, T., Murakami, T.N., Snaith, H.J. (2012). Efficient hybrid solar cells based on meso-superstructured organometal halide perovskites. *338*, 643–647.

Leijtens, T., Ding, I. K., Giovenzana, T., Bloking, J. T., McGehee, M. D., & Sellinger, A. (2012). Hole transport materials with low glass transition temperatures and high solubility for application in solid-state dye-sensitized solar cells. *ACS Nano*, *6*, 1455–1462.

Leijtens, T., Lim, J., Teuscher, J., Park, T., & Snaith, H.J. (2013). Charge density dependent mobility of organic hole-transporters and mesoporous TiO_2 determined by transient mobility spectroscopy: Implications to dye-sensitized and organic solar cells. *25*, 3227–3233.

Leyden, M. R., Jiang, Y., & Qi, Y. (2016). Chemical vapor deposition grown formamidinium perovskite solar modules with high steady state power and thermal stability. *Journal of Materials Chemistry A*, *4*, 13125–13132.

Leyden, M. R., Ono, L. K., Raga, S. R., Kato, Y., Wang, S., & Qi, Y. (2014). High performance perovskite solar cells by hybrid chemical vapor deposition. *Journal of Materials Chemistry A*, *2*, 18742–18745.

Li, C. Z., Chueh, C. C., Yip, H. L., Ding, F., Li, X., & Jen, A. K. Y. (2013). Solution-processible highly conducting fullerenes. *Advanced Materials*, *25*, 2457–2461.

Li, J.-F., Zhang, Z.-L., Gao, H.-P., Zhang, Y., & Mao, Y.-L. (2015). Effect of solvents on the growth of TiO_2 nanorods and their perovskite solar cells. *Journal of Materials Chemistry A*, *3*, 19476–19482.

Liang, L., Huang, Z., Cai, L., Chen, W., Wang, B., Chen, K., ... Fan, B. (2014). Magnetron sputtered zinc oxide nanorods as thickness-insensitive cathode interlayer for perovskite planar-heterojunction solar cells. *ACS Applied Materials & Interfaces*, *6*, 20585–20589.

Lim, K.-G., Kim, H.-B., Jeong, J., Kim, H., Kim, J.Y., & Lee, T.-W. (2014). Boosting the power conversion efficiency of perovskite solar cells using self-organized polymeric hole extraction layers with high work function. 26, 6461–6466.

Ling, X., Yuan, J., Liu, D., Wang, Y., Zhang, Y., Chen, S., ... Ma, W. (2017). Room-temperature processed Nb_2O_5 as the electron-transporting layer for efficient planar perovskite solar cells. *ACS Applied Materials & Interfaces*, *9*, 23181–23188.

Liu, C., Wang, K., Du, P., Meng, T., Yu, X., Cheng, S. Z., & Gong, X. (2015). High performance planar heterojunction perovskite solar cells with fullerene derivatives as the electron transport layer. *ACS Applied Materials & Interfaces*, *7*, 1153–1159.

Liu, M., Johnston, M. B., & Snaith, H. J. (2013). Efficient planar heterojunction perovskite solar cells by vapour deposition. *Nature*, *501*, 395–398.

Luo, D., Yang, W., Wang, Z., Sadhanala, A., Hu, Q., Su, R., … Xu, Z. (2018). Enhanced photovoltage for inverted planar heterojunction perovskite solar cells. *Science (New York, N.Y.)*, *360*, 1442–1446.

Mahjabin, S., Mahfuzul Haque, M., Khan, S., Selvanathan, V., Jamal, M. S., Bashar, M. S., … Akhtaruzzaman, M. (2021). Effects of oxygen concentration variation on the structural and optical properties of reactive sputtered WOx thin film. *Solar Energy*, *222*, 202–211.

Mahmood, K., Khalid, A., Ahmad, S. W., & Mehran, M. T. (2018). Indium-doped ZnO mesoporous nanofibers as efficient electron transporting materials for perovskite solar cells. *Surface and Coatings Technology*, *352*, 231–237.

Mahmood, K., Swain, B. S., & Amassian, A. (2014). Double-layered ZnO nanostructures for efficient perovskite solar cells. *Nanoscale*, *6*, 14674–14678.

Mahmood, K., Swain, B. S., & Jung, H. S. (2014). Controlling the surface nanostructure of ZnO and Al-doped ZnO thin films using electrostatic spraying for their application in 12% efficient perovskite solar cells. *Nanoscale*, *6*, 9127–9138.

Mahmood, K., Sarwar, S., & Mehran, M. T. (2017). Current status of electron transport layers in perovskite solar cells: Materials and properties. *RSC Advances*, *7*, 17044–17062.

Mahmud Hasan, A. K., Raifuku, I., Amin, N., Ishikawa, Y., Sarkar, D. K., Sobayel, K., … Akhtaruzzaman, M. (2020). Air-stable perovskite photovoltaic cells with low temperature deposited NiOx as an efficient hole-transporting material. *Optics Materials Express*, *10*, 1801–1816.

Malinkiewicz, O., Yella, A., Lee, Y. H., Espallargas, G. M., Graetzel, M., Nazeeruddin, M. K., & Bolink, H. J. (2014). Perovskite solar cells employing organic charge-transport layers. *Nature Photonics*, *8*, 128–132.

Marinova, N., Valero, S., & Delgado, J. L. (2017). Organic and perovskite solar cells: Working principles, materials and interfaces. *Journal of Colloid and Interface Science*, *488*, 373–389.

Maruyama, T., & Arai, S. (1992). Titanium dioxide thin films prepared by chemical vapor deposition. *Solar Energy Materials and Solar Cells*, *26*, 323–329.

Md, S., Kohei, Y., Yoshikazu, F., Takayuki, K., Kohshin, T., & Tetsuya, T. (2016). Shape-controlled CH_3 NH_3 PbI_3 nanoparticles for planar heterojunction perovskite solar cells. *Japanese Journal of Applied Physics*, *55*, 02BF05.

Md, S., Yoshikazu, F., Kohei, Y., Kyosuke, Y., Yosuke, A., Michinori, K., … Tetsuya, T. (2018). Influence of coating steps of perovskite on low-temperature amorphous compact TiO x upon the morphology, crystallinity, and photovoltaic property correlation in planar perovskite solar cells. *Japanese Journal of Applied Physics*, *57*, 03EJ06.

Mejía Escobar, M. A., Pathak, S., Liu, J., Snaith, H. J., & Jaramillo, F. (2017). ZrO_2/TiO_2 electron collection layer for efficient meso-superstructured hybrid perovskite solar cells. *ACS Applied Materials & Interfaces*, *9*, 2342–2349.

Meng, L., You, J., & Yang, Y. (2018). Addressing the stability issue of perovskite solar cells for commercial applications. *Nature Communications*, *9*, 5265.

Mohammadian-Sarcheshmeh, H., & Mazloum-Ardakani, M. (2018). Recent advancements in compact layer development for perovskite solar cells. *Heliyon*, *4*, e00912.

Murase, S., & Yang, Y. (2012). Solution processed MoO_3 interfacial layer for organic photovoltaics prepared by a facile synthesis method. *Advanced materials*, *24*, 2459–2462.

Nickel, N.H., Lang, F., Brus, V.V., Shargaieva, O., & Rappich, J. (2017). Unraveling the light-induced degradation mechanisms of $CH_3NH_3PbI_3$ perovskite films. *3*, 1700158.

Nie, W., Tsai, H., Blancon, J. C., Liu, F., Stoumpos, C. C., Traore, B., ... Tretiak, S. (2018). Critical role of interface and crystallinity on the performance and photostability of perovskite solar cell on nickel oxide. *Advanced Materials*, *30*, 1703879.

Noh, J. H., Im, S. H., Heo, J. H., Mandal, T. N., & Seok, S. I. (2013). Chemical management for colorful, efficient, and stable inorganic–organic hybrid nanostructured solar cells. *Nano Letters*, *13*, 1764–1769.

NREL. (2019). Best research-cell efficiency chart <https://www.nrel.gov/pv/cell-efficiency.html>.

Ogomi, Y., Morita, A., Tsukamoto, S., Saitho, T., Fujikawa, N., Shen, Q., ... Hayase, S. (2014). CH3NH3SnxPb(1−x)I_3 perovskite solar cells covering up to 1060 nm. *The Journal of Physical Chemistry Letters*, *5*, 1004–1011.

Ou, Q.-D., Li, C., Wang, Q.-K., Li, Y.-Q., Tang, J.-X. (2017). Recent advances in energetics of metal halide perovskite interfaces. *4*, 1600694.

Ozin, G.A. (1970). The single crystal Raman spectrum of orthorhombic PbCl2. *48*, 2931–2933.

Park, J.H., Seo, J., Park, S., Shin, S.S., Kim, Y.C., Jeon, N.J. ... Seok, S.I. (2015). Efficient CH3NH3PbI3 perovskite solar cells employing nanostructured p-Type NiO electrode formed by a pulsed laser deposition. *27*, 4013–4019.

Park, J. H., Seo, J., Park, S., Shin, S. S., Kim, Y. C., Jeon, N. J., ... Yoon, S. C. (2015). Efficient $CH_3NH_3PbI_3$ perovskite solar cells employing nanostructured p-type NiO electrode formed by a pulsed laser deposition. *Advanced Materials*, *27*, 4013–4019.

Park, N.-G. (2015). Perovskite solar cells: An emerging photovoltaic technology. *Materials Today*, *18*, 65–72.

Park, N.-G., Miyasaka, T., & Grätzel, M. (2016). *Organic-inorganic halide perovskite photovoltaics*. Cham: Springer.

Park, N.-G., & Segawa, H. (2018). Research direction toward theoretical efficiency in perovskite solar cells. *ACS Photonics*, *5*, 2970–2977.

Robertson, J., & Falabretti, B. (2006). Band offsets of high K gate oxides on III–V semiconductors. *Journal of Applied Physics*, *100*, 014111.

Rong, Y., Hu, Y., Mei, A., Tan, H., Saidaminov, M.I., Seok, S.I., ... Han, H. (2018). Challenges for commercializing perovskite solar cells. 361, eaat8235.

Salazar, R., Altomare, M., Lee, K., Tripathy, J., Kirchgeorg, R., Nguyen, N. T., ... Schmuki, P. (2015). Use of anodic TiO_2 nanotube layers as mesoporous scaffolds for fabricating $CH_3NH_3PbI_3$ perovskite-based solid-state solar cells. *ChemElectroChem*, *2*, 824–828.

Salhi, B., Wudil, Y. S., Hossain, M. K., Al-Ahmed, A., & Al-Sulaiman, F. A. (2018). Review of recent developments and persistent challenges in stability of perovskite solar cells. *Renewable and Sustainable Energy Reviews*, *90*, 210–222.

Saliba, M., Matsui, T., Seo, J.-Y., Domanski, K., Correa-Baena, J.-P., Nazeeruddin, M. K., ... Grätzel, M. (2016). Cesium-containing triple cation perovskite solar cells: Improved stability, reproducibility and high efficiency. *Energy & Environmental Science*, *9*, 1989–1997.

Salim, T., Sun, S., Abe, Y., Krishna, A., Grimsdale, A. C., & Lam, Y. M. (2015). Perovskite-based solar cells: Impact of morphology and device architecture on device performance. *Journal of Materials Chemistry A*, *3*, 8943–8969.

Samiul Islam, M., Sobayel, K., Al-Kahtani, A., Islam, M. A., Muhammad, G., Amin, N., ... Akhtaruzzaman, M. (2021). Defect study and modelling of SnX_3-based perovskite solar cells with SCAPS-1D. *11*, 1218.

Selvanathan, V., Yahya, R., Shahiduzzaman, M., Ruslan, M. H., Muhammad, G., Amin, N., & Akhtaruzzaman, M. (2021). Ionic liquid infused starch-cellulose derivative based quasi-solid dye-sensitized solar cell: Exploiting the rheological properties of natural polymers. *Cellulose*, *28*, 5545–5557.

Seo, S., Park, I. J., Kim, M., Lee, S., Bae, C., Jung, H. S., ... Shin, H. (2016). An ultra-thin, un-doped NiO hole transporting layer of highly efficient (16.4%) organic–inorganic hybrid perovskite solar cells. *Nanoscale*, *8*, 11403–11412.

Sha, W.E.I., Ren, X., Chen, L., & Choy, W.C.H. (2015). The efficiency limit of CH3NH3PbI3 perovskite solar cells. *106*, 221104.

Shahiduzzaman, M., Ashikawa, H., Kuniyoshi, M., Visal, S., Sakakibara, S., Kaneko, T., ... Tomita, K. (2018). Compact TiO_2/anatase TiO_2 single-crystalline nanoparticle electron-transport bilayer for efficient planar perovskite solar cells. *ACS Sustainable Chemistry & Engineering*, *6*, 12070–12078.

Shahiduzzaman, M., Fukaya, S., Muslih, E.Y., Wang, L., Nakano, M., Akhtaruzzaman, M., ... Taima, T. (2020). Metal oxide compact electron transport layer modification for efficient and stable perovskite solar cells. *13*, 2207.

Shahiduzzaman, M., Hamada, K., Yamamoto, K., Nakano, M., Karakawa, M., Takahashi, K., & Taima, T. (2019). Thermal control of PbI_2 film growth for two-step planar perovskite solar cells. *Crystal Growth & Design*, *19*, 5320–5325.

Shahiduzzaman, M., Hossain, M. I., Visal, S., Kaneko, T., Qarony, W., Umezu, S., ... Isomura, M. (2021). Spray pyrolyzed TiO_2 embedded multi-layer front contact design for high-efficiency perovskite solar cells. *Nano-Micro Letters*, *13*, 36.

Shahiduzzaman, M., Karakawa, M., Yamamoto, K., Kusumi, T., Yonezawa, K., Kuwabara, T., ... Taima, T. (2018). Interface engineering of compact-TiOx in planar perovskite solar cells using low-temperature processable high-mobility fullerene derivative. *Solar Energy Materials and Solar Cells*, *178*, 1–7.

Shahiduzzaman, M., Kulkarni, A., Visal, S., Wang, L., Nakano, M., Karakawa, M., ... Taima, T. (2020). Single-phase brookite TiO_2 Nanoparticle bridge enhances the stability of perovskite solar cells. *Sustainable Energy & Fuels*.

Shahiduzzaman, M., Kuwahara, D., Nakano, M., Karakawa, M., Takahashi, K., Nunzi, J.-M., & Taima, T., (2020). Low-temperature processed TiOx electron transport layer for efficient planar perovskite solar cells. *10*, 1676.

Shahiduzzaman, M., Muslih, E. Y., Hasan, A. K. M., Wang, L., Fukaya, S., Nakano, M., ... Taima, T. (2021). The benefits of ionic liquids for the fabrication of efficient and stable perovskite photovoltaics. *Chemical Engineering Journal*, 128461.

Shahiduzzaman, M., Sakuma, T., Kaneko, T., Tomita, K., Isomura, M., Taima, T., ... Iwamori, S. (2019). Oblique electrostatic inkjet-deposited TiO_2 electron transport layers for efficient planar perovskite solar cells. *Scientific Reports*, *9*, 19494.

Shahiduzzaman, M., Visal, S., Kuniyoshi, M., Kaneko, T., Umezu, S., Katsumata, T., ... Tomita, K. (2019). Low-temperature-processed brookite-based TiO_2 heterophase junction enhances performance of planar perovskite solar cells. *Nano Letters*, *19*, 598–604.

Shahiduzzaman, M., Wang, L., Fukaya, S., Muslih, E. Y., Kogo, A., Nakano, M., ... Taima, T. (2021). Ionic liquid-assisted $MAPbI_3$ nanoparticle-seeded growth for efficient and stable perovskite solar cells. *ACS Applied Materials & Interfaces*, *13*, 21194–21206.

Shahiduzzaman, M., Yamamoto, K., Furumoto, Y., Kuwabara, T., Takahashi, K., & Taima, T. (2015a). Enhanced photovoltaic performance of perovskite solar cells via modification of surface characteristics using a fullerene interlayer. *Chemistry Letters*, *44*, 1735–1737.

Shahiduzzaman, M., Yamamoto, K., Furumoto, Y., Kuwabara, T., Takahashi, K., & Taima, T. (2015b). Ionic liquid-assisted growth of methylammonium lead iodide spherical nanoparticles by a simple spin-coating method and photovoltaic properties of perovskite solar cells. *RSC Advances*, *5*, 77495–77500.

Shahiduzzaman, M., Yamamoto, K., Furumoto, Y., Yonezawa, K., Hamada, K., Kuroda, K., ... Taima, T. (2017). Viscosity effect of ionic liquid-assisted controlled growth of $CH_3NH_3PbI_3$ nanoparticle-based planar perovskite solar cells. *Organic Electronics*, *48*, 147–153.

Shahiduzzaman, M., Yonezawa, K., Yamamoto, K., Ripolles, T. S., Karakawa, M., Kuwabara, T., ... Taima, T. (2017). Improved reproducibility and intercalation control of efficient planar inorganic perovskite solar cells by simple alternate vacuum deposition of PbI_2 and CsI. *ACS Omega*, *2*, 4464–4469.

Shao, Y., Yuan, Y., & Huang, J. (2016). Correlation of energy disorder and open-circuit voltage in hybrid perovskite solar cells. *Nature Energy*, *1*, 15001.

Shi, S., Li, Y., Li, X., & Wang, H.J.M.H. (2015). Advancements in all-solid-state hybrid solar cells based on organometal halide perovskites. *2*, 378–405.

Shimada, K., Shahiduzzaman, M., & Taima, T. (2020). Platinum leaf counter electrodes for dye-sensitized solar cells. *Japanese Journal of Applied Physics*, *59*, SDDC07.

Shin, S.S., Lee, S.J., & Seok, S.I. (2019). Metal oxide charge transport layers for efficient and stable perovskite solar cells. *29*, 1900455.

Singh, T., Öz, S., Sasinska, A., Frohnhoven, R., Mathur, S., & Miyasaka, T. Sulfate-assisted interfacial engineering for high yield and efficiency of triple cation perovskite solar cells with alkali-doped TiO_2 electron-transporting layers, advanced functional materials. *0*, 1706287.

Singh, T., Singh, J., & Miyasaka, T. (2016). Role of metal oxide electron-transport layer modification on the stability of high performing perovskite solar cells. *9*, 2559–2566.

Son, D.-Y., Im, J.-H., Kim, H.-S., & Park, N.-G. (2014). 11% efficient perovskite solar cell based on ZnO nanorods: An effective charge collection system. *The Journal of Physical Chemistry C*, *118*, 16567–16573.

Song, J., Zheng, E., Bian, J., Wang, X.-F., Tian, W., Sanehira, Y., & Miyasaka, T. (2015). Low-temperature SnO_2-based electron selective contact for efficient and stable perovskite solar cells. *Journal of Materials Chemistry A*, *3*, 10837–10844.

Song, Z., Watthage, S., Phillips, A., & Heben, M. (2016). Pathways toward high-performance perovskite solar cells: Review of recent advances in organo-metal halide perovskites for photovoltaic applications. *Journal of Photonics for Energy*, 022001.

Stranks, S.D., Eperon, G.E., Grancini, G., Menelaou, C., Alcocer, M.J.P., Leijtens, T., ... Snaith, H.J. (2013). Electron-hole diffusion lengths exceeding 1 micrometer in an organometal trihalide perovskite absorber, *342*, 341–344.

Su, T.-S., Hsieh, T.-Y., Hong, C.-Y., & Wei, T.-C. (2015). Electrodeposited ultrathin TiO_2 blocking layers for efficient perovskite solar cells. *Scientific Reports*, *5*, 16098.

Sun, S., Salim, T., Mathews, N., Duchamp, M., Boothroyd, C., Xing, G., ... Lam, Y. M. (2014). The origin of high efficiency in low-temperature solution-processable bilayer organometal halide hybrid solar cells. *Energy & Environmental Science*, *7*, 399–407.

Sun, W., Li, Y., Ye, S., Rao, H., Yan, W., Peng, H., ... Chen, Z. (2016). High-performance inverted planar heterojunction perovskite solar cells based on a solution-processed CuO x hole transport layer. *Nanoscale*, *8*, 10806–10813.

Sun, X., Ji, L., Chen, W., Guo, X., Wang, H., Lei, M., ... Li, Y. (2017). Halide anion–fullerene π noncovalent interactions: N-doping and a halide anion migration mechanism in p–i–n perovskite solar cells. *Journal of Materials Chemistry A*, *5*, 20720–20728.

Tablero Crespo, C. (2019). The effect of the halide anion on the optical properties of lead halide perovskites. *Solar Energy Materials and Solar Cells*, *195*, 269–273.

Taima, T., Shahiduzzaman, M., Yamamoto, K., Furumoto, Y., Kuwabara, T., & Takahashi, K. (2016). Planar heterojunction type perovskite solar cells based on TiOx compact layer fabricated by chemical bath deposition. SPIE.

Tang, Z.-K., Xu, Z.-F., Zhang, D.-Y., Hu, S.-X., Lau, W.-M., & Liu, L.-M. (2017). Enhanced optical absorption via cation doping hybrid lead iodine perovskites. *Scientific Reports*, *7*, 7843.

Tavakoli, M. M., Gu, L., Gao, Y., Reckmeier, C., He, J., Rogach, A. L., ... Fan, Z. (2015). Fabrication of efficient planar perovskite solar cells using a one-step chemical vapor deposition method. *Scientific Reports*, *5*, 14083.

Wang, J. T.-W., Ball, J. M., Barea, E. M., Abate, A., Alexander-Webber, J. A., Huang, J., ... Nicholas, R. J. (2014). Low-temperature processed electron collection layers of graphene/TiO_2 nanocomposites in thin film perovskite solar cells. *Nano Letters*, *14*, 724–730.

Wang, J. T.-W., Ball, J. M., Barea, E. M., Abate, A., Alexander-Webber, J. A., Huang, J., ... Snaith, H. J. (2013). Low-temperature processed electron collection layers of graphene/TiO_2 nanocomposites in thin film perovskite solar cells. *Nano Letters*, *14*, 724–730.

Wang, L., Shahiduzzaman, M., Fukaya, S., Muslih, E. Y., Nakano, M., Karakawa, M., ... Taima, T. (2021). Low-cost molecular glass hole transport material for perovskite solar cells. *Japanese Journal of Applied Physics*, *60*, SBBF12.

Wang, L., Shahiduzzaman, M., Muslih, E. Y., Nakano, M., Karakawa, M., Takahashi, K., ... Taima, T. (2021). Double-layer CsI intercalation into an $MAPbI_3$ framework for efficient and stable perovskite solar cells. *Nano Energy*, *86*, 106135.

Wang, P., Zhang, X., Zhou, Y., Jiang, Q., Ye, Q., Chu, Z., ... You, J. (2018). Solvent-controlled growth of inorganic perovskite films in dry environment for efficient and stable solar cells. *Nature Communications*, *9*, 2225.

Wang, Q., Chen, B., Liu, Y., Deng, Y., Bai, Y., Dong, Q., & Huang, J. (2017). Scaling behavior of moisture-induced grain degradation in polycrystalline hybrid perovskite thin films. *Energy & Environmental Science*, *10*, 516–522.

Wang, Y., Su, Y., Wang, Z., Zhang, Z., Han, X., Dong, M., ... Chen, M. (2018). Reversible conductivity recovery of highly sensitive flexible devices by water vapor. *NPJ Flexible Electronics*, *2*, 31.

Weber, D. (1978). CH3NH3PbX3, ein Pb(II)-System mit kubischer Perowskitstruktur/ CH3NH3PbX3, a Pb(II)-system with cubic perovskite structure. *Journal of Zeitschrift für Naturforschung B*, *33*, 1443–1445.

Wehrenfennig, C., Eperon, G. E., Johnston, M. B., Snaith, H. J., & Herz, L. M. (2014). High charge carrier mobilities and lifetimes in organolead trihalide perovskites. *Advanced Materials*, *26*, 1584–1589.

Wu, C.-G., Chiang, C.-H., Tseng, Z.-L., Nazeeruddin, M. K., Hagfeldt, A., & Grätzel, M. (2015). High efficiency stable inverted perovskite solar cells without current hysteresis. *Energy & Environmental Science*, *8*, 2725–2733.

Wu, W. Q., Huang, F., Chen, D., Cheng, Y. B., & Caruso, R. A. (2015). Thin films of dendritic anatase titania nanowires enable effective hole-blocking and efficient light-harvesting for high-performance mesoscopic perovskite solar cells. *Advanced Functional Materials*, *25*, 3264–3272.

Wu, W.-Q., Zhong, J.-X., Liao, J.-F., Zhang, C., Zhou, Y., Feng, W., ... Kuang, D.-B. (2020). Spontaneous surface/interface ligand-anchored functionalization for extremely high fill factor over 86% in perovskite solar cells. *Nano Energy*, *75*, 104929.

Xiao, L., Xu, J., Luan, J., Zhang, B., Tan, Za, Yao, J., & Dai, S. (2017). Achieving mixed halide perovskite via halogen exchange during vapor-assisted solution process for efficient and stable perovskite solar cells. *Organic Electronics*, *50*, 33–42.

Xie, J., Zhao, Ce, Lin, Zq, Gu, Py, & Zhang, Q. (2016). Nanostructured conjugated polymers for energy-related applications beyond solar cells. *Chemistry–An Asian Journal*, *11*, 1489–1511.

Xu, J., Buin, A., Ip, A. H., Li, W., Voznyy, O., Comin, R., ... McDowell, J. J. (2015). Perovskite–fullerene hybrid materials suppress hysteresis in planar diodes. *Nature Communications*, *6*, 7081.

Xu, T., Chen, L., Guo, Z., & Ma, T. (2016). Strategic improvement of the long-term stability of perovskite materials and perovskite solar cells. *Physical Chemistry Chemical Physics*, *18*, 27026–27050.

Yamamoto, K., Furumoto, Y., Shahiduzzaman, M., Kuwabara, T., Takahashi, K., & Taima, T. (2016). Degradation mechanism for planar heterojunction perovskite solar cells. *Japanese Journal of Applied Physics*, *55*, 04ES07.

Yang, T. C.-J., Fiala, P., Jeangros, Q., & Ballif, C. (2018). High-bandgap perovskite materials for multijunction solar cells. *Joule*, *2*, 1421–1436.

Yang, W. S., Noh, J. H., Jeon, N. J., Kim, Y. C., Ryu, S., Seo, J., & Seok, S. I. (2015). High-performance photovoltaic perovskite layers fabricated through intramolecular exchange. *Science (New York, N.Y.)*, *348*, 1234–1237.

Yang, W. S., Park, B.-W., Jung, E. H., Jeon, N. J., Kim, Y. C., Lee, D. U., ... Seok, S. I. (2017). Iodide management in formamidinium-lead-halide–based perovskite layers for efficient solar cells. *Science (New York, N.Y.)*, *356*, 1376–1379.

Ye, S., Sun, W., Li, Y., Yan, W., Peng, H., Bian, Z., ... Huang, C. (2015). CuSCN-based inverted planar perovskite solar cell with an average PCE of 15.6%. *Nano Letters*, *15*, 3723–3728.

Yella, A., Heiniger, L.-P., Gao, P., Nazeeruddin, M. K., & Grätzel, M. (2014). Nanocrystalline rutile electron extraction layer enables low-temperature solution processed perovskite photovoltaics with 13.7% efficiency. *Nano Letters*, *14*, 2591–2596.

Yin, X., Chen, P., Que, M., Xing, Y., Que, W., Niu, C., & Shao, J. (2016). Highly efficient flexible perovskite solar cells using solution-derived NiOx hole contacts. *ACS Nano*, *10*, 3630–3636.

Yin, X., Yao, Z., Luo, Q., Dai, X., Zhou, Y., Zhang, Y., ... Wang, N. (2017). High efficiency inverted planar perovskite solar cells with solution-processed NiOx hole contact. *ACS Applied Materials & Interfaces*, *9*, 2439–2448.

Yongzhen, W., Xudong, Y., Han, C., Kun, Z., Chuanjiang, Q., Jian, L., ... Liyuan, H. (2014). Highly compact TiO_2 layer for efficient hole-blocking in perovskite solar cells. *Applied Physics Express*, *7*, 052301.

Yoo, J. J., Wieghold, S., Sponseller, M. C., Chua, M. R., Bertram, S. N., Hartono, N. T. P., ... Bawendi, M. G. (2019). An interface stabilized perovskite solar cell with high stabilized efficiency and low voltage loss. *Energy & Environmental Science*, *12*, 2192–2199.

You, J., Meng, L., Song, T.-B., Guo, T.-F., Yang, Y., Chang, W.-H., ... Yang, Y. (2016). Improved air stability of perovskite solar cells via solution-processed metal oxide transport layers. *Nature Nanotechnology*, *11*, 75–81.

Zhang, M., Lyu, M., Yun, J.-H., Noori, M., Zhou, X., Cooling, N. A., ... Wang, L. (2016). Low-temperature processed solar cells with formamidinium tin halide perovskite/fullerene heterojunctions. *Nano Research*, *9*, 1570–1577.

Zhang, M., Yun, J. S., Ma, Q., Zheng, J., Lau, C. F. J., Deng, X., ... Ho-Baillie, A. W. Y. (2017). High-efficiency rubidium-incorporated perovskite solar cells by gas quenching. *ACS Energy Letters*, *2*, 438–444.

Zhang, Q., Dandeneau, C. S., Zhou, X., & Cao, G. (2009). ZnO nanostructures for dye-sensitized solar cells. *Advanced Materials*, *21*, 4087–4108.

Zhang, Y., Grancini, G., Feng, Y., Asiri, A. M., & Nazeeruddin, M. K. (2017). Optimization of stable quasi-cubic $FAxMA_1-xPbI_3$ perovskite structure for solar cells with efficiency beyond 20%. *ACS Energy Letters*, *2*, 802–806.

Zhao, D., Sexton, M., Park, H.-Y., Baure, G., Nino, J.C., So, F. (2015). High-efficiency solution-processed planar perovskite solar cells with a polymer hole transport layer. *5*, 1401855.

Zhao, D., Yu, Y., Wang, C., Liao, W., Shrestha, N., Grice, C. R., ... Yan, Y. (2017). Low-bandgap mixed tin–lead iodide perovskite absorbers with long carrier lifetimes for all-perovskite tandem solar cells. *Nature Energy*, *2*, 17018.

Zhou, H., Chen, Q., Li, G., Luo, S., Song, T.-B., Duan, H.-S., ... Yang, Y. (2014). Interface engineering of highly efficient perovskite solar cells, *345*, 542–546.

Zhu, H. L., & Choy, W. C. (2018). Crystallization, properties, and challenges of low-bandgap Sn–Pb binary perovskites. *Solar RRL*, *2*, 1800146.

Zhu, Z., Bai, Y., Liu, X., Chueh, C. C., Yang, S., & Jen, A. K. Y. (2016). Enhanced efficiency and stability of inverted perovskite solar cells using highly crystalline SnO2 nanocrystals as the robust electron-transporting layer. *Advanced Materials*, *28*, 6478–6484.

Zhu, Z., Bai, Y., Zhang, T., Liu, Z., Long, X., Wei, Z., ... Yan, F. (2014). High-performance hole-extraction layer of sol–gel-processed NiO nanocrystals for inverted planar perovskite solar cells. *Angewandte Chemie*, *126*, 12779–12783.

SUBCHAPTER

Optics in high efficiency perovskite tandem solar cells 4.4

Mohammad Ismail Hossain[1,2,3], Wayesh Qarony[2,4,5], Md. Shahiduzzaman[6], Md. Akhtaruzzaman[7], Yuen Hong Tsang[2] and Dietmar Knipp[8]

[1]*Department of Materials Sciences and Engineering, City University of Hong Kong, Kowloon, Hong Kong, P.R. China*

[2]*Department of Applied Physics, Hong Kong Polytechnic University, Kowloon, Hong Kong, P.R. China*

[3]*Department of Electrical and Computer Engineering, University of California, Davis, CA, United States*

[4]*Materials Sciences Division, Lawrence Berkeley National Laboratory, Berkeley, CA, United States*

[5]*Department of Electrical Engineering and Computer Sciences, University of California Berkeley, CA, United States*

[6]*Nanomaterials Research Institute, Kanazawa University, Kanazawa, Japan*

[7]*Solar Energy Research Institute, Universiti Kebangsaan Malaysia, Bangi, Malaysia*

[8]*Geballe Laboratory for Advanced Materials, Department of Materials Science and Engineering, Stanford University, Stanford, CA, United States*

4.4.1 Introduction

A solar cell (SC) or photovoltaic (PV) is currently considered as a promising candidate to satisfy the world's energy crisis since it directly converts sunlight into electricity. Over the last decade, manufacturing, installation, and maintenance costs of SCs have been dramatically decreasing compared to all other energy resources (Alharbi & Kais, 2015; Duran Sahin, Dincer, & Rosen, 2007; Rau, Paetzold, & Kirchartz, 2014). Due to the inadequate energy conversion efficiency (ECE), PV technology has not yet been extensively utilized as a primary source in the electrical sector (Hossain, Qarony et al., 2019; Rau et al., 2014). Literature review reveals that several fundamental factors are responsible for restricting the ECE of SC, such as insufficient photon absorptions, low short-circuit current density (J_{SC}), low open-circuit voltage (V_{OC}), poor fill-factor (FF), device stability, etc. (Futscher & Ehrler, 2016; Honsberg, Corkish, & Bremner, 2001). The photon absorption ability, including ECE, entirely depends on the active material used to design SCs. Various materials (e.g., Si, GaAs, organic materials, CdTe, perovskite, InGaN) have been used in PV technology (Chen et al., 2019; Hossain et al.,

Comprehensive Guide on Organic and Inorganic Solar Cells. DOI: https://doi.org/10.1016/B978-0-323-85529-7.00013-X

2017; Hossain, Hasan et al., 2020; Jošt et al., 2018; Lee & Ebong, 2017; Qarony & Hossain, 2015; Qarony et al., 2015; Tamang et al., 2016). It is noted that PV materials absorb sunlight mostly in the visible and a fraction of the near-infrared regions, where photons beyond the infrared part solely contribute to heat energy generation. The standard AM 1.5G reference solar spectra (ASTM G173–03) is shown in Fig. 4.4.1A (Reference Solar Spectral Irradiance, n.d.). The present PV market is predominantly conquered by single-junction silicon (Si) SCs because of their earth abundancy and comparative higher ECE (Honsberg et al., 2001; Lee & Ebong, 2017). The laboratory crystalline silicon (c-Si) SC has already crossed 26% ECE; however, the ECE is still lower than their theoretical upper limit (~33%) as seen from the Shockley-Queisser (SQ) limit illustrated in Fig. 4.4.1B (Green et al., 2020; Shockley & Queisser, 1961). Based on their progress throughout the last decades, it can be considered that there is a tiny scope to enhance the ECE of single-junction Si SCs (Green et al., 2020; NREL Transforming Energy, 2020). So far, several light trapping strategies have been anticipated and demonstrated (e.g., nanostructure, plasmonic design, photonic crystals, interface engineering, etc.) to overcome the ECE limitation (Atwater & Polman, 2010; Haque et al., 2019, 2020; He, Zheng, Wang, Lin, & Lin, 2014; Hossain, Hongsingthong et al., 2019; Hossain, Hasan et al., 2020; Tamang et al., 2016; Wang et al., 2014). Nevertheless, researchers are, until now, far from reaching the SQ limit (Shockley & Queisser, 1961). So far, SC with tandem configuration is considered to be the most encouraging approach that can reach high ECEs (De Vos, 1980; Leijtens, Bush, Prasanna, & McGehee, 2018; Li & Zhang, 2020). Detailed balance calculations show that the ECE of a tandem SC (TSC) may go beyond 45% if optimal material bandgaps (Egs) are selected (De Vos, 1980; Hossain, Qarony et al., 2019; Shockley & Queisser, 1961) for the top and bottom cells. The ECE of the

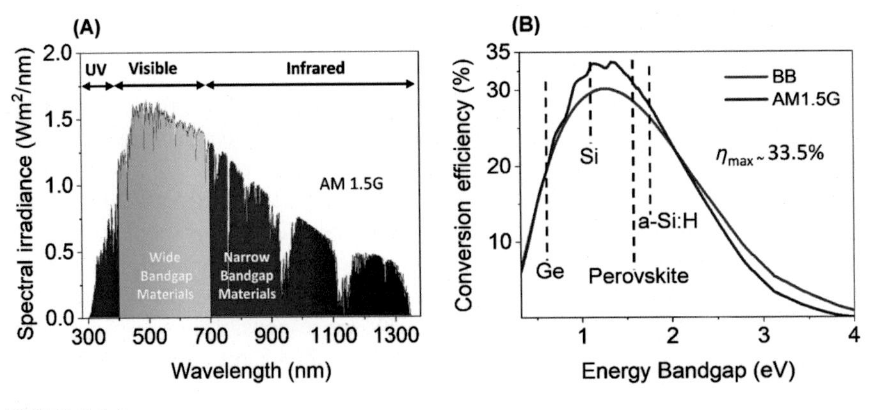

FIGURE 4.4.1

(A) The spectral density for standard solar reference spectra (AM 1.5G). (B) Detailed Balance (SQ) limit for the energy conversion efficiency of single-junction solar cells under AM 1.5G spectrum with blackbody (BB) spectrum at 6000K.

TSC is maximized only if the top and bottom cell *Eg* are in a range of 1.7 ~ 1.8 eV and 0.85 ~ 1.2 eV, respectively (Hossain, Qarony et al., 2019; Shockley & Queisser, 1961; Hossain, Shahiduzzaman, Saleque et al., 2021).

Hence, several material combinations have been studied to realize efficient TSCs. However, so far, a crystalline silicon (c-Si) (*Eg* of 1.12 eV) bottom cell is considered as a promising candidate for TSC development (Bush et al., 2017; Hossain, Qarony et al., 2019; Li & Zhang, 2020; Sahli et al., 2018). Several aspects have been considered while fabricating TSCs with the combination of c-Si and other materials. According to the detailed balance theory, amorphous silicon (a-Si:H) exhibits almost ideal *Eg* (~1.7 eV) as a top cell with the c-Si SC used as a bottom cell. Nevertheless, the high defect density of a-Si:H prevents realizing high V_{OC}, which is a prerequisite for efficient TSCs (Qarony et al., 2017; Rech & Wagner, 1999). Additionally, most material deposition involves high-temperature sintering that limits the incorporation while preparing the top cell on a processed bottom c-Si SC. Recently, perovskite material systems have attracted considerable attention to the PV community due to their advancement in ECEs (Green, Ho-Baillie, & Snaith, 2014; Green et al., 2020; Hossain, Hongsingthong et al., 2019; NREL Transforming Energy, 2020; Park, 2015; Rong et al., 2018; Zuo et al., 2016). The record ECE of single-junction perovskite SCs (PSCs) has been increased from 3.8% to over 25% within a few years (Haque et al., 2019; Hossain, Hasan et al., 2020; Kojima, Teshima, Shirai, & Miyasaka, 2009; NREL Transforming Energy, 2020). Over the few years, perovskite materials exhibited outstanding results because of their excellent optical and electronic properties, such as considerable diffusion length, high absorption coefficient, direct and tunable bandgap, and low-cost processing (Hossain, Hongsingthong et al., 2019; Hossain, Hasan et al., 2020; Hossain et al., 2021; Köhnen et al., 2019; Rong et al., 2018). Moreover, multibandgap property and low-temperature deposition methods of perovskite materials allow them to utilize the potential top cell on a bottom SC while fabricating TSCs (Bush et al., 2017; Green et al., 2014; Hossain, Qarony, Jovanov, Tsang, & Knipp, 2018; Hossain, Qarony et al., 2019; Sahli et al., 2018; Shahiduzzaman et al., 2018). To date, promising results have already been demonstrated for perovskite TSCs by several research groups (Al-Ashouri et al., 2020; Bush et al., 2017; Jošt, Kegelmann, Korte, & Albrecht, 2020; Leijtens et al., 2018; Li & Zhang, 2020; NREL Transforming Energy, 2020; Sahli et al., 2018). Perovskite/silicon (PVK/Si) TSCs with ECEs exceeding 23% was presented by Bush et al., while Oxford PV and Albrecht's research group have even surpassed the record ECE of single-junction c-Si SCs (Al-Ashouri et al., 2020; Bush et al., 2017; Green et al., 2020; Köhnen et al., 2019; NREL Transforming Energy, 2020). It is expected that the ECE of PVK/Si TSCs will exceed 30% very soon. On the other hand, perovskite/perovskite (PVK/PVK) or all-perovskite TSCs are substitutes for realizing cost-effective devices with high ECEs due to the primarily flexible, lightweight, and low-temperature monolithic fabrication process (Hossain et al., 2021; Jošt et al., 2020; Lin et al., 2019; Rajagopal et al., 2017; Zhao et al., 2018). This study only focuses on the PVK/Si

TSCs, which is still a new research area. The record ECEs of such PVK/Si TSCs can only be achieved if both subcells operate near the theoretical limit. Hence, a detailed understanding of device design, fabrication process, electronic properties, and underlying optics of such devices are essential for attaining TSCs with high ECEs. Most research in the literature emphasizes the fabrication process of PVK/Si TSCs and their electronic properties; only a few works have been illustrated the optics of SCs (Bush et al., 2017; Hossain et al., 2018; Leijtens et al., 2018; Oxford PV, 2020; Sahli et al., 2018). Furthermore, the importance of interface morphology while placing the top SC on a textured bottom SC and its influence on the device performance has not been fully explained.

In this study, a detailed investigation of PVK/Si TSCs has been provided high efficiency. In the first part (Section 4.4.2), a basic understanding of TSCs and their theoretical upper limits will be discussed. The theoretical upper ECE limits are calculated by considering realistic assumptions and well-established DB theory or SQ limit. As fabricating such TSCs is challenging, which requires several issues, Section 4.4.3 provides scientific challenges for realizing high-efficiency PVK/Si tandem devices. The PVK/Si TSC is different from the single-junction PSC, and it requires several modifications while depositing PSC on top of the bottom cells. Hence, discussions on various optimized device structures by considering realistic device interface morphologies are presented in Section 4.4.4. Section 4.4.5 broadly describes how optics and nanophotonics can be used to realize TSCs with better PV parameters. Finally, Section 4.4.6 provides a brief explanation of TSC fabrication techniques for determining high ECEs and their present status before delivering final remarks and prospects in Section 4.4.7.

4.4.2 Fundamentals of tandem solar cells

4.4.2.1 Tandem solar cells and working mechanism

Generally, PV is an electronic device that converts sunlight into electricity due to electron-hole pair (EHP) generation and separation. Most high-efficiency PV devices consider a p-i-n structure, where an absorber material is sandwiched between two doped (transport) layers that contribute to the electron and hole collections (Alharbi & Kais, 2015; Limpert, Bremner, & Linke, 2015). Fig. 4.4.2A shows a conventional single-junction SC structure, where EHP generates in the intrinsic layer later goes to electron transporting/hole blocking layer (ETL) and hole transporting/electron blocking layer (HTL). The placement of such ETL and HTL may vary on the SC design. In a single-junction SC, emitted photons from the sunlight are absorbed by the SC only if (A) photon energies are higher than that of the absorber's Eg and (B) incident photons have at least equal energies to the Eg of the absorber, where rest of photons are transmitted through the SC or gets reflected by the device (Cuevas & Yan, 2013; Limpert et al., 2015; Wurfel, Cuevas, & Wurfel, 2015). Absorbed photons develop an excited electronic state

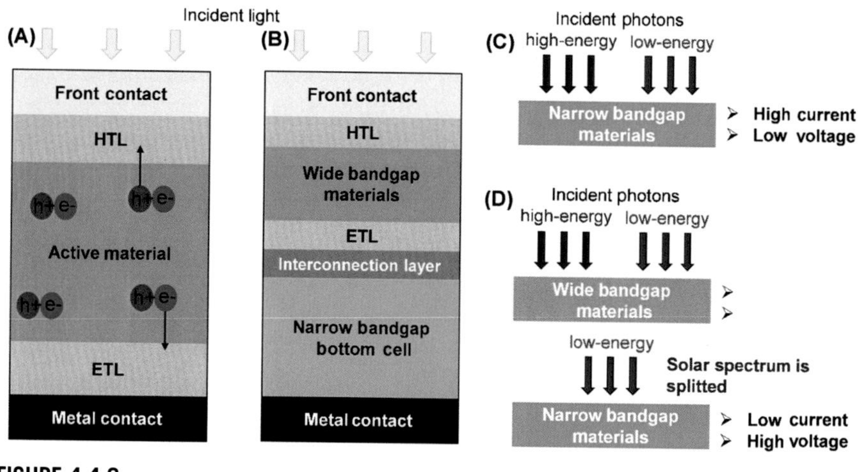

FIGURE 4.4.2

A schematic design of (A) the heterojunction thin-film solar cell, (B) the monolithic tandem solar cell. Arrows show the directions of electron and hole transportations. Schematic illustration of the working mechanism of (C) narrow bandgap solar cells, (D) tandem solar cells. Particular arrows show the photons with high and low energies.

at the absorber material, leading to the conversion from optical (electromagnetic radiation) into electrical (chemical products) by producing EHPs (Pridham, 1970; Smestad & Ries, 1992). The number of photons contributing to electrons' production to the external circuit can be calculated by considering absorption coefficient, mobility, transport parameters, and electron lifetime (Shah et al., 2004; Smestad & Ries, 1992). Moreover, Eg of the material also confirms the spectral range for the maximum photon absorption by the SC; for example, Si has Eg of 1.12 eV that allows absorbing photons up to 1100 nm. However, a single absorber is not able to utilize both high and low energy photons effectively due to (1) low energy photons than the absorber's Eg do not contribute to the EHP generations, and (2) higher energy photons exceeding the Eg are lost without contributing to the J_{SC} (Shah et al., 2004). Fig. 4.4.2C illustrates the working mechanism of a single-junction low Eg SC, where a fraction of high and low energy photons is only absorbed by the SC, resulting in low V_{OC} and high J_{SC}. The device can exhibit an opposite behavior if a high Eg absorber is selected. As a result, the upper limit of single-junction SCs is limited to only ~33.5%, as shown in Fig. 4.4.1B. For better utilization of incident photons, it is essential to consider more than one semiconductor absorber with identical Egs to guide photons to be absorbed by the device, which can be understood from the solar spectral irradiance shown in Fig. 4.4.1A.

Hence, the TSC concept is introduced, where a wide Eg and narrow Eg absorbers are stacked to form a single device, which leads to maximizing photon absorptions in the SC (Li & Zhang, 2020). In an ideal tandem device, the sunlight should first hit the top cell so that all high energy photons can absorb by the top cell and transmit low energy photons to the bottom cell. V_{OC} of a TSC is the addition of individual V_{OC} of top and bottom cells; however, J_{SC} either limited by top cell or bottom cell, resulting in an improved ECE. Henceforth, an efficient interconnection layer (IL), which connects two subcells, is required for adequate photon management. The working mechanism of TSC is schematically presented in Fig. 4.4.2D. Fundamentally, the tandem devices include two basic designs (1) 2-terminal that allows realizing monolithic deposition of a top cell on the bottom cell, and (2) 4-terminal that usually mechanically connect two separate SCs. As compared to the 4-terminal tandem device, monolithic TSC is considered an efficient design; nevertheless, it requires a matching of current from top to bottom cells.

4.4.2.2 Detailed balance theory and efficiency limit

It is essential to understand the fundamental ECE limits of SCs for realizing high-efficiency PV devices. The basic calculation of upper limits of SC efficiencies derived from the first and second thermodynamics laws (Hegedus & Luque, 2005; Kosyachenko, 2015). However, the realistic ECE limits of SCs were first described by Shockley and Queisser by applying thermodynamics to SC considered a semiconductor device (Shockley & Queisser, 1961). The model considered the Eg of PV material and possible radiative recombination. The derived limit is called a DB limit or SQ limit. In the SQ model, the following five processes have to be taken into account for determining a more realistic description of the ECE limit of SCs (Shockley & Queisser, 1961): (1) generation of EHPs through incident light, (2) radiative recombination, (3) nonradiative or thermal generation, (4) nonradiative recombination, and (5) extraction of EHPs as current flow. The SQ limit of single-junction SCs has already been shown in Fig. 4.4.1B. The DB limit is not constrained to single-junction SCs, which can be applied for determining ECE limits of TSCs (De Vos, 1980; Green, 2006). The ECE limits of 2- and 4-terminal TSCs are provided in Fig. 4.4.3A and B, respectively. In the case of a 4-terminal device, it is possible to achieve an ECE of ~40% for various combinations. Both subcells are electrically independent, where both diodes' output powers are calculated independently and later added while determining the overall ECE. The ECE of a serial-connected 2-terminal TSC is a complicated process as it mainly relies on the J_{SC} of the device. The J_{SC} is determined by the bottom cell if the bottom cell J_{SC} is lower than the J_{SC} of the top cell. The J_{SC} is equal to the top cell J_{SC}, when the bottom J_{SC} is higher than the top cell J_{SC}. If both subcells' J_{SC}s are matched, the device works at the optimal point, leading to high ECE (Hossain, Qarony et al., 2019). The matching of the right material Egs is essential to reach matched J_{SC} and high ECE; when appropriated Egs are selected and J_{SC}s are matched, the ECE of 2-terminal device will be equal to the ECE of the 4-terminal device. An upper limit of the ECE of TSCs can be determined from the formula: $E_{G\_top}=0.5 \times E_{G\_bot}+1.15$ eV.

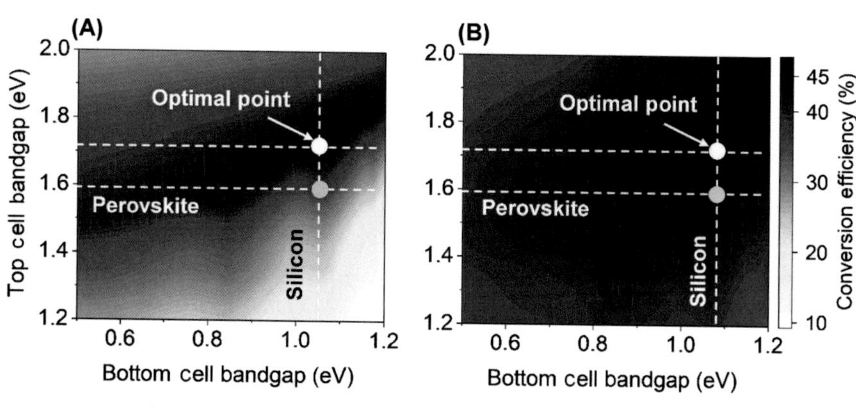

FIGURE 4.4.3

Detailed balance (SQ) limit of a (A) monolithic serial connected 2-terminal and (B) mechanically stacked 4-terminal tandem solar cells.

Reproduced with the permission from Hossain, M. I., Qarony, W., Ma, S., Zeng, L., Knipp, D., Tsang, Y. H. ... Tsang, Y. H. (2019). Perovskite/silicon tandem solar cells: From detailed balance limit calculations to photon management. Nano-Micro Letters, 11, *58. https://doi.org/10.1007/s40820–019–0287–8. Copyright 2020, Springer Nature.*

In the case of a Si bottom cell with an *Eg* of 1.12 eV, the top cell *Eg* must be ~1.73 eV for attaining maximum ECE of the TSC. The ECE of TSCs may go over 45% for the optimized condition, where the top cell bandgap must be within 1.7–1.8 eV and bottom cell *Eg* in a range of 0.9–1.2 eV. As the current study only focuses on the PVK/Si TSCs; hence, we will discuss the DB limit of PVK/Si TSCs and its implications in the following sections.

4.4.2.3 Photon management and quantum efficiency

The photon management improves the quantum efficiency (QE) by enhancing the photon absorption in the SC, which further increases the ECE. Photon management can be achieved by improving light incoupling and light trapping. According to the ray optics or geometrical limit ($A=4\alpha n^2 d$) derived by Yablonobitch et al., the upper limit of the optical path length enhancement is restricted to only $2n^2$, where n is the index of refraction, α is being absorption coefficient, and d is the thickness of the absorber material (Yablonovitch, 1982). Hence, the SC's QE is also limited to the geometrical absorption limit; nevertheless, the limit is valid if the absorber thickness is much larger than the incident wavelength. Therefore the assumptions might not fully satisfy in the case of thin-film SCs. The penetration depth of photons is usually smaller than $2d$ in the shorter wavelength region, where it is higher for the longer wavelength range. Hence, light trapping plays a vital role in the longer wavelengths, where light incoupling has to be improved for the shorter wavelengths for high QEs (Hossain, Qarony et al., 2019; Hossain, Shahiduzzaman, Ahmed et al., 2021). Both light trapping and light incoupling can be enhanced by utilizing textured interfaces that further improves the QE of the SC.

So far, several strategies have been applied to improve the QE of SCs or minimize material consumption during the device fabrication process. In the current study, the most widely used pyramid texture surface is considered to study the optics of SCs (Shahiduzzaman et al., 2021; Hossain, Shahiduzzaman, Ahmed et al., 2021). The behavior of optics in the SC can be explained by the three domains based on the texture size, as illustrated schematically in Fig. 4.4.4B (Hossain et al., 2018; Hossain, Qarony et al., 2019; Zhu, Yu, Fan, & Cui, 2010). Suppose the texture dimension is smaller than the incident wavelength. In that case, the optics can be explained by the effective medium theory, which can

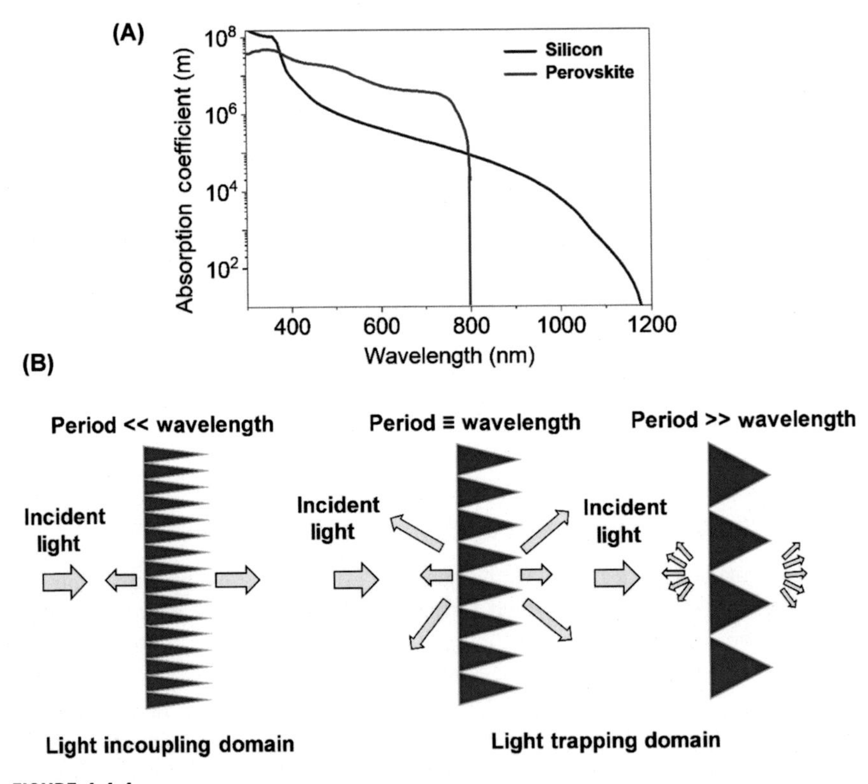

FIGURE 4.4.4

(A) Absorption coefficients of Si and $MAPbI_3$ perovskite materials. (B) Schematic illustration of optical wave propagation and photon management mechanisms in solar cells.

Reproduced with the permission from Ref. Hossain, M. I., Qarony, W., Ma, S., Zeng, L., Knipp, D., Tsang, Y. H. ... Tsang, Y. H. (2019). Perovskite/silicon tandem solar cells: From detailed balance limit calculations to photon management. Nano-Micro Letters, 11, *58. https://doi.org/10.1007/s40820-019-0287-8. Copyright 2020, Springer Nature.*

improve light incoupling and act as a broadband antireflection coating (ARC) by reducing reflections at the interface (Hossain, Qarony et al., 2019). If the texture size is equal or larger than the wavelength of the incident light, the optical wave mechanism can be derived by electromagnetic theory and geometrical optics theory through the diffraction of lights that are used to increase the optical path length and light trapping in the SC. The larger texture profile is often used in the case of silicon SCs due to their low absorption coefficient near the e.g., In this study, PVK/Si TSCs are investigated, where top PSC needs improved light incoupling because of high absorption coefficient and bottom Si SCs need improved photon absorption via light trapping. Fig. 4.4.4A shows absorption coefficients of silicon and perovskite materials. Hence, such SCs' optics are complicated as the incident light should efficiently couple to the SC and propagate to the device's back contact. As PSCs' optics is significantly different from wafer-based Si SCs, the structure must be efficiently designed so that the optical losses are minimized. The utilization of maximum photons can be achieved, leading to improved QE and J_{SC}.

4.4.3 Scientific challenges for monolithic 2-terminal tandem solar cells

This study focuses only on the 2-terminal PVK/Si TSCs, where top PSC is fabricated monolithically on top of bottom Si SCs. Such a serially connected multiabsorber device must address several vital factors while designing the SC (Jošt et al., 2020). The design challenges are briefly discussed in the following.

4.4.3.1 Front contact

The front contact of tandem plays a significant role in the device performance, which has several functions (Hossain, Hongsingthong et al., 2019; Hossain, Shahiduzzaman, Saleque et al., 2021; Hossain, Shahiduzzaman, Ahmed et al., 2021; Shahiduzzaman et al., 2021; Qarony et al., 2018). The front contact combined with the rear contact makes a junction; hence, the work function of used materials has to be selected wisely. The front contact should exhibit high lateral conductivity and better optical transparency so that incident photons can be transmitted to the SC effectively. The front contact must provide efficient coupling of light (may link to diffraction) to result in high QE, which is often attained by texturing the interface. Lastly, the absorption loss of the front contact or parasitic loss must be minimized as an increased doping concentration may increase free carrier absorption for longer wavelengths. Generally, doped metal oxides (MOs) are useful to consider as the front contact, fulfilling all necessary requirements (Hossain, Hongsingthong et al., 2019). Hence, the front contact of the TSC has to be efficient for realizing high QE and J_{SC}.

4.4.3.2 Interconnection layer

The IL is an essential parameter for efficient TSCs as it connects both subcells incorporated in the device. The IL should tunnel electrons and the holes cannot be effectively recombined; otherwise, device efficiency will suffer due to recombination losses. The IL must be highly transparent, provide good selectivity, and doped adequately to pass low energy photons to the bottom cell without pronouncing losses. Highly doped transparent conductive oxides are beneficial to act as an IL (Bush et al., 2017).

4.4.3.3 Conformal deposition of the top cell integration

Single-junction PSCs fabrication often uses the solution-processing technique, such as spin-coating, which is due to the low surface roughness of transparent conductive substrates. However, in PVK/Si TSCs, the commonly used spin-coating method might not work or require necessary modification. It depends on several factors such as temperature, greater surface roughness, and/or use of solvents. Conventional high-efficiency Si SC has high texture features with an etching angle of 54.7 degrees. As a result, the deposition process may need to alter by vapor deposition or atomic layer deposition (ALD), or other appropriate deposition techniques for attaining TSC with high performance, which can also contribute to the device stability (Jošt et al., 2020).

4.4.3.4 Current matching and optimization

In the monolithic TSC, both subcells should operate at their optimal PowerPoint to maximize the ECE, which requires matching their J_{SC}s. Furthermore, the matched J_{SC} must be maximized; therefore photon management is applied to the device. As a result, subcells are imbalanced, and in most cases, device efficiency often suffers because of the matching issue. Hence, the optimization of subcells for realizing matched J_{SC} is highly demanding for the TSC with high performance. Experimentally, such optimization is challenging as it requires longer times and higher costs. As an alternative, optical simulations become a powerful tool to perform such jobs (Haque et al., 2020; Hossain, Khan et al., 2020; Hossain et al., 2021; Qarony, Hossain et al., 2020; Qarony, Kozawa et al., 2020).

4.4.3.5 Material bandgap selection

Material selection is the critical part for realizing efficient tandem devices. As the optimum material combination can provide very high ECE (Fig. 4.4.3), it must consider suitable subcell semiconductors in terms of their at least optical and electrical properties. Furthermore, the material should have a multibandgap property to achieve the desired results by making necessary tuning. In the case of a PVK/Si TSC, the bottom cell is fixed to silicon with an Eg of 1.12 eV; hence, the top cell Eg has to be wide so that high ECE can be achieved. This is due to the tunability of the perovskite material system, which allows changing the *Eg* by compositional engineering. The *Eg* of perovskite can be varied from low to high (e.g., 1.15–3.1 eV) (Hossain, Khan et al., 2020; Qarony, Kozawa et al., 2020). In this study, the most widely used $MAPbI_3$ perovskite with an *Eg* of ~1.6 eV is considered.

4.4.4 Device structure, film growth, and interface morphology

Several research groups investigated the potential designs of PVK/Si TSC, where the top PSC is fabricated either on a flat or a textured Si SCs (Jošt et al., 2016;

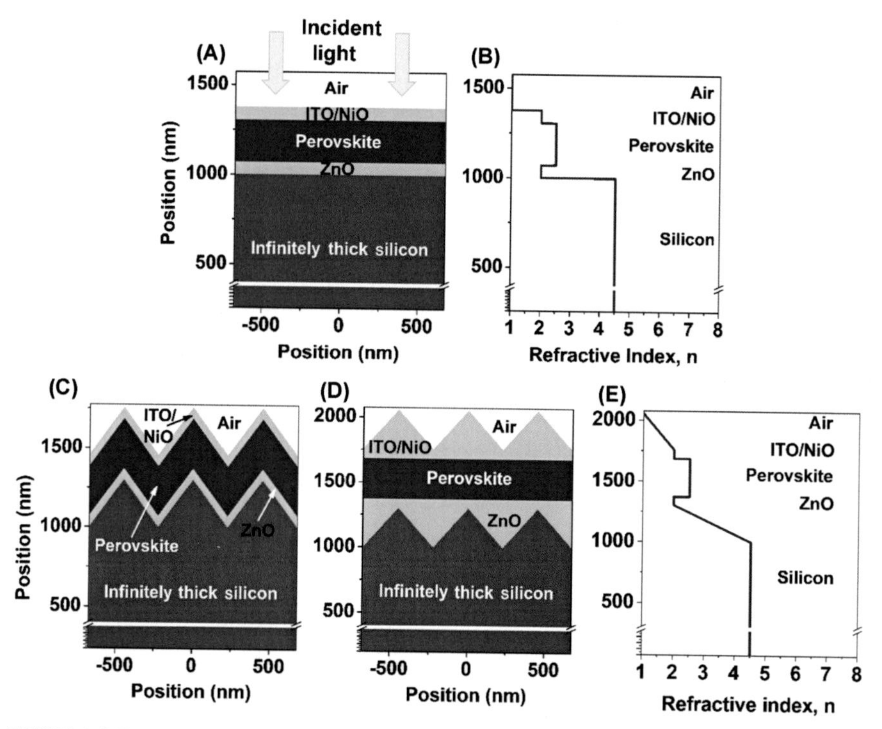

FIGURE 4.4.5

Schematic cross sections of the tandem solar cell (A) with flat top perovskite solar cell interfaces, (C) with the electrically and optically rough interface, and (D) with the electrically flat but optically rough interface. (B), (E) Corresponding refractive index profiles.

Reproduced with the permission from Qarony, W., Hossain, M. I., Jovanov, V., Salleo, A., Knipp, D., & Tsang, Y.H. (2020). Influence of Perovskite interface morphology on the photon management in perovskite/silicon tandem solar cells. ACS Applied Materials & Interfaces, 12, *15080–15086. https://doi.org/10.1021/acsami.9b21985. Copyright 2020, ACS.*

Jošt et al., 2018; Qarony, Hossain et al., 2020). Fig. 4.4.5 illustrates the summary of proposed device designs for PVK/Si TSCs, where MO (e.g., ZnO, ITO, NiO, etc.) transport contact materials are considered for simplicity (Qarony, Hossain et al., 2020). In the first design, as shown in Fig. 4.4.5A, the top PSC is deposited on a flat side of c-Si SC, leading to an ECE of 23.6% with a matched J_{SC} of 18.1 mA/cm^2 (Bush et al., 2017; Hossain, Qarony et al., 2019). Nevertheless, high optical losses due to the reflection at interfaces restrict the J_{SC} of the device, leading to a reduction of the ECE (Bush et al., 2017; Hossain et al., 2018). To overcome the limitation, two alternative device structures, as shown in Fig. 4.4.5C,D, are studied to enhance QE and J_{SC} through improved light incoupling and light trapping (Hossain, Hongsingthong et al., 2019; Qarony, Hossain, Salleo, Knipp, & Tsang, 2019). The device's electrical performance shown in Fig. 4.4.5D is better than the device depicted in Fig. 4.4.5C due to the flat interfaces between perovskite absorber and transport layers. The top cell of the tandem device in Fig. 4.4.5C is electrically and optically rough (EOR) and the top cell of the device in Fig. 4.4.5D is electrically flat but optically rough (EFOR) (Qarony et al., 2019). The optical performance of both structures is almost similar, where both top cells are deposited on top of a textured c-Si SCs. Both designs allow achieving efficient photon management and reduce optical losses. Fully textured monolithic PVK/Si TSCs were investigated by Sahli et al., where they have reached a matched J_{SC} of 20.1 mA/cm^2 with an ECE of 25.2%; the fabrication of such SCs will be discussed in Section 4.4.6. Moreover, the film growth and interface morphology have a significant impact on device performance, which has not been thoroughly studied so far. In PSC fabrication on top of textured c-Si bottom cells, two growth mechanisms are mostly followed, which are schematically presented

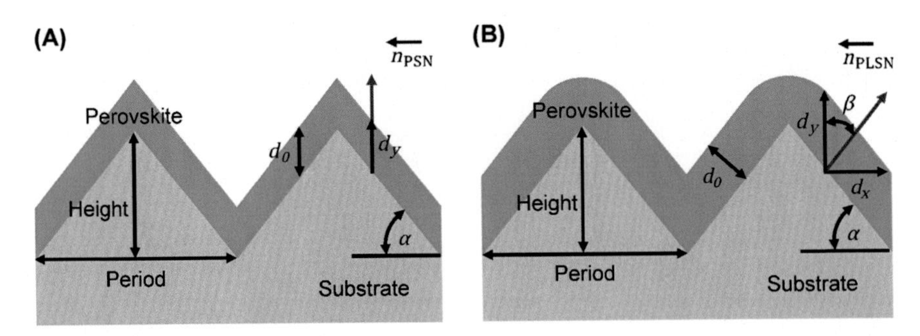

FIGURE 4.4.6

Schematic illustration of the perovskite film growth mechanisms on a textured c-Si SC in the direction of (A) substrate normal and (B) local surface normal.

Reproduced with the permission from Qarony, W., Hossain, M. I., Jovanov, V., Salleo, A., Knipp, D., & Tsang, Y.H. (2020). Influence of Perovskite interface morphology on the photon management in perovskite/silicon tandem solar cells. ACS Applied Materials & Interfaces, 12, *15080–15086. https://doi.org/10.1021/acsami.9b21985. Copyright 2020, ACS.*

in Fig. 4.4.6. Both cases have equal actual perovskite film thickness and deposition time; however, their growth processes are distinct (Qarony, Hossain et al., 2020).

The film growth is shown in Fig. 4.4.6A, named as substrate normal (SN) typically made by physical vapor deposition (PVD) techniques, which exhibits an inadequate step coverage. In contrast, local surface normal (LSN) film growth, as shown in Fig. 4.4.6B, offer excellent step coverage, which is representative of chemical vapor deposition (CVD) or ALD techniques (Bush et al., 2017; Jovanov et al., 2013; Köhnen et al., 2019; Plummer, Deal, & Griffin, 2000; Qarony, Hossain et al., 2020). Such growth mechanisms largely influence the effective film thickness, interface morphology, and optics of the device.

The c-Si SC consists of pyramid textures with an etching angle (α) of 54.7 degrees; hence, height to period ratio of the pyramid remains constant. The effective thickness (denoted by d_0) of both growth methods is comparable; however, their nominal thicknesses are different that is a product of deposition rate and deposition time. The effective thickness of the film in the y-direction can be defined by (Qarony, Hossain et al., 2020):

$$d_{eff} = \frac{\iint d_y(x,y)dxdy}{\iint dxdy} \tag{4.4.1}$$

In the case of SN growth, $d_y^{SN} = d_0$ and hence, $d_{eff}^{SN} = d_0$. On the other hand, $d_{eff}^{LSN} = d_0/\cos(\beta)$ for the LSN growth, while $\beta = \alpha$, so that the effective thickness is given by:

$$d_{eff}^{LSN} = \frac{d_0}{\cos(\alpha)} = d_0\sqrt{1 + \frac{4h^2}{p^2}} \tag{4.4.2}$$

Hence, the effective thickness of the LSN case, mainly depends on the period to height ratio of the pyramid texture.

4.4.5 Optics of perovskite/silicon tandem solar cells

The record Si SCs exhibits a J_{SC} of $\sim$42 mA/cm^2, which is almost their theoretical upper limit (46 mA/cm^2) (Green et al., 2020; Masuko et al., 2014; Shockley & Queisser, 1961; Yoshikawa et al., 2017). In the case of PVK/Si TSCs, the matched J_{SC} can be maximum 23 mA/cm^2 (half of the single-junction bottom SC's J_{SC}). Hence, it is essential to design an optimized structure so that high J_{SC} can be achieved under matching conditions. Furthermore, the conventional planar configuration of the device has to be changed for realizing high ECE. In this study, TSCs with both planar and texture structures are investigated, illustrated in Fig. 4.4.5. The planar structure is considered as a reference while comparing the performance of a textured device. The optics and optimization of SCs are studied using three-dimensional (3D) finite-difference time-domain (FDTD) simulations where a complex refractive index of materials is used as input parameters. The hybrid technique was adapted to characterize the tandem device's optical properties as wafer-based c-Si is much thicker than the thin perovskite film. In the simulation environment, the collection efficiency is considered to be 100%. A detailed discussion on the FDTD and hybrid techniques is provided in the literature (Hossain, Hasan et al., 2020; Hossain, Mohammad et al., 2020; Qarony, Kozawa et al., 2020). In this study, MOs (ITO, NiO, and ZnO) are considered as potential charge transporting materials, which are typically prepared by the sputtering process (Xu et al., 2015; Zhou et al., 2017).

Fig. 4.4.7A shows a planar device structure schematic, where PSC is deposited on the planar side of Si wafer. The $MAPbI_3$ perovskite absorber is placed between ZnO ETL and ITO/NiO HTL, where the double layer of ITO/NiO provides high work function and better conductivity (Hossain et al., 2018). In this case, the front contact of this device is planar and highly doped. The calculated QE of the TSC is presented in Fig. 4.4.7B, where top PSC contributes the photon absorption up to 800 nm only as the absorption near the perovskite Eg (1.6 eV) is 0. The planar SC gives top cell and bottom cell J_{SC} of 17.2 and 17.47 mA/cm^2; the matched J_{SC} (restricted to only 17.2 mA/cm^2) is a limited top cell. The result is very much comparable with the experimentally realized value published by Bush et al. (2017). QE and J_{SC} of the planar device are mainly limited due to higher reflection losses; hence, an advanced photon management strategy must be applied so that the SC can be improved.

A potential design for the PVK/Si TSC, as shown in Fig. 4.4.5C, is proposed so that the optical losses are minimized and optics are improved. In the proposed device structure, PSC is assumed to be fabricated on top of the textured Si SC while maintaining the Si wafer's etching angle; however, the conventional spin-coating method will not work as the surface roughness of the bottom SC is

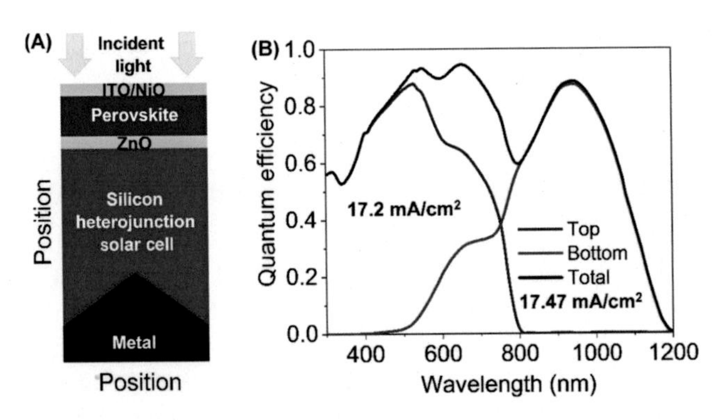

FIGURE 4.4.7

(A) Schematic cross-section and (B) corresponding QE of the PVK/Si planar tandem solar cell. The top perovskite solar cell is placed on top of a planar side of the c-Si SC.

Reproduced with the permission from Hossain, M. I., Qarony, W., Jovanov, V., Tsang, Y. H., & Knipp, D. (2018). Nanophotonic design of perovskite/silicon tandem solar cells. Journal of Materials Chemistry A, 6, *3625–3633. https://doi.org/10.1039/C8TA00628H. Copyright 2020, RSC.*

significantly larger. Hence, it is assumed that the preparation of top PSC is made by CVD or PVD techniques. Recently, promising results have been presented for perovskite films grown via the CVD technique (Leyden et al., 2014; Spina et al., 2016). In this study, it is assumed that the front and back contacts of the device are pyramidally textured with the same height to period ratio. Such a textured

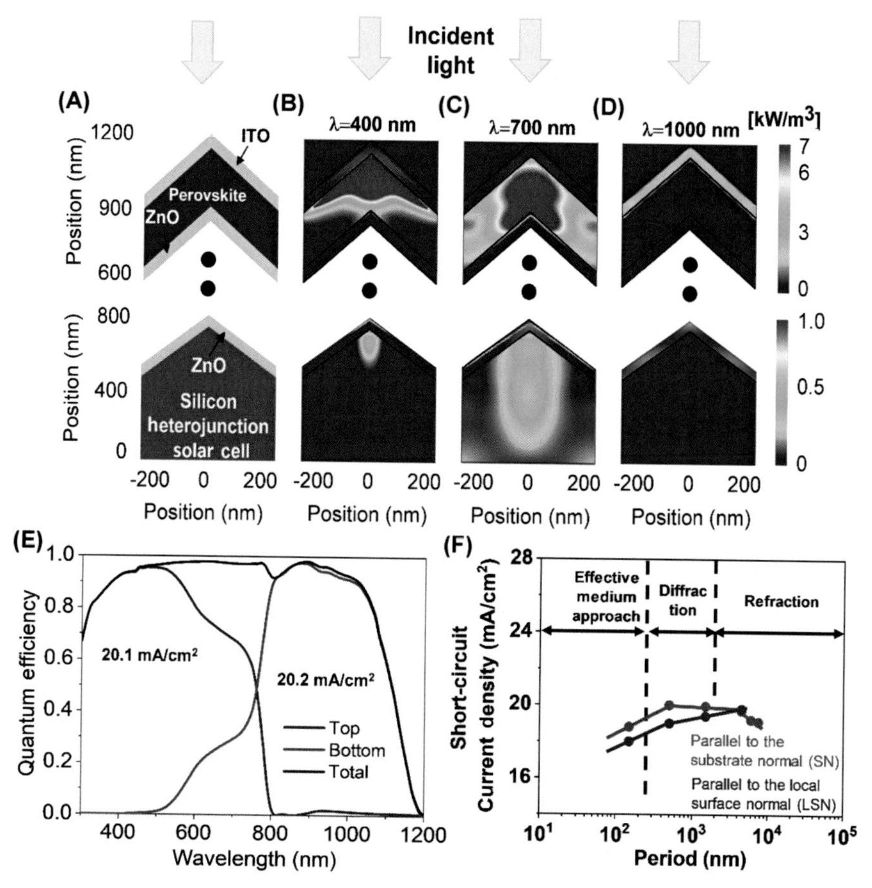

FIGURE 4.4.8

(A) The schematic cross-section of a fully textured PVK/Si TSC with a pyramid textured front contact. Corresponding power density of the device under monochromatic illumination of (B) 400 nm, (C) 700 nm, (D), and 1000 nm. (E) Calculated QE and J_{SC} of the EOR TSC under matching conditions. (F) Matched J_{SC} of the EOR TSC for various pyramid texture periods with SN and LSN film growth modes. *EOR*, electrically and optically rough; *LSN*, local surface normal; *TSC*, tandem solar cell.

Reproduced with the permission from Hossain, M. I., Qarony, W., Jovanov, V., Tsang, Y. H., & Knipp, D. (2018). Nanophotonic design of perovskite/silicon tandem solar cells. Journal of Materials Chemistry A, 6, *3625–3633. https://doi.org/10.1039/C8TA00628H. Copyright 2020, RSC.*

front contact allows achieving efficient light incoupling to the SC by minimizing reflection losses. The calculated power densities of the textured TSC with a period of 450 nm is illustrated in Fig. 4.4.8B–D under monochromatic wavelengths of 400, 750, and 1000 nm, along with a schematic of the device structure shown in Fig. 4.4.8A. The optics of perovskite and Si is different due to their refractive index and absorption coefficients; hence, power densities are visualized discretely.

For the shorter wavelengths (at 400 nm), most photons are absorbed by the top PSC, as shown in Fig. 4.4.8B. At 700 nm wavelength, most photons are yet absorbed by the top cell, and a fraction of light goes to the bottom cell and gets absorbed by the Si SC shown in Fig. 4.4.8C. For a wavelength of 1000 nm, a large fraction of light propagates to the bottom SC's deep, and a significant amount of light is absorbed by the ITO due to free carrier absorption, which does not contribute to the QE and J_{SC}. The influence of top PSC thickness on J_{SC} matching is investigated, which allows a realization of J_{SC} of 20.1 mA/cm^2 for the TSC. QE of the textured TSC is presented in Fig. 4.4.8E, where blue and red colors show the QE of top and bottom cells, respectively. The pyramid texture dimension is small (period: 450 nm and height: 300 nm); however, the influence of the large dimension of the pyramids is also studied, where the period is varied up to 7500 nm, and corresponding height varied up to 5000 nm. It has been found that increasing the profile dimension of the texture does not affect the optics of the SC primarily that can be observed from the realized J_{SC}s for both growth modes while varying the period of the pyramid texture revealed in Fig. 4.4.8F. Nevertheless, such a monolithic textured device's fabrication processing has to be highly accurate; otherwise, the

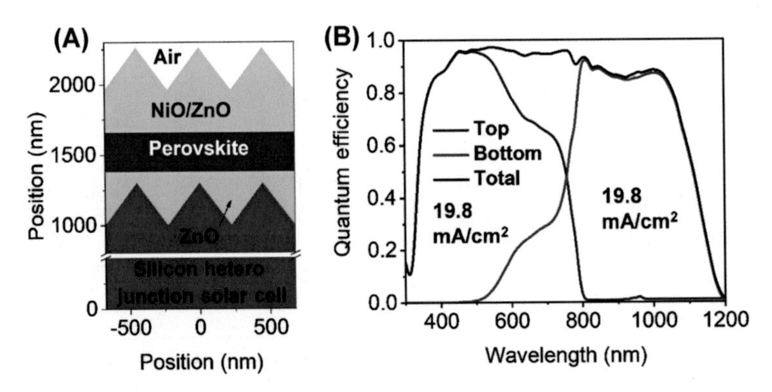

FIGURE 4.4.9

(A) Schematic cross-section and (B) corresponding QE of the EFOR PVK/Si TSC. The top PSC is placed on top of a textured side of the c-Si SC. *EFOR*, electrically flat but optically rough; *PSC*, perovskite solar cell; *TSC*, tandem solar cell.

Reproduced with the permission from Ref. Hossain, M. I., Hongsingthong, A., Qarony, W., Sichanugrist, P., Konagai, M., Salleo, A., ... Tsang, Y.H. (2019). Optics of Perovskite solar cell front contacts. ACS Applied Materials & Interfaces, 11, *14693–14701. https://doi.org/10.1021/acsami.8b16586. Copyright 2020, ACS.*

device performance may suffer due to low electrical properties. Sahli et al. investigated such structures and realized a matched J_{SC} of 20.1 mA/cm^2; however, their device ECE is limited to only 25.2% due to insufficient electrical parameters (Sahli et al., 2018); fabrication of fully textured PVK/Si TSC is presented in the next section. To overcome the limitation of fully textured TSC, the EFOR TSC is further investigated, which allows determining comparable electrical properties to the planar device. The proposed device structure is depicted in Fig. 4.4.9A. The investigation of such an arrangement improves the device performance without significantly compromising the optical performance. The QE of the matched EFOR-TSC is shown in Fig. 4.4.9B, which leads to achieving a J_{SC} of 19.8 mA/cm^2. By considering the experimentally realized best PSC with a V_{OC} of 1.182 V and FF of 84% along with the record Si SCs with a V_{OC} of 0.74 V and FF of 84.9%, the maximum ECE of the PVK/Si TSC can be estimated to be >32% (Fu et al., 2018; Green et al., 2020; Li et al., 2019).

4.4.6 Fabrication of perovskite/silicon tandem solar cells

In general, fabrication of monolithic PVK/Si TSC is challenging due to a combination of direct bandgap thin-film perovskite and indirect bandgap Si materials. Furthermore, matching of subcell currents is a challenging and exhausting job. Hence, it is essential to take necessary processing steps with the proper care. This section will discuss the experimental realization of PVK/Si TSC with the monolithic planar and fully textured configurations. One of the most efficient monolithic PVK/Si TSC with a planar top PSC was investigated by Bush et al. with an ECE of 23.6% (Bush et al., 2017). The CsFAPbIBr perovskite with a tuned Eg of 1.63 eV was used as an absorber for the top PSC. The fabrication of the TSC was started from the preparation of a single-junction top PSC. The p-i-n architecture was considered for the fabrication of the top cell. The perovskite layer is sandwiched between low work-function MOs/PCBM ETL (e.g., ZnO, TiO_2, etc.) and NiO HTL MO works as a window layer, and NiO provides high voltage and improved stability. An extremely thin LiF shunt blocking layer is used between PCBM and perovskite to improve the FF. A thin layer of ALD deposited SnO_2/ZTO was used on top of the PCBM layer. A comparatively thick ITO layer, along with LiF ARC, is used as an electrode. Then, the same procedures were followed while depositing top PSC on top of Si SCs, and Ag contact was thermally evaporated.

A schematic of the device is depicted in Fig. 4.4.10A, and SEM images of PSC and Si Cells are shown in Fig. 4.4.10C–E. The fabricated TSC exhibits a maximum ECE of 23.6% with a matched J_{SC} of 18.1 mA/cm^2, V_{OC} of 1.65 V, and FF of 79%, as shown from the J-V characteristic in Fig. 4.4.10F. Fig. 4.4.10G shows the QE of the TSC along with reflectance and parasitic absorption, which do not contribute to the QE and J_{SC}. From the QE plots, the J_{SC} under matching condition is slightly more realized. It is clear that the ECE of the investigated TSC by Bush et al. is mainly limited due to

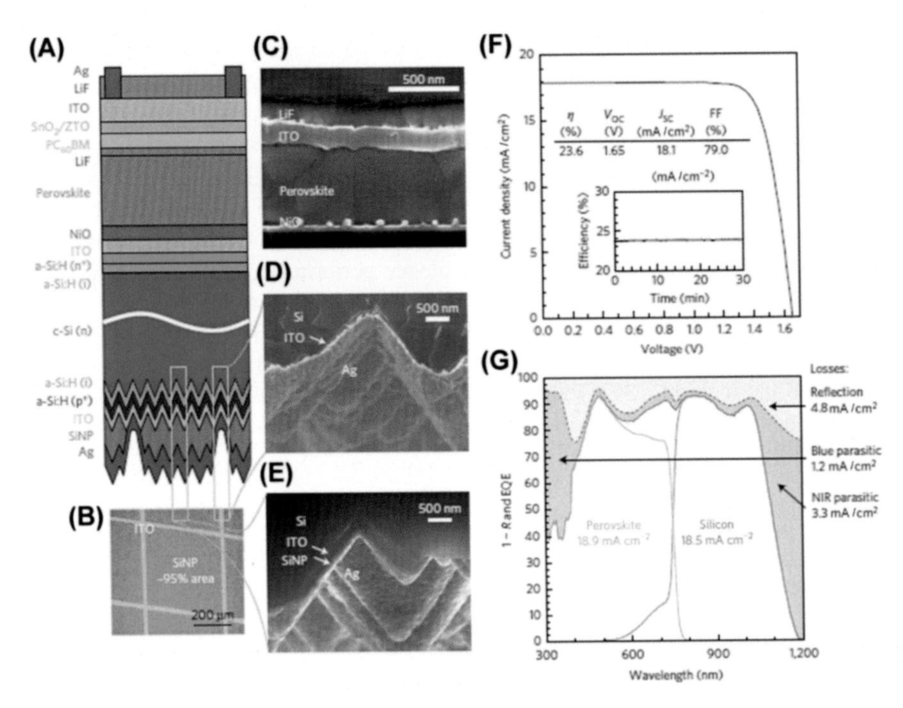

FIGURE 4.4.10

(A) Schematic of the investigated device structure. (B) Optical microscopic image of the Si nanoparticle-patterned rear side of the Si cell before silvering. Cross-sectional SEM image of (C) the top PSC, (D) rear side of the Si cell without nanoparticles, and (E) rear side of the Si cell with nanoparticles. (F) *J-V* curve of the champion device with extracted photovoltaic parameters. (G) QE of the top and bottom cells along with reflectance and parasitic losses.

Reproduced with the permission from Bush, K. A., Palmstrom, A. F., Yu, Z. J., Boccard, M., Cheacharoen, R., Mailoa, J. P., ... McGehee, M. D. (2017). 23.6%-efficient monolithic perovskite/silicon tandem solar cells with improved stability. Nature Energy, 2, *17009. https://doi.org/10.1038/nenergy.2017.9. Copyright 2020, Nature.*

the lower matched J_{SC}. To overcome this limitation, Sahli et al. proposed a fully textured monolithic PVK/Si TSC with an improved ECE of 25.2% and J_{SC} of 20.1 mA/cm^2, made due to efficient photon management (Sahli et al., 2018).

A double-sided textured SHJ SCs were considered, and an ITO recombination layer is placed between PSC and bottom cells. A Spiro-OMeTAD HTL layer is deposited through thermal evaporation, followed by a spin coating of CsFAPbIBr ($Eg \sim 1.6$ eV) perovskite layer. A LiF/C_{60} ETL layer subsequently prepared thermal evaporation. A stack of ALD deposited SnO_2 and sputtered IZO was used as a buffer layer. The evaporated Ag used a front metal contact, and MgF_2 acts as an ARC layer. The device structure's schematic and AFM and SEM images are portrayed in Fig. 4.4.11A–E. The investigated best TSC exhibits a high QE

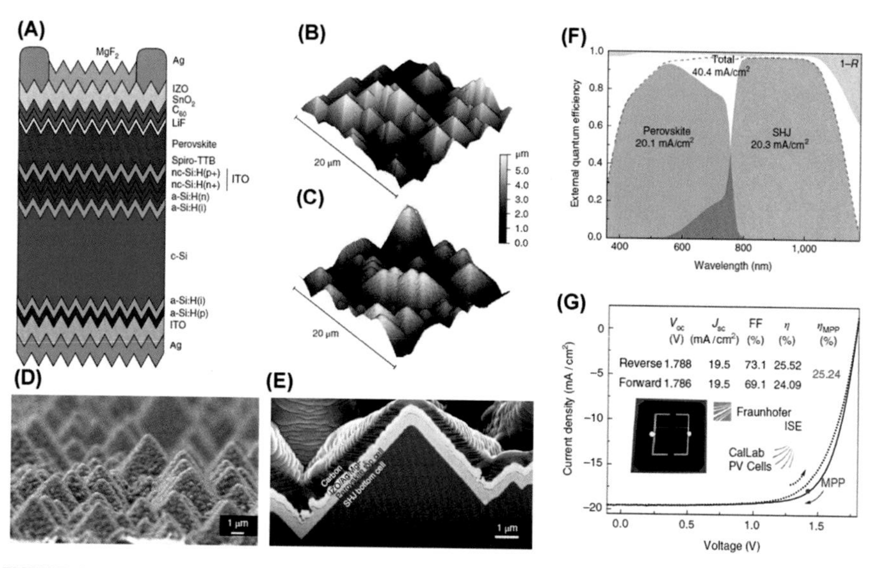

FIGURE 4.4.11

(A) Schematic of the investigated fully textured TSC structure. AFM images of (B) bare c-Si pyramids, (C) c-Si pyramids covered with perovskite layer. Cross-sectional SEM image of (D) perovskite film, and (E) full PSC deposited on SHJ SC. (F) QE of the top and bottom cells along with 1-R curve. (G) *J-V* curve of the champion device with extracted photovoltaic parameters. *PSC*, perovskite solar cell; *TSC*, tandem solar cells.

Reproduced with the permission from Ref. Sahli, F., Werner, J., Kamino, B. A., Bräuninger, M., Monnard, R., Paviet-Salomon, B., ... Ballif, C. (2018). Fully textured monolithic perovskite/silicon tandem solar cells with 25.2% power conversion efficiency. Nature Materials. *https://doi.org/10.1038/s41563-018-0115-4.*

presented in Fig. 4.4.11F, which allows the J_{SC} of 20.1 mA/cm^2 under matching conditions. However, a discrepancy in the J_{SC} has been seen from the J-V characteristic, as shown in Fig. 4.4.11G, where the J_{SC} is matched to 19.5 mA/cm^2. Although the device shows a V_{OC} of 1.78 V and high J_{SC}, the device ECE is restricted due to poor FF (73.1%). Hence, it is realized that the FF must be improved to achieve high ECE from TSC. Very recently, Amran et al. has demonstrated a monolithic PVK/Si TSC with a high ECE of 29.15% by improving the electrical performance of the device (V_{OC} of 1.9 V and FF of 79.52%) through enhanced hole extraction (Al-Ashouri et al., 2020). More details on the device fabrication and characterization can be found in the literature (Al-Ashouri et al., 2020). The summary of the present status of PVK/Si 2-terminal TSCs is presented in Table 4.4.1. It has been realized that the PVK/Si TSC has huge prospects for overcoming the efficiency limitation of single-junction SCs.

As shown in Table 4.4.1, the study of PVK/Si TSCs began in 2017 with an ECE of 23.6%, which has dramatically improved to 29.1% within the last three years.

Table 4.4.1 Summary of current status of monolithic 2-terminal PVK/Si efficient tandem solar cell.

Research group	Year	Area (cm^2)	J_{SC} (mA/cm^2)	V_{OC} (V)	FF (%)	Energy conversion efficiency (%)	References
U.Stan/ASU	2017	1	18.1	1.65	79	23.6	Bush et al. (2017)
EPFL	2018	1.42	19.5	1.79	73.2	25.5	Sahli et al. (2018)
U.Stan/ASU	2018	1	18.4	1.77	77	25	Bush et al. (2018)
HZB	2018	0.77	18.5	1.76	78.5	25.5	Jošt et al. (2018)
UNL/ASU	2019	0.42	17.8	1.8	79.4	25.4	Chen et al. (2019)
HZB/U.Ox/ Ox PV	2019	1.1	19	1.79	74.6	25.4	Mazzarella et al. (2019)
EPFL	2019	1.43	19.5	1.74	74.7	25.4	Nogay et al. (2019)
HZB	2019	0.77	19.2	1.77	76.6	26	Köhnen et al. (2019)
Ox PV	2018	1.03	19.8	1.8	78.7	28	Oxford PV (2020)
HZB	2020	1.06	19.26	1.9	79.52	29.1	Al-Ashouri et al. (2020)

According to theory, the ECE can be enhanced to greater than 35% by upgrading the electrical and optical performance parameters of the SCs. In fact, Oxford PV has just reported another record of 29.5% (Oxford scientists develop record-breaking solar power technology, n.d.) in PVK/Si TSCs, which could be implemented for the rooftop energy harvesting process as early as by the year of 2022.

4.4.7 Final remarks and prospects

The traditional Si-SC has reached its practical limit, whereas PSC technology is a relatively new research field which has emerged in less than a decade. The potential of this new technology has been investigated by numerous experts in the field. The single junction PSC has already reached an ECE of 25.5% (Best Research-Cell Efficiency Chart, n.d.). This fast and skyrocketing ECE achieved in PSC along with excellent optoelectronic properties of perovskites (PVK) was the encouraging point of interest in developing TSCs. Making SC technology even slightly more efficient than the conventional Si-SC would have a dramatic impact on the amount of renewable power generated in the world. An extra one percent of efficiency of the currently installed base of world solar power could power an additional million homes. The combination of a silicon and PSC is the only way to drive fast adoption of solar PV in the world. There are still many scopes to improve the technology by enhancing ECE in perovskite/silicon (PVK/Si) TSCs since it has not yet reached the SQ limit of the ECE. Besides improving the electrical performance parameters of the PVK/Si TSCs by ameliorating interface engineering, surface roughness in the PVK, especially at the interface of textured heterojunction silicon bottom SC and perovskite top SC, and transport layers of the cells. Then the optics of the SCs has to be improved as well, which is the key to increasing short circuit current density and efficiency.

The book chapter aims at providing guidelines for the realization of PVK/Si TSCs with high ECEs. PVK/Si TSCs have been designed with the J_{SC} exceeding 20 mA/cm^2 while taking realistic device structures into considering. The high J_{SC} allows for the realization of PVK/Si TSCs with ECEs exceeding 30%. The PVK/Si TSC has to be textured to minimize reflection losses. Hence the perovskite top cell has to be prepared on top of the textured silicon bottom cell. This requires the use of materials and deposition techniques compatible with the fabrication on a textured substrate. MOs like ITO/NiO and IOH/NiO hole and ZnO ETLs could be used, which can be prepared by sputtering. The textured silicon SC allows for an almost perfect incoupling of the longer wavelengths light in the bottom cell. Then the contact layers can be optimized by minimizing the free carrier absorption losses. The higher charge carrier mobility allows for using a lower doping concentration while keeping the sheet resistance of the contact layer constant.

The perovskite film growth has also distinct influence on the roughness of the SC, the effective thickness of the top PSC, and the optics of PVK/Si TSCs. In the case of PSN film growth, small surface textures of the underlying c-Si bottom SC are large enough to allow for efficient light coupling in the SC. For PLSN film growth, a flattening of the front surface is observed, which leads to an increased reflection of the SC. By increasing the size of the surface texture an improved light incoupling is achieved. Films grown with the PLSN exhibit a 73% larger effective thickness compared to PSN. In other words, the PLSN growth leads to a reduction of the deposition time by 42%. However, the diffraction and refraction of the light have a smaller effect on the J_{SC}. Both growth modes allow for reaching J_{SC} of approx. 20 mA/cm^2 and ECE of over 32%. In case of PSN growth, the maximal ECE is reached for periods of 500 nm or larger, while the maximal ECE of SCs with perovskite film grown with the PLSN is reached for periods of 5000 nm or larger. Depositing thin PSCs on the large dimensional highly efficient heterojunction silicon SC is very challenging, which limits the study of the monolithic fabrication as well as calculation of the TSCs. Moreover, the all-perovskite tandem is another way of simultaneously improving ECE closer to the SQ limit and reducing the processing cost, while the fabrication complexities experienced in the PVK/Si TSC can be reduced.

Acknowledgments

The research was supported by a grant from the Innovation and Technology Commission of Hong Kong (Project Number: ITS/461/18). This work was also supported by the Research Grants Council of Hong Kong, China (Project number: 152093/18E). The authors would also like to acknowledge the support from The National University of Malaysia under Grant LRGS/1/2019/UKM-UKM/6/1.

References

Al-Ashouri, A., Köhnen, E., Li, B., Magomedov, A., Hempel, H., Caprioglio, P., ... Albrecht, S. (2020). Monolithic perovskite/silicon tandem solar cell with >29% efficiency by enhanced hole extraction. *Science*, *370*, 1300–1309. Available from https://doi.org/10.1126/science.abd4016.

Alharbi, F. H., & Kais, S. (2015). Theoretical limits of photovoltaics efficiency and possible improvements by intuitive approaches learned from photosynthesis and quantum coherence. *Renewable and Sustainable Energy Reviews*, *43*, 1073–1089. Available from https://doi.org/10.1016/j.rser.2014.11.101.

Atwater, H. A., & Polman, A. (2010). Plasmonics for improved photovoltaic devices. *Nature Materials*, *9*, 205–213. Available from https://doi.org/10.1038/nmat2629.

Best research-cell efficiency chart (n.d.). *Photovoltaic research*. NREL.

Bush, K. A., Manzoor, S., Frohna, K., Yu, Z. J., Raiford, J. A., Palmstrom, A. F., ... McGehee, M. D. (2018). Minimizing current and voltage losses to reach 25% efficient monolithic two-terminal perovskite–silicon tandem solar cells. *ACS Energy Letters*, *3*, 2173–2180. Available from https://doi.org/10.1021/acsenergylett.8b01201.

Bush, K. A., Palmstrom, A. F., Yu, Z. J., Boccard, M., Cheacharoen, R., Mailoa, J. P., ... McGehee, M. D. (2017). 23.6%-efficient monolithic perovskite/silicon tandem solar cells with improved stability. *Nature Energy.*, *2*, 17009. Available from https://doi.org/10.1038/nenergy.2017.9.

Chen, B., Yu, Z., Liu, K., Zheng, X., Liu, Y., Shi, J., ... Huang, J. (2019). Grain engineering for perovskite/silicon monolithic tandem solar cells with efficiency of 25.4%. *Joule*, *3*, 177–190. Available from https://doi.org/10.1016/j.joule.2018.10.003.

Cuevas, A., & Yan, D. (2013). Misconceptions and misnomers in solar cells. *IEEE Journal of Photovoltaics*, *3*, 916–923. Available from https://doi.org/10.1109/JPHOTOV.2013.2238289.

De Vos, A. (1980). Detailed balance limit of the efficiency of tandem solar cells. *Journal of Physics D: Applied Physics*, *13*, 839–846. Available from https://doi.org/10.1088/0022-3727/13/5/018.

Duran Sahin, A., Dincer, I., & Rosen, M. A. (2007). Thermodynamic analysis of solar photovoltaic cell systems. *Solar Energy Materials and Solar Cells*, *91*, 153–159. Available from https://doi.org/10.1016/j.solmat.2006.07.015.

Fu, Y., Wang, G., Ming, X., Liu, X., Hou, B., Mei, T., ... Wang, X. (2018). Oxygen plasma treated graphene aerogel as a solar absorber for rapid and efficient solar steam generation. *Carbon*. Available from https://doi.org/10.1016/j.carbon.2017.12.124.

Futscher, M. H., & Ehrler, B. (2016). Efficiency limit of perovskite/Si tandem solar cells. *ACS Energy Letters*, *1*, 863–868. Available from https://doi.org/10.1021/acsenergylett.6b00405.

Green, M.A. (2006). *Third generation photovoltaics: Advanced solar energy conversion*. doi:10.1021/la100123q.

Green, M. A., Dunlop, E. D., Hohl-Ebinger, J., Yoshita, M., Kopidakis, N., & Hao, X. (2020). Solar cell efficiency tables (version 56). *Progress in Photovoltaics: Research and Applications*, *28*, 629–638. Available from https://doi.org/10.1002/pip.3303.

Green, M. A., Ho-Baillie, A., & Snaith, H. J. (2014). The emergence of perovskite solar cells. *Nature Photonics*, *8*, 506–514. Available from https://doi.org/10.1038/nphoton.2014.134.

Haque, S., Alexandre, M., Mendes, M. J., Águas, H., Fortunato, E., & Martins, R. (2020). Design of wave-optical structured substrates for ultra-thin perovskite solar cells. *Applied Materials Today.*, *20*, 100720. Available from https://doi.org/10.1016/j.apmt.2020.100720.

Haque, S., Mendes, M. J., Sanchez-Sobrado, O., Águas, H., Fortunato, E., & Martins, R. (2019). Photonic-structured TiO_2 for high-efficiency, flexible and stable Perovskite solar cells. *Nano Energy*, *59*, 91–101. Available from https://doi.org/10.1016/j.nanoen.2019.02.023.

He, M., Zheng, D., Wang, M., Lin, C., & Lin, Z. (2014). High efficiency perovskite solar cells: From complex nanostructure to planar heterojunction. *Journal of Materials Chemistry A.*, *2*, 5994. Available from https://doi.org/10.1039/c3ta14160h.

Hegedus, S. S., & Luque, A. (2005). *Status, trends, challenges and the bright future of solar electricity from photovoltaics. Handbook of photovoltaic science and engineering* (pp. 1–43). Chichester: John Wiley & Sons, Ltd. Available from http://doi.org/10.1002/0470014008.ch1.

Honsberg, C. B., Corkish, R. C., & Bremner, S. P. (2001). A new generalized detailed balance formulation to calculate solar cell efficiency limits. *Georgia Institute of Technology*, *1*, 3. Available from http://smartech.gatech.edu/handle/1853/26162.

Hossain, M. I., Hasan, A. K. M., Qarony, W., Shahiduzzaman, M., Islam, M. A., Ishikawa, Y., ... Tsang, Y. H. (2020). Electrical and optical properties of nickel-oxide films for efficient perovskite solar cells. *Small Methods*, *4*, 2000454. Available from https://doi.org/10.1002/smtd.202000454.

Hossain, M. I., Hongsingthong, A., Qarony, W., Sichanugrist, P., Konagai, M., Salleo, A., ... Tsang, Y. H. (2019). Optics of perovskite solar cell front contacts. *ACS Applied Materials & Interfaces*, *11*, 14693–14701. Available from https://doi.org/10.1021/acsami.8b16586.

Hossain, M. I., Khan, H. A., Kozawa, M., Qarony, W., Salleo, A., Hardeberg, J. Y., ... Knipp, D. (2020). Perovskite color detectors: Approaching the efficiency limit. *ACS Applied Materials & Interfaces.*, *12*, 47831–47839. Available from https://doi.org/10.1021/acsami.0c12851.

Hossain, M. I., Mohammad, A., Qarony, W., Ilhom, S., Shukla, D. R., Knipp, D., ... Tsang, Y. H. (2020). Atomic layer deposition of metal oxides for efficient perovskite single-junction and perovskite/silicon tandem solar cells. *RSC Advances*, *10*, 14856–14866. Available from https://doi.org/10.1039/D0RA00939C.

Hossain, M. I., Shahiduzzaman, M., Ahmed, S., Huqe, M. R., Qarony, W., Saleque, A. M., ... Zapien, J. A. (2021). Near field control for enhanced photovoltaic performance and photostability in Perovskite solar cells. *Nano Energy*, *89*(PA), 106388. Available from https://doi.org/10.1016/j.nanoen.2021.106388.

Hossain, M. I., Shahiduzzaman, M., Saleque, A. M., Huqe, M. R., Qarony, W., Ahmed, S., ... Zapien, J. A. (2021). Improved nanophotonic front contact design for high-performance Perovskite single-junction and Perovskite/Perovskite tandem solar cells. *Solar RRL*, 2100509. Available from https://doi.org/10.1002/solr.202100509.

Hossain, M. I., Qarony, W., Hossain, M. K., Debnath, M. K., Uddin, M. J., & Tsang, Y. H. (2017). Effect of back reflectors on photon absorption in thin-film amorphous silicon solar cells. *Applied Nanoscience*, *7*, 489–497. Available from https://doi.org/10.1007/s13204-017-0582-y.

Hossain, M. I., Qarony, W., Jovanov, V., Tsang, Y. H., & Knipp, D. (2018). Nanophotonic design of perovskite/silicon tandem solar cells. *Journal of Materials Chemistry A.*, *6*, 3625–3633. Available from https://doi.org/10.1039/C8TA00628H.

Hossain, M. I., Qarony, W., Ma, S., Zeng, L., Knipp, D., Tsang, Y. H., ... Tsang, Y. H. (2019). Perovskite/silicon tandem solar cells: From detailed balance limit calculations to photon management. *Nano-Micro Letters*, *11*, 58. Available from https://doi.org/10.1007/s40820-019-0287-8.

Hossain, M. I., Saleque, A. M., Ahmed, S., Saidjafarzoda, I., Shahiduzzaman, M., Qarony, W., ... Tsang, Y. H. (2021). Perovskite/perovskite planar tandem solar cells: A comprehensive guideline for reaching energy conversion efficiency beyond 30%. *Nano Energy*, *79*, 105400. Available from https://doi.org/10.1016/j.nanoen.2020.105400.

Jošt, M., Albrecht, S., Lipovšek, B., Krč, J., Korte, L., Rech, B., & Topič, M. (2016). Back- and front-side texturing for light-management in perovskite/silicon-heterojunction tandem solar cells. *Energy Procedia*, *102*, 43–48. Available from https://doi.org/10.1016/j.egypro.2016.11.316.

Jošt, M., Kegelmann, L., Korte, L., & Albrecht, S. (2020). Monolithic perovskite tandem solar cells: A review of the present status and advanced characterization methods toward 30% efficiency. *Advanced Energy Materials*, *10*, 1904102. Available from https://doi.org/10.1002/aenm.201904102.

Jošt, M., Köhnen, E., Morales-Vilches, A. B., Lipovšek, B., Jäger, K., Macco, B., ... Albrecht, S. (2018). Textured interfaces in monolithic perovskite/silicon tandem solar cells: Advanced light management for improved efficiency and energy yield. *Energy & Environmental Science*, *11*, 3511–3523. Available from https://doi.org/10.1039/C8EE02469C.

Jovanov, V., Xu, X., Shrestha, S., Schulte, M., Hüpkes, J., & Knipp, D. (2013). Predicting the interface morphologies of silicon films on arbitrary substrates: Application in solar cells. *ACS Applied Materials & Interfaces.*, *5*, 7109–7116. Available from https://doi.org/10.1021/am401434y.

Köhnen, E., Jošt, M., Morales-Vilches, A. B., Tockhorn, P., Al-Ashouri, A., Macco, B., ... Albrecht, S. (2019). Highly efficient monolithic perovskite silicon tandem solar cells: Analyzing the influence of current mismatch on device performance. *Sustain. Energy & Fuels: An American Chemical Society Journal*, *3*, 1995–2005. Available from https://doi.org/10.1039/C9SE00120D.

Kojima, A., Teshima, K., Shirai, Y., & Miyasaka, T. (2009). Organometal halide perovskites as visible-light sensitizers for photovoltaic cells. *Journal of the American Chemical Society*, *131*, 6050–6051. Available from https://doi.org/10.1021/ja809598r.

Kosyachenko, L.A. (2015). *Solar cells—new approaches and reviews*. InTech. doi:10.5772/58490.

Lee, T. D., & Ebong, A. U. (2017). A review of thin film solar cell technologies and challenges. *Renewable and Sustainable Energy Reviews*. Available from https://doi.org/10.1016/j.rser.2016.12.028.

Leijtens, T., Bush, K. A., Prasanna, R., & McGehee, M. D. (2018). Opportunities and challenges for tandem solar cells using metal halide perovskite semiconductors. *Nature Energy.*, *3*, 828–838. Available from https://doi.org/10.1038/s41560-018-0190-4.

Leyden, M. R., Ono, L. K., Raga, S. R., Kato, Y., Wang, S., & Qi, Y. (2014). High performance perovskite solar cells by hybrid chemical vapor deposition. *Journal of Materials Chemistry A*, *2*, 18742–18745. Available from https://doi.org/10.1039/C4TA04385E.

Li, H., & Zhang, W. (2020). Perovskite tandem solar cells: From fundamentals to commercial deployment. *Chemical Reviews, 120*, 9835–9950. Available from https://doi.org/10.1021/acs.chemrev.9b00780.

Li, Z., Wang, F., Liu, C., Gao, F., Shen, L., & Guo, W. (2019). Efficient perovskite solar cells enabled by ion-modulated grain boundary passivation with a fill factor exceeding 84%. *Journal of Materials Chemistry A., 7*, 22359–22365. Available from https://doi.org/10.1039/C9TA08081C.

Limpert, S., Bremner, S., & Linke, H. (2015). Reversible electron–hole separation in a hot carrier solar cell. *New Journal of Physics, 17*, 095004. Available from https://doi.org/10.1088/1367-2630/17/9/095004.

Lin, R., Xiao, K., Qin, Z., Han, Q., Zhang, C., Wei, M., ... Tan, H. (2019). Monolithic all-perovskite tandem solar cells with 24.8% efficiency exploiting comproportionation to suppress Sn(ii) oxidation in precursor ink. *Nature Energy*. Available from https://doi.org/10.1038/s41560-019-0466-3.

Masuko, K., Shigematsu, M., Hashiguchi, T., Fujishima, D., Kai, M., Yamaguchi, T., ... Okamoto, S. (2014). Achievement of more than 25%: Conversion efficiency with crystalline silicon heterojunction solar cell,. *Photovoltaics, IEEE Journal, 4*, 1433–1435. Available from https://doi.org/10.1109/JPHOTOV.2014.2352151.

Mazzarella, L., Lin, Y., Kirner, S., Morales-Vilches, A. B., Korte, L., Albrecht, S., ... Schlatmann, R. (2019). Infrared light management using a nanocrystalline silicon oxide interlayer in monolithic perovskite/silicon heterojunction tandem solar cells with efficiency above 25%. *Advanced Energy Materials, 9*, 1803241. Available from https://doi.org/10.1002/aenm.201803241.

Nogay, G., Sahli, F., Werner, J., Monnard, R., Boccard, M., Despeisse, M., ... Ballif, C. (2019). 25.1%-efficient monolithic perovskite/silicon tandem solar cell based on a p -type monocrystalline textured silicon wafer and high-temperature passivating contacts. *ACS Energy Letters, 4*, 844–845. Available from https://doi.org/10.1021/acsenergylett.9b00377.

NREL Transforming Energy. (2020). Best research-cell efficiency chart. <https://www.nrel.gov/pv/cell-efficiency.html>. Accessed 01.10.20.

Oxford PV-The Perovskite Company. (2020). Oxford PV perovskite solar cell achieves 28% efficiency. <https://www.oxfordpv.com/news/oxford-pv-perovskite-solar-cell-achieves-28-efficiency>. Accessed 11.12.18.

Oxford scientists develop record-breaking solar power technology (n.d.).

Park, N.-G. (2015). Perovskite solar cells: An emerging photovoltaic technology. *Materials Today, 18*, 65–72. Available from https://doi.org/10.1016/j.mattod.2014.07.007.

Plummer, J.D., Deal, M.D., & Griffin, P.B. (2000). *Silicon VLSI Technology*, p. 817.

Pridham, G. J. (1970). Physics of semiconductor devices. *Electron Power., 16*, 34. Available from https://doi.org/10.1049/ep.1970.0039.

Qarony, W., & Hossain, M. I. (2015). Organic solar cell: Optics in smooth and pyramidal rough surface. *IOSR Journal of Electrical and Electronics Engineering. Version III, 10*. Available from https://doi.org/10.9790/1676-10436772, 2278–1676.

Qarony, W., Hossain, M. I., Dewan, R., Fischer, S., Meyer-Rochow, V. B., Salleo, A., ... Tsang, Y. H. (2018). Approaching perfect light incoupling in perovskite and silicon thin film solar cells by moth eye surface textures. *Advanced Theory in Simulations., 1*, 1800030. Available from https://doi.org/10.1002/adts.201800030.

Qarony, W., Hossain, M. I., Hossain, M. K., Uddin, M. J., Haque, A., Saad, A. R., & Tsang, Y. H. (2017). Efficient amorphous silicon solar cells: Characterization, optimization, and optical loss analysis. *Results Physics*, *7*, 4287–4293. Available from https://doi.org/10.1016/j.rinp.2017.09.030.

Qarony, W., Hossain, M. I., Jovanov, V., Salleo, A., Knipp, D., & Tsang, Y. H. (2020). Influence of Perovskite interface morphology on the photon management in perovskite/silicon tandem solar cells. *ACS Applied Materials & Interfaces.*, *12*, 15080–15086. Available from https://doi.org/10.1021/acsami.9b21985.

Qarony, W., Hossain, M. I., Salleo, A., Knipp, D., & Tsang, Y. H. (2019). Rough vs planar interfaces: How to maximize the short circuit current of perovskite single and tandem solar cells. *Materials Today Energy.*, *11*, 106–113. Available from https://doi.org/10.1016/j.mtener.2018.10.001.

Qarony, W., Jui, Y. A., Das, G. M., Mohsin, T., Hossain, M. I., & Islam, S. N. (2015). Optical analysis in $CH_3NH_3PbI_3$ and $CH_3NH_3PbI_2Cl$ based thin-film perovskite solar cell. *American Journal of Energy Research*, *3*, 19–24. Available from https://doi.org/10.12691/ajer-3-2-1.

Qarony, W., Kozawa, M., Khan, H. A., Hossain, M. I., Salleo, A., Tsang, Y. H., … Knipp, D. (2020). Vertically stacked perovskite detectors for color sensing and color vision. *Advanced Materials Interfaces*, 2000459. Available from https://doi.org/10.1002/admi.202000459.

Rajagopal, A., Yang, Z., Jo, S. B., Braly, I. L., Liang, P.-W., Hillhouse, H. W., & Jen, A. K. Y. (2017). Highly efficient perovskite-perovskite tandem solar cells reaching 80% of the theoretical limit in photovoltage. *Advanced Materials*, *29*, 1702140. Available from https://doi.org/10.1002/adma.201702140.

Rau, U., Paetzold, U. W., & Kirchartz, T. (2014). Thermodynamics of light management in photovoltaic devices. *Physical Review B: Condensed Matter and Materials Physics.*, *90*. Available from https://doi.org/10.1103/PhysRevB.90.035211.

Rech, B., & Wagner, H. (1999). Potential of amorphous silicon for solar cells. *Applied Physics A: Materials Science and Processing*, *69*, 155–167. Available from https://doi.org/10.1007/s003390050986.

Reference solar spectral irradiance: ASTM G-173 (n.d.). <http://rredc.nrel.gov/solar/spectra/am1.5/astmg173/astmg173.html>. Accessed 19.03.18.

Rong, Y., Hu, Y., Mei, A., Tan, H., Saidaminov, M. I., Il Seok, S., … Han, H. (2018). Challenges for commercializing perovskite solar cells. *Science*, *361*. Available from https://doi.org/10.1126/science.aat8235, eaat8235.

Sahli, F., Werner, J., Kamino, B. A., Bräuninger, M., Monnard, R., Paviet-Salomon, B., … Ballif, C. (2018). Fully textured monolithic perovskite/silicon tandem solar cells with 25.2% power conversion efficiency. *Nature Materials*. Available from https://doi.org/10.1038/s41563-018-0115-4.

Shah, A. V., Schade, H., Vanecek, M., Meier, J., Vallat-Sauvain, E., Wyrsch, N., … Bailat, J. (2004). Thin-film silicon solar cell technology. *Progress in Photovoltaics: Research and Applications*, *12*, 113–142. Available from https://doi.org/10.1002/pip.533.

Shahiduzzaman, M., Ashikawa, H., Kuniyoshi, M., Visal, S., Sakakibara, S., Kaneko, T., … Tomita, K. (2018). Compact TiO_2/anatase TiO_2 single-crystalline nanoparticle electron-transport bilayer for efficient planar perovskite solar cells. *ACS Sustainable Chemistry & Engineering*, *6*, 12070–12078. Available from https://doi.org/10.1021/acssuschemeng.8b02406.

Shahiduzzaman, M., Ismail Hossain, M., Otani, S., Wang, L., Umezu, S., Kaneko, T., … Taima, T. (2021). Low-temperature treated anatase TiO2 nanophotonic-structured contact design for efficient triple-cation Perovskite solar cells. *Chemical Engineering Journal*, *426*, 131831. Available from https://doi.org/10.1016/j.cej.2021.131831.

Shockley, W., & Queisser, H. J. (1961). Detailed balance limit of efficiency of p-n junction solar cells. *Journal of Applied Physics*, *32*, 510–519. Available from https://doi.org/10.1063/1.1736034.

Smestad, G., & Ries, H. (1992). Luminescence and current-voltage characteristics of solar cells and optoelectronic devices. *Solar Energy Materials and Solar Cells.*, *25*, 51–71. Available from https://doi.org/10.1016/0927-0248(92)90016-I.

Spina, M., Bonvin, E., Sienkiewicz, A., Náfrádi, B., Forró, L., & Horváth, E. (2016). Controlled growth of CH3NH3PbI3 nanowires in arrays of open nanofluidic channels. *Scientific Reports*, *6*, 19834. Available from https://doi.org/10.1038/srep19834.

Tamang, A., Sai, H., Jovanov, V., Hossain, M. I., Matsubara, K., & Knipp, D. (2016). On the interplay of cell thickness and optimum period of silicon thin-film solar cells: Light trapping and plasmonic losses. *Progress in Photovoltaics: Research and Applications*, *24*, 379–388. Available from https://doi.org/10.1002/pip.2718.

Wang, H.-P., Lien, D.-H., Tsai, M.-L., Lin, C.-A., Chang, H.-C., Lai, K.-Y., & He, J.-H. (2014). Photon management in nanostructured solar cells. *Journal of Materials Chemistry C*, *2*, 3144. Available from https://doi.org/10.1039/c3tc32067g.

Wurfel, U., Cuevas, A., & Wurfel, P. (2015). Charge carrier separation in solar cells. *IEEE Journal of Photovoltaics*, *5*, 461–469. Available from https://doi.org/10.1109/JPHOTOV.2014.2363550.

Xu, X., Liu, Z., Zuo, Z., Zhang, M., Zhao, Z., Shen, Y., ... Wang, M. (2015). Hole selective NiO contact for efficient perovskite solar cells with carbon electrode. *Nano Letters*, *15*, 2402–2408. Available from https://doi.org/10.1021/nl504701y.

Yablonovitch, E. (1982). Statistical ray optics. *Journal of the Optical Society of America*, *72*, 899. Available from https://doi.org/10.1364/JOSA.72.000899.

Yoshikawa, K., Kawasaki, H., Yoshida, W., Irie, T., Konishi, K., Nakano, K., ... Yamamoto, K. (2017). Silicon heterojunction solar cell with interdigitated back contacts for a photoconversion efficiency over 26%. *Nature Energy.*, *2*, 17032. Available from https://doi.org/10.1038/nenergy.2017.32.

Zhao, D., Chen, C., Wang, C., Junda, M. M., Song, Z., Grice, C. R., ... Yan, Y. (2018). Efficient two-terminal all-perovskite tandem solar cells enabled by high-quality low-bandgap absorber layers. *Nature Energy.*, *3*, 1093–1100. Available from https://doi.org/10.1038/s41560-018-0278-x.

Zhou, J., Meng, X., Zhang, X., Tao, X., Zhang, Z., Hu, J., ... Yang, S. (2017). Low-temperature aqueous solution processed ZnO as an electron transporting layer for efficient perovskite solar cells. *Materials Chemistry Frontiers.*, *1*, 802–806. Available from https://doi.org/10.1039/C6QM00248J.

Zhu, J., Yu, Z., Fan, S., & Cui, Y. (2010). Nanostructured photon management for high performance solar cells. *Materials Science and Engineering R*, *70*, 330–340. Available from https://doi.org/10.1016/j.mser.2010.06.018.

Zuo, C., Bolink, H. J., Han, H., Huang, J., Cahen, D., & Ding, L. (2016). Advances in perovskite solar cells. *Advancement of Science*, *3*, 1500324. Available from https://doi.org/10.1002/advs.201500324.

CHAPTER

Commercial viability of different photovoltaic technologies

5

Mohammad Aminul Islam[1], Md. Akhtaruzzaman[2], Nowshad Amin[3] and Kamaruzzaman Sopian[2]

[1]*Department of Electrical Engineering, Faculty of Engineering, Universiti Malaya, Jalan Universiti, Kuala Lumpur, Malaysia*

[2]*Solar Energy Research Institute, Universiti Kebangsaan Malaysia, Bangi, Malaysia*

[3]*Institute of Sustainable Energy, Universiti Tenaga Nasional (@UNITEN; The Energy University), Kajang, Malaysia*

5.1 Introduction

The energy demand is increasing concurrently with the increase of the world's population. To satisfy the growing energy demand, solar photovoltaics (PV) are considered an indispensable source of energy all over the world and has attracted significant attention in recent times. As a consequence, the PV industry is experiencing huge development potential and large-scale development to replace traditional energy. Also, a significant increase in energy conversion efficiency and a decrease in the price of the PV modules along with various national policies over the world enhanced the solar PV-based energy generation with the lowest levelized-cost-of-energy. As a consequence, with over 600 GWp of accumulated installed capacity as of 2019, solar PV systems have become a major contributor to global electricity production (Abdullah, Osman, Ab Kadir, & Verayia, 2019). PV electricity is now cheaper than electricity from conventional sources such as coal in many regions of the world, making PV the most widely deployed new generation source, along with wind (Ech-Charqaouy, Saifaoui, Benzohra, & Lebsir, 2020). Among the several PV technologies available in the commercial market, crystalline silicon (c-Si) PV modules are used in most current PV installations (Fu et al., 2015; Gielen et al., 2019).

Despite the current success, the degradation of PV modules under the stress of natural and environmental factors has recently been observed, which has great impacts on the commercial market and makes solar PV inefficient compared to traditional energy sources. Particularly, the degradation rate of PV modules depends greatly on numerous factors, including manufacturer, technology, local environment, system mounting, failure modes and mechanism, and measurement uncertainty, etc. Thus it is very important to define the causes of degradation and

Comprehensive Guide on Organic and Inorganic Solar Cells. DOI: https://doi.org/10.1016/B978-0-323-85529-7.00014-1

failure modes and finding suitable PV technology for a specific site for increasing their commercial viability. An in-depth investigation and detail statistical data are indispensable to ensure optimal efficiency and maximum energy yield of PV modules for a specific location and climate condition.

As observing failure modes in the field need a long time, accelerated lifetime testing has been developed. However, the accelerated test method should be matched with the observed environmental factors and associated failure modes in the field. Therefore an important part of designing lifetime service tests is to detail and measure failure modes and power loss from field observation. The PV industries rely on qualification testing named IEC 61215 and IEC 61646 for product lifetime prediction. This chapter is dedicated to the reliability and various failure modes that have been documented for various PV technologies over the last few years via field and laboratory test investigation.

5.2 Photovoltaic performance

PV performance is measured in efficiency, which quantifies the amount of incident solar irradiation on a PV module surface and converted into electricity. Particularly, PV module performance is determined by two main factors; the PV solar cell efficiency, based on the cell design and material type, and the total module efficiency, based on the module layout, configuration and size. Because of the recent advancements in PV technology in recent years, average module conversion efficiency has improved from 15% to 20%. There are three major types of solar cells available in the market: monocrystalline or single crystalline silicon (sc-Si) solar cells, polycrystalline (pc-Si) silicon solar cells, and thin-film. Each type has its unique advantages and disadvantages, which are shown in Table 5.1.

Generally, c-Si and p-Si solar cells, both are made of silicon wafers. The c-Si solar cells are cut out of a single pure silicon crystal. On the other hand, p-Si solar cells are made up of silicon crystal fragments that are fused in a mold before being cut into wafers. The c-Si solar cells usually have the highest conversion efficiencies and power generation capacity among all types of solar cells. The c-Si PV module efficiency can be higher than 20%, while the efficiencies of the p-Si PV module typically between 15% to 17% only. Also, the power capacity of the recent commercialized c-Si PV module is more than 300 watts (W) of power capacity, interestingly, some of even exceeding 400 W. Alternatively, p-Si PV module tend to have lower wattage. Thus even both c-Si and p-Si PV modules have come into the commercial market with 60, 72, or 96 solar cell variants (usually for large-scale installations); however, c-Si modules can produce more electricity. The highest market share occupied by the c-Si module is shown in Fig. 5.1 (Xu et al., 2018).

Table 5.1 Characteristics of different types of commercial photovoltaic modules or solar cells.

Solar module type	Advantages	Disadvantages
Mono or single crystalline (sc-Si)	• High efficiency • High performance • Better performance in high light conditions • Esthetic	• Expensive • Circular (square-round) cell creates wasted space on the module • High-temperature coefficient
Multi or polycrystalline (pc-Si)	• Less expensive • Square shape cells fit into module efficiently using the entire space	• Lower efficiency • Lower performance • High-temperature coefficient
Thin-film	• Low cost • Could be flexible • Less susceptible to shading problems • Better performance in low light conditions • Lightweight • Esthetic • Low-temperature coefficient	• Lower efficiency • Sometimes difficult to synthesis • Some may have toxic material

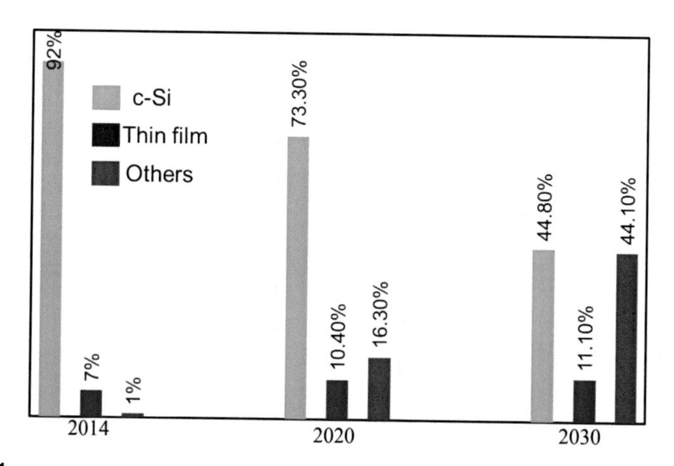

FIGURE 5.1

Market share of PV modules by technology type (2014–2030).

Image generated from data by Xu et al., 2018. Xu, Y., Li, J., Tan, Q., Peters, A. L., & Yang, C. (2018). Global status of recycling waste solar panels: A review. Waste Management, 75, 450–458.

However, thin-film solar cells are made from various types of materials. The prevalent type is cadmium telluride (CdTe) thin-film solar cells. Additionally, Copper Indium Gallium Selenide (CIGS) is another popular thin-film solar cell. The physical structure of both solar cells is very similar, but the CdTe is a superstrate type and CIGS is substrate type. Thin-film solar cells are also made from Si, which is known as amorphous silicon (a-Si) solar cells. Even though these a-Si solar cells are made of Si; they are not composed of solid silicon wafers. Rather, they are made on glass, metal, or plastic substrates, similar to other thin-film solar cells. The efficiency and/or power capacity of thin-film solar cells are lower than that of c-Si and p-Si solar cells. First Solar reported an average efficiency of 16.1% at the end of 2015, which is the highest reported CdTe module efficiency. Besides, Avancis GmbH reported the highest efficiency of CIGS PV module up to 15.5% on $30 \times 30\ cm^2$ and 13.9% on 1 m^2 size (Karg, 2012).

Regardless of the power conversion efficiency, c-Si PV modules are most expensive than any other type. The higher cost largely comes from the manufacturing process cost comes from the wafer processing. Particularly, making a single-crystalline Si wafer is energy-intensive and results in wasted silicon. Due to this, p-Si modules are much cheaper than c-Si, because the cells are made from silicon fragments (sometimes wasted of the c-Si wafer) rather than a pure and single crystalline silicon. But thin-film PV modules are typically cheaper than c-Si and p-Si modules, and particularly, CdTe is the cheapest PV module to produce, while CIGS solar modules are highly expensive to manufacture than both a-Si and CdTe. Moreover, the PV module installation cost is much lower than the c-Si and p-Si PV modules due to the fewer labor requirements. Particularly, thin-film PV modules are lighter weight, and easy to handle, transport, and secure in place (Sica et al., 2018).

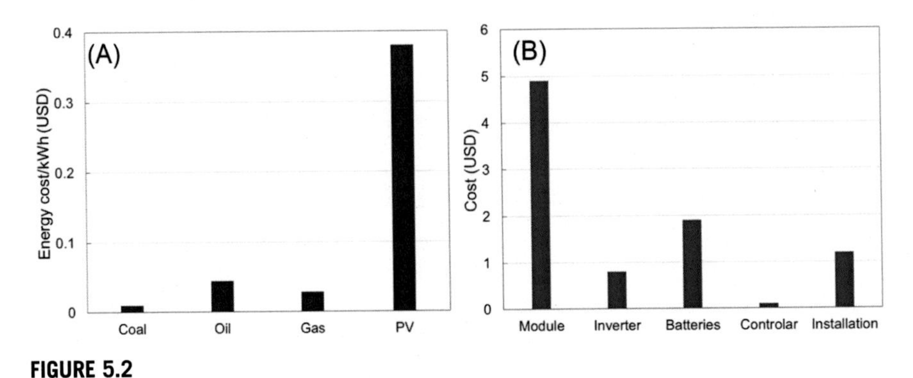

FIGURE 5.2

(A) Comparisons of energy cost per kilowatt-hour from different sources and (B) cost of solar energy per watt belongs to the system components.

Image generated from data from https://greenecon.net/understanding-the-cost-of-solar-energy/energy_economics.html.

Average energy cost per kilowatt-hour (kWh) and PV system cost per-watt energy generation is shown in Fig. 5.2A and B. It could be seen that the cost of solar energy in producing electricity is considerably higher compared to traditional hydrocarbon fuels such as coal, oil, or gas. It could also be seen in Fig. 5.2B that the highest cost is contributed by the PV module. Thus current research and manufacturers are mostly giving focus on cost reduction via reducing cell fabrication costs for which c-Si with a variety of structures are seen in recent years. However, as this is not the concern of this chapter, we will not discuss it here. Even that PV has exhibited a gradual cost reduction and seen a rapid incline after 2005 (Fig. 5.3), more cost reduction is expected for minimizing the fossil fuel dependency. The PV module cost can be reduced further by decreasing the cost of material production per watt and labor cost using new technologies rather than traditional methods. Also, the minimum efficiency of the PV module should increase to 25% and the lifetime should be increased to 30 years (Trube et al., 2018). Particularly, if the PV module efficiency and lifetime increase, the cost related to the balance of the system will also be reduced. The key cost reductions facts for the production of electricity from PV systems are as follows (Weckend, Wade, & Heath, 2016):

1. Higher energy conversion efficiency.
2. Less material consumption.

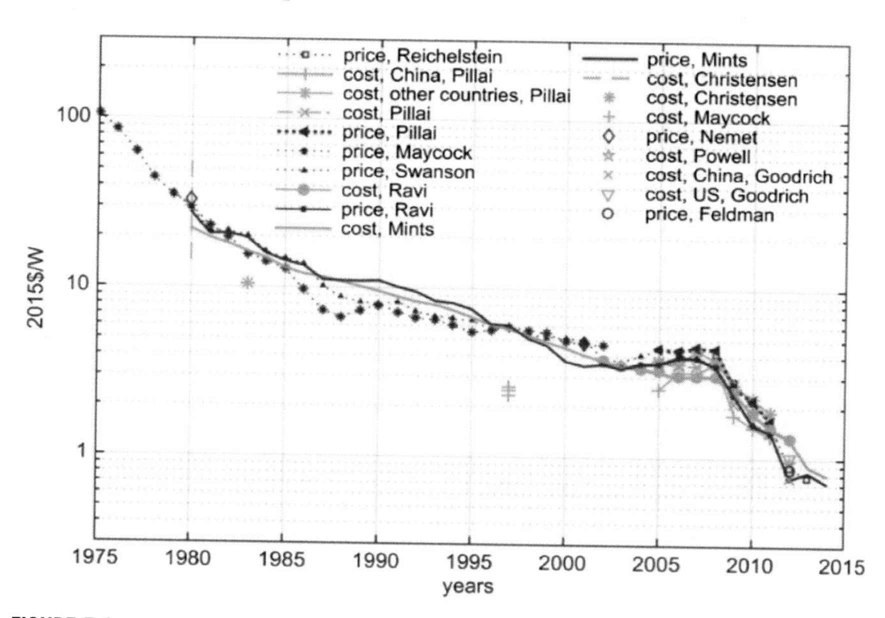

FIGURE 5.3

Costs and prices of modules after 1975. The cost is shown in orange and the prices are shown in purple. Values are averages across different c-Si PV technologies (Kavlak, McNerney, & Trancik, 2018).

Table 5.2 Pathways of cost reduction potential for photovoltaic systems (Cengiz & Mamiş, 2015).

Characteristics	Value or qualifier
1. Module	
a. Efficiency	≥25%
b. Substrate	Lower cost and lower weight than glass
c. Reliability	≥30 years or can be substituted for minimum labor
d. Materials	Earth-abundant, nontoxic, established, or possible to recycle
2. Balance of the system/ installation	
a. Labor	Can be installed by nonspecialized labor
b. Process	Lightweight, easy to handle, and no need for any special equipment
Assembly	Snap together mechanical and electrical
3. Power electronics	
a. Efficiency	≥95%, optimized module-peak power management
b. Reliability	≥30 years
c. Assembly	Wiring integration, components with minimum electrical connections

3. Cost-effective materials.
4. Optimized manufacturing and mass production.
5. Improved PV module technologies.
6. Optimized integration of the grid (smart grids).
7. New design concepts for energy conversion using PV technologies.

Research and development work in PV sectors show that PV module and system prices will be reduced soon, and using PV systems will be increase exponentially. Besides, as an aim of cost reduction in PV modules and systems, researchers in PV determined a variety of pathways as shown in Table 5.2.

5.3 Stability and reliability

As the installation capacity of global PV approaches 900 GW by 2025, the concern about the PV module stability and reliability is becoming crucial. Particularly, investors are looking for confidence in the long-term stability of PV performance the investment will be profitable only through reliable operation for a decade. The stability and reliability in PV is a key factor in the design of PV modules or systems, which are mostly involved in cost approximation. The research on PV module reliability and stability is well recognized and has led to significant contributions to the quality improvement of PV system components and modules. Several inherent system causes may influence the reduced stability and deviation of PV system performance from the expected output (Bosman & Darling, 2018). The primary cause is related to the PV module and system design. It has been reported that 40% of PV module failures are related to microscopic cracks (Komoto et al., 2018). This failure mode is often for newer modules manufactured following 2008 when the fabrication of thinner cells started (Haque, Bharath, Khan, Khan, & Jaffery, 2019). The module encapsulation materials, especially degradation of ethylene-vinyl acetate (EVA) influence the module performance and stability. The failure modes such as cell cracks, discoloration, and delamination depend mostly on the EVA degradation (Dias, Javimczik, Benevit, Veit, & Bernardes, 2016). But the solar cells in a PV module are connected in series-parallel topology and several PV modules are connected parallel in an array which are then connected in series for achieving expected system power and voltage. Therefore any mismatch in the connected strings can affect the total system performance and ultimately performance stability. Second, PV module installation can be flawed. The module installation angle and face should comply with the climate zone. The module better is installed at a suitable angle for which it could less affect by wind and snow and its surface is clean the dust and debris automatically via rainfall. Particularly, the accumulation of dirt and debris for a long time on the module surface degrade the module performance (Bouraiou et al., 2018) and led to the other failure modes, such as potential induced degradation (PID). Third, the PV system is connected to the inverter for DC to AC conversion. The incorrect sizing of the inverter may cause lower system performance. Also, incorrect inverter less robust and increased the possibility of failure. Also, other electrical types of equipment include junction boxes, fuse boxes, cabling, and/or grounding have a significant impact on the stability and reliability of a PV system.

Table 5.3 shows the degradation rates of different PV technologies that are commercially available now where a general scenario of the degradation behavior of the different PV technologies could be seen. It should be mentioned that the degradation rates of PV modules depend on numerous factors including the climate zone, temperature, humidity, and UV light including system size and design. However, every technology is improving over time, and each has the opportunity to make further progress. For making more reliable PV technology, advancement in materials science and process engineering is essential.

5.3.1 Climate and technology

The technological choices from the different PV available in the market are very critical in the sense of the lack of guide and forecasting tools suited to the climates and environment of the installation sites. There have been many PV system projects going to fail due to the bad choice of PV technology where failure causes are influenced by the environmental parameters, such as heat, humidity, partial

Table 5.3 Outdoor performance and degradation rates for photovoltaic technologies described in various reports.

Different photovoltaic (PV) technologies	The efficiency of PV modules as reported by 2015		Reliability	
	Most widely shipped module	Efficiencies offerings from various brands	Performance degradation rate (95% of confidence interval) (Jordan & Kurtz, 2013; Jordan, Kurtz, VanSant, & Newmiller, 2016)	Total range [B]
sc-Si	16%	(14%–22%)	0.5%–0.8%/year	0.4%–2.8%/year
pc-Si	16%	(13%–19%)	0.5%–0.8%/year	0.4%–3.2%/year
CdTe	16%	(13%–18%)	0.7%–1.0%/year	0.0%–1.5%/year
CIS or CIGS	14%	(12%–17%)	0.7%–1.0%/year	0.4%–3.0%/year
a-Si	8%	(6%–10%)	1.0%–1.3%/year	0.2%–4.5%/year

shadow and dust, etc. Manufacturers provide a characteristic of PV modules measured in standard test conditions, however, the performance cannot reach that level in real operating conditions. Performing a PV module in the installation sites depends on the weather (Huld & Amillo, 2015), the cell/module design (Idoko, Anaya-Lara, & McDonald, 2018), the solar spectrum, and the spectral response of each technology (Okullo, Munji, Vorster, & Van Dyk, 2011). For example, crystalline silicon (Si) technology could be a good choice for a high level of irradiation whereas amorphous Si and chalcopyrite (CdTe and CIGS) based modules behaved opposite and show voltage and current mismatch due to materials' metastable phenomenon (Okullo et al., 2011). Besides, the PV module's performances and aging strongly depend on the climate and the surrounding environment of the installation site.

Several studies have focused on the knowledge and impact of the environment on the performance of the PV technology that delivers the best trade-off between the performance and cost of the PV module. Akhmad et al. (1997) have been compared the performance of polysilicon (pc-Si) and amorphous silicon (a-Si) at Kobe, Japan, and found a-Si modules are better for this region. Nishioka et al. (2003) compared sc-Si, pc-Si module, and heterojunction silicon at Nara Institute of Science and Technology under Japanese climate. They reported that the HIT technology is better suited for this region due to its low-temperature dependency. Poissant (2009) have evaluated the temperature effect on four novel PV module technologies, (1) heterojunction silicon, (2) back-contact crystalline silicon, (3) amorphous silicon triple-junction, and (4) laser-grooved buried junction crystalline Silicon. His study confirmed that the heterojunction silicon and triple-junction amorphous silicon technologies are less affected by temperature than other two crystalline silicon technologies. Cañete, Carretero, and Sidrach-de-Cardona (2014) also performed a comparative study four PV module technologies, (1) amorphous silicon (a-Si), (2) tandem structure of amorphous silicon-microcrystalline silicon (a-Si/mc-Si), (3) polycrystalline silicon module (pc-Si) and (4) cadmium telluride (CdTe). Their results show that the performance of thin-film modules is better than that of pc-Si modules for the location of Southern Spain. Sharma, Kumar, Sastry, & Chandel (2013) performed a comparative study of the performances of a-Si, p-Si, and HIT PV module technologies under Indian climatic conditions and found that a-Si and HIT have better performance than p-Si at this location. Eric Maluta (Maluta & Sankaran, 2011) evaluated the performances of c-Si and a-Si PV modules under South Africa climate conditions. His findings showed that both technologies give a similar and apposite performance for the climate of this region. Three different PV technologies (monocrystalline, polycrystalline, and amorphous silicon) have been evaluated under desert climate by Shaltout, El-Hadad, Fadly, Hassan, & Mahrous (2000). They reported that the polycrystalline silicon cells are more suitable in such a climate. The effect of spectral irradiance distribution on the performance of a-Si/mc-Si stacked PV modules has been analyzed by Minemoto et al. (2007) installed at Kusatsu-city (Japan). Their study revealed these stacked PV modules are extremely spectrally sensitive compared to pc-Si PV modules installed on the same site. Aste, Del Pero, and Leonforte (2014) have been performed a comparative analysis of c-Si, a-Si/mc-Si stack cell, and heterojunction under different temperate climates of Italy. Their analysis shows that the a-Si/mc-Si stack cell showed performance higher than the other technologies tested in this study, which may be due to its low-temperature coefficient and thermal annealing. All these above-mentioned studies indicate the difficulty for choosing the appropriate PV technology for a site. Thus the prediction of PV energy potentials before installation helps us to understand the economic advantages associated with it and for policy regulation for electric utilities.

Earlier literature has been indicated that some specific factors are influencing the performance evaluation of solar PV plants. García-Domingo, Aguilera, De la Casa, & Fuentes (2014) has been reported the effect of insulation on the performance of PV modules. It is important to mention that the output efficiency could be severely affected at relatively lower or higher solar densities due to the improper insulation. Besides the solar insulation effect, sunshine duration is one of the most important factors that affect PV plant output. The longer sunshine hours provide better annual average electricity outputs (Tsao, Lewis, & Crabtree, 2006). Above and beyond, other variables including wind speed, clouds, shading, and soiling also well recognized as the primary factors that influence PV power generation efficiency.

Conversely, the efficiency loss of a PV plant due to aging is also closely related to the environment of the installation site. Due to the complex and changeable weather in which PV modules are working susceptible to some factors such as dust, gravel, haze, wind, snow, high temperature, and their high potential. These harsh working environments bring faults like "cracks," "hot spots," "PID," and other electrical equipment failures that often appeared in age, which are prone to reduce the efficiency and lifetime of PV systems. They not only lead to a large energy loss but can also make the entire PV system paralyzed in severe cases. Therefore a scientific and accurate fault detection method of PV modules is also projected in this proposal with the measured relevant electrical and environmental parameters. The impact factors that affect system performance will be analyzed, which will help to understand the working conditions of the PV system. Finally, the maximum potential of its performance and benefits is achieved, which lays a good foundation for the further development and promotion of PV generation technology and has a great strategic significance for adjusting the energy structure.

5.4 Failure and degradation modes

The understanding of the synergistic nature of the various failure modes via the underlying chemical and physical processes is indispensable. However, research exploration on this field yet not to be completed and is demanding more research because PV modules are multifaceted and can show a variety of failure modes. Moreover, the cost and size of PV modules typically hinder for testing of many samples. Besides, the surrounding environment of an installed PV module that is seasonal climate-dependent can be varied in many ways, leading to various failure modes.

As the field failure occurred after a long time, thus the accelerated test has been developed for predicting the stability and reliability of PV modules. The acceleration test can be duplicate the failure modes observed in the field. Particularly, Information on failure modes that occurred in the field can be acquired empirically through warranty returns (Wohlgemuth, Cunningham, Monus, Miller, & Nguyen, 2006), site visits (DeGraaff, Lacerda, Campeau, & Xie, 2011), and/or maintenance records (Golnas, 2012). However, most of the time it becomes very difficult to link between time elapse and module failure modes due to several reasons. Practically different modules, even similar types of PV modules produced by different manufacturers often show different failure modes at different times in their life cycle. PV modules most often show multiple failure modes at the same time, making it difficult to isolate single failure mode causes for a specific power loss. Thus PV failures that alternatively identify the stability classify by mechanism, modes, and detectability and failure modes are generally specified by effects that led to observed failure (Pentti & Atte, 2002). The observed PV module failures are summarized in Table 5.4 in terms of failure mechanism, mode, onset, severity, and detectability (Jordan & Kurtz, 2017). Particularly, visual inspection is a primary failure observation on which another

Table 5.4 Commonly observed PV module failure modes, mechanism, severity, time onset, and detectability.

Failure mode	Failure mechanism	Severity	Onset	Detectability
Discoloration (EVA)	Chemical alteration of cross-linking	Low	Gradual	Evident, visual
Delamination (front side)	Adhesion loss, loss of elastomeric properties	Low to high	Gradual	Evident, visual
Back sheet failure	Adhesion loss, fracture	Low to high	Gradual or sudden	Evident, visual
Hot spot	Fracture, electrical mismatch, partial shading	Low to high	Gradual	Hidden, need IR
Burn mark	Fracture, electrical mismatch	High	Gradual	Evident, visual
Potential induced degradation	Electrical stress, Na migration	High	Gradual	Hidden, need IR, EL, performance data
Glass breakage	Fracture	High	Sudden	Evident, visual
Cell breakage	Fracture	Low to high	Sudden	Evident, visual
Internal circuitry discoloration	Corrosion	Low to high	Gradual	Evident, visual
Internal circuitry interruption or disconnect (solder bonds, ribbon, etc.)	Corrosion, fatigue	High	Sudden	Visual, detail inspection required
External circuitry disruption (j-box, bypass diode, cables)	Wear, electrical stress	Low to high	Gradual	Evident, visual
Soiling	Wear, etching of glass	Low	Gradual	Evident, visual
Structural failure	Deformation (frame from snow load)	High	Sudden	Evident, visual
AR coating delamination	Adhesion loss	Low	Sudden	Evident, visual
LID	Chemical, oxygen-boron complex	Low	Gradual	Hidden, need performance data

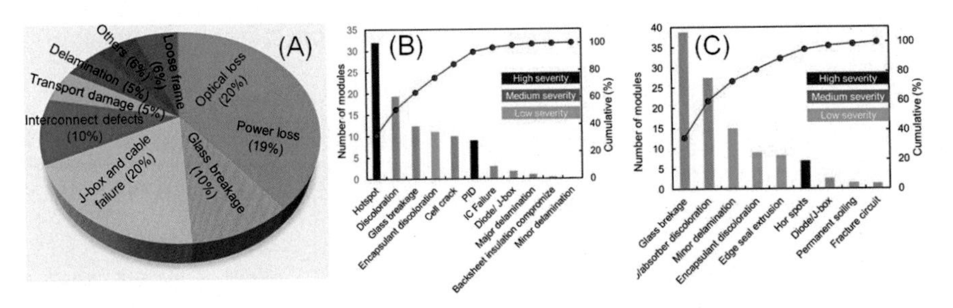

FIGURE 5.4

PV module failure rates according to customer complaints (A) (Chowdhury et al., 2020; Komoto et al., 2018), Pareto chart of the most significant degradation modes in c-Si (B) and thin-film modules (C) (Jordan, Silverman, Wohlgemuth, Kurtz, & VanSant, 2017).

inspection relies, however, it certainly depends on the person's knowledge and experience going for inspection (Walston et al., 2015). According to the customer complaint, the highest loss comes from the optical loss and j-box failure. Fig. 5.4A shows the PV failure rate according to the customer complaints (Chowdhury et al., 2020; Komoto et al., 2018) and Fig. 5.4B and C shows the different failure modes in c-Si and thin-film PV modules according to the number of modules and severity. The details on the failure modes will be discussed in the next section.

5.4.1 Solar cell degradation

Mechanisms of performance deterioration may involve either a gradual decrease of PV module output power overtime or an overall decrease in power because of failure that occurred in an individual solar cell in the module. Particularly, solar cell performance and quality degradation results from various operating condition, includes temperature cycle, humidity, and installation condition. PV modules installed in desert regions are subjected to various temperature cycles with significant gradients that accelerated the cell degradation.

It has been reported that the open-circuit voltage (V_{oc}) is mostly affected by the temperature cycle (Amelia et al., 2016). Also, it was found that amorphous silicon (a-Si) type solar cells degrade to 38% while operating in open circuit conditions; however, only 14% degradation is observed while operating in maximum power conditions (Lund, Sinclair, Pryor, Jennings, & Cornish, 1999).

Particularly, performance degradation at cell levels are caused of:

1. Increases in series resistance (R_{S}) as a cause of decreased adherence of contact fingers or corrosion in the finger level. Water vapor entering through the back sheet is the main cause of finger corrosion and an increase of R_s.
2. Decreases in shunt resistance (R_{SH}) owing to metal ion migration, locally shorting the cell, and/or reduced the p-n junction quality. Particularly, the high

voltage in the PV module frame influence the Na^+ ion migration from the module glass to the cell leads to the deterioration of R_{sh}, the process simply known by PID (Islam, Noguchi, Nakahama, & Ishikawa, 2016a). The details on PID will be discussed in the next section.

3. Deterioration of antireflection coating as an impact of encapsulant decomposition. As an impact of UV light in the incident radiation, the encapsulant, such as EVA decomposition to acetic acid (Islam, Noguchi, Nakahama, & Ishikawa, 2016b) that led to the deterioration of the anti-reflection coating. It is also well known that EVA decomposition influences the PID mechanism.

Other than the above causes, the individual cell could be shorted as illustrated in Fig. 5.5. Even this type of failure is not common for wafer-based solar cells, however, very common to occur in thin film-based solar cells. Particularly, the front and rear contacts are much closer in thin-film solar cells for which it has more chance to shorted two consecutive solar cells via pin-holes, corroded region, and/or damaged solar cell material.

Another common failure in cell level is cell cracking as illustrated in Fig. 5.5B. Here, busbars hold the cracked part and prevent open circuit failure. The cell seems to function as a normal solar cell; however, the ultimate output is reduced, especially cell I_{sc} is reduced. Cracked cell failure has become a very common recent time in the c-Si PV module because the thinned wafer is using for cell fabrication. Particularly, cell cracking is an early stage failure observed in all types of PV technologies while it could occur during the processing and assembly, transporting of a module. Some typical situations leading to cell cracks are (Köntges et al., 2014):

1. PV modules falling.
2. During transportation, the rigid pallet unevenly touching the lowest PV module in the stack.

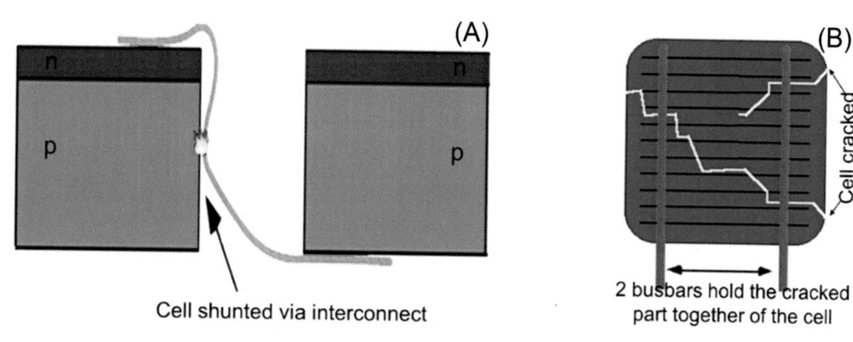

FIGURE 5.5

(A) Cell failure via interconnect shorting and (B) cell cracking (holding cracked part by busbars can help to prevent open-circuit failure).

3. Too tight transport corners in the transport stack. The second highest module is also lifted during the de-stacking of the top module of the stack and unexpectedly drops down.
4. Someone walking on top of the PV module.
5. The cells in PV modules can crack during "natural" transport, even in well-designed transport containers.

In the c-Si module fabrication stage, the cracked cell could be identified by "snail tails" in which encapsulant shows a different color near the crack than elsewhere as shown in Fig. 5.6A and B and reported that the substantial power loss is occurred due to these cracked cells (Dolara, Leva, Manzolini, & Ogliari, 2014). Usually these cracks cannot detect by visual inspection during the manufacturing process, however, they visible sometime later. In that case, electroluminescence (EL) and infrared thermal (IR) imaging are appropriate tools for identifying the cracked cells in a module in Fig. 5.6C and E (Kurtz et al., 2015). Also, cell cracking can occur in many technologies as an impact of thermal stress, since different module materials may not behave comparable way to the thermal cycle.

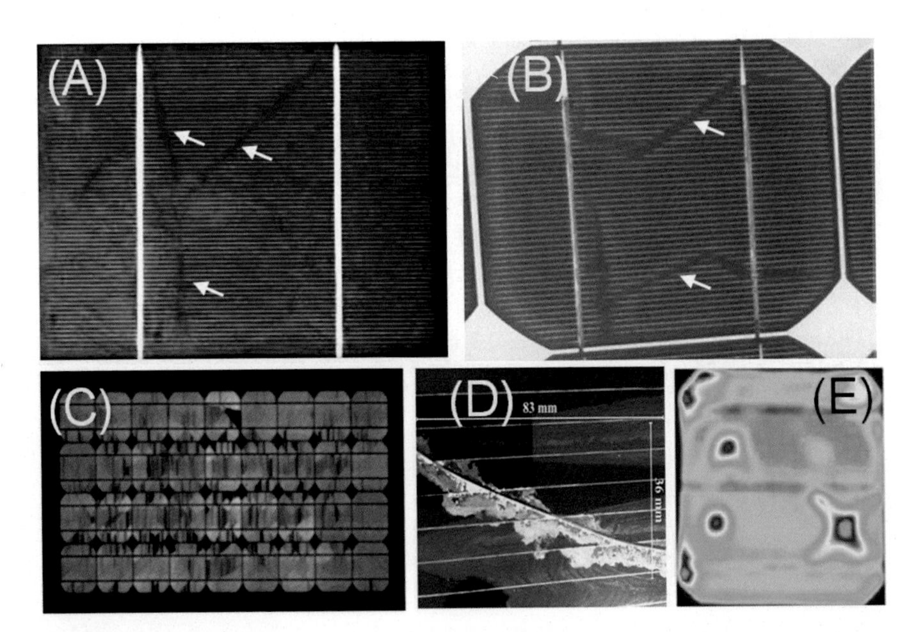

FIGURE 5.6

(A) Image showing 'snail trails' that decorate cracked cells in c-Si PV modules (Duerr, Bierbaum, Metzger, Richter, & Philipp, 2016), (B) the discoloration reveals the underlying cracks, (C) Electroluminescence image of a module with many cracked cells in c-Si PV modules (Jordan & Kurtz, 2017), (D) optical image of a cracked thin-film PV module due to the stressed induced by ice load (Demirtaş, Tamyürek, Erol, Çetinbaş, & Öztürk, 2019) and (E) hot spots detected by IR camera along the crack (Dong et al., 2018).

Sometimes, thermal stress lead to microcracking which cannot be detected via visual inspection. These microcracks could influence the ion migration and accelerate the PID mechanism (Dong, Islam, Ishikawa, & Uraoka, 2018). Cell cracking could be occurred in the field aged module due to hail and ice load as shown in Fig. 5.6D. In severe cracking failure, sometimes fatigue could occur in the busbar due to wind loading and cyclic thermal stress leading to the open circuit failure.

5.4.2 Photovoltaic module failure

PV modules consist of several layers as illustrated in Fig. 5.7. Deterioration and damage of every component individually or combined with another led to the module power degradation and/or failure. The details of the different failure modes that occur at the module level have been discussed in this section.

5.4.2.1 Delamination

The module parts are sealed together by EVA encapsulation via a lamination process. However, due to the lack of manufacturing process and as an impact of hot and humid environment, the module layers are lost their adhesion and locally detached which recognized as delamination. The delamination may occur between the EVA and the cells or between the front glass and cells or between cell and back sheet. PV modules essentially are water and air-tight and the lamination work should be done under vacuum for ensuring this. Quick delamination occurred due to inadequate lamination where these causes may involve:

1. Mismatched and unclean glass and/or plastics are used.
2. Back sheet plastic is of substandard quality.
3. The thermal properties of EVA are not adequate for which it does not melt precisely over the module.

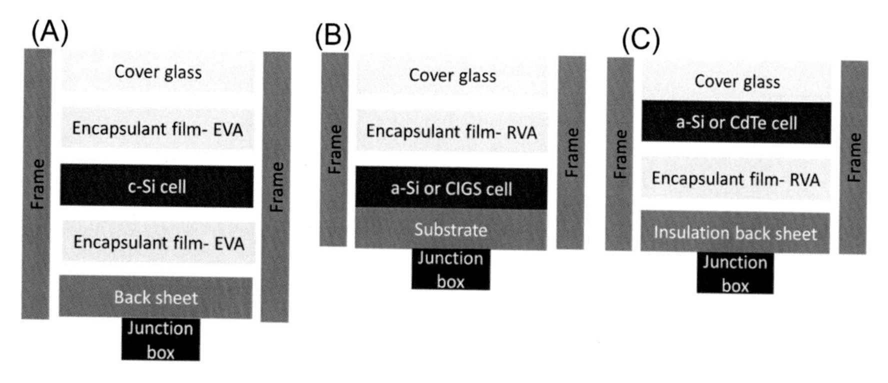

FIGURE 5.7

PV module structure; (A) c-Si PV module, (B) a-Si or CIGS V module (cell is substrate type), and (C) a-Si or CdTe PV module (cell is superstrate type). *PV*, Photovoltaic.

4. The temperature and pressure are poorly regulated during the lamination.

Generally, because air bubbles at or near the front and back surface can be seen, as shown in Fig. 5.8C, the detachment of the layer is evident. The delamination is most commonly occurred along the busbar, frame, and/or at the level of the ribbons. It is therefore easy, even without technical instruments, to recognize a delaminated module as seen in Fig. 5.8(A–D). Delamination affects the module I_{sc} at the initial stage because of light transmission loss and the development of an additional interface. It recognized as a major failure because as an impact of delamination, the water ingress inside the module structure is increases and the incident light reflection is increased (Yedidi et al., 2014). Unfortunately, the delamination problem is not recoverable, and it would be costly to recover the module's insulating ability. Here, it is important to swap the modules when the problem occurs, to prevent both more output losses and electrical hazards. If delamination persists and becomes more serious, more water vapor ingress that led to internal circuit corrosion and reduced the module FF. It has been reported that delamination is most critical if it arises on the module edges because it not

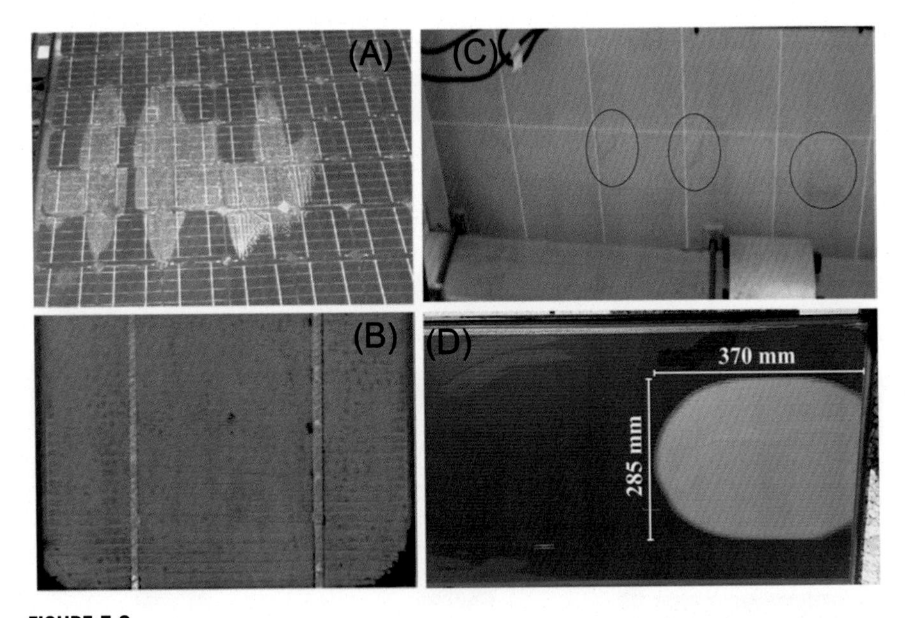

FIGURE 5.8

(A) Large-scale delamination across multiple c-Si solar cells (Alonso-Abella, Chenlo, Alonso, & González, 2014), severe delamination that occurred over an entire cell (Skoczek, Sample, Dunlop, & Ossenbrink, 2007), (C) delamination of the back sheet causing bubbles (Jordan & Kurtz, 2017), and (D) delamination in thin-film PV modules (Demirtaş et al., 2019).

only reduces the power production capability but also increases electrical risks for the module and also for the entire system (Ndiaye et al., 2013).

The delamination of a module can be understood via the acceleration stress test in the laboratory either conducting a dump heat test (85°C, 85% RH, 1000 hours), thermal cycle test, and UV-test according to the regulation IEC 61215. Skoczek et al. (2008) analyzed the degradation of PV modules associated with the delamination via acceleration test following IEC 61215 (Skoczek, 2008).

5.4.2.2 Corrosion

Corrosion is particularly preceded by delamination where ingress water vapor causes corrosion (Kempe, 2005). The retained water vapor in the module housing and react with the metal fingers of PV cells in presence of oxygen (Osterwald, Anderberg, Rummel, & Ottoson, 2002). Corrosion also reduces the adhesion between metal fingers and solar cells. Other than solar cells, corrosion could occur in the junction box in high humid conditions (Yedidi et al., 2014). Corrosion could easily detect via visual inspection as shown in Fig. 5.9. The power loss associated with corrosion is the reduction of FF through the increase of series RS and leakage current. According to Vázquez and Rey-Stolle, 2008; Realini, 2003; and Quintana et al., 2002, corrosion and discoloration are the major factor of PV modules' power degradation.

Wohlgemuth and Kurtz in 2011 (Wohlgemuth & Kurtz, 2011) reported that corrosion appeared after 1000 h of dump heat test (85°C, 85% RH), according to IEC 61215 standard. It has been reported that the major cause of the corrosion that occurred in PV modules edges is the reaction between sodium (Na) that is ample in module glass and ingresses water vapor (Rathore, Panwar, Yettou, & Gama, 2019). On the other hand, Osterwald et al. (2002) reported that the main cause of silicon PV module degradation is oxygen which influences the corrosion of the silicon junction (Osterwald et al., 2002). Kempe (2005) reported that there

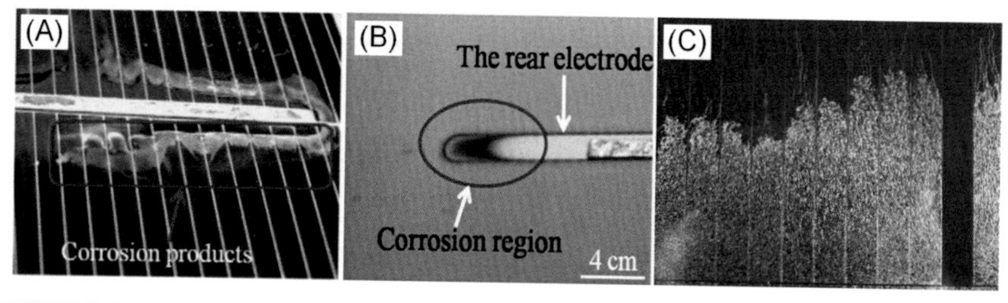

FIGURE 5.9

(A) Severe corrosion of the metallization (the busbar and fingers) (Xiong et al., 2017), (B) corrosion at the rear electrode of c-Si PV module after aging test conducted in indoor environment with 25°C, and 45% RH, and (C) TCO corrosion in a-Si thin-film PV module (Osterwald, McMahon, & Del Cueto, 2003).

is a correlation between the rate of degradation and the moisture in the PV module, especially for a hot and humid region like Miami, Florida (Kempe, 2005). On the other hand, except for high humidity, corrosion in thin-film PV modules occurred as seen in Fig. 5.9C when they operate in high voltage (Wohlgemuth et al., 2015). Hot and humid environments accelerate the water vapor diffusion in the EVA and corrosion occurred via EVA decomposition. It has been suggested that the only way to prevent corrosion is to reduce water vapor ingress into the module.

5.4.2.3 Hot-spot failures

In a PV module, individual solar cells are connected in series and current flow through the cells are identical. If the module is partially shaded due to surface fouling, soiling, and/or foreign objects on the surface then the cells in the shaded area become reverse biased and dissipate power that leading to heating effects. Bypass diodes are connected to the PV cells in parallel and the opposite direction for limiting the power dissipating effect as illustrated in Fig. 4.10A. Also, broken glass, cracked cells, and broken/bent frames can lead to hot-spots because they act similarly to the partial shading effect.

Partial shading, however, cannot lead to one or more cells being reverse biased. A current-generation mismatch between different cells in a substring may have a similar effect. The current generation mismatch may increase during long term field exposure where the different cells are degraded and/or affected differently, worsening the effect. Also, if the bypass diode somehow fails or broken, the heating will increase rapidly and will exacerbate the effect. Also, after a long time being exposed to the field, the connection between the internal module circuitry, such as high series RS, shunts, local delamination, weak solder joints, etc., and becomes weak due to the thermal cycling effect. Since the elevation in the hot-spot area, significant damage can occur in the module's front and back as shown in Fig. 5.10C and D, even it has a fire risk, thus safety becomes the primary concern. A robust O&M plan is needed that allows for early detection can minimize safety and revenue risks. Particularly, hot-spots in a module or a system can observe onsite using IR imaging technique as shown in Fig. 5.10C. With thin films, glass breakage causes hot spots along the crack and the risk of burning as seen in Fig. 5.10E and F.

5.4.2.4 Potential induced degradation

PID was first reported on high-efficiency n-type Si modules concerning the location in the string (Swanson et al., 2005); however, the cause was unknown at that time. A similar effect was discovered just a few years later for p-type c-Si modules put under negative bias (Pingel et al., 2010). The higher the voltage, the more power losses are more pronounced, and this PV module failure mode is thus called "potential induced degradation" (PID). The initial confounding result was there was severe degradation in modules that were subjected to the highest

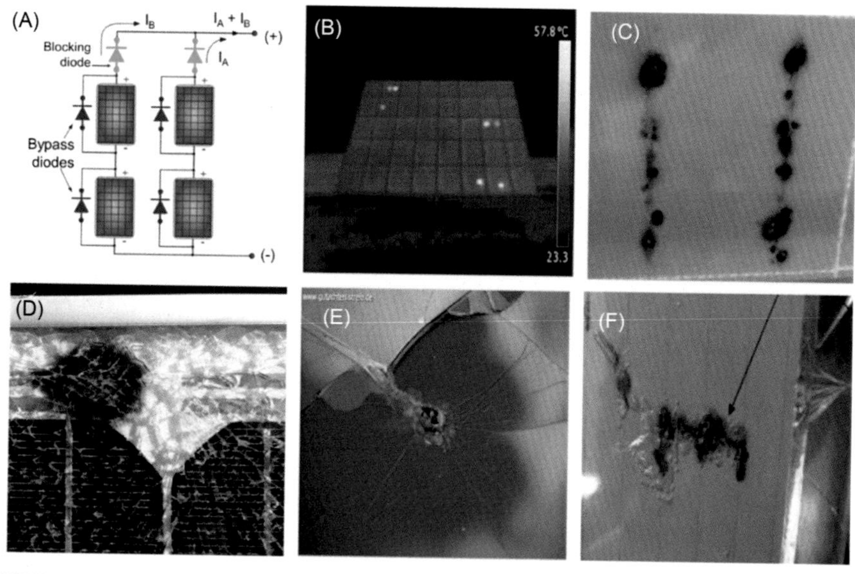

FIGURE 5.10

(A) By-pass diode in PV arrays, (B) hot-spots observable in infrared imaging (García, Marroyo, Lorenzo, Marcos, & Pérez, 2014), (C) and (D) damage caused by hotspots on the backside and front side of a module (DeGraaff, Lacerda, & Campeau, 2011), and (E) and (F) shows the hot-spot and burning in thin-film PV module lead by glass breakage and partial shading (Friesen, 2013). *PV*, Photovoltaic.

negative potential, as illustrated in Fig. 5.11. More complex thing is that it is very difficult to predict where and when PID might occur in a PV system.

Due to the rapid and severe effect on the efficiency and the reliability of PV systems (PV systems, PID has gained considerable attention last few years). So far it has been revealed in this rapidly evolving area that the migration of sodium ions (Na +) into the semiconductor junction as illustrated in Fig. 5.11(*middle*) appears to cause the effect. The migration of Na + ions at the interface of antireflection coating and Si has been detected utilizing multiple in-depth analysis tools by Islam, Matsuzaki, Okabayashi, and Ishikawa (2019) which is shown in Fig. 5.12. It has also been confirmed that PID influences the delamination between the encapsulant and the solar cells as seen in Fig. 5.12B.

These Na^+ ions at the Si surface increase the surface charge concentration (induced negative charge) and inverted the junction structure as illustrated by Naumann et al. (Naumann, Hagendorf, Grosser, Werner, & Bagdahn, 2012) as shown in Fig. 5.13A. This inversion intensity may depend on the density of the induced negative charge and the emitter's charge carrier density. This phenomenon has been confirmed employing SCM image analysis considering DC bias dependence and differential capacitance, as shown in Fig. 5.13B and C (Islam

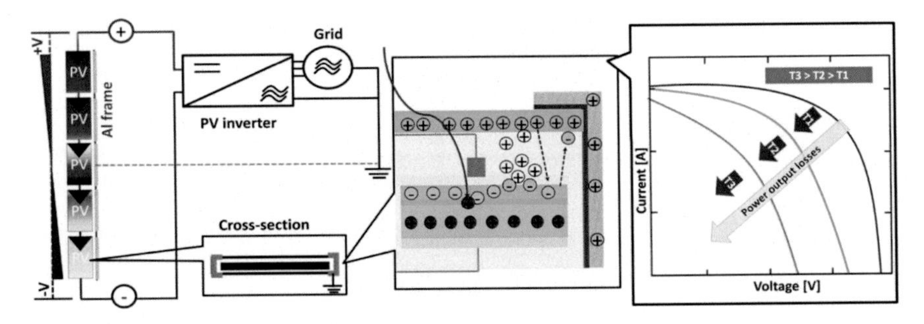

FIGURE 5.11

(*left*) Schematics of a PV array connected to the grid employing a transformerless inverter, (*middle*) module cross-section that elucidating the Na ion migration due to PID and (*right*) impact of PID on cell or module performance. PID, potential induced degradation; *PV*, Photovoltaic.

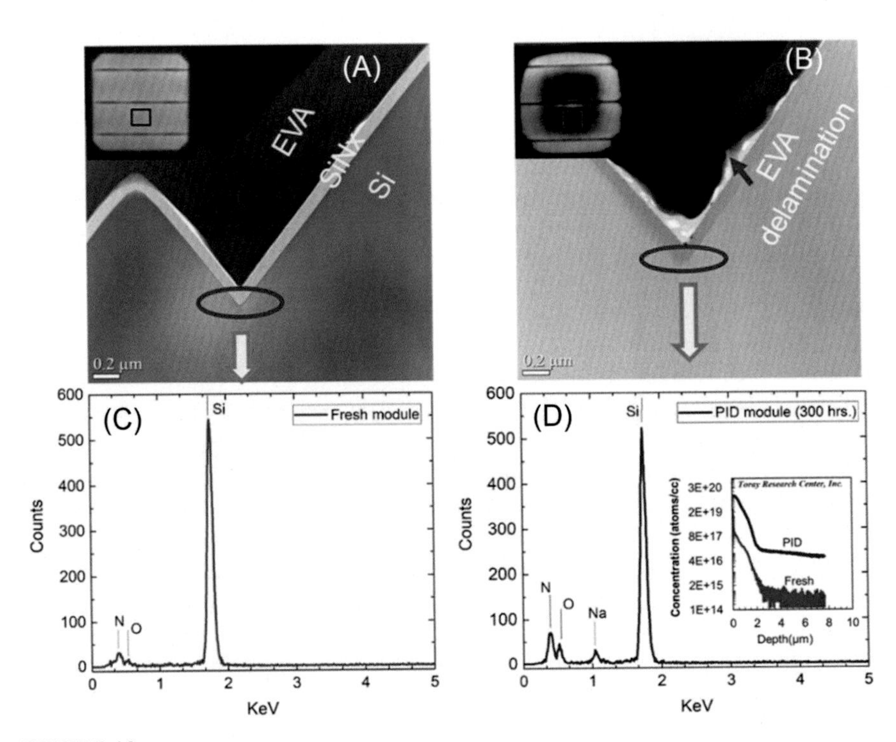

FIGURE 5.12

(Color online) (A), (B) BF STEM images of fresh and PID modules (inset show the examined areas). (C), (D) showing EDX spectra of the elemental composition in the solar cell at a selected area [inset of (D) shows the SIMS profile of Na in the modules] (Islam et al., 2018).

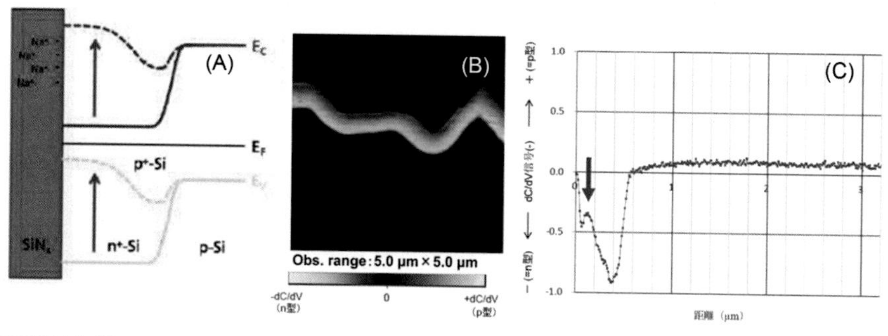

FIGURE 5.13

(A) Schematic of original p–n junction band structure, and changed by PID (dashed lines) (Naumann et al., 2012), (B) DC bias dependence SCM image of p–n junction and (C) variation of capacitance over the thickness [arrow in (C) shows the carrier concentration of n + layer near the surface seems to decrease] (Islam et al., 2018).

et al., 2018). Particularly, these accumulated Na^+ ions cause recombination in the junction and causing lattice damage by move over silicon nitride due to external electrical potential.

Similar to the c-Si modules, PID has also been commonly observed in the field aged CIGS and CdTe thin-film modules. Hack et al. studied details on PID that occurred in CdTe thin-film solar cells (Hacke et al., 2016). In thin-film PV modules, Na ions migrate to the ZnO buffer layer and CdS window layer through TCO and reduce shunt RS by degrading the CdS layer. Also, Na ions move to the laser scribes and decrease the shunt RS. TCO corrosion is also lead by PID in thin-film PV module (Wohlgemuth et al., 2015) and substantial power loss is observed. PID cannot detect via visual inspection. Lock-in-thermal (LIT) imaging is an opposite technique in the field measurement. The image should be taken when the array is operating at maximum power generation in highly illuminated conditions. Also, PID affected module could analyze by bringing back them into the lab. The LIT and EL image of a PID affected (accelerated tested) single-cell PV module are shown in Fig. 5.14. Particularly, shunt RS decreases as a cause of PID, thus the cell FF and V_{oc} are severely affected as illustrated in Fig. 5.11. The different factors that influence the PID in PV modules are summarized in Table 5.5.

Even several articles reported that the PID effects are reversible and affected modules begin to recover under the proper mitigation activities, however, this is not practically always true. Some modules might show improvement, however, improvement depends on the severity of the power loss while recovery work is started. Particularly, modules with minor power losses ($\leq 10\%$) are recoverable, but those with serious power losses ($\geq 30\%$) are not expected to completely recover (Islam, Oshima, Nakahama, & Ishikawa, 2017). The policy for resolving the PID effects can have a major impact on future PV plant performance.

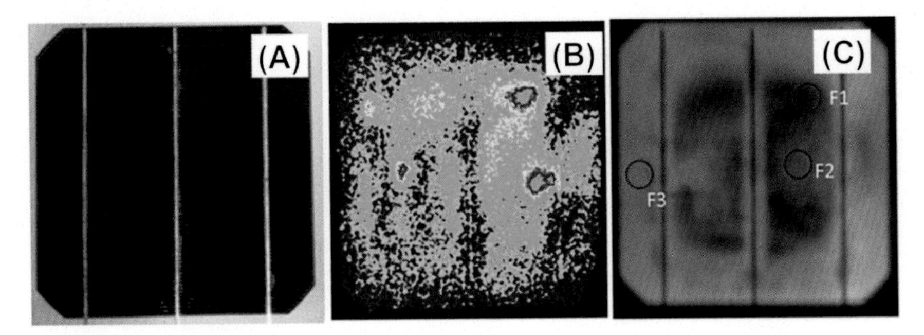

FIGURE 5.14

Visual image, lock-in-thermal image, and corresponding electroluminescence image of PID affected single cell c-Si PV module (Islam et al., 2019). *PID*, potential induced degradation; *PV*, Photovoltaic.

Recently, module manufacturers usually mark their products are "PID free" or "PID resistant;" however, it is not clear what studies they carried out for confirming that mark. Yet, there is no approved industry standard for certifying PID RS modules. Particularly, the PV industry has defined a series of tests, denoted by IEC/TS 62804, which is worthwhile in measuring PV module reliability.

Various efforts from the cell level to the system level have been developed to decrease or reduce PID. SiNx films typically used as an antireflection coating on the surface of a solar cell are modified for reducing its resistivity (Mishina et al., 2014). The conventional ethylene vinyl acetate (EVA) encapsulant has been replaced by different types of polymer with high volume resistivity, such as polyolefin or ionomer (Kapur, Stika, Westphal, Norwood, & Hamzavytehrany, 2014). The cover glass is also chemically treated for reducing the PID (Kambe et al., 2013). On the other hand, the PV industry also proposed two techniques for PID mitigation at the system level. They are, namely, (1) charge equalizers, and (2) high impedance grounding. The choice of the appropriate method depends on the system design, compatibility, and cost. In the case of the charge equalizer technique, a PV offset box is used to transform less inverter that is not grounded. Because of a bipolar or ungrounded system, PID occurs in day time due to the high potential relative to the ground, and at night, the charge equalizer neutralizes the PID by applying reverse charge via offset box. However, this technique cannot work for the full recovery of the PV module, a 5% permanent power loss at the module level can remain. However, at the system level, this power loss often becomes unnoticeable. This technique has shown success in recovering marginally PID affected modules for an extended period, however, the technique is not workable with severely PID affected module. In the case of the high impedance grounding technique, the negative pole of a transformerless system is grounded via a high-value resistor, ($\approx 22\ \mathrm{k\Omega}$). However, to allow for ground fault detection, additional hardware needed to be mounted. This technique reduces the

Table 5.5 Factors influence the potential induced degradation (Köntges et al., 2014).

Level	Design factors	Influence on/accelerating factor
Environmental conditions (Micro-, macroclimate)	Temperature and humidity	Surface conductivity, leakage current, ion mobility, chemical reactivity
	Humidity, rain, and condensation	Surface and encapsulant bulk conductivity, leakage current
	Insulation(-distribution)	The energy yield fraction at low light
	Aerosols	Surface conductivity, leakage current
Cell (manuf.) level	Si/N ratio, thickness and homogeneity in Anti-reflective coating (ARC)	No ARC—no potential induced degradation (PID), conductive coating on ARC arrests PID, higher refracting index of SiNx lowers PID (but increases reflective losses)
	Surface morphology	Reduction of "attractive" K-centers
	Depth of emitter	Emitter sheet resistivity influences PID
	Doping level and semiconductor type (p or n)	Wafer base resistivity influences PID
Module-level	Mounting orientation (angle, portrait or landscape orientation)	Wetness, number of cells next to the lower edge with higher surface conductivity, soiling, temperature, and leakage current
	Frame and mounting on a structure	Conductivity of electrical path, leakage current
	Encapsulant material and thickness	Bulk resistivity, ion mobility, leakage Current
	Back-sheet material	Water vapor transmission rate (WVTR), encapsulant's water content, bulk resistivity, chemical reactivity, leakage current
	Front cover material	Electrical conductivity, sodium ion concentration, ion mobility, leakage current
	Front cover surface treatment and coating	Surface conductivity, soiling, leakage current
System-related factors	Operating and open circuit system voltage	Leakage current
	Inverter topology and array potentials	Array polarity levels (DC + AC content), leakage current, and polarity
	Reverse array polarity during nighttime	Recovery
	Grounding concept	Conductivity of electrical path, leakage current

voltage potential on the string, thereby allow to reduce or prevent PID. However, recovery of the PID affected module as like charge equalizer technique is not possible in this technique.

5.4.2.5 Encapsulant failure or discoloration

Encapsulant (EVA or any other adhesive materials) failure of degradation usually leads to discoloration. Due to the structural advantages, thin-film PV modules are rarely affected by encapsulant discoloration. The discoloration of the module is a shift of material color that turns yellow and occasionally brown as shown in Fig. 5.15A–C. Discoloration reduces the transmittance of light reaching PV cells and thus power generation by the cell is decreased. Module short circuit current is reduced as an impact of discoloration (Jordan, Wohlgemuth, & Kurtz, 2012). The discoloration is suspected as a key contributor to the slow degradation (0.5%/year) which is usually seen for c-Si PV modules. It has been reported that UV rays combined with water vapor of temperatures higher than 50°C are the main causes of EVA degradation (Oreski & Wallner, 2009).

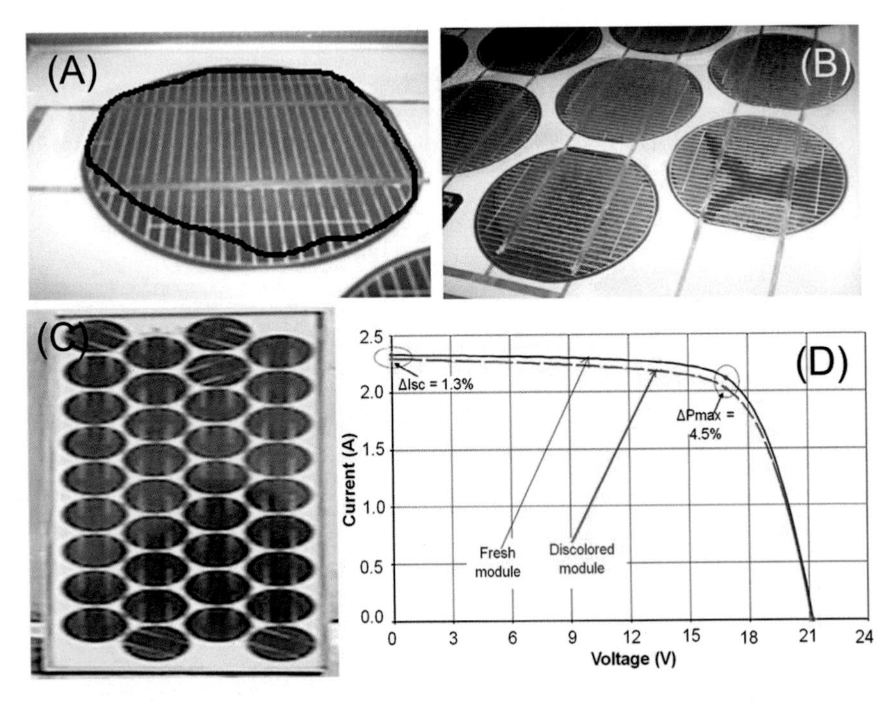

FIGURE 5.15

(A) and (B) Partial discoloration occurred in c-Si PV modules (Kaplanis & Kaplani, 2011), (C) discoloration occurred over the whole surface in a c-Si module (Sinha, Sastry, & Gupta, 2016), and (D) current-voltage characteristics of PV module healthy (solid line) and discolored (dash line) (Realini, 2003).

Table 5.6 Comparison of c-Si and thin photovoltaic modules in terms of common failure modes.

	c-Si PV module			Thin-film photovoltaic module		
Failure mode	**Frequency (relative)**	**Impact on power**	**Safety issue**	**Frequency (relative)**	**Impact on power**	**Safety issue**
Yellowing/ browning	High	High	Low	Low	Low	Low
EVA delamination	High	High to low	High	Low	Low	Low
TCO delamination	N/A	N/A	N/A	High to low	High	Low
PID	High	Very high	Low	High	Very high	Low
Glass breakage	High to low	Very high	Very high	Very high	Very high	Very high

It is commonly seen that discoloration does not occur over the whole area, which is because EVA may not spread similarly in all the areas of the modules during module preparation. Kojima and Yanagisawa in 2004 reported that rapid discoloration has occurred at 400 h in acceleration test when radiation of 4000 Wm^2 is applied (Kojima & Yanagisawa, 2004). There are numerous paper has been reported on the degradation of silicon crystalline PV modules were focused on the EVA degradation (Kempe, 2010). Between 1982 and 2003, Realini conducted an experiment in which he correlates the electrical characteristics of the modules with their encapsulant discoloration (Realini, Burà, Cereghetti, Chianese, & Rezzonico, 2002). He reported that for a partial discoloration, I_{sc} may reduce from 6% to 8% and I_{sc} may decrease from 10% to 13% for discoloration occurred over the whole area of a module. Discoloration is still an important topic for research in the PV community, even it could be detected by visual inspection. From the several reports as mentioned in the above section, the failure mechanism that occurred in the c-Si and thin film has been compared in Table 5.6.

5.4.2.6 Permanent soiling

The term "soiling" means the accumulation of leaves, dirt, snow, dust, pollen, bird droppings, and other particles on PV modules. Soiling reduces the sunlight entering the solar cells and leads to significant PV performance and power loss. The impact depends on the quantity of soil that covers the module surface. In the comparison of other factors, the dust has the highest contribution to soiling. Fig. 5.16 shows dust levels at the different of the world. The dark blue regions have the highest density of dust.

It could be seen in Fig. 5.16 (Maghami et al., 2016) that North Africa and the Middle East have the highest dust accumulation zones in the world

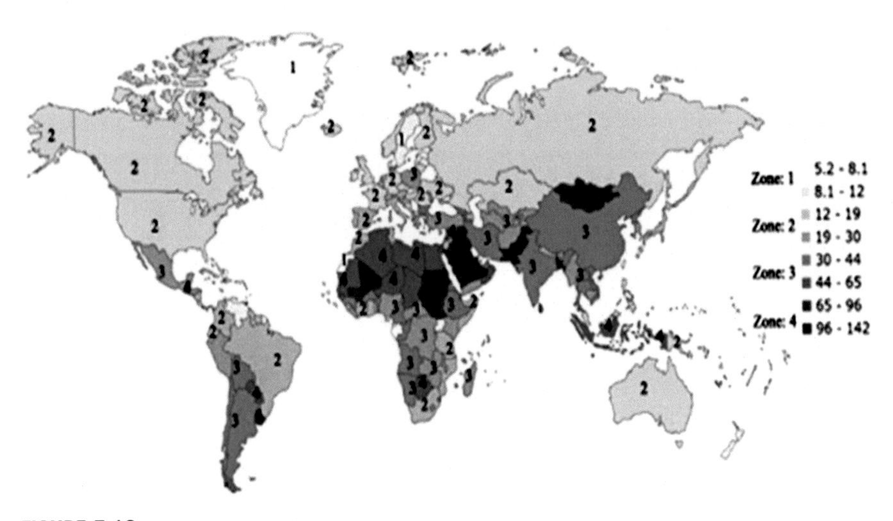

FIGURE 5.16

Dust level at different places in the world (Ghazi, Sayigh, & Ip, 2014).

(Ghazi et al., 2014). A detailed study on the effect of dust has been conducted by Sarver et al. in 2012 (Sarver, Al-Qaraghuli, & Kazmerski, 2013), which introduced key input to the understanding, effect on performance, and easing of this problem. Other than dust, the impact of snowfall on the PV module surface sometimes may become worst and no energy is generated when it covers the whole surface (Sonnenenergie, 2013).

The dust accumulation on PV modules depends on two independent parameters: the dust properties and site environment. Components of dust property include size (usually less than 10 μm in diameter), weight, and shape (Mani & Pillai, 2010) however, all the parameters again depends on the site and its environment. Also, dust in the environment comes from many sources such as vehicular movements, pedestrian, wind pollution, and volcanic eruptions. The site environment refers to the system's surroundings that consist of human activities, vegetation, built environment, and weather condition. The details of the cause of dust accumulation are shown in Fig. 5.17.

Besides, surface and/or glass module properties are also a significant contributing factor in the process of soiling. If the surface is rough, sticky and fury rather than smooth, indeed it allows more dust to accumulate. On the other hand, if the wind speed is higher, it able to clear the dust particles from the PV module and reduce the soiling (El-Shobokshy & Hussein, 1993). However, a slow breeze assists in soiling. Besides, the position and angle of the module and wind flow through the module surface are also important in the dust accumulation or soiling process. The airspeed may not be uniform over the solar module and/or array, thus wherever airspeed is higher, the possibility of soling is less. Also, more dust could be accumulated on the horizontal surface. The impact of soiling on PV

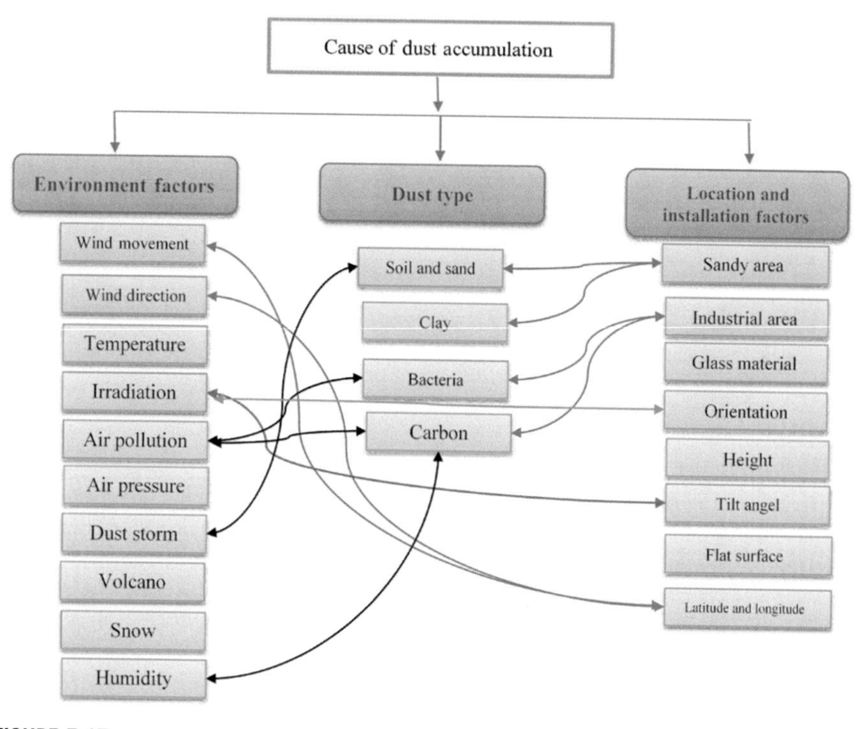

FIGURE 5.17

Cause of dust accumulating on the surface of solar arrays (Maghami et al., 2016).

modules are quite comparable to the impact of discoloration and/or partial shading.

5.4.2.7 Glass breakage

Glass breakage is one of the most significant failures that occurred in the field for all types of PV technologies. Richter (2011) summarized different types of failure including glass breakage according to the customer complaints as shown in Fig. 5.18A. The statistic has been made based on nearly 2 million PV modules that have been delivered from 2006 to 2010 by a German distributor. Unknown defects are for some modules where failure categories were not found. Also, DeGraaff et al. illustrated similar failure modes for field-aged modules that were produced by 21 manufactures and installed eight years in the field as shown in Fig. 5.18B. The rate of failure shown in the figure is relative to the total number of failures. From both studies, it could be seen that a significantly high rate of failure due to the glass breakage occurred in the field.

The mechanical stresses that developed due to the rapid increase of module temperature typically boost glass breakage along with delamination in the case of the conventional module which built with a front-glass and back-sheet as shown

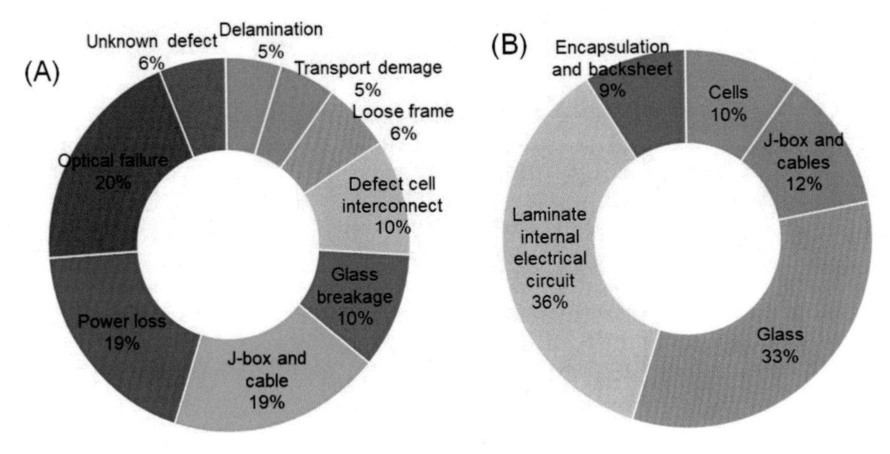

FIGURE 5.18

(A) The rate of failure calculated from the customer complaints within two years after module delivery , and (B) PV module failures found in the field for various PV modules come from 21 manufactures installed for 8 years. PV, Photovoltaic.

Data from Richter, A. (2011). Schadensbilder nach Wareneingang und im Reklamationsfall, 8. Workshop "Photovoltaik-Modultechnik", 24/25. November 2011, TÜV Rheinland, Köln; DeGraaff, D., et al., How do qualified modules fail—What is the root cause? SunPower Corp, NREL, 2011.

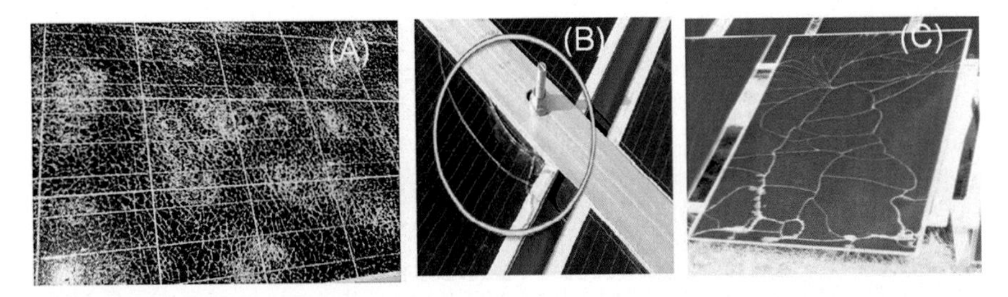

FIGURE 5.19

Cracked front cover glass for silicon (A) and thin-film (B) and (C) modules (Köntges et al., 2014; Muehleisen et al., 2018; Wohlgemuth et al., 2015) [Figure (B) shows glass breakage caused by too tight screws].

in Fig. 5.19A and C. In case of poor lamination, the excess encapsulant is pushed out on the sides of a module due to the temperature cycle and creates significant tensile stress that tries to bend the glass. On the other hand, the tempered glass used in PV modules is not perfectly flat, and thus the thickness of EVA over the whole module is not distributed uniformly. This imperfection brings the residual mechanical stresses and increases the probability of glass breakage as well as delamination which can lead to serious issues of performance and safety including hot-spot and burning.

Particularly, glass breakage in thin-film PV modules is more common than in silicon-based PV modules caused by clamps as shown in Fig. 5.19B and glass properties (Fig. 5.19C). Most of the Si PV modules use heat-strengthened or tempered glass, whereas thin-film modules use annealed glass, which is susceptible to the heat cycle. On the other hand, the glass breakage could occur due to several causes, such as, (1) the bad clamping positions, that is, positions are not being chosen in compliance with the manual of the manufacturer, (2) used of clamps are too narrow and/or too short and/or (3) sharp edge or improper clamp geometry (Dietrich, Pander, Ebert, & Bagdahn, 2008). Also, excessively tightened screws can lead to glass breakage (Heinstein, Ballif, & Perret-Aebi, 2013). Besides, glass breakage is attributed to inappropriate maintenance, poor handling practices, improper structure design, and lastly built-up defects. Glass breakage allows to ingress water vapor and oxygen that lead to electrical circuit corrosions, delamination, and hot-spot. Electrical safety concerns are big problems triggered by glass breakage because the modules' insulation is no longer guaranteed.

5.4.2.8 Frame deformation

Frame deformation or bending is a common phenomenon in weathers with snow and ice. The ice that accumulated in the surface of the modules Fig. 5.20A may partially melt then refreeze near the bottom edge of the module and put mechanical stress to the frame due to the gravitational force, ultimately frame bent as shown in Fig. 5.20B. Longer frame length PV modules are more vulnerable to easy bending. If the frame is separated from the glass, the PV module is damaged

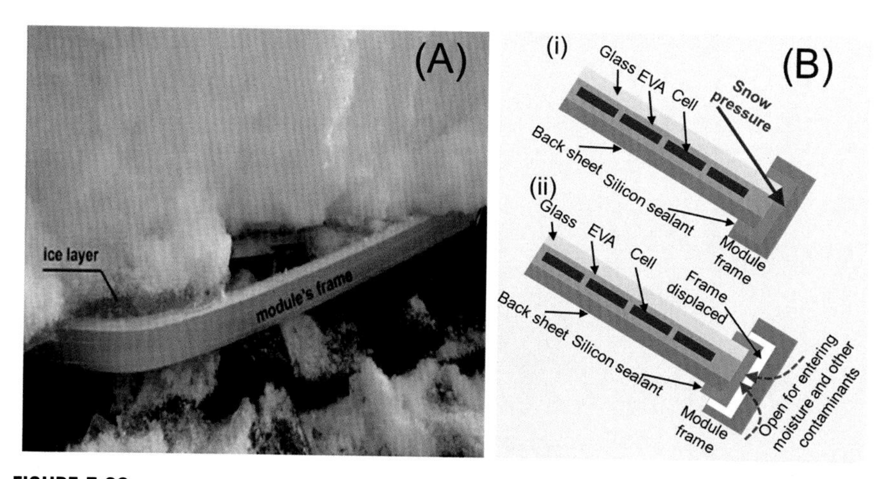

FIGURE 5.20

(A) Ice accumulation on top of the PV module (Köntges et al., 2014) and (B) Schematically shown the PV module frame displacement or damage due to the snow load (Reil et al., 2012); (i) without displacement and (ii) frame displaced from the glass which may lead to other failure mechanism.

and influences the other failure mechanism, thus the frame must be replaced as soon as possible. In recent times, for heavy snow load sites, several PV modules have been developed and implemented. Particularly, the IEC 61215 load test is performed to test and certify the PV module's stability and reliability for heavy snow load.

It has been reported that modules that are used silicone-based adhesive can withstand loads of up to almost 500 kg (~3 kPa) without any permanent damage and frame bending. On the other hand, the similar modules with tape-based adhesives tolerate frame bending and glass breakage at lower loads of 230–360 kg (1.4–2.3 kPa) (Köntges et al., 2014). Modules are generally installed with an inclined angle which allows the moving of melted ice to the lower sections of the module and often causes torque at the clamped locations. The torque is further amplified by the gravitational force and thus bottom part feels stressed much more than the center part of the module. The mechanism of how ice accumulation influences the module surface and frame is shown in Fig. 5.21.

Observed Fig. 5.21, ice load on PV modules can be summarized primarily in four features that are used for developing a new test method.

1. The ice load F_g acting on the inclined module surface can be separated into two force components (1) the downhill force F_H and (2) the normal force F_N. Particularly, the downhill force F_H is the primary cause of frame bending.
2. Ice gliding down on the surface of the module is commonly not homogeneously distributed.
3. The lower part of the module is subjected to moments and torque due to the inhomogeneous ice loads and significantly influences the adhesive bonds between the glass and the frame.

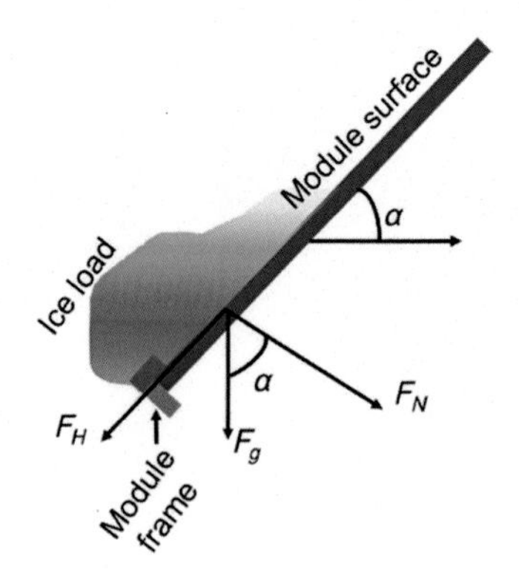

FIGURE 5.21

Introduction of snow loads, the difference between load vectors.

Redrawn from Aarseth, B. L. et al. (2018). Mitigating snow on rooftop PV systems for higher energy yield and safer roofs.

4. Low temperatures ($<0°C$) can cause the decrease of adhesive bonds and accelerate the failure mechanisms. Creeping at higher temperatures may occur.

5.4.3 System failure

A PV system either c-Si or thin film is typically composed of several solar modules or arrays combined with a junction box, an inverter, and other electrical and mechanical hardware as shown schematically in Fig. 5.22. Particularly, PV modules and arrays are connected through a diode to confirm the unidirectional current flow. All the PV arrays are connected in a junction box before connected to an inverter. The failure mechanism that occurred in the module and/or cell level has been discussed in the previous section. In this section, the system failure will be discussed in terms of diode failure and junction box failure.

It should be noticed that very few reports have been found where PV module and system long term performance were studied together, thus it is still not clear how modules performance variation by time relates to system performance. It has been reported that the system performance degradation is similar or slightly higher than PV module degradation (Jordan & Kurtz, 2017). Observing a c-Si based

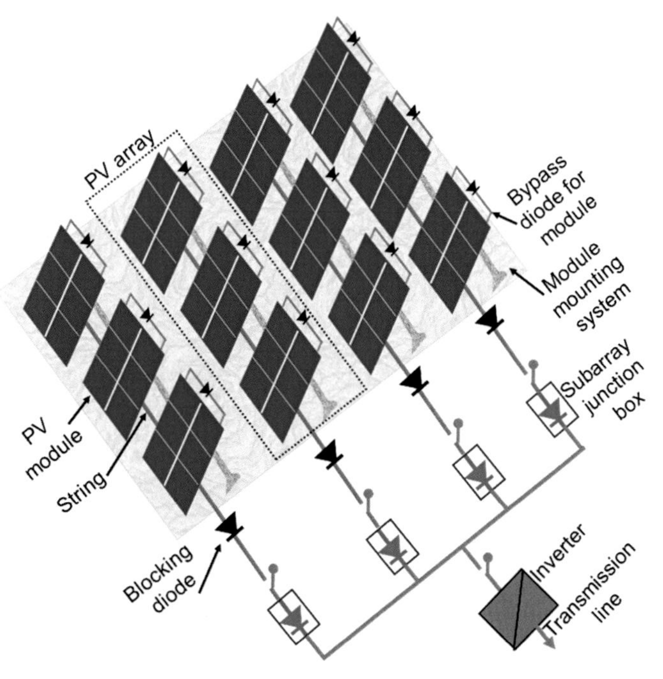

FIGURE 5.22

Schematic of a photovoltaic system.

PV system for 20 years, it has been reported that the highly degraded modules affect the string performance and lead to reduce the system performance (Jordan, Sekulic, Marion, & Kurtz, 2015). The PV system performance and failure are related to several factors, such as PV type, voltage selection, electrical protection system, grounding system, inverter type, and Field arrangement, etc. Particularly, the protection system, such as bypass and blocking diode and junction box is critical and failure is much more common in a PV system that will be discussed in the next section.

5.4.3.1 ByPass diode failure

To understand the impact of bypass diode failure, it is essential to know why the bypass diode has to be mounted in the PV module and device. The purpose of the bypass diodes is to remove the hot-spot phenomenon that can damage PV cells and even cause a fire if the incident light is not uniform at the cell surface in a module. Particularly, bypass diodes are used in solar PV systems to protect partially shaded PV cells from fully operating cells in the full sun within the same module where they are connected in series. The bypass diodes are mounted externally across (in parallel) but in opposite polarity that could be seen in Figs. 5.22 and 4.23B, also the mounting could be different depends on the PV system design. Also, a diode in series installed with a string of modules for connecting the arrays as shown in Fig. 5.23A and B.

If a PV module is partially shaded and there is no bypass diode but have a blocking diode as shown in Fig. 5.23A, in that case, the cells are in the shaded area will not produce any current, and current will flow through the shaded cell due to the voltage potential. The shaded cell then works as a load or semiconductor RS; thus it will seem to be a hotspot, and once upon a time, the cell will be completely damaged. However, if there is a bypass diode as shown in Fig. 5.23B, the bypass diode will conduct the current instead of shaded cells and save them. However, if there is no bypass and blocking diode as shown in Fig. 5.23C, if a module is shaded in string A, then the current from string B could flow through the string A due to the voltage potential, by which all the cell in string A could be affected by reverse current. If there is a bypass diode across the PV modules as shown in Fig. 5.23D, then reverse current flow through string A will be blocked until the voltage potential applied by string B is reached to the knee voltage of the diode across the shaded PV modules. Usually, excessive forward current and a high reverse voltage are the usual causes for a diode failure. High reverse voltage typically leads to a short diode, while overcurrent causes it to not open. The failure of a single bypass diode in a string usually switched off three consecutive modules and output suddenly dropped by one third. In the yield curve of a string with good solar radiation conditions, a decrease in the yield of this magnitude and many modules can be observed.

Despite the critical role of bypass diodes for power and safety, there are relatively few comprehensive failure-analysis studies that evaluate bypass diodes. There have been records of failures of bypass diodes including externally applied

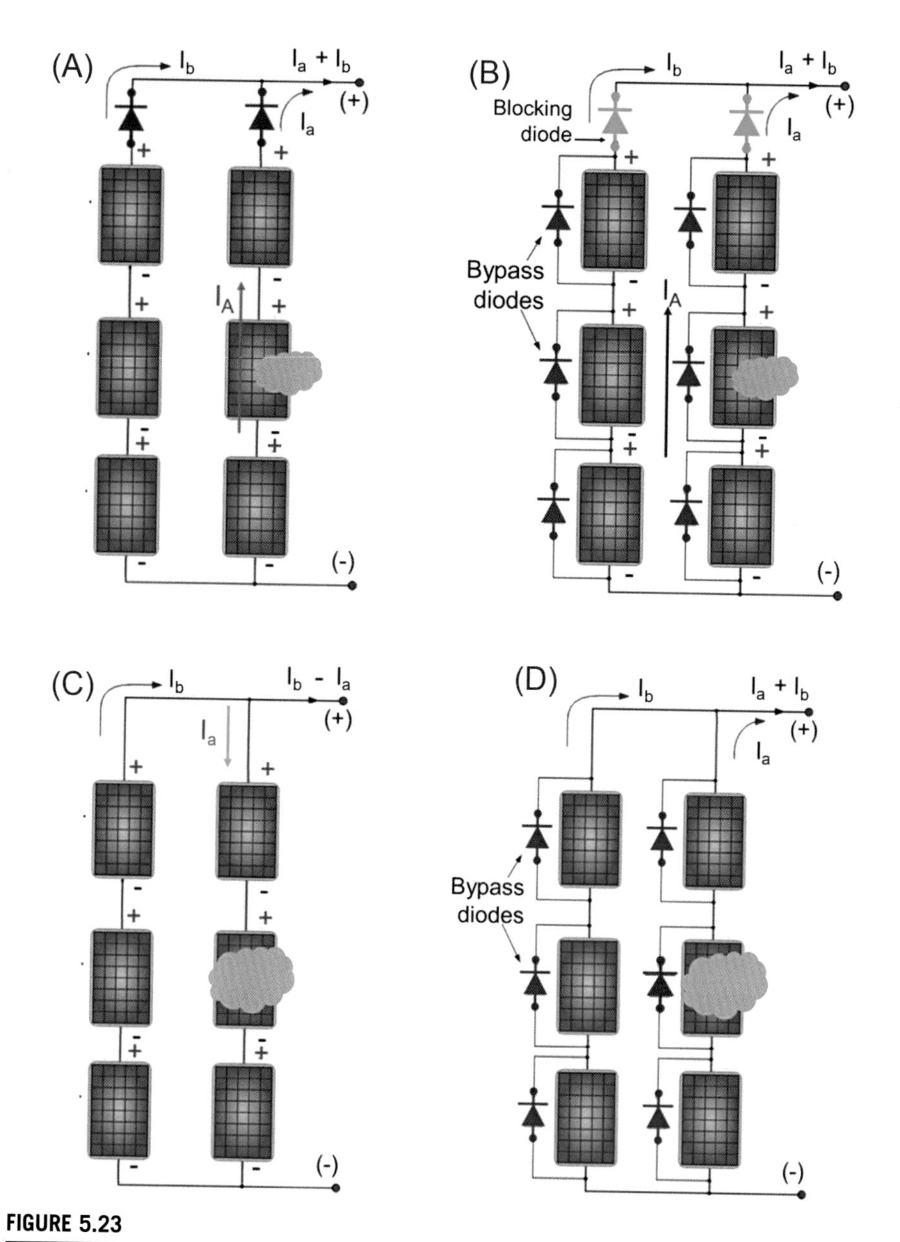

FIGURE 5.23

Schematic of bypass diode and blocking diode application in PV array for preventing partial shading effect.

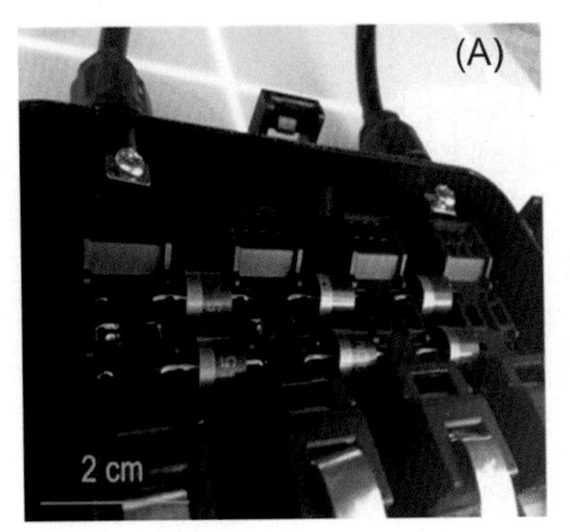

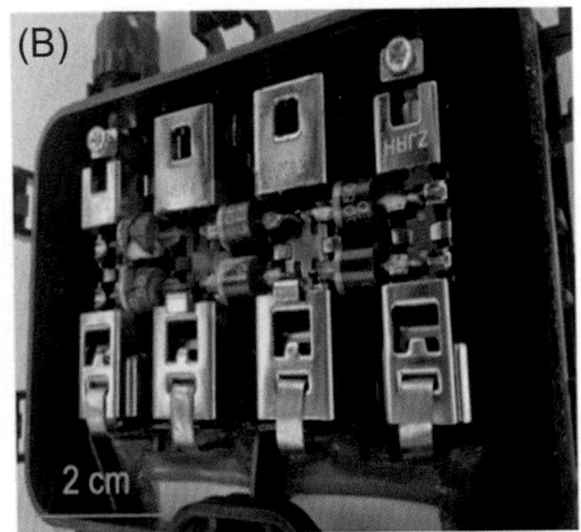

FIGURE 5.24

(A) Image of open junction box showing six exposed diodes free of any visible damage in rooftop modules and (B) inside a junction box showing cracked diodes (Xiao et al., 2020).

stress (Islam, 2019), like electrostatic discharge, lightning strikes, thermomechanical fatigue, thermal runaway, and single-event or continuous long-term overstress in forwarding bias. Apart from the external stress factors, failures may result from defects including impurities and crystalline defects intrinsic to the device itself. Local faults can focus the current in such a specific area causing local melting of the diode. The junction box with fresh and cracked bypass diodes are shown in Fig. 5.24.

In practice, defective bypass diodes are sometimes found to cause a tremendous power loss in PV modules. For instance, one study stated that bypass-diode deterioration in some environments was the largest single factor in a power outage, 11% per year in a hot and dry environment and 25% per year in moderate climates (Xiao et al., 2020). Bypass diode failures could contribute to power loss in many aspects: firstly, shunting would short-circuit the cells' substring in the module that it is aimed to protect and as a result, more than 33% of the module output is lost. Secondly, open-circuit bypass diode failure drives reverse-bias current through the modules in the connected string series as they are shaded. Lastly, hot spots contributing to encapsulant yellowing, solder bond failure, module breakage, and fire may result.

Woo et al. (Shin et al., 2018) investigated the origin of bypass-diode failure in c-Si PV modules in working in the field. They found that when the ambient temperature increases, the temperature of the inner junction box where the bypass diode is located increases. On sunny days in summer, the module surface and junction box temperature increase to over 70°C. As the junction box temperature increases from 25°C to 70°C, under a reverse voltage of 15 V, the leakage current increases up to 35 times. In damaged diodes obtained from abnormally operating PV modules, the meltdown of the junction barrier between the metal and semiconductor has been observed as a result of the high leakage current of the bypass diode at high temperatures.

On the other hand, using a blocking diode is very important if multiple solar modules or arrays are connected in parallel. In that case, blocking the diode will be helped to prevent the current flow through the shaded modules from the sunny modules. In a day, PV modules or arrays are partially or fully shaded by soil, falling leaves, shades of trees, and/or by cloud. A blocking diode in series with each string of the parallel-connected PV modules or arrays allows the sunny modules or arrays to put all their power to the output via disconnect the shaded modules or arrays. Also, blocking the diode is prevents discharging of batteries backward through solar modules at night. However, between the battery and the PV module, most PV systems use a charge controller recent time which has a system to prevent the backflow of electricity and removing the use of a blocking diode. It should be remembered that there is a slight voltage loss, about 0.5 V, using a diode in the system.

5.4.3.2 Inverter failure

Studies demonstrate that the inverter is the PV plant portion with the biggest number of service requests and the highest cost burden of operation and maintenance, as a result, significantly reducing the power output of PV system (Hacke et al., 2018), which is responsible for up to 36% of the energy loss (Golnas, 2012). The failure or damage of different components or parts in the inverter is the biggest failure cause related to the inverter loss. However, software or firmware control failures are important enough to be the first or second biggest source of inverter power loss in PV systems. According to the SunEdison data, parts and materials lumped together is attributable to at least 57% of failures where 16% of failures from software (Klise & Balfour, 2015).

Electric and thermal stresses experienced by the inverter are the primary cause of failure in a PV system. On the other hand, the most durable part is the solar module, which can last over 20 years. In terms of inverter failure, many sources can cause the inverter to fail. Components of the inverter such as a capacitor, fan, Insulated-gate bipolar transistor (IGBT), printed-circuit boards (PCBs), and solder joints and fuses, etc., are considered a source that can induce failures of the inverter. Analysis of 255 residential solar PV systems over 4 years was performed by Pecan Street (Formica, 2017), a firm that gathers information on energy needs and water supply. Within the period, 54 of these solar PV systems registered minor maintenance problems. Of the 54 reports, 13 suffered PV inverter failures because of blown AC fuses inside the inverters.

Besides, electromechanical wear on capacitors is another cause for inverter failure. Inverters depend on capacitors to provide a stable output power at different current levels, but electrolytic condensers have a limited lifetime and age faster than dry components. Capacitors are highly sensitive to temperature. The life of the component can be shortened by temperatures above the specified operating conditions which are mostly induced by the elevated current. However, as the electrolytes evaporate faster at

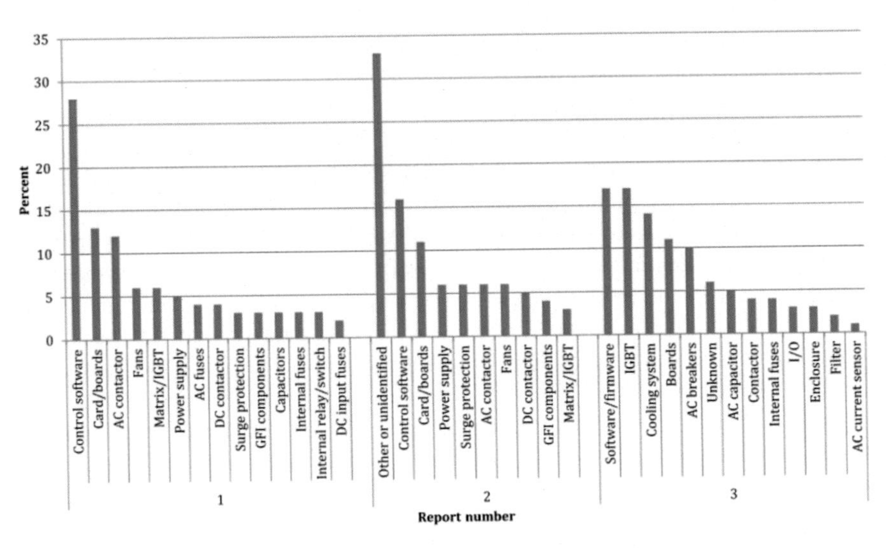

FIGURE 5.25

Three reports detailing the relative frequency (in percent) of inverter component failures, primarily for central inverters (IGBTs are insulated gate bipolar transistors; GFIs are ground fault interrupters) (Hacke et al., 2018).

higher temperatures, capacitor life increases when they are run at lower than operating temperature (Luo, Wang, Dooner, & Clarke, 2015).

Furthermore, Failure of PCBs ranks in the top four for larger inverters as shown in Fig. 5.25. Tilt or warpage between board and parts and stretched solder joints tend to be causes of failure that can be elevated by extra stress. Additional PCB failures are associated with vibration, shocks, and stress which bring to package cracking, hole plating cracks, die cracking, and wire bond breaks (Hacke et al., 2018). Additionally, the observed IGBT module failures originating from drive-board failures (18%), diode failures (11%), defective components (9%), loose connections (5%), overcurrent (5%), other wire management issues (4%), capacitor failure (4%), and several other electronic equipment failures. The root cause of failure cannot always readily be determined and 27% of the failure causes remain unknown, but some fraction of these may be the IGBT device within the module (Formica, Khan, & Pecht, 2017).

5.5 Scale-up possibilities towards lifetime and reliability

PV production and deployment have been significantly increasing since the last few years, including different PV projects ongoing. These increasing trends of PV are a result of the cost reduction of PV installation. The International Energy Agency (IEA) forecasts a 43% increase in the capacity for renewable electricity before 2022. Currently, PV technology is in the middle of the reform and played a significant role towards energy transition at the individual level. Particularly, the lifetime and stability of PV modules could realize in two ways: (1) develop new technologies that can directly slow down deterioration and extend lifetimes, and (2) development of understanding, evaluations, and standards that help reduce the confusion around PV reliability for PV clients and investors. It should be noticed that, for achieving the objectives in scaling-up of stability and reliability in PV technologies in the real world, the numerous study in failure mechanism and modes should be conducted in various location over the world and finding out the limitation and suitability of each technology for a specific site or region. The summarized results certainly could help to optimize the PV technologies as well as increase the PV deployment. We summarized some research findings in Table 5.7 as an example works for finding out the best PV technology by location and its climate. It should be noted that the PV device performance and reliability mostly depend on its spectral response. Which in turn depends on its absorber material properties and quality, as well as characteristics of the installation sites, for instance, the spectral distribution, climate, environment, latitude, longitude, albedo, etc., of the location. Besides, the performance and reliability of PV systems for a specific sites again depend on the cloudiness, water vapor and aerosol content in the sky of that location. The analysis considering all the above factors certainly will be too difficult, thus the researcher most of the time considers only some of the factors for simplifying their work.

Current research topics that address the reliability (stability and lifetime) of PV modules are include optimizing the module fabrication process for which solder bond failure, broken ribbon, and cell cracking will be minimized, fabricating encapsulant materials with higher resistivity that could hinder the PID; developing new encapsulant materials that will not yellow by the aging of PV modules; and developing new lamination and/ packaging techniques that will reduce the delamination or chemical corrosion. Also, multiple research efforts have been given to p-type c-Si module design that can reduce the ultra-violet radiation damage as well as heat and light-induced degradation targeting the module lifetime of 40–50 years. More research is required, in particular, to optimally monitor the process window either PV cells or PV modules, and prevent the shipment associated failure. Typically, PV products are qualified with inexpensive and easy checks, and production processes are managed with little need for comprehensive product testing to ensure reliability by maintaining the temperature and/or other process parameters within the required process window.

As research sheds light on the science behind PV failure mechanisms, the PV community certainly befitted via developing standard ways of test for these failures. Indeed the results reported by the various researcher on PV module reliability and stability in Table 5.7 are very difficult to compare because the work has been conducted focusing on different locations and various time scales (instantaneous, monthly, annual), different energy effects, and even the works are different by used metrics and calculation. Thus a proper and universal testing method is

Table 5.7 Summary of few reported works for finding out the best photovoltaic technology by location and climate.

Author(s)	Location	Environmental parameters	Tested technologies	Best perform technology
Zdyb and Gulkowski (2020)	Lublin, Poland	Temperate climate, 950–1250 (kWh/ m^2)/year, 15°C–48°C	pc-Si, a-Si, CIGS, and CdTe	pc-Si and CIGS
Romero-Fiances, Muñoz-Cerón, Espinoza-Paredes, Nofuentes, and De la Casa (2019)	1. Arequipa, Peru 2. Tacna, Peru 3. Lima, Peru	Diverse climates 1. 2380 kW/m^2, 3.81°C–32°C 2. 2280 kW/m^2, 13.4°C–31.5°C 3. 1740 kW/m^2, 18.8°C–18.9°C	1. sc-Si 2. ps-Si 3. a-Si/uc-Si	a-Si/μc-Si
Cotfas and Cotfas (2019)	Brasov, Romania	Temperate-continental climate, 2.1–1.82 Wh/m^2/ day, − 4.0°C–24°C	sc-Si and a-Si	sc-Si
Gulkowski, Zdyb, and Dragan (2019)	Lublin, Poland	Temperate climate, 950–1250 (kWh/ m^2)/year, 15°C–48°C	CdTe, CIGS, and pc-Si	CIGS
Bora et al. (2018)	Different parts of India	0.82–0.87 kW/m^2/ day not mentioned	a-Si, HIT, and pc- Si	All (cold and sunny zone)
Makrides, Zinsser, Phinikarides, Schubert, and Georghiou (2012)	Cyprus	Mediterranean climate, 1988–2054 kWh/ m^2, 10°C–40°C	sc-Si, pc-Si, a-Si, CIGS and CdTe	a-Si
(Louwen, Van Sark, Faaij, & Schropp, 2016; Gaglia, Lykoudis, Argiriou, Balaras, & Dialynas, 2017)	Utrecht, Netherlands	Oceanic climate, 20.5°C–29.5°C, 950–1050 W/m^2	SHJ, a- Si, sc-Si, pc-Si, CIGS, CIS and CdTe	sc-Si and SHJ
Dirnberger, Blackburn, Müller, and Reise (2015)	Breisgau, Germany	Maritime climate, 5°C–25°C, 1,117 kW/m^2/year (approx.)	a- Si, sc-Si, CIGS and CdTe	a-Si
(Edalati, Ameri, & Iranmanesh, 2015; Ghitas, 2012)	Kerman, Iran	Dry climate 68.64–198.72 kW/ m^2, 20°C	sc-Si, and pc-Si	pc-Si

(Continued)

Table 5.7 Summary of few reported works for finding out the best photovoltaic technology by location and climate. *Continued*

Author(s)	Location	Environmental parameters	Tested technologies	Best perform technology
Cañete et al. (2014)	Southern Spain	Dry Mediterranean climate, 3.7–7.4 kWh/m^2/day, 15°C–30°C	a-Si, a-Si/μc-Si CdTe, and pc-Si	a-Si and CdTe
Aste et al. (2014)	Milan, Italy	Temperate climatic, 1270 kW/m^2/year, − 5°C–32°C	c-Si, a-Si/uc-Si, HIT	HIT
Poissant (2009)	Montreal, Canada	Continental climate, 950–1050 W/m^2, max. 20°C to 22°C	SHJ, IBC, a-Si/uc-Si, and c-Si	a-Si/uc-Si
Minemoto et al. (2007)	Kusatsu-city, Japan	Subtropical climate, 200 kW/m^2, 9°C–33°C	pc-Si, and a-Si	pc-Si

indispensable for future technological development. There is a number of the technical committee working over the world for developing a standard PV reliability testing methodology. For example, the International PV Quality Assurance Task Force works for implementing and adopting repeatable and reproducible manufacturing following the ISO 9001 standard. The Technical Committee 82 (TC82) of the International Electrotechnical Commission (IEC) has established guidelines for testing PV modules in terms of certification (IEC 61215) and protection (IEC 61730). International Electrotechnical Commission of Renewable Energy works to establish a set of guidelines to check the correct implementation and efficiency of the device. Furthermore, in the United States, organizations such as the North American Board of Certified Energy Practitioners promote training and verification of essential skills in PV installation. Certainly, PV installation training will be a stepping stone for solar and storage installers for dominating the energy market. It should be mentioned that the skill of installer should be a higher level in recent times, and he must be familiar with building electrical wiring, various protocols for wireless communications, mobile phone, laptop and desktop devices, and hundreds of configuration options for inverter and/or battery.

5.6 Conclusions

The reliability and stability of PV modules and plants have been and/or will remain be a major concern for the manufacturer, investors, and owners. The power output, as well as reliability of PV modules, are highly dependent on the technology, site environment, and system design. The failures that have occurred in different PV technologies in the same location are different, and the failure causes are affecting the stability of different PV technologies in a very dissimilar way. There are numerous factors involved that influence the PV module performance and reliability, and analysis considering all factors is too difficult. Typically, most researchers only consider some of the factors for simplifying their work, which again limits the realization of the actual failure mode. Particularly, encapsulant and back sheet delamination, corrosion, and hot spots are major failure modes in PV technology. Besides, discoloration in EVA is a common failure mode in the case of c-Si PV module.

Strategic routes and future patterns should be developed for studying PV modules mounted in the various geographical locations focusing to detect the unalike degradation and failure modes as well as finding out the most efficient PV technology for specific climate conditions. The IEC standard and other test standards should also be modified following the climatic zones. Furthermore, a global network is necessary to improve the quality, stability, and reliability of PV modules or systems and their components via gathering, analyzing, and spreading information in terms of technological and financial performance. In the near future, PV cells, modules, and systems will be modified by utilizing new materials, cell or module or system design, manufacturing methods, and new integrated components focusing on cost reduction and increase of reliability. Certainly, new concepts in PV cells, modules, or systems are a crucial need for improving lifetime and reliability. This chapter expected to be help researchers, students, consumers, and manufacturers for improving PV technologies by increasing the material quality and maintenance to reduce the failure causes. At the same time, this chapter will help to realize the best-suited PV technology for a specific site.

Acknowledgments

This work was supported by the Malaysian ministry of higher education through FRGS grant FRGS/1/2020/TK0/UM/02/33. The authors would like to acknowledge the Institute of Sustainable Energy (ISE), Universiti Tenaga Nasional (@The National Energy University) for partial supports through the BOLD2025 Program. The authors also acknowledge the support from the Faculty of Engineering, Universiti Malaya (@UM) for other supports. The authors would also like to acknowledge the support from The National University of Malaysia and Ministry of Higher Education of Malaysia (MoHE) under Grant LRGS/1/2019/UKM-UKM/6/1.

References

Abdullah, W. S. W., Osman, M., Ab Kadir, M. Z. A., & Verayia, R. (2019). The potential and status of renewable energy development in Malaysia. *Energies*, *12*(12), 2437.

Akhmad, K., Kitamura, A., Yamamoto, F., Okamoto, H., Takakura, H., & Hamakawa, Y. (1997). Outdoor performance of amorphous silicon and polycrystalline silicon PV modules. *Solar Energy Materials and Solar Cells*, *46*(3), 209–218.

Amelia, A., Irwan, Y. M., Leow, W. Z., Irwanto, M., Safwati, I., & Zhafarina, M. (2016). Investigation of the effect temperature on photovoltaic (PV) panel output performance. *International Journal on Advanced Science Engineering Information Technology*, *6*(5), 682–688.

Aste, N., Del Pero, C., & Leonforte, F. (2014). PV technologies performance comparison in temperate climates. *Solar Energy*, *109*, 1–10.

Bora, B., Kumar, R., Sastry, O. S., Prasad, B., Mondal, S., & Tripathi, A. K. (2018). Energy rating estimation of PV module technologies for different climatic conditions. *Solar Energy*, *174*, 901–911.

Bosman, L. B., & Darling, S. B. (2018). Performance modeling and valuation of snow-covered PV systems: examination of a simplified approach to decrease forecasting error. *Environmental Science and Pollution Research*, *25*(16), 15484–15491.

Bouraiou, A., Hamouda, M., Chaker, A., Neçaibia, A., Mostefaoui, M., Boutasseta, N., . . . Lachtar, S. (2018). Experimental investigation of observed defects in crystalline silicon PV modules under outdoor hot dry climatic conditions in Algeria. *Solar Energy*, *159*, 475–487.

Cañete, C., Carretero, J., & Sidrach-de-Cardona, M. (2014). Energy performance of different photovoltaic module technologies under outdoor conditions. *Energy*, *65*, 295–302.

Cengiz, M. S., & Mamiş, M. S. (2015). Price-efficiency relationship for photovoltaic systems on a global basis. *International Journal of Photoenergy*, *2015*.

Chowdhury, M. S., Rahman, K. S., Chowdhury, T., Nuthammachot, N., Techato, K., Akhtaruzzaman, M., . . . Amin. (2020). An overview of solar photovoltaic panels' end-of-life material recycling. *Energy Strategy Reviews*, *27*, 100431.

Cotfas, D. T., & Cotfas, P. A. (2019). Comparative study of two commercial photovoltaic panels under natural sunlight conditions. *International Journal of Photoenergy*, 2019.

DeGraaff, D., Lacerda, R., & Campeau, Z. (2011). February. Degradation mechanisms in Si module technologies observed in the field; their analysis and statistics. In NREL 2011 Photovoltaic Module Reliability. Workshop, (Vol. 20).

DeGraaff, D., Lacerda, R., Campeau, Z., & Xie, Z. (2011). How do qualified modules fail—What is the root cause? *SunPower Corp, NREL*.

Alonso-Abella, M., Chenlo, F., Alonso, A., & González, D. (2014). Toledo PV plant 1MWp—10 years of operation. In European Photovoltaic Solar Energy Conference, 20th EUPVSEC, Amsterdam, NL.

Demirtaş, M., Tamyürek, B., Erol, K., Çetinbaş, İ., & Öztürk, M. K. (2019). Effects of aging and environmental factors on performance of CdTe and CIS thin-film photovoltaic modules. *Journal of Electronic Materials*, *48*(11), 6890–6900.

Dias, P., Javimczik, S., Benevit, M., Veit, H., & Bernardes, A. M. (2016). Recycling WEEE: Extraction and concentration of silver from waste crystalline silicon photovoltaic modules. *Waste Management*, *57*, 220–225.

Dirnberger, D., Blackburn, G., Müller, B., & Reise, C. (2015). On the impact of solar spectral irradiance on the yield of different PV technologies. *Solar Energy Materials and Solar Cells*, *132*, 431–442.

Dolara, A., Leva, S., Manzolini, G., & Ogliari, E. (2014). Investigation on performance decay on photovoltaic modules: Snail trails and cell microcracks. *IEEE Journal of Photovoltaics*, *4*(5), 1204–1211.

Dong, N. C., Islam, M. A., Ishikawa, Y., & Uraoka, Y. (2018). The influence of sodium ions decorated micro-cracks on the evolution of potential induced degradation in p-type crystalline silicon solar cells. *Solar Energy*, *174*, 1–6.

Duerr, I., Bierbaum, J., Metzger, J., Richter, J., & Philipp, D. (2016). Silver grid finger corrosion on snail track affected PV modules–Investigation on degradation products and mechanisms. *Energy Procedia*, *98*, 74–85.

Ech-Charqaouy, S. S., Saifaoui, D., Benzohra, O., & Lebsir, A. (2020). Integration of decentralized generations into the distribution network-smart grid downstream of the meter. *International Journal of Smart Grid-ijSmartGrid*, *4*(1), 17–27.

Edalati, S., Ameri, M., & Iranmanesh, M. (2015). Comparative performance investigation of mono-and poly-crystalline silicon photovoltaic modules for use in grid-connected photovoltaic systems in dry climates. *Applied Energy*, *160*, 255–265.

El-Shobokshy, M. S., & Hussein, F. M. (1993). Degradation of photovoltaic cell performance due to dust deposition on to its surface. *Renewable Energy*, *3*(6–7), 585–590.

Dietrich, S., Pander, M., Ebert, M., & Bagdahn, J. (2008, September). Mechanical assessment of large photovoltaic modules by test and finite element analysis. In *23rd European Photovoltaic Solar Energy Conference. Valencia, Spain.*

Formica, T. J. (2017). *The effectiveness of warranties in the solar photovoltaic and automotive industries.*

Formica, T. J., Khan, A., & Pecht, M. G. (2017). The effect of inverter failures on the return on investment of solar photovoltaic systems. *IEEE Access*, *5*, 21336–21343.

Fu, R., James, T. L., Chung, D., Gagne, D., Lopez, A., & Dobos, A. (2015). *Economic competitiveness of US utility-scale photovoltaics systems in 2015: Regional cost modeling of installed cost ($/W) and LCOE ($/kWh)* (pp. 1–11). IEEE.

Gaglia, A. G., Lykoudis, S., Argiriou, A. A., Balaras, C. A., & Dialynas, E. (2017). Energy efficiency of PV panels under real outdoor conditions—An experimental assessment in Athens, Greece. *Renewable Energy*, *101*, 236–243.

Friesen, T. (2013). Failures of thin film PV modules: Field experience, https://www.pvqat.org/pdfs/2013_thinfilm_wkshp_friesen.pdf.

García, M., Marroyo, L., Lorenzo, E., Marcos, J., & Pérez, M. (2014). Observed degradation in photovoltaic plants affected by hot-spots. *Progress in Photovoltaics: Research and Applications*, *22*(12), 1292–1301.

García-Domingo, B., Aguilera, J., De la Casa, J., & Fuentes, M. (2014). Modelling the influence of atmospheric conditions on the outdoor real performance of a CPV (Concentrated Photovoltaic) module. *Energy*, *70*, 239–250.

Ghazi, S., Sayigh, A., & Ip, K. (2014). Dust effect on flat surfaces–A review paper. *Renewable and Sustainable Energy Reviews*, *33*, 742–751.

Ghitas, A. E. (2012). Studying the effect of spectral variations intensity of the incident solar radiation on the Si solar cells performance. *NRIAG Journal of Astronomy and Geophysics*, *1*(2), 165–171.

Golnas, A. (2012). PV system reliability: An operator's perspective. In *Proceedings of the IEEE thirty-eighth photovoltaic specialists conference (PVSC) PART* 2. IEEE.

Gielen, D., Boshell, F., Saygin, D., Bazilian, M. D., Wagner, N., & Gorini, R. (2019). The role of renewable energy in the global energy transformation. *Energy Strategy Reviews*, *24*, 38–50.

Gulkowski, S., Zdyb, A., & Dragan, P. (2019). Experimental efficiency analysis of a photovoltaic system with different module technologies under temperate climate conditions. *Applied Sciences*, *9*(1), 141.

Hacke, P., Lokanath, S., Williams, P., Vasan, A., Sochor, P., TamizhMani, G., … Kurtz. (2018). A status review of photovoltaic power conversion equipment reliability, safety, and quality assurance protocols. *Renewable and Sustainable Energy Reviews*, *82*, 1097–1112.

Hacke, P., Spataru, S., Johnston, S., Terwilliger, K., VanSant, K., Kempe, M., … Propst. (2016). Elucidating PID degradation mechanisms and in situ dark I–V monitoring for modeling degradation rate in CdTe thin-film modules. *IEEE Journal of Photovoltaics*, *6*(6), 1635–1640.

Haque, A., Bharath, K. V. S., Khan, M. A., Khan, I., & Jaffery, Z. A. (2019). Fault diagnosis of photovoltaic modules. *Energy Science & Engineering*, *7*(3), 622–644.

Heinstein, P., Ballif, C., & Perret-Aebi, L.-E. (2013). Building integrated photovoltaics (BIPV): Review, potentials, barriers and myths. *Green*, *3*(2), 125–156.

Huld, T., & Amillo, A. M. G. (2015). Estimating PV module performance over large geographical regions: The role of irradiance, air temperature, wind speed and solar spectrum. *Energies*, *8*(6), 5159–5181.

Idoko, L., Anaya-Lara, O., & McDonald, A. (2018). Enhancing PV modules efficiency and power output using multi-concept cooling technique. *Energy Reports*, *4*, 357–369.

Islam, M. (2019). *A comparative study and performance analysis of poly and mono Si photovoltaic modules*. Brac University.

Islam, M. A., Matsuzaki, H., Okabayashi, Y., & Ishikawa, Y. (2019). Transient carrier recombination dynamics in potential-induced degradation p-type single-crystalline Si photovoltaic modules. *Progress in Photovoltaics: Research and Applications*, *27*(8), 682–692.

Islam, M. A., Noguchi, K., Nakahama, H., & Ishikawa, Y. (2016). *Localized defect study of laboratory PID tested module* (pp. 0885–0888). IEEE.

Islam, M. A., Noguchi, K., Nakahama, H., & Ishikawa, Y. (2016b). Investigation of the EVA degradation and prediction of reliability by the Raman spectroscopy. *Group*, *900*, 1.

Islam, M. A., Oshima, T., Kobayashi, D., Matsuzaki, H., Nakahama, H., & Ishikawa, Y. (2018). Carrier dynamics in the potential-induced degradation in single-crystalline silicon photovoltaic modules. *Japanese Journal of Applied Physics*, *57*(8S3), 08RG14.

Jordan, D., & Kurtz, S. (2017). *Photovoltaic module stability and reliability. The performance of photovoltaic (PV) systems* (pp. 71–101). Elsevier.

Jordan, D. C., & Kurtz, S. R. (2013). Photovoltaic degradation rates—An analytical review. *Progress in Photovoltaics: Research and Applications*, *21*(1), 12–29.

Jordan, D. C., Kurtz, S. R., VanSant, K., & Newmiller, J. (2016). Compendium of photovoltaic degradation rates. *Progress in Photovoltaics: Research and Applications*, *24*(7), 978–989.

Jordan, D. C., Sekulic, B., Marion, B., & Kurtz, S. R. (2015). Performance and aging of a 20-year-old silicon PV system. *IEEE Journal of Photovoltaics*, *5*(3), 744–751.

Jordan, D. C., Silverman, T. J., Wohlgemuth, J. H., Kurtz, S. R., & VanSant, K. T. (2017). Photovoltaic failure and degradation modes. *Progress in Photovoltaics: Research and Applications*, *25*(4), 318–326.

Jordan, D. C., Wohlgemuth, J. H., & Kurtz, S. R. (2012). *Technology and climate trends in PV module degradation*. Golden, CO: National Renewable Energy Lab.(NREL).

Islam, M. A., Oshima, T., Nakahama, H., & Ishikawa, Y. (2017). Study on potential-induced degradation and recovery of bifacial n-type single crystalline Si photovoltaic modules. In 33rd European Photovoltaic Solar Energy Conference and Exhibition. Amsterdam, Netherlands (pp. 1730-1733).

Kambe, M., Hara, K., Mitarai, K., Takeda, S., Fukawa, M., Ishimaru, N., & Kondo, M. (2013). *Chemically strengthened cover glass for preventing potential induced degradation of crystalline silicon solar cells* (pp. 3500–3503). IEEE.

Kaplanis, S., & Kaplani, E. (2011). Energy performance and degradation over 20 years performance of BP c-Si PV modules. *Simulation Modelling Practice and Theory*, *19*(4), 1201–1211.

Kapur, J., Stika, K. M., Westphal, C. S., Norwood, J. L., & Hamzavytehrany, B. (2014). Prevention of potential-induced degradation with thin ionomer film. *IEEE Journal of Photovoltaics*, *5*(1), 219–223.

Karg, F. (2012). High efficiency CIGS solar modules. *Energy Procedia*, *15*, 275–282.

Kavlak, G., McNerney, J., & Trancik, J. E. (2018). Evaluating the causes of cost reduction in photovoltaic modules. *Energy Policy*, *123*, 700–710.

Kempe, M.D. (2005). Control of moisture ingress into photovoltaic modules. In *Conference record of the thirty-first IEEE photovoltaic specialists conference*, IEEE.

Kempe, M. D. (2010). Ultraviolet light test and evaluation methods for encapsulants of photovoltaic modules. *Solar Energy Materials and Solar Cells*, *94*(2), 246–253.

Klise, G. & Balfour, J. (2015). *A best practice for developing availability guarantee language in photovoltaic (PV) O&M agreements*. SAND2015–10223.

Kojima, T., & Yanagisawa, T. (2004). The evaluation of accelerated test for degradation a stacked a-Si solar cell and EVA films. *Solar Energy Materials and Solar Cells*, *81*(1), 119–123.

Köntges, M., Kurtz, S., Packard, C. E., Jahn, U., Berger, K. A., Kato, K., ... & Friesen, G. (2014). Review of failures of photovoltaic modules.

Komoto, K., Lee, J. S., Zhang, J., Ravikumar, D., Sinha, P., Wade, A., & Heath, G. A. (2018). *End-of-life management of photovoltaic panels: Trends in PV module recycling technologies*. Golden, CO: National Renewable Energy Lab. (NREL).

Kurtz, S., Sample, T., Wohlgemuth, J., Zhou, W., Bosco, N., Althaus, J., & Kondo, M. (2015). *Moving toward quantifying reliability-the next step in a rapidly maturing PV industry* (pp. 1–8). IEEE.

Louwen, A., Van Sark., Faaij, A. P., & Schropp, R. E. (2016). Re-assessment of net energy production and greenhouse gas emissions avoidance after 40 years of photovoltaics development. *Nature Communications*, *7*(1), 1–9.

Lund, C., Sinclair, M., Pryor, T., Jennings, P., & Cornish, J. (1999). Degradation studies of A: SI: H solar sell modules under different loads in the field.

Luo, X., Wang, J., Dooner, M., & Clarke, J. (2015). Overview of current development in electrical energy storage technologies and the application potential in power system operation. *Applied Energy*, *137*, 511–536.

Maghami, M. R., Hizam, H., Gomes, C., Radzi, M. A., Rezadad, M. I., & Hajighorbani, S. (2016). Power loss due to soiling on solar panel: A review. *Renewable and Sustainable Energy Reviews*, *59*, 1307–1316.

Makrides, G., Zinsser, B., Phinikarides, A., Schubert, M., & Georghiou, G. E. (2012). Temperature and thermal annealing effects on different photovoltaic technologies. *Renewable Energy*, *43*, 407–417.

Maluta, E., & Sankaran, V. (2011). Outdoor testing of amorphous and crystalline silicon solar panels at Thohoyandou. *Journal of Energy in Southern Africa*, *22*(3), 16–22.

Mani, M., & Pillai, R. (2010). Impact of dust on solar photovoltaic (PV) performance: Research status, challenges and recommendations. *Renewable and Sustainable Energy Reviews*, *14*(9), 3124–3131.

Minemoto, T., Toda, M., Nagae, S., Gotoh, M., Nakajima, A., Yamamoto, K., ... Hamakawa. (2007). Effect of spectral irradiance distribution on the outdoor performance of amorphous Si//thin-film crystalline Si stacked photovoltaic modules. *Solar Energy Materials and Solar Cells*, *91*(2–3), 120–122.

Mishina, K., Ogishi, A., Ueno, K., Doi, T., Hara, K., Ikeno, N., ... Masuda. (2014). Investigation on antireflection coating for high resistance to potential-induced degradation. *Japanese Journal of Applied Physics*, *53*(3S1), 03CE01.

Muehleisen, W., Eder, G. C., Voronko, Y., Spielberger, M., Sonnleitner, H., Knoebl, K., & Hirschl, C. (2018). Outdoor detection and visualization of hailstorm damages of photovoltaic plants. *Renewable energy*, *118*, 138–145.

Naumann, V., Hagendorf, C., Grosser, S., Werner, M., & Bagdahn, J. (2012). Micro structural root cause analysis of potential induced degradation in c-Si solar cells. *Energy Procedia*, *27*, 1–6.

Ndiaye, A., Charki, A., Kobi, A., Kébé, C. M., Ndiaye, P. A., & Sambou, V. (2013). Degradations of silicon photovoltaic modules: A literature review. *Solar Energy*, *96*, 140–151.

Nishioka, K., Hatayama, T., Uraoka, Y., Fuyuki, T., Hagihara, R., & Watanabe, M. (2003). Field-test analysis of PV system output characteristics focusing on module temperature. *Solar Energy Materials and Solar Cells*, *75*(3–4), 665–671.

Okullo, W., Munji, M. K., Vorster, F. J., & Van Dyk, E. E. (2011). Effects of spectral variation on the device performance of copper indium diselenide and multi-crystalline silicon photovoltaic modules. *Solar Energy Materials and Solar Cells*, *95*(2), 759–764.

Oreski, G., & Wallner, G. (2009). Evaluation of the aging behavior of ethylene copolymer films for solar applications under accelerated weathering conditions. *Solar Energy*, *83* (7), 1040–1047.

Osterwald, C. R., McMahon, T. J., & Del Cueto, J. A. (2003). Electrochemical corrosion of SnO2: F transparent conducting layers in thin-film photovoltaic modules. *Solar Energy Materials and Solar Cells*, *79*(1), 21–33.

Osterwald, C. R., Anderberg, A., Rummel, S., & Ottoson, L. (2002, May). Degradation analysis of weathered crystalline-silicon PV modules. In Conference Record of the Twenty-Ninth IEEE Photovoltaic Specialists Conference, 2002. (pp. 1392-1395). IEEE.

Patel, A. P., Sinha, A., & Tamizhmani, G. (2019). Field-aged glass/backsheet and glass/glass PV modules: Encapsulant degradation comparison. *IEEE Journal of Photovoltaics*, *10*(2), 607–615.

Pentti, H., & Atte, H. (2002). Failure mode and effects analysis of software-based automation systems. *VTT Industrial Systems, STUK-YTO-TR*, *190*, 190.

Poissant, Y. (2009). Field assessment of novel PV module technologies in Canada. In *Proceedings of the fourth Canadian solar buildings conference*, June.

Pingel, S., Frank, O., Winkler, M., Daryan, S., Geipel, T., Hoehne, H., & Berghold, J. (2010). *Potential induced degradation of solar cells and panels* (pp. 002817–002822). IEEE.

Quintana, M. A., King, D. L., McMahon, T. J., & Osterwald, C. R. (2002, May). Commonly observed degradation in field-aged photovoltaic modules. In Conference Record of the Twenty-Ninth IEEE Photovoltaic Specialists Conference, 2002. (pp. 1436–1439). IEEE.

Rathore, N., Panwar, N. L., Yettou, F., & Gama, A. (2019). A comprehensive review of different types of solar photovoltaic cells and their applications. *International Journal of Ambient Energy*, 1–18.

Realini, A., Burà, E., Cereghetti, N., Chianese, D., & Rezzonico, S. (2002). Mean time before failure of photovoltaic modules (MTBF-PVm). *Annual Report.*

Realini, A. (2003). Mean time before failure of photovoltaic modules. *Final Report (MTBF Project), Federal Office for Education and Science Tech. Rep.*, BBW, 99.

Reil, F., Mathiak, G., Strohkendl, K., Raubach, S., Schloth, C., Wangenheim, B.v., et al., "Experimental testing of PV modules under inhomogeneous snow loads," 27th EUPVSEC, Frankfurt, Germany, 2012, pp. 3414-3417.

Richter, A. (2011). Schadensbilder nach Wareneingang und im Reklamationsfall, 8. Workshop "Photovoltaik-Modultechnik", 24/25. November 2011, TÜV Rheinland, Köln.

Romero-Fiances, I., Muñoz-Cerón, E., Espinoza-Paredes, R., Nofuentes, G., & De la Casa, J. (2019). Analysis of the performance of various pv module technologies in Peru. *Energies*, *12*(1), 186.

Sarver, T., Al-Qaraghuli, A., & Kazmerski, L. L. (2013). A comprehensive review of the impact of dust on the use of solar energy: History, investigations, results, literature, and mitigation approaches. *Renewable and Sustainable Energy Reviews*, *22*, 698–733.

Shaltout, M. M., El-Hadad, A. A., Fadly, M. A., Hassan, A. F., & Mahrous, A. M. (2000). Determination of suitable types of solar cells for optimal outdoor performance in desert climate. *Renewable Energy*, *19*(1–2), 71–74.

Sharma, V., Kumar, A., Sastry, O. S., & Chandel, S. S. (2013). Performance assessment of different solar photovoltaic technologies under similar outdoor conditions. *Energy*, *58*, 511–518.

Shin, W. G., Ko, S. W., Song, H. J., Ju, Y. C., Hwang, H. M, & Kang, G. H. (2018). Origin of bypass diode fault in c-Si photovoltaic modules: Leakage current under high surrounding temperature. *Energies*, *11*(9), 2416.

Sica, D., et al. (2018). Management of end-of-life photovoltaic panels as a step towards a circular economy. *Renewable and Sustainable Energy Reviews*, *82*, 2934–2945.

Skoczek, A., Sample, T., & Dunlop, E. D. (2009). The results of performance measurements of field-aged crystalline silicon photovoltaic modules. *Progress in Photovoltaics: Research and applications*, *17*(4), 227–240.

Sinha, A., Sastry, O. S., & Gupta, R. (2016). Nondestructive characterization of encapsulant discoloration effects in crystalline-silicon PV modules. *Solar Energy Materials and Solar Cells*, *155*, 234–242.

Sonnenenergie, D. G. F. (2013). *Planning and installing photovoltaic systems: a guide for installers, architects and engineers.* Routledge.

Skoczek, A., Sample, T., Dunlop, E. D., & Ossenbrink, H. A. (2007, September). Electrical performance results from long-term outdoor weathered modules. In *Proceedings of the 22nd European Photovoltaic Solar Energy Conference* (pp. 2458-2466).

Skoczek, A., Sample, T., Dunlop, E. D., & Ossenbrink, H. A. (2008). Electrical performance results from physical stress testing of commercial PV modules to the IEC 61215 test sequence. Solar Energy Materials and Solar Cells, 92(12), 1593–1604.

Swanson, R., Cudzinovic, M., DeCeuster, D., Desai, V., Jürgens, J., Kaminar, N., ... & Wilson, K. (2005, October). The surface polarization effect in high-efficiency silicon solar cells. In 15th PVSEC. Shanghai, China.

Trube, J., Fischer, M., Erfert, G., Li, C. C., Ni, P., Woodhouse, M., & Wang, Q. (2018). *International technology roadmap for photovoltaic (ITRPV). VDMA photovoltaic equipement*, *24*, 77–97.

Tsao, J., N. Lewis, & G. Crabtree (2006). *Solar FAQs*. United States Department of Energy. 13.

Vázquez, M., & Rey-Stolle, I. (2008). Photovoltaic module reliability model based on field degradation studies. *Progress in photovoltaics: Research and Applications, 16*(5), 419–433.

Walston, L.J., Rollins, K.E., Smith, K.P., LaGory, K.E., Sinclair, K., Turchi, C., ... & Souder, H. (2015). A review of avian monitoring and mitigation information at existing utility-scale solar facilities.

Weckend, S., Wade, A., & Heath, G. A. (2016). *End of life management: Solar photovoltaic panels*. Golden, CO: National Renewable Energy Lab.(NREL).

Wohlgemuth, J., Silverman, T., Miller, D. C., McNutt, P., Kempe, M., & Deceglie, M. (2015). *June). Evaluation of PV module field performance* (pp. 1–7). IEEE.

Wohlgemuth, J. H. & Kurtz, S. (2011). Reliability testing beyond qualification as a key component in photovoltaic's progress toward grid parity. In *Proceedings of the international reliability physics symposium*. IEEE.

Wohlgemuth, J. H., Cunningham, D. W., Monus, P., Miller, J., & Nguyen, A. (2006). *May). Long term reliability of photovoltaic modules* ((Vol. 2, pp. 2050–2053). IEEE.

Xiao, C., Hacke, P., Johnston, S., Sulas-Kern, D. B., Jiang, C. S., & Al-Jassim, M. (2020). Failure analysis of field-failed bypass diodes. *Progress in Photovoltaics: Research and Applications, 28*(9), 909–918.

Xu, Y., Li, J., Tan, Q., Peters, A. L., & Yang, C. (2018). Global status of recycling waste solar panels: A review. Waste Management, 75, 450–458.

Xiong, H., Gan, C., Yang, X., Hu, Z., Niu, H., Li, J., & Luo, X. (2017). Corrosion behavior of crystalline silicon solar cells. *Microelectronics Reliability, 70*, 49–58.

Yedidi, K., Tatapudi, S., Mallineni, J., Knisely, B., Kutiche, J., & TamizhMani, G. (2014). *Failure and degradation modes and rates of PV modules in a hot-dry climate: Results after 16 years of field exposure* (pp. 3245–3247). IEEE.

Zdyb, A., & Gulkowski, S. (2020). Performance assessment of four different photovoltaic technologies in Poland. *Energies, 13*(1), 196.

Conclusion

In conclusion, the field of photovoltaics is constantly evolving with research and development work dedicated to optimizing the device towards commercialization. To realize global aspirations of producing clean energy from renewable sources, every aspect of photovoltaic devices, including efficiency, stability, reliability, shelf-life, manufacturing cost, and material sustainability have been scrutinized. The conception of this book was intended to be a one-stop resource for information on different types of solar cells, from inorganic to organic to hybrids. This approach was adopted to benefit prospective students and researchers in solar cell technology to understand the similarities and differences as well as the pros and cons of each device. Hence, we hope the content of this book has provided the fundamental concepts of different generations of photovoltaics for readers who wish to be acquainted with the various categories of solar cells before transitioning into a more complex representation of each category. We sincerely hope that you find this book beneficial for your future endeavors.

We would like to take this opportunity to thank all the authors who have contributed to this book for their diligence, especially amidst the challenging pandemic situation. We are also happy to extend our gratitude to the editorial team from the publishers for their professional support throughout this project.

Index

Note: Page numbers followed by "*f*" and "*t*" refer to figures and tables, respectively.

D

E

Q

R

S

T

V

W

Z